Real-Time Applications of Advanced Electrochemical Sensing Devices

Online at: https://doi.org/10.1088/978-0-7503-5377-9

IOP Series in Sensors and Sensor Systems

The IOP Series in Sensors and Sensor Systems includes books on all aspects of the science and technology of sensors and sensor systems. Spanning fundamentals, fabrication, applications and processing, the series aims to provide a library for instrument and measurement scientists, engineers and technologists in universities and industry.

The series seeks (but is not restricted to) publications in the following topics:

- Advanced materials for sensing
- Biosensors
- Chemical sensors
- Industrial applications
- Internet of Things (IoT)
- Lab-on-a-chip
- Localization and object tracking
- Manufacturing and packaging
- Mechanisms, modelling and simulations
- Microelectromechanical systems/nanoelectromechanical systems
- Micro and nanosensors
- Non-destructive testing
- Optoelectronic and photonic sensors
- Optomechanical sensors
- Physical sensors
- Remote sensors
- Sensing for health, safety and security
- Sensing principles
- Sensing systems
- Sensor arrays
- Sensor devices
- Sensor networks
- Sensor technology and applications
- Signal processing and data analysis
- Smart sensors and monitoring
- Telemetry

Authors are encouraged to take advantage of electronic publication through the use of colour, animations, video and interactive elements to enhance the reader experience.

A full list of titles published in this series can be found here: https://iopscience.iop.org/bookListInfo/iop-series-in-sensors-and-sensor-systems.

Real-Time Applications of Advanced Electrochemical Sensing Devices

Edited by
Jamballi G Manjunatha
Department of Chemistry, Field Marshal KM Cariappa College, Constituent College of Mangalore University, Madikeri, Karnataka, India

IOP Publishing, Bristol, UK

ISBN 978-0-7503-5377-9 (ebook)
ISBN 978-0-7503-5375-5 (print)
ISBN 978-0-7503-5378-6 (myPrint)
ISBN 978-0-7503-5376-2 (mobi)

DOI 10.1088/978-0-7503-5377-9

Version: 20241201

IOP ebooks

British Library Cataloguing-in-Publication Data: A catalogue record for this book is available from the British Library.

Published by IOP Publishing, wholly owned by The Institute of Physics, London

IOP Publishing, No.2 The Distillery, Glassfields, Avon Street, Bristol, BS2 0GR, UK

US Office: IOP Publishing, Inc., 190 North Independence Mall West, Suite 601, Philadelphia, PA 19106, USA

*Dedicated to my mother J Sharadamma Jamballi,
Professor GMH and the IOP Publishing staff.*

Contents

Preface

This book parades the advancements in electrochemical devices for various real-time applications. This book encloses the research trend and utilization of advanced electrochemical sensing devices with essential applications in different fields such as environmental, biomedical, agricultural, pharmaceutical, forensic, energy storage and management, food analysis, water analysis, and so on. The important purpose of this book is to attract scientific researchers from the field of electrochemistry, and to produce more researchers in this field to develop highly sensitive devices.

In the current era, progress and research in electrochemical sensing devices are becoming widespread in chemistry, physics, electronics, biology, engineering, and material science. Real-time applications of electrochemical sensing devices deliver prospects for the construction of a new generation of sensing devices based on novel sensor technologies. Nanomaterials in sensing devices improve some important properties such as electrochemical, mechanical, magnetic, and optical properties of the sensing devices. The combination of electrochemical sensing technology with nanotechnology provides a green and cost-effective electrochemical sensing device for various real-time applications.

Dr J G Manjunatha
Editor

Acknowledgements

I gratefully acknowledge my sincere gratitude to my mother Sharadamma Jamballi, wife Sinchana Manjunath, daughters Nirvikaa J M, and Vruddhi J M and research colleagues.

Editor biography

Jamballi G Manjunatha

J G Manjunatha is working as an Assistant Professor in Chemistry at FMKMC College, a constituent college of Mangalore University, Madikeri, India. He received his Ph.D. degree in Chemistry from Kuvempu University and Postdoc from the University of Kebangsaan Malaysia. He has received various awards and published more than 193 research articles in reputed international journals. He has been an editor for 21 books (RSC, Springer, Willy, ACS, IOP, Elsevier and Bentham Science Publishers), and special issues (IOP Science Publisher, Frontiers in Sensors, MDPI). He is also an editorial board member for many reputed journals and Editor-in-Chief for the Sensing Technology journal (Taylor and Francis).

List of contributors

Prashanth S Adarakatti
Department of Chemistry, SVM Arts, Science and Commerce College, affiliated to Rani Channamma University, Belagavi, India

Vinayak Adimule
Angadi Institute of Technology and Management, Belagavi, Karnataka, India

Nida Aydogdu
Department of Analytical Chemistry, Afyonkarahisar Health Sciences University, Afyonkarahisar, Turkiye

Gözde Aydoğdu Tığ
Department of Chemistry, Ankara University, Ankara, Turkiye

Hemanth B S
Department of Chemistry, Sri Jayachamarajendra College of Engineering, JSS Science and Technology University, Mysuru, India

Beejaganahalli Sangameshwar Madhukar
Department of Chemistry, Sri Jayachamarajendra College of Engineering, JSS Science and Technology University, Mysuru, India

Gangadhar Bagihalli
Department of Chemistry, KLE Institute of Technology, Hubli, India

Chinnapiyan Vedhi
Department of Chemistry, V.O. Chidambaram College, Thoothukudi, India

Bruna Coldibeli
Departamento de Química, Centro de Ciências Exatas, Universidade Estadual de Londrina, Londrina, Brazil

S Alwin David
Department of Chemistry, V.O. Chidambaram College, Thoothukudi, India

Ersin Demir
Department of Analytical Chemistry, Afyonkarahisar Health Sciences University, Afyonkarahisar, Turkiye

Maya Devi
Department of Physics, School of Applied Sciences, KIIT Deemed to be University, Odisha, India

Cem Erkmen
Department of Analytical Chemistry, Ankara University, Ankara, Turkiye

Gustavo Fix
Departamento de Química, Centro de Ciências Exatas, Universidade Estadual de Londrina, Londrina, Brazil

Parvin Abedi Ghobadloo
Food and Drug Safety Research Center, Tabriz University of Medical Sciences, Tabriz, Iran

Surya Gopidas
Amrita School of Physical Sciences, Amritapuri, India

G Amala Jothi Grace
Department of Chemistry, St. Mary's College (Autonomous), Thoothukudi, India

Honnegowdanahalli Shivabasappa Nagendra Prasad
Department of Chemistry, Sri Jayachamarajendra College of Engineering, JSS Science and Technology University, Mysuru, India

Samin Hamidi
Research Center of Psychiatry and Behavioral Sciences, Tabriz University of Medical Sciences, Tabriz, Iran

Yousef Javadzadeh
Biotechnology Research Center, Tabriz University of Medical Sciences, Tabriz, Iran

P Karpagavinayagam
Department of Chemistry, V.O. Chidambaram College, Thoothukudi, India

M Kavitha
Department of Chemistry, V.O. Chidambaram College, Thoothukudi, India

Rangappa Keri
Department of Chemistry, Centre for Nano and Material Sciences, Jain University, Bangalore, India

İpek Kucuk
Department of Analytical Chemistry and The Graduate School of Health Sciences, Ankara University, Ankara, Turkiye

Sevinc Kurbanoglu
Department of Analytical Chemistry, Faculty of Pharmacy, Ankara University, Akara, Turkiye

Murudagalli Basavaraju Shivaswamy
Department of Chemistry, Sri Jayachamarajendra College of Engineering, JSS Science and Technology University, Mysuru, India

Gopika Meenakumari Gopakumar
Amrita School of Physical Sciences, Amritapuri, India

Mosale Jagadeesharadya Deviprasad
Department of Chemistry, Sri Jayachamarajendra College of Engineering, JSS Science and Technology University, Mysuru, India

Murat Misir
Department of Chemical Engineering, Kırşehir Ahi Evran University, Kırşehir, Turkiye

Milan Z Momčilović
University of Niš, Faculty of Sciences and Mathematics, Department of Chemistry, Višegradska 33, 18000 Niš, Serbia.

Santosh Nandi
Department of Humanity and Basic Science, KLE Tech University, Belagavi, India

Geethanjali Parivara Appaji
Department of Microbiology, Field Marshal K M Cariappa College (A constituent College of Mangalore University), Madikeri, India

Swetapadma Praharaj
Department of Physics, School of Applied Sciences, KIIT Deemed to be University, Odisha, India

V Prasad
International and Inter University Centre for Nanoscience and Nanotechnology, Mahatma Gandhi University, Kottayam, Kerala, India

P Rajakani
Department of Chemistry, V.O. Chidambaram College, Thoothukudi, India

Dibyaranjan Rout
Department of Physics, School of Applied Sciences, KIIT Deemed to be University, Odisha, India

Ashoka Siddaramanna
Department of Chemistry, School of Applied Science, REVA University, Bengaluru, India

Somashettihalli Gangadharappa Manjushree
Department of Chemistry, Sri Siddhartha Institute of Technology, Tumkur, India

Selenay Sadak
Department of Analytical Chemistry and The Graduate School of Health Sciences, Ankara University, Ankara, Turkiye

Bhama Sajeevan
Amrita School of Physical Sciences, Amritapuri, India

Beena Saraswathyamma
Department of Chemistry, Amrita School of Physical Sciences, Amritapuri, India

Elen Romão Sartori
Departamento de Química, Centro de Ciências Exatas, Universidade Estadual de Londrina, Londrina, Brazil

Özge Selcuk
Department of Analytical Chemistry, Ankara University, Ankara, Turkiye, and, Department of Analytical Chemistry, Mersin University, Mersin, Turkiye

N Senthilkumar
Department of Chemistry, Govt. Arts and Science College, Cheyyaru, India

Hulya Silah
Department of Chemistry, Bilecik Seyh Edebali University, Bilecik, Turkiye

Nazlı Şimşek
Graduate School of Natural and Applied Sciences, Department of Chemistry, Ankara University, Ankara, Turkiye

R Baby Sunnetha
Department of Chemistry, V.O. Chidambaram College, Thoothukudi, India

Arezou Taghvimi
Biotechnology Research Center, Tabriz University of Medical Sciences, Tabriz, Iran

Kübra Turan
Department of Chemistry, Ankara University, Ankara, Turkiye

Bengi Uslu
Department of Analytical Chemistry, Ankara University, Ankara, Turkiye

B Vijaya
Department of Chemistry, V.O. Chidambaram College, Thoothukudi, India

Monireh Zamani-Kalajah
Food and Drug Safety Research Center, Tabriz University of Medical Sciences, Tabriz, Iran

IOP Publishing

Chapter 1

A brief review on basic principles of electrochemistry and electrochemical sensing devices

S G Manjushree, S Ashoka and Prashanth S Adarakatti

The first half of this text is concerned with the basic concepts of electrochemistry including instrumental methods of quantitative analysis. It enumerates the fundamental ideas behind contemporary voltammetric analysis methods. This text's central idea is to give fundamentals of capacitive and faradaic currents which determines how sensitive a technique is and its electrochemical reversibility and irreversibility. Finally, a brief explanation of the benefits and limitations of voltammetric analysis is provided. The realm of electrochemical sensors, including potentiometric, amperometric, impedimetric, conductometric, and immunosensors, is covered in the second section of this text. These sensors have a variety of interesting uses in environmental, industrial, and clinical analyses. Here, we give a broad review of the several primary categories of electrochemical sensors, outlining their essential characteristics, recent advancements, and contributions to the field of analytical chemistry, while also discussing pertinent developments in the field.

1.1 Introduction

Voltammetric techniques of electrochemical analysis are unquestionably less significant in industry and research than chromatographic, electrophoretic, and spectroscopic methods. However, for particular analytical tasks, they provide superior solutions, thus it is crucial to become familiar with their key characteristics and the underlying principles. Voltammetric techniques of analysis are quite diverse, and the underlying theory delves deeply into the thermodynamics and kinetics of electrochemistry, a field of physical chemistry that also receives insufficient attention in many curricula, making it difficult to provide an overview of them. This text aims

doi:10.1088/978-0-7503-5377-9ch1

to provide a thorough overview of voltammetric methods. There are also electroanalytical methods like cyclic voltammetry, chronocoulometry, and impedance spectroscopy that are extremely useful for researching the electrochemical characteristics of substances and systems but not very significant for quantitative analysis. These methods won't be covered in this text.

Analytical chemistry research shows that electrochemical sensors are the class of chemical sensors that is developing the fastest. A gadget that continuously delivers data about its surroundings is referred to as a chemical sensor. The ideal response from a chemical sensor is one that is directly proportional to the concentration of a given chemical species. All chemical sensors are made up of a transducer, which converts the analyte's reaction into a signal that can be detected by modern instruments, and a chemically selective layer, which separates the analyte's reaction from its immediate surroundings. Electrical, optical, mass, and thermal sensors can be used to detect and react to an analyte in the gaseous, liquid, or solid form, depending on the property to be measured [1]. Electrochemical sensors are particularly appealing in comparison to optical, mass, and thermal sensors due to their exceptional detectability, ease of experimentation, and inexpensive price. They hold a dominant position among the sensors that are now on the market, have passed the commercialization stage, and have a wide variety of significant applications in the sectors of clinical, industrial, environmental, and agricultural analysis.

Electrochemical sensors come in varieties: potentiometric, amperometric, impedimetric, conductometric, and immunosensors. For potentiometric sensors, a local equilibrium is created at the sensor interface, where either the electrode potential or membrane potential is recorded. From the potential difference between two electrodes, information about the composition of a sample is derived. In order to make an electroactive species oxidize or reduce, a voltage is provided between a working electrode and a reference electrode in amperometric sensors. The resulting current is then measured. Contrarily, conductometric sensors are used to detect conductivity across a range of frequencies. It would be hard to list every development in the field of electrochemical sensors in the context of this review, given the field's tremendous advancement and its rising influence on analytical chemistry. There are more literature reviews on electrochemical sensors [2–19]. The major goal of this study is to offer a broad overview of electrochemical sensors while also highlighting key concepts and recent advancements in a subject that has attracted a lot of research attention.

1.2 Basic principle of electrochemistry

1.2.1 Electrochemistry

An effective method for examining reactions involving electron transfers is electrochemistry. Electrochemistry links changes in chemical composition to electron flow. The ensuing chemical shift in inorganic chemistry frequently involves the oxidation or reduction of a metal complex. Consider the reduction of ferrocenium $[Fe(Cp)_2]^+$ (Cp = cyclopentadienyl), abbreviated as Fc^+, to ferrocene $[Fe(Cp)_2]$, abbreviated as

Fc, to comprehend the distinction between a chemical reduction and an electrochemical reduction:

- Through a chemical reducing agent: $Fc^+ + [Co(Cp^*)_2] \leftrightarrow Fc + [Co(Cp^*)_2]^+$
- At an electrode: $Fc^+ + e^- \leftrightarrow Fc$

What causes $[Co(Cp^*)_2]$ to diminish Fc^+ (Cp^* = pentamethylcyclopentadienyl)? The lowest unoccupied molecular orbital (LUMO) of Fc^+ is at a lower energy than the electron in the highest occupied molecular orbital (HOMO) of $[Co(Cp^*)_2]$, according to the most basic explanation, therefore an electron moves from $[Co(Cp^*)_2]$ to Fc^+. The difference in energy levels is what initiates the reaction, and the transfer of an electron between the two molecules in solution is thermodynamically advantageous.

What drives this process in an electrochemical reduction where Fc^+ is reduced by heterogeneous electron transfer from an electrode? An electrode is a material that conducts electricity, usually made of glassy carbon, platinum, gold, or mercury. To modify the energy of the electrons in the electrode, voltage can be provided using an external power source (such a potentiostat). An electron from the electrode is transported to Fc^+ when the energy of the electrons in the electrode is higher than the LUMO of Fc^+. The energy differential between the electrode and the LUMO of Fc^+ is once again the catalyst for this electrochemical reaction.

The identification of the molecule utilized as the reductant must be changed in order to alter the driving force of a chemical reduction [20]. The simplicity with which the driving force of a reaction may be regulated and the convenience with which thermodynamic and kinetic parameters can be measured are at the heart of electrochemistry [21].

1.2.2 Electroanalytical techniques based on measuring the charge transport properties of ions

Voltammetric analysis occupies the position of a subdivision if electroanalysis is understood to include all electrical applications for chemical analysis: it is possible to study (a) conduction (charge transport) in the electrolyte and (b) the charge transfer at the electrode/electrolyte interface when electrodes (electron conductors, like metals, such as Hg, Au, Pt) or carbon (graphite, glassy carbon) are added to an electrolyte solution (or a solid electrolyte, i.e., ion conductor). Conductometry, which measures the ionic conductivity of solutions, and electrophoresis [22], which uses changes in charge transport to separate ions, are the two basic methods based on charge transport. Although it uses ion migration in an electric field and might therefore be considered an electrochemical technique, electrophoresis is often studied within the context of separation techniques. Both the equivalency point of titrations and the fundamental measuring concept of detectors in ion chromatography are widely known applications of conductometry.

The term 'electroanalysis' is frequently associated with voltammetric techniques of analysis, despite the fact that electroanalytical techniques based on measuring the transport properties of ions are frequently used. This is due to the fact that voltammetric techniques are more prevalent than all other electroanalytical techniques combined. Naturally, potentiometric methods, particularly pH measurements using glass electrodes, as well as other potentiometric measures with ion-selective (or ion-sensitive) electrodes are extensively used in labs, industry, and environmental monitoring.

1.2.3 Electroanalytical techniques based on measuring charge-transfer processes at interfaces

The subsequent actions take place when an electrochemical cell receives a current: (A) The metal wires that connect a current (or voltage) source with the electrodes must convey electrons. The free electrons in the metals provide this assurance (metal conduction). (a) A charge transfer procedure must ensure that charge can pass through the electrode/electrolyte interface once the electrons have arrived there. The electron may in some circumstances be transferred to a species in solution, such as an iron(III) ion:

$$Fe^{3+} + e^- \rightarrow Fe^{2+}$$

In some situations, a cation may attach to the metal electrode (such as a lead piece or a mercury droplet) and absorb electrons to reduce:

$$Pb^{2+} + 2e^- \rightarrow Pb$$

Because the cation that joins the electrode 'neutralizes' the electrons, this reaction is similar to electrons passing across the contact.

Because they adhere to Faraday's law, the currents circulating in these two reactions are known as faradaic currents.

The electric charge that passes through the electrode, q, determines how many moles of a material, m, are created or used during an electrode operation. $q = nF\text{m}$, where n and F are the number of electrons occurring in the electrode reaction equation and the Faraday constant, respectively, assuming that there are no parallel processes [23].

The charge of one mole of electrons, or $9.648\ 533\ 83(83) \times 10^4\ \text{C mol}^{-1}$, is the Faraday constant. The discovery of this law's background is nicely laid up in reference [24].

Another way for current to flow through an electrochemical cell is when the electrode/electrolyte interfaces' two sides are charged and discharged in a similar fashion to how a plate condenser's two plates are charged and discharged. These currents are also known as charging currents or capacitive currents. They will be described in more depth below.

Static and dynamic electroanalytical approaches based on charge-transfer processes at electrode/electrolyte interfaces can be distinguished. When using static techniques, the currents should be zero and there should be no net reaction. The system is dynamic in the sense that a net reaction occurs when using the dynamic approaches because measurable currents flow through the system.

Potentiometry, which compares the potential of an electrode to that of a reference electrode, is the most significant static technique. This method is also known as an equilibrium method or an $I = 0$ method. Two reference electrodes are used to detect the potential dips at the inner and outer glass membrane solution interfaces in the context of pH glass electrodes. The most significant kind of ion-sensitive electrodes are membrane electrodes.

Controlled potential and controlled current techniques are two subcategories of dynamic techniques (sometimes known as transient techniques because they record time-dependent events) used to measure processes at electrode/electrolyte interfaces. In controlled potential approaches, potentiostats—devices that allow manipulating the potential in a desired manner, such as keeping it constant at a specific value or moving it in a programmed manner—are utilized. The system's response to the potential is then determined by measuring the current and voltammetric procedures are this category of methods.

As an alternative, galvanostats, which regulate current and gauge the voltage of the working electrode, can be used to determine the current. Although controlled current and controlled potential approaches can theoretically provide the same information, controlled potential techniques have taken the lead thanks to the invention of potentiostats.

1.2.4 Voltammetric techniques

Voltammetry is a term that derives from voltamperometry and expresses that the current is expressed as a function of voltage, or electrode potential. Any electrochemical cell has two electrodes, therefore if both electrodes determined the size of the flowing current, it would be difficult to collect unambiguous analytical information. As a result, one electrode is made much smaller than the other so that it is the only one that can limit the flow of current. The other, larger electrode is known as the auxiliary electrode, and this one is referred to as the working electrode.

Even though an electrochemical cell only requires two electrodes to function, adding a third electrode—a reference electrode, which has a constant and known electrode potential—is tremendously advantageous. This electrode should never be subjected to current since doing so might alter its potential and possibly cause injury. By measuring the voltage between these two electrodes, the reference electrode is utilized to regulate the potential of the working electrode. A three-electrode cell is one that has a working electrode (WE), an auxiliary electrode (AE), and a reference electrode (RE). A potentiostat is used to manage the potential difference at the electrolyte/electrode contact of the working electrode. The majority of potentiostats employ a feedback circuit to apply a voltage between the auxiliary and working electrodes that precisely matches the value that the electrode's potential should have. The working electrode is often grounded, which effectively controls the solution's potential.

The word 'voltammetry' is now used to refer to all methods for measuring current as a function of electrode potential. Jaroslav Heyrovsky, the creator of a voltammetric technique he called polarography, was responsible for its early development.

Heyrovsky's working electrode was a mercury electrode that was dropping. Voltammetry and polarography have long been used interchangeably, but today polarography is reserved only for 'voltammetry with a dropping mercury electrode'.

1.2.5 Faradaic currents

Measuring the faradaic currents that a molecule produces when it is oxidized or reduced at electrodes is the easiest approach to figure out its concentration. The molecules or ions must get to the electrode surface in order to produce faradaic currents. Only mass transport toward the electrode surface will be able to do this. They can only be transported by the following three mechanisms: (a) diffusion in a concentration gradient, (b) ion migration in a potential gradient, and (c) convection. When the substance is oxidized or reduced at the electrode surface, a concentration gradient will always develop at the electrode/electrolyte interface.

The diffusion profile is a function of time and is dependent on the diffusion coefficients of diffusing ions. With time, the thickness of the diffusion layer increases, and the steepness of the concentration versus distance relationship, or the concentration gradients, will flatten out over time.

The electrodes' size and geometry also have an impact on the diffusion profile. When the electrode is a disc, the diffusion layer may assume a spherical shape, and edge effects become prominent when electrode dimensions' shrink into the one metre range. When the solution is stirred (or the electrode is moved, such as rotated when it is a disc or wire), the diffusion layer thins out as the mixture is mixed) [25].

1.2.6 Capacitive currents

Because charge can build up on both sides of an electron conductor, the interface between an electron conductor and a solution creates a capacitor. The side of an electron conductor can be charged either negatively or positively by the accumulation or deficiency of electrons. A surplus of either anions or cations might charge the solution side. Dipole solvent molecules that are positioned on the interface can contribute to the accumulated charge in addition to the ions. The interface is additionally referred to as an electrochemical double layer (EDL) due to its capacitive characteristics.

Although both sides of the EDL's structure are more complex in reality, this has no bearing on our understanding of electroanalytical procedures. Within the compact layer and all the way to the outer Helmholtz plane, the potential in the EDL varies linearly. In the diffuse layer, it then experiences exponential decline. This is crucial because there is no electric field outside the EDL as a result of the charged electrode, meaning that ions outside the EDL are unaffected by the electrode potential and cannot migrate. The diffuse layer develops as a result of the system's thermal energy, which prevents the ions from being placed in a fixed distance from the metal but somewhat randomizes their position. Their location is fixed by the electrical field, but the temperature movement prevents this fixing. The electrode potential, the charge density on the metal side, and the number of ions per volume of solution (i.e., their concentration) are the main determinants of the EDL thickness. A third element is the solution's dielectric constant. Due to the dipoles' non-random orientation, the dielectric constant fluctuates in the EDL. There is no extra charge on either side of the EDL at a specific potential. The 'potential of zero charge' is the name given to this potential (pzc). It depends on the metal, the kind and quantity of the electrolyte, and the solvent.

1.2.7 Electrochemical reversibility and irreversibility

We discuss the reversibility or irreversibility of electrochemical reactions in the sections that follow. So it is necessary to provide some fundamental knowledge about the kinetics of electrode reactions [26, 27].

Transporting the educt to the interface and the product away from the interface are necessary for an electrode reaction to proceed. The charge transfer at the interface happens at a rate of v_{ct}. The overall rate of the electrode reaction can be limited (i) by the mass transport of educt to the electrode, (ii) by the mass transport of product to the bulk of solution, or (iii) by the rate of charge transfer. We shall discuss here only case (i) and (iii), as case (ii) is less frequently found. The ratio of oxidized species to reduced species at the interfaces will take on exactly the value that follows the Nernst equation for this potential when the equilibrium between the oxidized and reduced forms of the electroactive compound is established at the interface very quickly, i.e., when v_{ct} is faster than v_{mt}. The Nernst equation depicts an equilibrium situation, and as a result, a system is said to be electrochemically reversible when the ratio of reduced species to oxidized species at the interfaces corresponds to the potential indicated by the Nernst equation. The equilibrium cannot be reached with a sufficient rate when the charge-transfer rate v_{ct} is less than the mass transport rate v_{mt}, and the ratio of oxidized species and reduced species at the interface will depart from the ratio predicted by the Nernst equation for this potential. The electrochemical system must thus be considered to be essentially irreversible. You may alternatively put it like this: the current is the observable that lets us know an electrochemical reaction is happening. If that current is less than anticipated, it is clear that the reaction is not taking place at equilibrium and is not reversible.

It is significant to remember that, within certain bounds, the goal can alter the rates of mass transfer and charge transfer in voltammetric analysis. The hydrodynamic circumstances can modify the transfer rate (rate of stirring the solution, rate of rotating the electrode, drop time of the DME, etc). Only by altering the chemistry can the rate of charge transmission be modified. This means that categorizing an electrochemical system as reversible or irreversible is not an absolute, but rather depends on the circumstances of the experiment.

1.2.8 The advantages and disadvantages of voltammetric techniques of analysis

This section sought to show that the primary problem facing voltammetric techniques (VTs) of analysis is the discriminating of capacitive currents in order to achieve the most favourable ratios of the faradaic to capacitive currents. Capacitive currents are only occasionally employed for analytical calculations, in which case the faradaic currents must be distinguished.

To reach low detection limits, a variety of approaches that each discriminate capacitive currents are very successful. Examples of such pairings include:

- Pre-concentration techniques (electrochemical or adsorptive) + differential pulse (or square-wave, or phase-sensitive ac techniques) + static mercury electrode (or mercury film electrode).
- Pre-concentration techniques (electrochemical or adsorptive) + catalytic currents + differential pulse (or square-wave, or phase-sensitive ac techniques) + static mercury electrode (or mercury film electrode) [28].

1.3 Electrochemical sensing devices

The IUPAC [29] defines a chemical sensor as 'a device that translates chemical data into an analytically useful signal, ranging from the concentration of a single sample component to comprehensive composition analysis'. A chemical sensor is typically made up of two key functional components: a receptor and a physicochemical transducer. The receptors vary and can include complex (macro)molecules that interact with the analyte (figure 1.1) in highly specific ways as well as activated or doped surfaces.

A device is referred to as a biosensor if the receptor is biological in origin (e.g., DNA, antibodies, and enzymes). The recognition event is transformed into a pre-set output signal by the receptor's interaction with the analyte. In order to prevent false-positive results, sensors must retain a high level of selectivity for the intended analyte in the presence of potentially interfering chemical species. The transducer, which transforms the signal produced by the receptor–analyte interaction into a value that can be read, is another crucial part of sensors. So, catalytic or affinity-based devices can be used as chemical or biological sensors. While affinity-based devices rely on highly specific interactions between the receptor and analyte, such as using the specific affinity of nucleic acids (i.e., ssDNA and aptamers), antibodies–antigens, or host–guest interactions, catalytic sensors use catalytic activity to generate the signal, as in the case of enzymatic, DNAzyme, or functionalized surfaces that can perform

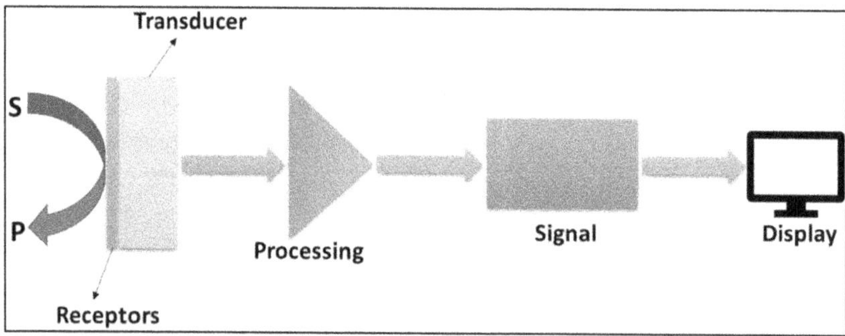

Figure 1.1. A schematic diagram illustrating the major components of a standard sensor. Adapted from [31], copyright (2019), with permission from Elsevier.

redox reactions under certain conditions. Depending on the type of transducer used, several techniques (such as optical, gravimetric, or electrochemical) can be used to monitor the recognition events [30].

Due to its advantages of having low detection limits (as low as picomoles), rapidity, and inexpensive sensing equipment, electrochemical sensors—the market leaders—are by far the most widely used type of sensor. There are several different form factors for electrochemical sensors, from top-bench to completely integrated wearables [31]. A chemical sensor's ability to provide precise, real-time information on the chemical composition of its environment is what makes it useful. In a perfect world, such a gadget would be able to react continuously and in both directions without affecting the sample. Such devices have a transduction element coated with a biological or chemical identifying layer. In electrochemical sensors, the electrical signal generated by the interaction of the target analyte and the recognition layer is used to derive the analytical information. Depending on the analyte, sample matrix, and sensitivity or selectivity requirements, a variety of electrochemical devices can be used for environmental monitoring. The majority of these devices can be classified for example as electrochemical amperometric and potentiometric sensors [32, 33]. Amperometric sensors are used to find electroactive species engaged in chemical or biological identification.

1.3.1 Types of electrochemical sensors

Amperometric, potentiometric, impedimetric, photoelectrochemical, Conductometric are some of the several types of electrochemical sensors. When no current is permitted to flow in the system providing information about the analyte's concentration, a Nernstian equilibrium forms at the sensor interface for potentiometric sensors as a result of unique sensor–analyte interactions. According to the Cottrell equation, amperometric sensors use a voltage applied between a reference and working electrode to start electrochemical oxidation or reduction, and then measure the resulting current as a quantitative indicator of the analyte's concentration.

$$i = \frac{nFAc_j^0\sqrt{D_j}}{\sqrt{\pi t}}$$

i = current (in ampere); n = number of electrons; F = Faraday constant (96 485 C mol^{-1}); A = area of the electrode in cm^2; c_j^0 = initial concentration of the reducible analyte; D_j = diffusion coefficient for species; and t = time in seconds.

Contrarily, conductometric sensors—also known as impedimetric sensors—measure variations in surface impedance to find and quantify analyte-specific recognition events on the electrode. It is challenging to discuss all of the accomplishments in the field of electrochemical sensor research within the confines of this review due to the field's extraordinary success and growing influence on analytical chemistry, so our goal was to highlight the field's diversity rather than go in-depth on any particular type of electrochemical sensor [34].

1.3.2 Basic components and fabrication

1.3.2.1 Basic setups of electrochemical devices

We'll start by briefly describing the fundamental parts, configurations, and tools used in electrochemical devices. A three-electrode setup, consisting of the working, reference, and auxiliary electrodes, is usually used for amperometry and voltammetry. In some cases, two working electrodes are utilized for redox cycling, for example. The potentiostat controls the three-electrode system. The working electrode is subjected to a fixed potential or a waveform-varying potential applied with respect to the reference electrode, and the current produced by a redox reaction is monitored. The working and auxiliary electrodes both have current flowing through them while no current passes through the reference electrode. A kind of amperometry is coulometry. The charge, which is the integral of current, is the output in this case.

Two electrodes are used in the simplest potentiometry configuration: an indication electrode to detect an analyte and a reference electrode. Using an electrometer, the potential between the first and second is determined. According to the Nernst or Nikolsky–Eisenman equations, the potential varies linearly with respect to the logarithm of the analyte's concentration.

The three-electrode system discussed earlier can be the fundamental setup for impedimetry, and variations in impedance around the working electrode are measured. However, a pair of electrodes is often utilized in the majority of practical sensing systems. An impedance analyzer is used to examine the impedance between the electrodes.

Many commercial sources now offer thin- and thick-film electrochemical microdevices, such as three-electrode systems and microelectrodes. As a result, several electrochemical microdevices are built using these readily accessible commercial devices. In order to concentrate on the development of new devices rather than the advancements in detecting chemistry, we chose not to mention these devices in the discussion that follows.

1.3.2.2 Integration of reference electrodes

In electroanalytical systems, the reference electrode is crucial. Conventional reference electrodes come in a wide variety of designs. However, the shrinking of sensors today necessitates more compact and straightforward reference electrodes. The reference electrode is a necessary component whose performance directly affects that of the complete electrochemical device, despite the fact that it is typically overlooked because it is an uninteresting part. The level of precision and stability needed depends on the mode of operation. The working electrode's potential is typically fixed in amperometry at the diffusion-limiting zone where the generated current is independent of electrode potential. As seen in the case of many microfabricated amperometric devices, the shift in the reference electrode potential in this instance is not significant, and even polarizable electrodes constructed of platinum, gold, and carbon can be viable alternatives. However, with potentiometry, a change in the reference electrode's potential immediately influences how the indicator electrode's potential is read and results in a significant inaccuracy. As a result, many miniature potentiometric devices today use commercially available reference electrodes (in the order of cm).

As was already noted, amperometric sensing does not have as stringent criteria for the stability of the reference electrode potential. Therefore, under some circumstances, the straightforward so-called pseudo reference electrodes (that are in direct contact with an analyte solution) are frequently used. Even polarizable electrodes like platinum, gold, and carbon (which are produced concurrently with the working and auxiliary electrodes) are utilized in place of non-polarizable electrodes like Ag/AgCl. This strategy does not apply to potentiometric sensors, which require a separate structure to maintain the concentration of the ions that determine the primary potential, as is the case with commercialized liquid-junction electrodes.

The traditional liquid-junction Ag/AgCl electrode can be directly miniaturized with an internal filling solution to provide dependable small reference electrodes for microfabricated electrochemical devices. The Nernst equation states that the potential of the Ag/AgCl electrode is dependent on the concentration of Cl^- ions. In addition, measurement error may be brought on by the potential difference created at the liquid junction. A concentrated potassium chloride (KCl) solution is typically employed to keep the concentration of internal Cl^- ions at the same level, provide adequate outflow of internal KCl, and reduce the liquid junction potential in order to avoid problems caused by this phenomenon. But it gets more difficult to fulfil these demands when the internal solution's overall volume drops. However, the situation is not as bad because the chip needs to conform to a specific size to guarantee easy handling, with the exception of situations when a very thin probe is utilized or the available space is very limited—as in the case of chips of the sub mm scale. It gets simpler to include a liquid-junction electrode into a device and achieve appropriate performance as its size grows.

1.3.2.3 Potentiometric sensors

Since the early 1930s, potentiometric sensors have been the most often used in practical applications because of their affordability, familiarity, and simplicity. Ion-selective electrodes (IES) are the main types of potentiometric devices (FETS).

An indicator electrode with the ability to selectively measure the activity of a certain ionic species is the ion-selective electrode. Such electrodes are typically membrane-based, ion-conducting devices in the traditional configuration, which isolate the sample from the electrode's interior. The functioning electrode is one electrode, and its potential is determined by its surroundings. A solution that has the desired ion present at a constant activity fixes the potential of the second electrode, which serves as a reference electrode. The value of the potential difference (cell potential) can be attributed to the concentration of the dissolved ion because the potential of the reference electrode is constant. There are more sources where you can find the detailed theory of the membrane contact mechanisms that produce potential [35, 36].

The nature and composition of the membrane material determine various methods for creating an electrode that is selective to one species. The sole limitation is the choice of dopant and ionophore matrix of the membrane, although research in this field has opened up a broad variety of applications to a virtually infinite number of analytes. ISEs can be categorized into one of three types, depending on the kind of membrane: glass, liquid, or solid electrodes. More than two dozen ISEs are commercially available from companies like Orion, Radiometer, Corning, Beckman, Hitachi, and others. These devices are frequently used for the production and monitoring of drugs as well as the analysis of organic ions and cationic or anionic species from various effluents [37].

The pH electrode, which has been in use for many years, is the potentiometric instrument that is most frequently employed. Its success is credited to a number of undeniable benefits, including ease of use, speed, no destructiveness, affordability, application to a wide concentration range, and, in particular, its extraordinarily high selectivity for hydrogen ions. The most popular electrodes are made of glass and come in a variety of sizes and shapes. They are based on a thin, ion-sensitive glass membrane.

Nevertheless, different kinds of potentiometric sensors can also be used to monitor pH. There have also been reports [38] of the use of glass electrodes for other monovalent cations, such as sodium, lithium, [39, 40] ammonium, and potassium sensors. Although measuring the pH of solutions using a glass membrane electrode has been quite successful, this technique is only used for aqueous media. Corrections must be made when determining hydrogen ions in non-aqueous solutions [41].

1.3.2.4 Amperometric sensors

The phrase 'amperometric sensor' is an anomaly in and of itself. Amperometric measurements are conducted using electroanalytical techniques by observing the cell's current flow at a single applied voltage. The potential difference across an electrochemical cell is instead scanned from one preset value to another for a voltammetric measurement, which records the cell current as a function of the applied potential. The transport of electrons to or from the analyte is the fundamental operating component of voltammetric or amperometric devices in both situations. Two electrodes are placed in an appropriate electrolyte as part of an electrochemical cell, and controlled-potential equipment is needed for the fundamental instrumentation. A three-electrode cell with one of the electrodes acting as a reference electrode is a more complex and common layout. The reference electrode (such as Ag/AgCl or Hg/Hg_2Cl_2) offers a stable potential in comparison to the working electrode, whereas the working electrode is the electrode at which the reaction of interest occurs. Typically, an inert conducting substance is utilized as an auxiliary electrode (such as platinum or graphite). In controlled-potential investigations, a supporting electrolyte is necessary to prevent electromigration effects, reduce the solution's resistance, and keep the ionic strength constant. Theoretical elements and experimental techniques have a lot of documentation [42, 43].

This work will omit investigations on the mechanistic elements including electron-transfer reactions and electrode process in order to keep the definition of an amperometric sensor to a suitable range. Our discussion will only focus on the most recent developments in quantitative amperometric sensors, despite the fact that these studies are crucial to the evolution of amperometric sensors.

The working electrode material has a significant impact on the performance of amperometric sensors. As a result, creating and maintaining electrodes has taken a lot of work. Although Heyrovsky's invention of the dropping mercury electrode, for which he was awarded the Nobel Prize in 1922, marked the beginning of classical electrochemical measurements of analytes, solid electrodes made of noble metals and different types of carbon have recently become the preferred sensors. More recently, this field has made a remarkable amount of development, and its influence on electroanalytical chemistry is rising. There are many paths in which electrochemical sensor research is going.

For a long time, mercury was a very desirable electrode material due to its large cathodic potential range window, high repeatability, and renewable surface. The most often used working electrode for stripping analysis was the hanging mercury drop electrode or mercury film electrode [44]. In order to determine metals, anions, organometallics, and organic compounds by stripping analysis at concentration levels as low as 1×10^{-10} mol l^{-1} utilizing a straightforward pre-concentration step, numerous methods have been devised. The main drawbacks of the approach are the restricted anodic potential of mercury electrodes and its toxicity.

Due to their wide potential window, minimal background current, low cost, chemical inertness, and suitability for a variety of sensing and detection applications, solid electrodes (such as carbon, platinum, gold, silver, nickel, copper, and dimensionally stable anions) have become very popular as electrode materials. A more contemporary approach to electrode systems is created by the growth of chemically modified electrodes (QMEs), in which a suitable surface modifier is used to intentionally modify the electrode surface. Even though there was significant advancement in the field of amperometric sensors in the early 1970s, the majority of these devices could only be used under tightly controlled laboratory circumstances.

Fortunately, there has been a lot of interest in the shrinking of the working electrode, and microeletrodes (ME) with dimensions no more than 2 m have been produced, improving the potential of *in vivo* and *in vitro* measurements with small apparatus. This has the expected positive impact on the advancement of amperometric sensors for real-world sample analysis. The creation of biosensors is one example of this progression. They have the ability to allow a biospecific reagent to transform the biological recognition process into a quantifiable amperometric response when immobilized or held at a suitable electrode. The integration of sensitive amperometric sensors with flow injection and liquid chromatography systems is particularly pertinent. Finally, the development of screen-printed electrodes has produced contemporary amperometric sensors that are possibly portable in response to the growing demand to conduct decentralized analytical determinations outside of the laboratory. The incorporation of microelectrodes, microelectrode arrays, and chemically modified electrodes into various extremely sensitive sensor systems (biosensors) has boosted industrial and clinical interest in the novel possibilities of electrode design and manufacture. Below is a summary of each of these electroanalysis contributions.

1.3.2.5 Immuno sensors

For the identification of biochemical targets related to health issues ranging from cancer antigens in patient serum to bacterial species in food, immunosensors that conduct immunoassays based on antigen and antibody recognition have become essential [45]. The capacity to effectively detect pertinent immunological chemicals without jeopardizing the bioactivity of the immunoactive species on the electrode is a constant challenge with immunosensor development. A survey of current developments in flow-based immunosensors was published by Hartwell *et al* [46]. Nanomaterials are being used more and more by researchers to support immunoactive drugs and improve the electrochemical and analytical capabilities of the electrode.

Recent developments in the manufacture of immunosensors heavily include carbon nanotubes (CNTs). For accurate detection, the bioactive species must be immobilized, and CNTs provide a simple method for doing so. These composite immunosensors have demonstrated good analytical properties and have showed tremendous promise for therapeutic applications, along with other nanomaterials. A review of CNT-based sensors for biomolecule detection by Jacobs *et al* [47] has been released. CNTs have been used in immunosensors for the purpose of cancer detection.

CNTs have enabled the creation of innovative immunosensors to improve biomolecule screening capabilities in addition to cancer diagnosis. For instance, Serafin *et al* [48] recently created an immunosensor for testosterone that is practically applicable. Improved sensory abilities are required due to the illegal usage of testosterone boosters to increase athletic ability and the resulting health issues. These researchers discovered that a superior electrode composition made possible by the combination of gold nanoparticles, MWCNTs, and Teflon ensured the stability of the immunoactive substance, which were monoclonal antitestosterone antibodies. The resulting immunosensor required minimum setup to quantify testosterone in serum.

Due to its wide range of uses and comparatively simple operation, electrochemiluminescence has grown in popularity in analytical chemistry. Recent research by Jie *et al* [49] describes a novel electrochemiluminescence immunosensor that combines chitosan, CdS quantum dots, and CNTs with gold nanoparticles. Their sensor has excellent analytical properties for biological settings and successfully immobilized antibodies.

Interleukin-6 and beta-fetoprotein are two recent examples of CNT-based immunosensors for cancer biomarkers. Squamous cell carcinomas can result from immunological mechanisms that involve interleukin-6. Malhotra *et al* [50] developed and evaluated a sensor based on a SWCNT forest platform with tiered capture antibody functionalization to improve the detection of interleukin-6. This innovative immunosensor worked effectively in a range of physiologically relevant doses and correlated favourably with enzyme-linked immunosorbent tests.

1.3.2.6 Impedimetric sensors

Conductometric sensors are used to detect a medium's or an electrolyte solution's propensity to permit the passage of an electrical current between a working electrode and counter electrodes or a reference electrode. Impedimetric sensors are divided into conductometric approaches. Due to the fact that variations in capacitance are analysed using conductometric sensing technique [51]. Impedance spectroscopy, which is proportional to and dependent upon the concentration of analyte and application of the sensor, is used to analyse the impedance of electrochemical cell surfaces. Techniques for measuring capacitance are continuously examined, and electrochemical spectroscopy is a potential combination method [52]. Only the typical resistance value can be determined using a DC current. In impedimetric sensing, AC current is typically used to measure changes in the electrode's capacitance value [53].

Among the most popular electrochemical methods are cyclic voltammetry (CV), differential pulse voltammetry (DPV), bulk electrolysis (BE), square voltammetry (SWV), and AC impedance spectroscopy. A three-electrode system configuration is typically used in standard EIS systems. The system's three electrodes are the working cell, auxiliary/counter cell, and reference electrode. When measuring impedance, the working electrode receives an AC voltage signal that periodically switches between the polarities of +vely and −vely. Equal and opposite polarity to the working electrode are always given for the counter electrode. The reference electrode is kept close to the working electrode and has a constant voltage set for it to offset the effects of unequal cell charging. The first identification of germs using impedance spectroscopy involved the observation of a change in the electrical/electrochemical impedance of the solution as a result of bacterial growth and an increase in the chemicals that the bacteria produced. Electrochemical impedance spectroscopy is the term used to describe this method of detection.

Impedance is a complicated resistance that results when AC current drifts from a network made up of resistors, capacitors, and inductors, or compositions of them. The electrical components and the frequency sweep range of the AC signal applied to that electrode determine the magnitude and phase of the impedance that results. We only address the resistive (real) and capacitive behaviour of electrochemical cells because inductive behaviour is not demonstrated in electrochemistry [54].

By applying an excitation voltage with a time function $V(t)$ and detecting the induced response signal with a current–time function I, the impedance value may be calculated (t). After that, the impedance is computed.

Excitation voltage signal $= Vm.\sin(\omega t)$,

Response current signal $= Im.\sin(\omega t + \phi)$

And calculated impedance is

$$Z = \frac{Vm.\ \sin(\omega t)}{Im.\sin(\omega t + \varphi)} = e^{-j\varphi} = |Z|$$

Impedance is a complex quantity so Z can be written as $= \mathrm{Re}(Z) + j.\mathrm{Img}(Z)$.

1.3.2.7 Conductometric sensors

The electric conductivity of a film or a bulk material, whose conductivity is impacted by the analyte present, is changed by the presence of the sensor in this category. Fundamentally, conductimetric techniques lack selectivity. These have only recently become more practical ways for constructing sensors with the introduction of redesigned surfaces for selectivity and greatly enhanced instrumentation. Conductimetric methods are appealing due to a number of very practical factors, including their low cost and ease of use because reference electrodes are not required. Improved technology has made it possible to quickly and simply identify analytes using merely a conductivity measurement. We'll start by looking at the materials that make up these sensors the most frequently. Following a discussion of the techniques for numerically processing the analytical signal, several phenomena or properties that are conducive to sensing will be covered.

Thin films are mostly utilized as gas sensors since surface chemisorption modifies the conductivity of these materials. For instance, CdS films can be employed as oxygen sensors because of oxygen chemisorption [55]. Porous films of $MnWO_4$ can work as a humidity sensor [56], thin semiconducting Ga_2O_3 films can detect CH_4 and oxides doped with copper or copper oxide can be extremely sensitive to gases containing H_2S [57]. Additionally, polymers that are conductive on their own or when modified are frequently used. Volatile amines are detectable via polypyrrole [58] or when doped with ClO_4^- and tosylate, they can be made into a sensor for NH_3 [58].

Resistance is typically measured from a DC current. Most frequently, AC current (impedance) is used for the measurement, which also makes it possible to track changes in capacitive impedance [59]. For instance, zeolite layers' capacitance can fluctuate in reaction to combustion gas presence [60, 61]. Impedance discrimination between conductivity and capacitance can increase the range of applications for sensing materials. Impedance sensors for NO_2 and smoke from cigarettes were described in [62], odour detection in [63], or determination of water in an oil-in-water emulsion in [64]. Intelligent materials are frequently developed for this purpose [65].

Mixed oxide conductivity sensors are frequently used to measure humidity, but can also cause undesirable interference [66]. Sintered bismuth tungstate can be used as the foundation for a humidity-insensitive gas sensor for ethanol and acetone vapours [67]. There are additional reports on the creation of a multichannel scent sensor [63, 68]. One can identify antibody–antigen binding by impedance measurements at various frequencies [69]. A flow injection conductometric device suitable for assessing acidity in industrial hydrated ethyl alcohol has also been demonstrated by Fatibello and Borges [70].

1.4 Conclusion

Exploring electron transfer reactions that underlie fundamental mechanistic inorganic chemistry and renewable energy systems is made possible by quantitative electrochemical analysis. Emerging researchers are becoming more and more interested in using electrochemistry to further their work. We wanted to highlight these researchers with an approachable theory on using voltammetry in research and educational activities through this text. This text represents current best practice for applying voltammetric techniques by integrating the theoretical foundations of electrochemical analysis and offering introductory modules.

The numerous potential uses for electrochemical sensors have continued to spark interest in them. Electrochemical sensors are important in chemistry and related fields, as evidenced by the numerous journals that specialize in the field, big professional associations that have electrochemical sensor divisions, and frequent international conferences on the topic. Since chemical sensors have been incorporated into the analytical curricula at many universities, their influence can also be seen in chemical education. Because they meet the growing demand for quick, easy, and affordable methods of determining a variety of analytes, electrochemical sensors have evolved over the past 20 years into a widely acknowledged component of analytical chemistry. The practical obstacle of producing repeatable, affordable procedures and user-friendly routine analysis equipment has always been a problem for academics creating these crucial instruments. It is simple to understand the significance and usefulness of electrochemical sensors to the development of analytical chemistry given the wide range of options.

Acknowledgments

SGM is thankful to the Department of Science and Technology for the financial support via Women Scientists Scheme-A (WOS-A) SR/WOS-A/CS-153/2018, and also acknowledges Chemistry department, Sri Siddhartha Institute of Technology, Tumkur for their constant support and encouragement. The author SA thanks REVA University for continuous support. SAP greatly thanks Karnataka Science and Technology Academy (KSTA) for their financial support and also acknowledges SVM Arts, Science and Commerce College, ILKAL for their constant support and encouragement.

References

[1] Janata J 2001 *Peer Reviewed: Centennial Retrospective on Chemical Sensors* (ACS Publications)

[2] Hitchman M L and Berlouis L E A 1995 Electrochemical methods *Process Analytical Chemistry* ed F McLennan and B R Kowalski (Dordrecht: Springer) pp 217–58

[3] Janata J 1992 Chemical sensors *Anal. Chem.* **64** 196–219

[4] Widrig C A *et al* 2002 Dynamic electrochemistry: methodology and application *Anal. Chem.* **62** 1–20

[5] 2004 Chemical sensors *Handbook of Modern Sensors: Physics, Designs, and Applications* ed J Fraden (New York: Springer) pp 499–532

[6] Wang J 1988 *Electroanalytical Techniques in Clinical Chemistry and Laboratory Medicine* (New York: Wiley)

[7] Hareesha N *et al* 2021 Electrochemical analysis of indigo carmine in food and water samples using a poly (glutamic acid) layered multi-walled carbon nanotube paste electrode *J. Electron. Mater.* **50** 1230–8

[8] Tigari G and Manjunatha J 2020 Optimized voltammetric experiment for the determination of phloroglucinol at surfactant modified carbon nanotube paste electrode *Instrum. Exp. Tech.* **63** 750–7

[9] Manjunatha Charithra M and Manjunatha J G 2020 Electrochemical sensing of paracetamol using electropolymerised and sodium lauryl sulfate modified carbon nanotube paste electrode *ChemistrySelect* **5** 9323–9

[10] Pushpanjali P A and Manjunatha J G 2020 Development of polymer modified electrochemical sensor for the determination of alizarin carmine in the presence of tartrazine *Electroanalysis* **32** 2474–80

[11] Tigari G and Manjunatha J 2020 Poly (glutamine) film-coated carbon nanotube paste electrode for the determination of curcumin with vanillin: an electroanalytical approach *Monatsh. Chem.* **151** 1681–8

[12] Manjunatha J *et al* 2020 Electrochemical fabrication of poly (niacin) modified graphite paste electrode and its application for the detection of riboflavin *Open Chem. Eng. J.* **14** 90–8

[13] Manjunatha J G 2020 A promising enhanced polymer modified voltammetric sensor for the quantification of catechol and phloroglucinol *Anal. Bioanal. Electrochem.* **12** 893–903

[14] Manjunatha J 2018 Highly sensitive polymer based sensor for determination of the drug mitoxantrone *J. Surf. Sci. Technol.* **34** 74–80

[15] Prinith N S, Manjunatha J G and Hareesha N 2021 Electrochemical validation of L-tyrosine with dopamine using composite surfactant modified carbon nanotube electrode *J. Iran. Chem. Soc.* **18** 3493–503

[16] Raril C and Manjunatha J G 2020 Fabrication of novel polymer-modified graphene-based electrochemical sensor for the determination of mercury and lead ions in water and biological samples *J. Anal. Sci. Technol.* **11** 1–10

[17] Manjunatha J *et al* 2010 Sensitive voltammetric determination of dopamine at salicylic acid and TX-100, SDS, CTAB modified carbon paste electrode *Int. J. Electrochem. Sci.* **5** 682–95

[18] Manjunatha J 2020 A surfactant enhanced graphene paste electrode as an effective electrochemical sensor for the sensitive and simultaneous determination of catechol and resorcinol *Chem. Data Collect.* **25** 100331

[19] Charithra M M, Manjunatha J G G and Raril C 2020 Surfactant modified graphite paste electrode as an electrochemical sensor for the enhanced voltammetric detection of estriol with dopamine and uric acid *Adv. Pharm. Bull.* **10** 247

[20] Connelly N G and Geiger W E 1996 Chemical redox agents for organometallic chemistry *Chem. Rev.* **96** 877–910

[21] Elgrishi N *et al* 2018 A practical beginner's guide to cyclic voltammetry *J. Chem. Educ.* **95** 197–206

[22] Ndimba B K and Ngara R 2013 Sorghum and sugarcane proteomics *Genomics of the Saccharinae* ed A H Paterson (New York: Springer) pp 141–68

[23] Donne S W 2013 General principles of electrochemistry *Supercapacitors* (Wiley) pp 1–68

[24] Wendt H and Kreysa G 1999 Basic principles and laws in electrochemistry *Electrochemical Engineering: Science and Technology in Chemical and Other Industries* ed H Wendt and G Kreysa (Berlin: Springer) pp 8–16

[25] Bard A J, Faulkner L R and White H S 2022 *Electrochemical Methods: Fundamentals and Applications* (New York: Wiley)

[26] Brajter-Toth A 2002 *Electroanalytical Methods: Guide to Experiments and Applications* ed F Scholz (Berlin: Springer)

[27] Hendel S J and Young E R 2016 Introduction to electrochemistry and the use of electrochemistry to synthesize and evaluate catalysts for water oxidation and reduction *J. Chem. Educ.* **93** 1951–6

[28] Scholz F 2015 Voltammetric techniques of analysis: the essentials *ChemTexts* **1** 17
[29] Hulanicki A, Glab S and Ingman F 1991 Chemical sensors: definitions and classification *Pure Appl. Chem.* **63** 1247–50
[30] Miri P S *et al* 2021 MOF-biomolecule nanocomposites for electrosensing *Nanochem. Res.* **6** 213–22
[31] Shetti N P *et al* 2019 Graphene–clay-based hybrid nanostructures for electrochemical sensors and biosensors *Graphene-Based Electrochemical Sensors for Biomolecules* ed A Pandikumar and P Rameshkumar (Amsterdam: Elsevier) ch 10 pp 235–74
[32] Meti M D *et al* 2021 Nanostructured Au-graphene modified electrode for electrosensing of chlorzoxazone and its biomedical applications *Mater. Chem. Phys.* **266** 124538
[33] Neiva E G C *et al* 2014 PVP-capped nickel nanoparticles: synthesis, characterization and utilization as a glycerol electrosensor *Sens. Actuators* B **196** 574–81
[34] Rahman M A *et al* 2008 Electrochemical sensors based on organic conjugated polymers *Sensors* **8** 118–41
[35] Mell L D and Maloy J T 1975 Model for the amperometric enzyme electrode obtained through digital simulation and applied to the immobilized glucose oxidase system *Anal. Chem.* **47** 299–307
[36] Weetall H H 1974 Immobilized enzymes. Analytical applications *Anal. Chem.* **46** 602A–15a
[37] Meyerhoff M E and Opdycke W N 1986 Ion-selective electrodes *Advances in Clinical Chemistry* ed H E Spiegel (Amsterdam: Elsevier) pp 1–47
[38] Benet L Z 1968 *Glass Electrodes for Hydrogen and Other Cations* ed G Eisenman (New York: Marcel Dekker)
Benet L Z 1968 Review: Glass electrodes for hydrogen and other cations *J. Pharm. Sci.* **57** 1271
[39] Gadzekpo P Y *et al* 1985 Lipophillic lithium ion carrier in a lithium ion selective electrode *Anal. Chem.* **57** 493–5
[40] Metzger E *et al* 1987 Lithium/sodium ion concentration ratio measurements in blood serum with lithium and sodium ion selective liquid membrane electrodes *Anal. Chem.* **59** 1600–3
[41] Shao Y *et al* 1997 Nanometer-sized electrochemical sensors *Anal. Chem.* **69** 1627–34
[42] Woermann D and Janata J 1990 Principles of chemical sensors (New York: Plenum Press)
Woermann D 1990 J. Janata: Principles of Chemical Sensors. Plenum Press, New York and London 1989. 317 Seiten, Preis in Europa *Ber. Bunsenges. Phys. Chem.* **94** 543
[43] Wang J 1988 *Electroanalytical Techniques in Clinical Chemistry and Laboratory Medicine* (Weinheim: Wiley)
[44] Wang J 1999 Amperometric biosensors for clinical and therapeutic drug monitoring: a review *J. Pharm. Biomed. Anal.* **19** 47–53
[45] Mahato K and Wang J 2021 Electrochemical sensors: from the bench to the skin *Sens. Actuators* B **344** 130178
[46] Hartwell S K and Grudpan K 2010 Flow based immuno/bioassay and trends in micro-immuno/biosensors *Microchim. Acta* **169** 201–20
[47] Jacobs C B, Peairs M J and Venton B J 2010 Review: carbon nanotube based electrochemical sensors for biomolecules *Anal. Chim. Acta* **662** 105–27
[48] Jurado-Sánchez B *et al* 2021 Janus particles and motors: unrivaled devices for mastering (bio) sensing *Microchim. Acta* **188** 416

[49] Jie G *et al* 2010 Electrochemiluminescence immunosensor based on nanocomposite film of CdS quantum dots-carbon nanotubes combined with gold nanoparticles-chitosan *Electrochem. Commun.* **12** 22–6
[50] Malhotra R *et al* 2010 Ultrasensitive electrochemical immunosensor for oral cancer biomarker IL-6 using carbon nanotube forest electrodes and multilabel amplification *Anal. Chem.* **82** 3118–23
[51] Grieshaber D *et al* 2008 Electrochemical biosensors-sensor principles and architectures *Sensors* **8** 1400–58
[52] Yilong Z, Dean Z and li D 2015 Electrochemical and other methods for detection and determination of dissolved nitrite: a review *Int. J. Electrochem. Sci.* **10** 1144–68
[53] Stradiotto N R, Yamanaka H and Zanoni M V B 2003 Electrochemical sensors: a powerful tool in analytical chemistry *J. Braz. Chem. Soc.* **14** 159–73
[54] Park S-M and Yoo J-S 2003 Peer reviewed: electrochemical impedance spectroscopy for better electrochemical measurements *Anal. Chem.* **75** 455A–61A
[55] Smyntyna V *et al* 1995 Influence of chemical composition on sensitivity and signal reproducibility of CdS sensors of oxygen *Sens. Actuators* B **25** 628–30
[56] Wenmin Q and Jörg-Uwe M 1997 Thick-film humidity sensor based on porous material *Meas. Sci. Technol.* **8** 593
[57] Fleischer M and Meixner H 1995 Sensitive, selective and stable CH4 detection using semiconducting Ga_2O_3 thin films *Sens. Actuators* B **26** 81–4
[58] Index pages *Analyst* 1996 **121** A001–36
[59] Pänke O, Balkenhohl T, Kafka J, Schäfer D and Lisdat F 2007 Impedance Spectroscopy and Biosensing. In: Renneberg, R., Lisdat, F. (eds) Biosensing for the 21st Century. Advances in Biochemical Engineering/Biotechnology, Springer, Berlin, Heidelberg. vol 109 195–237
[60] Plog C *et al* 1995 Combustion gas sensitivity of zeolite layers on thin-film capacitors *Sens. Actuators* B **25** 403–6
[61] Kurzweil P, Maunz W and Plog C 1995 Impedance of zeolite-based gas sensors *Sens. Actuators* B **25** 653–6
[62] Souto J *et al* 1996 A.c. conductivity of gas-sensitive Langmuir-Blodgett films of ytterbium bisphthalocyanine *Thin Solid Films* **284–5** 888–90
[63] Toko K 1998 Electronic sensing of tastes *Sensors Update* **3** 131–60
[64] García-Golding F *et al* 1995 Sensor for determining the water content of oil-in-water emulsion by specific admittance measurement *Sens. Actuators* A **47** 337–41
[65] Fleischer M and Meixner H 1995 A selective CH_4 sensor using semiconducting Ga_2O_3 thin films based on temperature switching of multigas reactions *Sens. Actuators* B **25** 544–7
[66] Profijt H B *et al* 2011 Plasma-assisted atomic layer deposition: basics, opportunities, and challenges *J. Vac. Sci. Technol.* A **29** 050801
[67] Fan J *et al* 2012 Photocatalytic degradation of azo dye by novel Bi-based photocatalyst Bi_4TaO_8I under visible-light irradiation *Chem. Eng. J.* **179** 44–51
[68] Fujimoto C, Hayakawa Y and Ono A 1996 Evaluation of the efficiency of deodorants by semiconductor gas sensors *Sens. Actuators* B **32** 191–4
[69] DeSilva M S *et al* 1995 Impedance based sensing of the specific binding reaction between Staphylococcus enterotoxin B and its antibody on an ultra-thin platinum film *Biosens. Bioelectron.* **10** 675–82
[70] Fatibello-Filho O and Borges M T M R 1998 Flow-injection conductometric determination of acidity in industrial hydrated ethyl alcohol *Anal. Chim. Acta* **366** 81–5

IOP Publishing

Real-Time Applications of Advanced Electrochemical Sensing Devices

Jamballi G Manjunatha

Chapter 2

Types of electrochemical sensing and its utility

S Alwin David, R Baby Sunnetha, P Rajakani, P Karpagavinayagam and C Vedhi

2.1 Introduction

Electroanalytical chemistry can play a very important role in the protection of our environment. Electrochemical sensors and detectors, in particular, are indeed very appealing for on-site monitoring of major pollutants as well as tackling other needs and requirements. Such devices satisfy many of the requirements for on-site environmental analysis. In recent years, a great revival in the use of carbon electrodes for analytical and sensing purposes was observed in relation to the variety of novel carbon forms accessible to analytical electrochemists. They include advanced carbon nanomaterials, such as carbon nanotubes, carbon nano-onions, mesoporous carbon, graphene and graphene oxide, as well as other carbon forms available both as bulk material or ultrathin- or nano-layers, carbon black, doped diamond, etc. These materials present remarkable electroanalytical properties both for the direct detection of electroactive species as well as being functionally tailored to develop sensitive and specific electrochemical sensors and biosensors. Synthesis of these nanomaterials with modified carbon electrodes has been discussed along with their functionalization methods. The latest proposals for all these nanoparticles in sensor applications are spotlighted for the primary areas of application, as are the future outlook and potentials.

A chemical sensor is defined as 'a device that converts chemical data, ranging from the concentration of a single sample component to complete composition analysis, into an analytically usable signal' [1]. Due to simplicity, higher sensitivity and selectivity as well as cost efficiency, electrochemical sensors were fully investigated in recent decades. Recent advancements in nanomaterial design and synthesis have resulted in highly reliable sensing systems with superior analytical performance. The incorporation of nanomaterials into sensors has accelerated the discovery of new pathways and possibilities for the identification of analytes or target molecules. A receptor that binds the sample, the sample or analyte, and a

doi:10.1088/978-0-7503-5377-9ch2

transducer that converts the reaction into a measurable electrical signal are the three essential components of an electrochemical sensor. The electrode serves as the transducer in the case of electrochemical sensors. The site of the reaction in most electrochemical sensors is an electrode surface. The analyte of interest will either be oxidized or reduced by the electrode. The current generated by the reaction is monitored and used to compute important data such as concentration of samples (figure 2.1).

For the most part, a chemical sensor is constituted of two essential functional units: a receptor and a physicochemical transducer. The receptors vary and therefore can vary from activated or overdosed substrates to complicated (macro) particles that interact with the analyte in extremely specialized ways. If the receptor is of biological origin (e.g., DNA, antibodies, and enzymes), the device is referred to as a biosensor. The receptor interacts with the analyte, converting the recognition event into a predetermined output signal. To prevent wrong results, detectors must preserve a high degree of specificity for the intentional electrolyte in the presence of level close reactive species. The transducer, which is responsible for the conversion of the signals generated by the receptor–analyte interplay into readable value, is also another essential element of sensors. Thus, both chemical and biosensors can be classified into catalytic or affinity-based devices. While catalytic sensors apply catalytic activity to generate the signal, as in the case of enzymatic, DNAzyme, or functionalized surfaces that can perform redox reactions under certain conditions, affinity-based devices rely on highly specific interactions between the receptor and analyte, e.g., using the specific affinity of nucleic acids, antibodies–antigens, or host–guest interactions. Depending on the type of transmitter used, tracking of everything

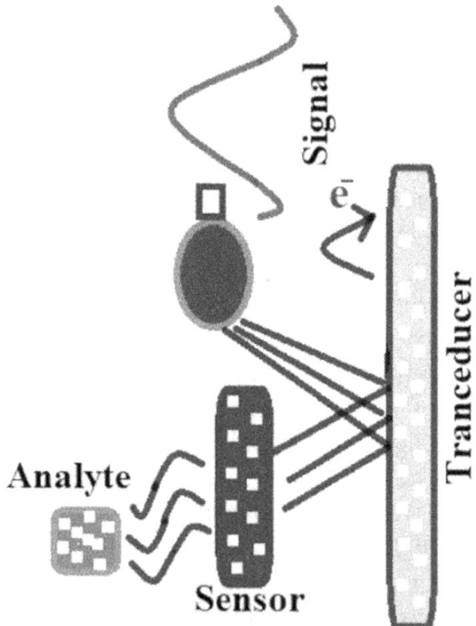

Figure 2.1. Schematic representation of electrochemical sensor.

can always be performed by a variety of methodologies (e.g., optical, gravimetric, or electrochemical) [2].

Electrochemical sensors come in a variety of form factors ranging from the top-bench to fully integrated wearable devices [3]. It was shown in the figure 2.1. The usefulness of a chemical sensor is to bring exact real-time information regarding the chemical composition of its atmospheres. In an ideal scenario, such a device would be able to respond constantly and reversibly without interfering with the sample. In such devices, a biological or chemical identification layer is coated on a transduction element. In electrochemical sensors, the analytical information is taken from the electrical signal produced by the interaction of the target analyte and the recognition layer.

2.2 Biosensors

Biosensing has developed rapidly since Clark and Lyons created the use of glucose oxidase (GOD) for the electrochemical detection of glucose in 1962 [4]. It can be usually understood as a device that can transform a definite biological indicator into an understandable one. Biosensors have many ways to convert signals, mainly in optical [5, 6] and electrochemical [7–9] approaches. Optical biosensors, as the name implies, convert biological signals into optical signals that can be read by technologies, devices, and people, thus enabling the analysis of biochemical information in biochemical reactions, biological toxins, food safety, and pharmaceuticals. Bioelectrochemical sensors can convert biological signals into readable electrical signals involving voltages, currents, frequencies, and amplitudes. To date, researchers have developed a wide range of biosensors based on metals [10], metal oxides [11], organics [12], and other materials [13, 14] that can detect a wide range of human and environmental conditions. The most progressive physical biosensors are often built on a hard substrate with common electronic recording hardware. Hence, the monitoring sensors are usually coupled to the skin via straps/tapes with wired interfaces.

2.3 Electrochemical biosensors

Electrochemical biosensors based on immobilized bio-recognition elements are one of the most popular and commercially successful groups of biosensors. The development of screen-printed planar electrodes, which are cheap, sensitive and capable of miniaturization, augments the appeal of these biosensors [15]. In the context of pathogen detection and diagnosis, electrochemical biosensors are highly advantageous as they have enabled rapid and multiplexed detection of pathogens, as well as their direct detection from samples without prior processing. Electrochemical detection platforms are usually low-cost and portable with wireless actuation and data acquisition formats. As a result, their use has been widely investigated for medical diagnostic, food and water safety, environmental monitoring and for biological-threat applications [16, 17]. Therefore, electrochemical biosensors are being increasingly used for the detection of pathogenic bacteria and other targets [18–21]. In order to design a high-performance biosensor, it is crucial that its bio-recognition element possesses excellent analyte selectivity [22]. For instance, it is worth considering aptamers, which are short oligonucleotides isolated *in vitro* from randomized libraries that can bind to specific molecules with high affinity [23]. They are excellent bioreceptor candidates, as they offer a number of advantages over antibodies, such as superior stability, versatility and lower production costs. The iterative selection process by which an aptamer is isolated is called systematic evolution of ligands by exponential enrichment (SELEX). When living cells, such as live bacteria, are used to select aptamers that bind specifically to a single target, or to multiple targets, present on the membrane surface of the same type of cell, it is termed cell-SELEX. The main advantage of cell-SELEX is that aptamers can be selected towards their targets, which are in their native form [24, 25]. Aptamers selected by cell-SELEX have been widely used as receptors in biosensors for the detection of bacteria [26–29]. So far, only one study reporting aptamers selected by cell-SELEX against *Yersinia enterocolitica* is available [30]. In that study, three aptamers were selected against cells harvested at different growth stages and were further characterized. However, concrete cell surface target molecules were not analyzed or identified. Moreover, those aptamers have not been further evaluated in detection systems. Electrochemical detection systems incorporating aptamers as sensory molecules have been developed for a number of other pathogens, including *Listeria monocytogenes* [31], *Escherichia coli* O157:H7 [32] and *Salmonella enterica* serovar Typhimurium [33], and have proved to be highly efficient [34]. Selection of a suitable target or analyte is also important to consider while designing a biosensor. When aptamers are selected towards a target by cell-SELEX, the target is usually exposed on the outer surface of the cell, such as a protein. Certain bacteria, such as *Y. enterocolitica*, assemble protein complexes called adhesins on their surface, which recognize specific molecular receptors of the host during colonization and are suitable markers for detection of the bacteria. During an infection, YadA mediates the binding to epithelial and polymorphonuclear cells, and is also essential for autoagglutination [35].

2.4 Types of electrochemical sensing

An examination of the evolution of analytical methods reveals that electrochemical sensors are the fastest expanding category of chemical sensors. A chemical sensor is an instrument that provides continuous details about its surroundings. A chemical sensor should in an ideal world provide such a specialized form of response that is primarily connected to the quantity of a particular chemical species. All chemical sensors are made up of a transducer, which converts the response into a detectable signal on modern instrumentation, and a chemically selective layer, which isolates the analyte's response from its immediate surroundings. Electrochemical sensors are particularly appealing when compared to optical, mass, and thermal sensors due to their exceptional detection range and ease of experimentation are relatively inexpensive. Electrochemical sensors are commonly used for absorption of nutrients investigation and natural process guidance in a variety of areas of assembly, such as vehicular, environmental, and medical monitoring. They are well-established and related materials for obtaining constant cycle control data without investigating the central nervous system *in situ*. This technology started in the mid-1950s. It is at the forefront of currently accessible sensors that have attained the commercial stage and also have found a wide range of crucial applications in diagnostic, manufacturing, ecologic, and agribusiness studies.

Electrochemical sensors are classified into three types viz., potentiometric, amperometric and conductometric.

2.4.1 Potentiometric

A potentiometric sensor is an electrochemical sensor that can calculate the informative gathering of logically separate gas or arrangement segments. When there is no flow, such sensors measure an anode's electrical potential. Potentiometric sensors use the effect of focus on the equilibrium of redox responses occurring at the anode–electrolyte interface of an electrochemical cell. Potentiometric technologies have enabled concentration of an analyte, primarily by measuring the variation of potential difference between working and reference electrodes at various analyte concentrations. The pH meter is the one example of potentiometric sensors. Potentiometric methods, employed since the early 1930s, have many advantages, such as a quick response time, a broad linear working range, lower power consumption, competitive prices, and simplicity of preparation, fast response, and low specific deterrence of various ionic species [36–40]. Potentiometric devices are classified into three types: ion-selective electrodes (IES), coated wire electrodes (CWES), and field-effect transistors (FETS).

2.4.2 Amperometric

An amperometric sensor is based on measuring current produced at the surface of the electrode by an enzyme—with mediated or bioaffinity response at a constant working potential with respect to the reference electrode. Electroanalytical procedures are used to make amperometric estimations by recording the current flow in the cell at a single applied potential. Then the voltammetric is obtained once more when the potential effect is tested, beginning with one preset value and progressing to that following, an electrochemical cell being used, and the mobile current is enlisted as a component of the probable applied. The basic operational excellence of voltammetric or amperometric instruments in the the transmission of electrons to or from the analyte occurs in two cases. The basic instrumentation requires controlled-potential hardware, and the electrochemical cell consists of two terminals immersed in a suitable electrolyte. A more complicated and commonplace strategy incorporates the utilization of a three-terminal cell, with one of the cathodes acting as a source of perspective anode. The reference cathode (e.g., Ag/AgCl, Hg/Hg_2Cl_2) has a constant likely comparative with the working anode, whereas the working terminal is the terminal at which the interest response occurs. A dormant leading material (e.g., platinum, graphite) is frequently used as a helper terminal [41–49].

2.4.3 Conductometric

The conductometric method assesses an analyte's or medium's capability to conduct an electrical current between electrodes or reference nodes. Only the sensors in this group rely on modifications in the electrical conductivity of a film or mass material that are influenced by the current analyte's conductivity. Conductometric procedures are essentially generic. With the appearance of refreshed surfaces for selectivity and significantly better instrumentation, these have simply become more accessible techniques for sensor planning. Because no reference anodes are required, there are some extremely useful considerations, for example, ease and effortlessness, which make conductimetric techniques appealing. Improved method, which is only concerned with estimating conductivity, has resulted in the fast and basic guarantee of solute [50, 51].

2.4.4 Biomolecule electrochemical detection

Biomolecules of a small size (e.g., hormones, nucleic acids, and enzymes) are detected based on their physiological and biological roles, which include transferring regulating biological activity, genetic information, and catalyzing cellular processes [52–54]. Nonetheless, creating biomolecule-sensing technology continues to be a difficult task [55]. Biomolecular methods, such as Western blot, polymerase chain reaction (PCR), and gel electrophoresis, have been developed for the analysis of biomolecules [56]. Despite their precision, they are limited by constraints such as large reagent needs, laboriousness, and long-time requirements [57]. Various studies on electrochemical approaches for identifying biomolecules as an early diagnostic tool have been published [58–60]. Mohan *et al* [61] developed an integrated electrochemical biosensor that could detect biomarkers in urine. This may help improve the effectiveness of clinical disease management and indicates that pathogen identification in combination with quantitative detection of lactoferrin can provide important information for the diagnosis of urinary tract infections.

2.4.5 Electrochemical biosensing for viral infections

Electrochemical biosensors are robust, easy to use, portable, and inexpensive analytical systems that can operate in turbid media and provide highly sensitive readouts [62]. DNA and RNA electrochemistry have been utilized to diagnose viral illnesses such as hepatitis E, coronavirus, HIV, influenza virus, bacterium, malaria, and Zika virus [63–67].

Aptamers are short, single-stranded oligonucleotides (i.e., DNA or RNA) that range in size from 10 to 100 nucleotides They are created using the SELEX method [68] which stands for the systematic evolution of ligands by exponential enrichment between the targets and the aptamers attached on the electrode surface of the sensor, the electrochemical apta-sensor determines the concentration of the interested targets [69]. Another example is the use of electrochemical techniques for detecting enzymes and hormones to monitor for pregnancy-related disease and cancer [70, 71]. In comparison to traditional procedures, such as Western blot and PCR, in terms of the time and cost, an electrochemical approach is a preferable option [72]. Nevertheless, its effectiveness is dependent on the conductivity characteristics of the sensing surface [73, 74]. Electrochemical performance with complex samples necessitates preventing signal overlapping due to the fact of interference [75].

2.4.6 Enzyme-based electro sensor applications

Enzymes are organic catalytic molecules created by living organisms. They accelerate biological processes by decreasing activation energy, and they can accelerate the conversion of substrates to products in cellular metabolism by a factor of at least 10 million [76]. Enzyme-mediated substrate conversion is very specific. Numerous enzymes are selective for a single substrate, whereas another type of enzyme can affect multiple structurally similar substrates. In order to begin an enzyme-catalyzed reaction, the enzyme must form a complex with its substrate. Enzymes are unaltered by the processes they catalyze and are recyclable and effective in minute quantities. Equally, the enzyme catalyzes either the forward or reverse process [77]. Enzymatic activity monitoring is in great demand. For measuring enzymatic activity, many analytical techniques have been reported, e.g., mass spectrometry [78], spectrophotometry, Raman spectroscopy, and electrochemical techniques. Because of their ease of use, cheap cost, and speed, electrochemical procedures are favored over other analytical techniques [79], which may need sophisticated pretreatment, filtering, and a knowledgeable operator. Enzymatic sensors are created by immobilizing an enzyme on an electrode and then used to determine the concentration of the matching substrate. The primary distinction between enzyme-based biosensors is the immobilization technique and the mediator used [80]. In a recent study, the authors constructed an amperometric Glc biosensor with Gox immobilized on multi-walled carbon nanotubes (MWCNTs), as the bio-recognition element, and RuO_2 acting as the mediator. To boost the sensor's stability, the enzyme was coated with a Nafion® membrane. The designed sensor was used to determine the concentrations of hydrogen peroxide and glycol. The developed sensor was employed as an electroanalytical technique for studying the inhibition of the enzyme's function, and the influence of the heavy metal cations (i.e., Cd^{2+}, Hg^{2+}, and Ag^{+}) on the activity of the Gox enzyme was examined [81].

2.4.7 Biosensors' distinct characteristics in health services

Diabetes prevalence and diabetes patients' use of biosensors are significant contributors to worldwide business profitability. Rapid and preventive diabetes detection is becoming increasingly popular. Biosensor developments have made it possible to detect blood glucose in the presence of various intervening substances throughout a wide temperature range. Using ZnO nanorods to detect glucose is a low-cost, accurate, rapid, and safe method [82]. The sensitivity and accuracy of biosensors within a minute sample volume are improving, and they are now widely employed in the diabetes domain, with significant market demand projected in the coming years. Portable electronic gadgets are an important part of the overall healthcare system because of their high capacity for monitoring, therapy, diagnosis, fitness, and well-being. They will increase preventative measures and obtain a better perspective on well-being by combining therapeutic technologies accessible in hospitals and emergency care centers. Technological advances and the increasing use of biosensors in several applications are driving the market [83]. People's lives have been enhanced by wearable biosensors [84].

2.4.8 Pathogen detection with electrochemical biosensors

One of the most common functions of chemical sensors is to generate a signal that can be analyzed [85]. Electrochemical sensors discern their objectives in a special manner. Pathogen detection strategies based on biosensors and the electrochemical approach share several characteristic features of specimen handling and device actions. The following section discusses electrical and chemical biosensing transmission, bio-recognition, and quantification layouts. Microbes can usually be defined by the presence of antibodies produced in an organism, which can be visible both during and after infectious disease. Both the bio-recognition element and the aim in such assays are antibodies. Electrochemical biosensors combine an analyte-receiving mechanism and an electrochemical transducer, where the interaction between the targeted analyte and the transducer generates an electrochemical signal in current, potential, resistance, or impedance format [86, 87]. There is a wide range of electrochemical biosensor schemes with different signal mechanisms, e.g., differential pulse voltammetry (DPV), voltammetric cyclic voltammetry (CV), polarography, square wave voltammetry (SWV), stripping voltammetry, alternating current voltammetry (ACV), and linear sweep voltammetry (LSV). Furthermore, electrochemical biosensors can use diverse types and forms of nanomaterials, nanoparticles, and nanocomposites to increase the sympathy of the detection mechanisms and to provide better detection limits through different strategies [88, 89].

2.4.9 Power supply for biosensor system

A significant part of the current research on wearable systems requires external power supply systems to carry out, which include computers, electrochemical workstations, and multimeters. Nowadays, most of the commercial wearable devices use traditional batteries such as lithium batteries and button cells, which limit the miniaturization and application of devices. Hence, efforts have been made on the power supply. Thus far, wearable systems can be expected to be powered by supercapacitors, solar cells, bioelectricity technology, physical hair technology, NFC technology, etc. Flexible self-healing supercapacitors, a competitive power supply system with high energy density, high charging and discharging efficiency, and excellent mechanical flexibility, meet substantially all the requirements for powering wearable sensing devices. Vu *et al* demonstrated a self-healing flexible supercapacitor based on a conductive composite electrode composed of polyurethane and carbon black (PU/CB) using a sandwich structure that provided excellent electrical performance and mechanical flexibility. The device has an electrical energy density of 5.8 μWh cm^{-2} at 1 mA cm^{-2} and 91% capacity retention during 10 000 charge/discharge cycles after breaking/healing [90]. Two-dimensional materials also emerge in this area. Kumar *et al* used graphene as a printing ink combined with 3D technology to produce a flexible supercapacitor without any additives [91]. Physical power generation techniques generally exploit the piezoelectric and Seebeck effects to power a thermodynamic generator by using ambient-skin-temperature differences. Lu *et al* designed a flexible piezoelectric nanogenerator through 3D nanoBCZT@Ag heterostructures. The device can be powered by human walking and can deliver 5.85 μA of current (38.6 V) [92]. Triboelectric nanogenerators (TENGs) are used to power devices by converting friction into electrical energy. Kim *et al* showed us a wearable ECG system based on a wearable thermoelectric generator (w-TEG) that can provide more than 13 $\mu W\ cm^{-2}$ of power for more than 22 h through temperature differences [93]. Guo *et al* presented a prototype of an all-in-one shape-adaptive self-charging power unit that can be used for scavenging random body motion energy under complex mechanical deformations. A kirigami paper-based supercapacitor (KP-SC) was designed to work as a flexible energy storage device (stretchability up to 215%). A stretchable and shape-adaptive silicone rubber triboelectric nanogenerator (SR-TENG) was utilized as a flexible energy harvesting device. By combining them with a rectifier, a stretchable, twistable, and bendable self-charging power package was achieved for sustainably driving wearable electronics. This work provides a potential platform for flexible self-powered systems [94]. Bioelectric conversion technologies are generally generated through redox reactions of electrolytes, enzymes, proteins, and other substances in body fluids. Falk *et al* constructed a self-powered glucose-sensing contact lens using ascorbate and oxygen from human tear fluid as the fuel and oxidant, able to deliver a stable current for up to 6 h [95]. Organic solar cells (OSCs) have attracted a lot of attention as a clean energy source due to their low cost, light weight, and flexibility. The large-area nonfullerene organic solar cells and modules are based on a flexible low-work function composite electrode (Ag grid/AgNWs:PEI-Zn). Large power conversion efficiencies (PCEs) with 13.1% and 12.6% are obtained with the solar cell areas of 6 and 10 cm^2, respectively, while the flexible module of 54 cm^2 achieves a PCE of 13.2% [96]. Remarkably, few integrated wearable sensing devices have been reported with both an excellent power supply system and excellent sensing capabilities; therefore, a collaborative effort by researchers from multiple research fields is required.

2.5 Immunosensors

Immunosensors are small, portable analytical tools that detect the development of antigen–antibody complexes and turn that event into an electrical signal that may be processed, stored, and displayed. In immunological biosensors, various transducing processes are used depending on whether a signal is generated (such as an optical or electrochemical signal) or a change in a property (such as mass changes) occurs after the creation of an antigen–antibody complex.

Yalow and Berson's use of a radiolabel in the late 1950s [97] to identify insulin using radioimmunoassay (RIA), that was a crucial step in the evolution of highly quantitative techniques, was a milestone in immunoassay technology. Since many clinical analyses may be automated and prescribed protocols deliver results of known quality, RIA remains the gold standard in many clinical analyses. In order to avoid the problems associated with dealing with radioactivity and to make the technological field portable, enzyme labels were first introduced in the 19th century. Nanomaterials are now being used as extremely sensitive labels for intricate matrices.

Immunoassay technology, which was initially created as a method for big biological molecules, was effectively used to detect pesticides in the 1960s and industrial chemicals, such as polychlorinated biphenyls and dioxins, a few years later [98, 99]. The development of an antigen–antibody complex is turned into an electrical signal in electrochemical immunosensors in the form of an electrical current (amperometric immunosensors), a voltage differential (potentiometric immunosensors), or a change in resistivity (conductimetric immunosensors).

The most popular kind of amperometric immunosensors can indeed be compared to ELISA tests with electrochemical sensing, in which the redox species produced by such a redox enzyme (enzymatic label) are transformed into a measurably large current. Piezoelectric transducers, including quartz crystal microbalances and microcantilevers, which vibrate at a given frequency sensitive, can be used in piezoelectric immunosensors to monitor the mass changes that occur after the creation of antigen–antibody formation. Piezoelectric devices can immobilize antigens or antibodies, and a vibration frequency shift caused by the development of the antigen–antibody combination can be detected with excellent sensitivity [100].

In optical immunosensors, the biologically sensitive component is immobilized on the transducer's surface and reacts to the interaction with the target analyte by either emitting a fluorescent signal or changing its optical characteristics, such as its absorption, reflectance, emission, refractive index, and optical path. A photodetector collects the optical signals, which are then transformed into electrical signals and analyzed electronically. Due to their ease of integration with electronic instruments, electrochemical immunosensors appear to be leading the way, whereas optical and piezoelectric transducers are more difficult to integrate. However, optical and piezoelectric transducers may soon find a wide range of practical applications due to their high sensitivity.

Sensors have the advantage of continuously and selectively detecting analytes, which produces a real-time response. In the pharmaceutical and food industries, sensors help quality control procedures for quick online analysis. By offering kinetic details on antibody–antigen reactions, they also aid proteomic research. A growing application for immunosensors is the monitoring of phosphorylated acetylcholinesterase as a sign of exposure to organophosphorous chemicals and chemical warfare agents [101].

The original purpose of environmental immunoassay methods was to lower the expenses associated with remedial initiatives through field analysis. The cost of personnel and equipment can be decreased by identifying hot spots on the ground to guide cleanup efforts at hazardous waste sites. On-site screening can identify samples that need additional laboratory testing and weed out negative samples, saving money on both shipping and analytical costs. Small sample sizes can be accommodated using immunoassay techniques, allowing for a dense sampling grid and an economical fingerprinting of a hazardous waste site. Monitoring studies for soil, sediment, surface and ground water, air, vegetation, and other matrices have used immunoassays for various substances of environmental significance (such as dioxins, pentachlorophenol, PCBs, paraquat, and other pesticides) [102]. Monitoring pollutants in urban areas has also found applications for field-deployable technologies.

The method was initially developed to reduce analytical costs at hazardous waste sites, but it has since been expanded to include the study of biomarkers of exposure. Analytical problems might arise from exposure monitoring studies since they usually call for the production of data for target analytes within predetermined time frames, exposure pathways, and exposure routes. In monitoring studies, immunoassay techniques have been used to identify occupational and non-occupational exposures to environmental contaminants in a wide range of matrices and exposure scenarios (such as urine, clothing patches, serum, air filters, building calking, dust, soil, and foods) (e.g., homes, orchards, contaminated waste sites, and day-care centers). In order to conduct high-throughput proteomic investigations to understand the structure and functional connections between proteins, multiplexed immunoassays are developing into robust and trustworthy instruments.

Development of cantilever-based immunosensors with distinctive enantio-selective antibodies has received a lot of attention recently [103]. These tools are mostly utilized for diagnostic biosensing for medical analysis as well as quality and process control in manufacturing. Furthermore, extremely flexible and valuable sensor platforms are created using mass-sensitive magnetoelastic immunosensors [104]. Since they have been around for a while, magnetoelastic sensors have an advantage over surface acoustic wave (SAW) sensors in terms of mass sensitivity. They might be considerably more affordable and smaller than SAW devices, though.

In order to achieve the highly sensitive detection of *E. coli*, Ruan *et al* presented a mass-sensitive magnetoelastic immunosensor based on the immobilization of affinity-purified antibodies on the surface of a micrometer-scale magnetoelastic cantilever [105]. Additionally, image ellipsometry (IE) has been created as a brand-new type of immunosensor, specifically for the detection of *Y. enterocolitica* pathogens [106]. Another illustration is a label-free multi-sensing immunosensor that is based on the integration of IE and protein chips and has been shown to be capable of simultaneously detecting several analytes and even monitoring multiple biological interaction processes *in situ* and in real time [107].

2.6 DNA sensors

Modern living greatly benefits from new biosensor research and advances. Nucleic acid recognition techniques-based DNA biosensors are being created for the quick, easy, and affordable diagnosis of viral and genetic illnesses. Additionally, the identification of a particular DNA sequence is important in a variety of contexts, including clinical, environmental, and dietary analysis. The use of biosensors is growing in the fields of bioterrorism [108], food analysis [109], environmental monitoring [110], and human health diagnostics [111].

A biosensor is often a compact device that uses biological recognition properties for a focused bio-analysis [112]. For these devices to translate biological signals into electrical signals or other signals proportionate to the concentration of analytes, a biological identification element must be intimately coupled with a physical transducer [113]. Since there is no longer a need for sample preparation, biosensors hold tremendous promise for a number of on-site analytical applications requiring quick and affordable measurements [114]. DNA is incredibly reliable, affordable, simple to change, and amenable to combinatorial selection.

The immobilized oligonucleotide (probe) and its counterpart strand (target) connect with affinity to create a stable double helix as a result of base pairing, which is the basis of how DNA sensors work. Typically, a moiety that makes it possible to identify and quantify the complimentary strand is attached to it. The increased emphasis has been brought about by the great sensitivity and quick reaction of biosensors. For the biosensing of DNA that is sequence-specific, electrochemical devices are highly helpful. Monitoring a current at a fixed voltage is typically required for electrochemical detection of DNA hybridization [115].

Using a differential pulse voltametric method and methylene blue as a DNA hybridization indicator, the avidin modified polyaniline electrochemically deposited onto a Pt disc electrode allowed for the direct detection of *E. coli* [116]. Similar to this, a polypyrrole-polyvinyl sulfonate–coated Pt disc electrode-based electrochemical DNA biosensor was created employing biotin–avidin binding [117].

The invention of an electrochemical DNA biosensor has a significant impact on the use of CNTs in DNA analysis. The invention of CNT-based biosensors and the application of arrayed CNT into DNA chips play key roles in DNA-based diagnostics in hospitals or at home [118]. The rise in current signal caused by a redox indicator (which detects the DNA duplex) or other hybridization-induced changes in electrochemical parameters is a typical way to detect hybridization (e.g. conductivity or capacitance). Improved single-strand (ss) and double-strand (ds) DNA discrimination are thanks to new redox indicators [119–121].

The identification of a particular DNA sequence is important in a variety of contexts, including dietary, environmental, and clinical analysis [122]. A crucial part of the rapid identification of genetic mutations is played by the analysis of gene sequences and the research of gene polymorphisms, which allows for an accurate diagnosis to be made even before any disease symptoms manifest. The detection of certain DNA sequences in the environmental and food sciences can be used to identify harmful bacteria or genetically modified organisms (GMOs). Single-stranded (ss) DNA probes can be immobilized on various electrodes and electroactive indicators can be used to assess the hybridization between the DNA probes and their complementary DNA strands to create DNA sensors [123–125]. By using DNA, it is frequently possible to selectively detect many significant metals at low parts-per-billion levels. Alkali metal Ions (sodium, potassium and cesium), alkaline earth metal Ions (magnesium, calcium, strontium and barium), lanthanide and actinide ions, post-transition metal Ions (lead, thallium and aluminum), transition metals (copper, zinc, mercury, cadmium, cobalt, manganese, nickel, Iron, molybdenum, tungsten and chromium), noble metals (silver, gold, platinum and palladium) can be detected by DNA sensors [126–139].

2.7 Conclusion

A biosensor is often a small device that uses biological recognition properties for a focused bio-analysis. For these devices to convert biological signals into electrical signals or other signals proportionate to the concentration of analytes, a biological identification element must be intimately coupled with a physical transducer. Since there is no longer a requirement for sample preparation, biosensors hold great promise for many on-site analytical applications requiring quick and affordable readings. A thorough introduction of electrochemical sensors, electrochemical sensing methods, gas sensors, biosensors, immunosensors, and DNA sensors is provided in this chapter. There is a description of the materials utilized in the sensor systems. These electrochemical gas sensors, biosensors, immunosensors, and DNA sensors have all been used in practical applications. Numerous studies that have been published in the literature and effectively used to detect a large variety of significant analytes show that electrochemical gas sensors, biosensors, immunosensors, and DNA sensors can all be used.

References

[1] Hulanicki A, Glab S and Ingman F O L K E 1991 Chemical sensors: definitions and classification *Pure Appl. Chem.* **63** 1247–50

[2] Miri P S, Khosroshahi N, Darabi Goudarzi M and Safarifard V 2021 MOF-biomolecule nanocomposites for electrosensing *Nanochem. Res.* **6** 213–22

[3] Shetti N P, Nayak D S, Reddy K R and Aminabhvi T M 2019 Graphene–clay-based hybrid nanostructures for electrochemical sensors and biosensors *Graphene-Based Electrochemical Sensors for Biomolecules* (Amsterdam: Elsevier) pp 235–74

[4] Oh S H, Altug H, Jin X, Low T, Koester S J, Ivanov A P, Edel J B, Avouris P and Strano M S 2021 Nanophotonic biosensors harnessing van der Waals materials *Nat. Commun.* **12** 1–18

[5] Bhadoria R and Chaudhary H S 2011 Recent advances of biosensors in biomedical sciences *Int. J. Drug Deliv.* **3** 571

[6] Chen C and Wang J 2020 Optical biosensors: an exhaustive and comprehensive review *Analyst* **145** 1605–28

[7] Sun M, Pei X, Xin T, Liu J, Ma C, Cao M and Zhou M 2022 A flexible microfluidic chip-based universal fully integrated nanoelectronic system with point-of-care raw sweat, tears, or saliva glucose monitoring for potential noninvasive glucose management *Anal. Chem.* **94** 1890–900

[8] Chaibun T, Puenpa J, Ngamdee T, Boonapatcharoen N, Athamanolap P, O'Mullane A P, Vongpunsawad S, Poovorawan Y, Lee S Y and Lertanantawong B 2021 Rapid electrochemical detection of coronavirus SARS-CoV-2 *Nat. Commun.* **12** 1–10

[9] Singh A, Sharma A, Ahmed A, Sundramoorthy A K, Furukawa H, Arya S and Khosla A 2021 Recent advances in electrochemical biosensors: applications, challenges, and future scope *Biosensors* **11** 336

[10] Tan P, Li H, Wang J and Gopinath S C B 2020 Silver nanoparticle in biosensor and bioimaging: clinical perspectives *Biotechnol. Appl. Biochem.* **68** 1236–42

[11] Zhao Z, Lei W, Zhang X, Wang B and Jiang H 2010 ZnO-based amperometric enzyme biosensors *Sensors* **10** 1216–31

[12] Wang X, Yang S, Shan J and Bai X 2022 Novel electrochemical acetylcholinesterase biosensor based on core–shell covalent organic framework@multi-walled carbon nanotubes (COF@MWCNTs) composite for detection of malathion *Int. J. Electrochem. Sci.* **17** 220543

[13] Strehlitz B, Nikolaus N and Stoltenburg R 2008 Protein detection with aptamer biosensors *Sensors* **8** 4296–307

[14] Fu H, Zhang S, Chen H and Weng J 2015 Graphene enhances the sensitivity of fiber-optic surface plasmon resonance biosensor *IEEE Sens. J.* **15** 5478–82

[15] Kucherenko I S, Soldatkin O O, Dzyadevych S and Soldatkin A P 2020 Electrochemical biosensors based on multienzyme systems: main groups, advantages and limitations—a review *Anal. Chim. Acta* **1111** 114–31

[16] Furst A L and Francis M B 2019 Impedance-based detection of bacteria *Chem. Rev.* **119** 700–26

[17] Cesewski E and Johnson B N 2020 Electrochemical biosensors for pathogen detection *Biosens. Bioelectron.* **159** 112214

[18] Ropero-Vega J L, Redondo-Ortega J F, Galvis-Curubo Y J, Rondón-Villarreal P and Flórez-Castillo J M 2021 A bioinspired peptide in TIR protein as recognition molecule on electrochemical biosensors for the detection of *E. coli* O157:H7 in an aqueous matrix *Molecules* **26** 2559

[19] Pandey R, Lu Y, Osman E, Saxena S, Zhang Z, Qian S, Pollinzi A, Smieja M, Li Y, Soleymani L *et al* 2022 DNAzyme-immobilizing microgel magnetic beads enable rapid, specific, culture-free, and wash-free electrochemical quantification of bacteria in untreated urine *ACS Sens.* **7** 985–94

[20] Salimiyan Rizi K, Hatamluyi B, Darroudi M, Meshkat Z, Aryan E, Soleimanpour S and Rezayi M 2022 PCR-free electrochemical genosensor for Mycobacterium tuberculosis complex detection based on two-dimensional Ti_3C_2 Mxene-polypyrrole signal amplification *Microchem. J.* **179** 107467

[21] Li G, Qi X, Wu J, Xu L, Wan X, Liu Y, Chen Y and Li Q 2022 Ultrasensitive, label-free voltammetric determination of norfloxacin based on molecularly imprinted polymers and Au nanoparticle-functionalized black phosphorus nanosheet nanocomposite *J. Hazard. Mater.* **436** 129107

[22] Morales M A and Halpern J M 2018 Guide to selecting a biorecognition element for biosensors *Bioconjug. Chem.* **29** 3231–9

[23] Rodrigues J L, Ferreira D and Rodrigues L R 2017 Synthetic biology strategies towards the development of new bioinspired technologies for medical applications *Bioinspired Materials for Medical Applications* ed L Rodrigues and M Mota (Kidlington: Woodhead Publishing) pp 451–97

[24] Yu H, Alkhamis O, Canoura J, Liu Y and Xiao Y 2021 Advances and challenges in small-molecule DNA aptamer isolation, characterization, and sensor development *Angew. Chem. Int. Ed.* **60** 16800–23

[25] Sola M, Menon A P, Moreno B, Meraviglia-Crivelli D, Soldevilla M M, Cartón-García F and Pastor F 2020 Aptamers against live targets: is *in vivo* SELEX finally coming to the edge? *Mol. Ther. Nucleic Acids* **21** 192–204

[26] Gupta R, Kumar A, Kumar S, Pinnaka A K and Singhal N K 2021 Naked eye colorimetric detection of Escherichia coli using aptamer conjugated graphene oxide enclosed gold nanoparticles *Sens. Actuators* B **329** 129100

[27] Savory N, Lednor D, Tsukakoshi K, Abe K, Yoshida W, Ferri S, Jones B and Ikebukuro K 2013 In silico maturation of binding-specificity of DNA aptamers against *Proteus mirabilis Biotechnol. Bioeng.* **110** 2573–80

[28] Wang C H, Wu J J and Lee G B 2019 Screening of highly-specific aptamers and their applications in paper-based microfluidic chips for rapid diagnosis of multiple bacteria *Sens. Actuators* B **284** 395–402

[29] Kaur H, Shorie M and Sabherwal P 2020 Electrochemical aptasensor using boron-carbon nanorods decorated by nickel nanoparticles for detection of *E. coli* O157:H7 *Microchim. Acta* **187** 461

[30] Shoaib M, Shehzad A, Mukama O, Raza H, Niazi S, Khan I M, Ali B, Akhtar W and Wang Z 2020 Selection of potential aptamers for specific growth stage detection of *Yersinia enterocolitica RSC Adv.* **10** 24743–52

[31] Sheng K, Jiang H, Fang Y, Wang L and Jiang D 2022 Emerging electrochemical biosensing approaches for detection of allergen in food samples: a review *Trends Food Sci. Technol.* **121** 93–104

[32] Razmi N, Hasanzadeh M, Willander M and Nur O 2020 Recent progress on the electrochemical biosensing of *Escherichia coli* O157:H7: material and methods overview *Biosensors* **10** 54

[33] Cinti S, Volpe G, Piermarini S, Delibato E and Palleschi G 2017 Electrochemical biosensors for rapid detection of foodborne Salmonella: a critical overview *Sensors* **17** 1910

[34] Subjakova V, Oravczova V, Tatarko M and Hianik T 2021 Advances in electrochemical aptasensors and immunosensors for detection of bacterial pathogens in food *Electrochim. Acta* **389** 138724

[35] Linke D, Riess T, Autenrieth I B, Lupas A and Kempf V A J 2006 Trimeric autotransporter adhesins: variable structure, common function *Trends Microbiol.* **14** 264–70

[36] Thévenot D R, Toth K, Durst R A and Wilson G S 1999 Electrochemical biosensors: recommended definitions and classification *Pure Appl. Chem.* **71** 2333

[37] Clark L C, Lyons C and Ann N Y 1962 Electrode systems for continuous monitoring in cardiovascular surgery *Acad. Sci.* **102** 29

[38] Updike S J and Hicks G P 1967 The enzyme electrode *Nature* **214** 986

[39] Kulys J J, Samalins A S and Svirmickas G J S 1980 Electron exchange between the enzyme active center and organic metal *FEBS Lett.* **144** 7

[40] Garcia C A B, Neto G D and Kubota L T 1998 New fructose biosensors utilizing a polypyrrole film and d-fructose 5-dehydrogenase immobilized by different processes *Anal. Chim. Acta* **374** 201

[41] Nunes G S, Skladal P, Yamnaka H and Barcelo D 1998 Determination of carbamate residues in crop samples by cholinesterase-based biosensors and chromatographic techniques *Anal. Chim. Acta* **362** 59

[42] Alfaya A A S and Kubota L T 2002 A utilização de materiais obtidos pelo processo de sol-gel na construção de biossensores *Quim. Nova* **25** 835–41

[43] Freire R S, Duran N, Kubota L T and Braz J 2002 Electrochemical biosensor-based devices for continuous phenols monitoring in environmental matrices *J. Braz. Chem. Soc.* **13** 456

[44] Brett A M O, Serrano S H P, Gutz I G and LaScalea M A 1997 Electrochemical reduction of metronidazole at a DNA-modified glassy carbon electrode *Bioeletrochem. Bioenerg.* **42** 175

[45] Brett A M O, Serrano S H P, Gutz I, LaScalea M A and Cruz M L 1997 Voltammetric behavior of nitroimidazoles at a DNA-biosensor *Electroanalysis* **9** 1132

[46] Fatibello-Filho O and Vieira I D 2002 Uso analítico de tecidos e de extratos brutos vegetais como fonte enzimática *Quim. Nova* **25** 455

[47] Vieira L C, Lupetti K O and Fatibello-Filho O 2002 Sweet potato (ipomoea batatas (L.) lam.) tissue as a biocatalyst in a paraffin/graphite biosensor for hydrazine determination in boiler feed water *Anal. Lett.* **35** 2221

[48] Fatibello-Filho O, Vieira L C and Lupetti K O 2001 Chronoamperometric determination of paracetamol using an avocado tissue (Persea americana) biosensor *Talanta* **55** 685

[49] Vieira I C, Lupetti K O and Fatibello-Filho O 2003 Determination of paracetamol in pharmaceutical products using a carbon paste biosensor modified with crude extract of zucchini (Cucurbita pepo) *Quim. Nova* **26** 39–43

[50] Vieira I D, Fatibello-Filho O and Angnes L 1999 Zucchini crude extract-palladium-modified carbon paste electrode for the determination of hydroquinone in photographic developers *Anal. Chim. Acta* **398** 145

[51] Lima A W O, Vidsiunas E K, Nascimento V B and Angnes L 1998 Vegetable tissue from *latania* sp.: an extraordinary source of naturally immobilized enzymes for the detection of phenolic compounds *Analyst* **123** 2377

[52] Knutson S D, Sanford A A, Swenson C S, Korn M M, Manuel B A and Heemstra J M 2020 Thermoreversible control of nucleic acid structure and function with glyoxal caging *J. Am. Chem. Soc.* **142** 17766–81

[53] Hannocks M J, Zhang X, Gerwien H, Chashchina A, Burmeister M, Korpos E, Song J and Sorokin L 2019 The gelatinases, MMP-2 and MMP-9, as fine tuners of neuroinflammatory processes *Matrix Biol.* **75–76** 102–13

[54] Lu M, Flanagan J U, Langley R J, Hay M P and Perry J K 2019 Targeting growth hormone function: strategies and therapeutic applications *Signal Transduct. Target. Ther.* **4** 3

[55] Kurbanoglu S, Ozkan S A and Merkoçi A 2017 Nanomaterials-based enzyme electrochemical biosensors operating through inhibition for biosensing applications *Biosens. Bioelectron.* **89** 886–98

[56] Wang C-F, Sun X-Y, Su M, Wang Y-P and Lv Y-K 2020 Electrochemical biosensors based on antibody, nucleic acid and enzyme functionalized graphene for the detection of disease-related biomolecules *Analyst* **145** 1550–62

[57] Asal M, Özen Ö, Sahinler M and Polato ğlu I 2018 Recent developments in enzyme, DNA and immuno-based biosensors *Sensors* **18** 1924

[58] Maduraiveeran G, Sasidharan M and Ganesan V 2018 Electrochemical sensor and biosensor platforms based on advanced nanomaterials for biological and biomedical applications *Biosens. Bioelectron.* **103** 113–29

[59] Ezzati Nazhad Dolatabadi J and de la Guardia M 2014 Nanomaterial-based electrochemical immunosensors as advanced diagnostic tools *Anal. Methods* **6** 3891–900

[60] Cheng M, Cuda G, Bunimovich Y, Gaspari M, Heath J, Hill H, Mirkin C, Nijdam A, Terracciano R and Thundat T 2006 Nanotechnologies for biomolecular detection and medical diagnostics *Curr. Opin. Chem. Biol.* **10** 11–9

[61] Mohan R, Mach K E, Bercovici M, Pan Y, Dhulipala L, Wong P K and Liao J C 2011 Clinical validation of integrated nucleic acid and protein detection on an electrochemical biosensor array for urinary tract infection diagnosis *PLoS One* **6** e26846

[62] Campuzano S, Yáñez-Sedeño P and Pingarrón J M 2017 Electrochemical biosensing for the diagnosis of viral infections and tropical diseases *ChemElectroChem.* **4** 753–77

[63] Faria H A M and Zucolotto V 2019 Label-free electrochemical DNA biosensor for zika virus identification *Biosens. Bioelectron.* **131** 149–55

[64] Salimian R, Shahrokhian S and Panahi S 2019 Enhanced electrochemical activity of a hollow carbon sphere/polyaniline-based electrochemical biosensor for HBV DNA marker detection *ACS Biomater. Sci. Eng.* **5** 2587–94

[65] Shabaninejad Z, Yousefi F, Movahedpour A, Ghasemi Y, Dokanehiifard S, Rezaei S, Aryan R, Savardashtaki A and Mirzaei H 2019 Electrochemical-based biosensors for microRNA detection: nanotechnology comes into view *Anal. Biochem.* **581** 113349

[66] Cui F, Zhou Z and Zhou H S 2020 Molecularly imprinted polymers and surface imprinted polymers based electrochemical biosensor for infectious diseases *Sensors* **20** 996

[67] Chowdhury A D, Takemura K, Li T-C, Suzuki T and Park E Y 2019 Electrical pulse-induced electrochemical biosensor for hepatitis E virus detection *Nat. Commun.* **10** 3737

[68] Ellington A D and Szostak J W 1990 *In vitro* selection of RNA molecules that bind specific ligands *Nature* **346** 818–22

[69] Lou B, Liu Y, Shi M, Chen J, Li K, Tan Y, Chen L, Wu Y, Wang T, Liu X *et al* 2022 Aptamer-based biosensors for virus protein detection *TrAC, Trends Anal. Chem.* **157** 116738

[70] Khunseeraksa V, Kongkaew S, Thavarungkul P, Kanatharana P and Limbut W 2020 Electrochemical sensor for the quantification of iodide in urine of pregnant women *Microchim. Acta* **187** 591

[71] Crulhas B P, Basso C R, Castro G R and Pedrosa V A 2021 Review—recent advances based on a sensor for cancer biomarker detection *ECS J. Solid State Sci. Technol.* **10** 047004

[72] Mishra G, Barfidokht A, Tehrani F and Mishra R 2018 Food safety analysis using electrochemical biosensors *Foods* **7** 141

[73] Luong J H T, Narayan T, Solanki S and Malhotra B D 2020 Recent advances of conducting polymers and their composites for electrochemical biosensing applications *J. Funct. Biomater.* **11** 71

[74] Sedlackova E, Bytesnikova Z, Birgusova E, Svec P, Ashrafi A M, Estrela P and Richtera L 2020 Label-free DNA biosensor using modified reduced graphene oxide platform as a DNA methylation assay *Materials* **13** 4936

[75] Suhito I R, Koo K-M and Kim T-H 2020 Recent advances in electrochemical sensors for the detection of biomolecules and whole cells *Biomedicines* **9** 15

[76] Mäntsälä P and Niemi J 2009 Enzymes: the biological catalysts of life *in Physiology and Maintenance—Volume II* ed O O P Hanninen and M Atalay (Abu Dhabi: EOLSS Publications) pp 1–22

[77] Aledo J C, Lobo C and del Valle A E 2003 Energy diagrams for enzyme-catalyzed reactions: concepts and misconcepts *Biochem. Mol. Biol. Educ.* **31** 234–6

[78] Mohammed A M, Rahim R A, Ibraheem I J, Loong F K, Hisham H, Hashim U and Al-Douri Y 2014 Application of gold nanoparticles for electrochemical DNA biosensor *J. Nanomater.* **2014** 1–7

[79] Kavita V 2017 DNA biosensors—a review *J. Bioeng. Biomed. Sci.* **7** 2

[80] Ashrafi A M, Sýs M, Sedláčková E, Farag A S, Adam V, Přibyl J and Richtera L 2019 Application of the enzymatic electrochemical biosensors for monitoring non-competitive inhibition of enzyme activity by heavy metals *Sensors* **19** 2939

[81] Marie M, Mandal S and Manasreh O 2015 An electrochemical glucose sensor based on zinc oxide nanorods *Sensors* **15** 18714–23

[82] Haleem A, Javaid M, Singh R P, Suman R and Rab S 2021 Biosensors applications in medical field: a brief review *Sensors* **2** 100100

[83] Sciutto G, Zangheri M, Anfossi L, Guardigli M, Prati S, Mirasoli M, Di Nardo F, Baggiani C, Mazzeo R and Roda A 2018 Miniaturized biosensors to preserve and monitor cultural heritage: from medical to conservation diagnosis *Angew. Chem.* **130** 7507–11

[84] Caygill R L, Blair G E and Millner P A 2010 A review on viral biosensors to detect human pathogens *Anal. Chim. Acta* **681** 8–15

[85] Thévenot D R, Toth K, Durst R A and Wilson G S 2001 Electrochemical biosensors: recommended definitions and classification *Biosens. Bioelectron.* **16** 121–31

[86] Kaya H O, Cetin A E, Azimzadeh M and Topkaya S N 2021 Pathogen detection with electrochemical biosensors: advantages, challenges and future perspectives *J. Electroanal. Chem.* **882** 114989

[87] Dyussembayev K, Sambasivam P, Bar I, Brownlie J C, Shiddiky M J A and Ford R 2021 Biosensor technologies for early detection and quantification of plant pathogens *Front. Chem.* **9** 636245

[88] Topkaya S N, Azimzadeh M and Ozsoz M 2016 Electrochemical biosensors for cancer biomarkers detection: recent advances and challenges *Electroanalysis* **28** 1402–19

[89] Srimaneepong V, Rokaya D, Thunyakitpisal P, Qin J and Saengkiettiyut K 2020 Corrosion resistance of graphene oxide/silver coatings on Ni–Ti alloy and expression of Il-6 and Il-8 in human oral fibroblasts *Sci. Rep.* **10** 3247

[90] Vu V-P, Mai V-D, Nguyen D C T and Lee S-H 2022 Flexible and self-healable supercapacitor with high capacitance restoration *ACS Appl. Energy Mater.* **5** 2211–20

[91] Kumar S, Goswami M, Singh N, Soni P, Sathish N and Kumar S 2022 Pristine graphene-ink for 3D-printed flexible solid-state supercapacitor *Carbon Lett.* **32** 979–85

[92] Lu H, Shi H, Chen G, Wu Y, Zhang J, Yang L, Zhang Y and Zheng H 2021 High-performance flexible piezoelectric nanogenerator based on specific 3D nano BCZT@Ag hetero-structure design for the application of self-powered wireless sensor system *Small* **17** 2101333

[93] Kim C S, Yang H M, Lee J, Lee G S, Choi H, Kim Y J, Lim S H, Cho S H and Cho B J 2018 Self-powered wearable electrocardiography using a wearable thermoelectric power generator *ACS Energy Lett.* **3** 501–7

[94] Guo H, Yeh M-H, Lai Y-C, Zi Y, Wu C, Wen Z, Hu C and Wang Z L 2016 All-in-one shape-adaptive self-charging power package for wearable electronics *ACS Nano* **10** 10580–8

[95] Falk M, Andoralov V, Silow M, Toscano M D and Shleev S 2013 Miniature biofuel cell as a potential power source for glucosesensing contact lenses *Anal. Chem.* **85** 6342–8

[96] Qin F, Sun L, Chen H, Liu Y, Lu X, Wang W, Liu T, Dong X, Jiang P, Jiang Y *et al* 2021 54 cm^2 large-area flexible organic solar modules with efficiency above 13% *Adv. Mater.* **33** 2103017

[97] Yalow R S and Berson S A 1960 Immunoassay of endogenous plasma insulin in man *J. Clin. Invest.* **39** 1157–75

[98] Chuang J C, Van Emon J M and Schrock M E 2009 High-throughput screening of dioxins in sediment and soil using selective pressurized liquid extraction with immunochemical detection *Chemosphere* **77** 1217–23

[99] Van Emon J M and Lopez-Avila V 1992 Immunochemical methods for environmental analysis *Anal. Chem.* **64** 79–88

[100] Raiteri R, Grattarola M, Butt H-J and Skládal P 2001 Micromechanical cantilever-based biosensors *Sensrors Actuators* B **79** 115–26

[101] Gonzalez-Martinez M A, Puchades R and Maquieira A 2007 Optical immunosensors for environmental monitoring: how far have we come? *Anal. Bioanal. Chem.* **387** 205–18
[102] Van Emon J M 2001 Immunochemical applications in environmental science *J. Assoc. Off. Anal. Chem. Int.* **84** 125–33
[103] Lee J H, Hwang K S, Park J, Yoon K H, Yoon D S and Kim T S 2005 Immunoassay of prostatespecific antigen (PSA) using resonant frequency shift of piezoelectric nanomechanical microcantilever *Biosens. Bioelectron.* **20** 2157–62
[104] Grimes C A, Mungle C S, Zeng K, Jain M K, Dreschel W R, Paulose M and Ong K G 2002 Invited paper: wireless magnetoelastic resonance sensors: a critical review *Sensors* **2** 289–308
[105] Ruan C M, Zeng K F, Varghese O K and Grimes C A 2003 Magnetoelastic immunosensors: amplified mass immunosorbent assay for detection of *Escherichia coli* O157:H7 *Anal. Chem.* **75** 6494–8
[106] Bae Y M, Oh B K, Lee W, Lee W H and Choi J W 2004 Immunosensor for detection of *Yersinia enterocolitica* based on imaging ellipsometry *Anal. Chem.* **76** 1799–803
[107] Wang Z H and Jin G 2003 A label-free multisensing immunosensor based on imaging ellipsometry *Anal. Chem.* **75** 6119–23
[108] Burkle F M 2003 Measures of effectiveness in large scale bioterrorism events *Prehosp. Disaster Med.* **18** 258–62
[109] Eden-Firstenberg R and Schaertel B J 1988 Biosensor in the food industry: present and future *J. Food Prot.* **51** 811–20
[110] Maseini M 2001 Affinity electrochemical biosensors for pollution control *Pure Appl. Chem.* **73** 23–30
[111] Malhotra B D and Chaube A 2003 Biosensors for clinical diagnostics industry *Sens. Actuators* B **9** 117–26
[112] Prummond T G 2003 Electrochemical DNA sensors *Nat. Biotechnol.* **21** 1192–9
[113] Berdat D, Marin A, Herrera F and Gijs M A M 2006 DNA biosensors using fluorescence microscopy andimpedance spectroscopy *Sens. Actuators* B **118** 53–9
[114] Sharma K, Sehgal N and Kumar A 2003 Biomolecules for development of biosensors and their application *Curr. Appl. Phys.* **3** 307–16
[115] Yau H C M, Chan H L and Yang M 2003 Electrochemical properties of DNA intercalating doxorubicin andmethylene blue on n-hexadecylmercaptandoped 5′-thiol-labeled DNA-modified gold electrodes *Biosens. Bioelectron.* **18** 873–9
[116] Arora K, Prabhakar N, Chand S and Malhotra B D 2007 Ultrasensitive DNA hybridization biosensor based on polyaniline *Biosens. Bioelectron.* **23** 613–20
[117] Yang W, Ozsoz M, Hibbert D B and Gooding J J 2002 Evidence for the direct interaction between methylene blue and guanine bases using DNA-modified carbon paste electrodes *Electroanalysis* **14** 1299–302
[118] Gu J, Lu X and Ju H 2002 DNA sensor for recognition of native yeast DNA sequence with methylene blue as an electrochemical hybridization indicator *Electroanalysis* **14** 949–53
[119] Reddy R R K, Chadha A and Bhattacharya E 2001 Porous silicon based potentiometric triglyceride biosensor *Biosens. Bioelectron.* **16** 313–7
[120] Kara P, Kerman K, Ozkan D, Meric B, Erdem A *et al* 2002 Electrochemical genosensor for the detection of interaction between methylene blue and DNA *Electrochem. Commun.* **4** 705–9

[121] Erdem A, Kerman K, Meric B and Ozsoz M 2001 Methylene blue as a novel electrochemical hybridization indicator *Electroanalysis* **13** 219–23
[122] Berney H, West J, Haefele H, Alderman J, Lane W *et al* 2000 DNA diagnostic biosensor development, characterization and performance *Sens. Actuators* B **68** 100–8
[123] Campbell C N, Gal D, Cristler N, Banditrat C and Heller A 2002 Enzyme amplified amperometric sandwich test for RNA and DNA *Anal. Chem.* **74** 158–62
[124] Millan K M and Mikkelsen S R 1993 Sequence selection biosensor for DNA based on electro active hybridization indicators *Anal. Chem.* **65** 2317–24
[125] Prabhakar N, Arora K, Singh S P, Singh H and Malhotra B D 2007 DNA entrapped polypyrrole-polyvinyl sulfonate film for application to electrochemical biosensor *Anal. Biochem.* **366** 71–9
[126] Zhang J F, Zhou Y, Yoon J and Kim J S 2011 Recent progress in fluorescent and colorimetric chemosensors for detection of precious metal ions (silver, gold and platinum ions) *Chem. Soc. Rev.* **40** 3416–29
[127] Hareesha N, Manjunatha J G, Amrutha B M, Pushpanjali P A, Charithra M M and Prinith Subbaiah N 2021 Electrochemical analysis of indigo carmine in food and water samples using a poly (glutamic acid) layered multi-walled carbon nanotube paste electrode *J. Electron. Mater.* **50** 1230–8
[128] Tigari G and Manjunatha J G 2020 Optimized voltammetric experiment for the determination of phloroglucinol at surfactant modified carbon nanotube paste electrode *Instrum. Exp. Tech.* **63** 750–7
[129] Manjunatha J G 2020 Poly (adenine) modified graphene-based voltammetric sensor for the electrochemical determination of catechol, hydroquinone and resorcinol *Open Chem. Eng. J.* **14** 52–62
[130] Manjunatha Charithra M and Manjunatha J G 2020 Electrochemical sensing of paracetamol using electropolymerised and sodium lauryl sulfate modified carbon nanotube paste electrode *ChemistrySelect* **5** 9323–9
[131] Pushpanjali P A and Manjunatha J G 2020 Development of polymer modified electrochemical sensor for the determination of alizarin carmine in the presence of tartrazine *Electroanalysis* **32** 2474–80
[132] Tigari G and Manjunatha J G 2020 Poly (glutamine) film-coated carbon nanotube paste electrode for the determination of curcumin with vanillin: an electroanalytical approach *Monatsh. Chem.* **151** 1681–8
[133] Manjunatha J G, Raril C, Hareesha N, Charithra M M, Pushpanjali P A, Tigari G, Ravishankar D K, Mallappaji S C and Gowda J 2020 Electrochemical fabrication of poly (niacin) modified graphite paste electrode and its application for the detection of riboflavin *Open Chem. Eng. J.* **14** 90–8
[134] Manjunatha J G and Hussain C M 2022 *Carbon Nanomaterials-Based Sensors: Emerging Research Trends in Devices and Applications* (Elsevier)
[135] Raril C, Manjunatha J G, Ravishankar D K, Fattepur S, Siddaraju G and Nanjundaswamy L 2020 Validated electrochemical method for simultaneous resolution of tyrosine, uric acid, and ascorbic acid at polymer modified nano-composite paste electrode *Surf. Eng. Appl. Electrochem.* **56** 415–26
[136] Prinith N S, Manjunatha J G and Hareesha N 2021 Electrochemical validation of L-tyrosine with dopamine using composite surfactant modified carbon nanotube electrode *J. Iran. Chem. Soc.* **18** 3493–503

[137] Pushpanjali P A, Manjunatha J G and Srinivas M T 2020 Highly sensitive platform utilizing poly (l-methionine) layered carbon nanotube paste sensor for the determination of voltaren *FlatChem.* **24** 100207
[138] Raril C and Manjunatha J G 2020 Fabrication of novel polymer-modified graphene-based electrochemical sensor for the determination of mercury and lead ions in water and biological samples *J. Anal. Sci. Technol.* **11** 1–10
[139] Manjunatha J G 2020 A surfactant enhanced graphene paste electrode as an effective electrochemical sensor for the sensitive and simultaneous determination of catechol and resorcinol *Chem. Data Coll.* **25** 100331

IOP Publishing

Real-Time Applications of Advanced Electrochemical Sensing Devices

Jamballi G Manjunatha

Chapter 3

A brief experimental protocol to carry out electroanalysis

G Amala Jothi Grace, P Karpagavinayagam, N Senthilkumar and C Vedhi

3.1 Introduction

The method for determining the amount of analyte in a specimen is known as electroanalysis. The current or the potential difference between the two electrodes is measured. Metals have played an important role in history. Accurate results are obtained for electroanalytical methods. These techniques are low cost and are easy to handle. Electrochemistry is an important tool in the medicinal field, enhancing research and significant in all fields. By this technique, the potential and current of an analyte is measured [1–4]. Because boron can oxidize or decrease other compounds in solution, it can provide superior performance when used in conjunction with boron-doped diamond electrodes. Electrical measurements are used in industry for quality control and environmental monitoring [5].

The impact of electroanalysis in industrial applications is currently significant. Conductometry, potentiometry, coulometry, voltammetry, amperometry, electrogravimetry, fluorometry, chromatography, spectrophotometry and photothermometry are the important electroanalytical methods. By the electroanalysis technique, we can measure conductance, potential and current of an analyte. The majority of the chemical compounds were discovered to be electrochemically active. There has been a remarkable evolution of advancement in the discovery, synthesis, and sensitive electrochemical analysis over the last few years. The purpose of this chapter is to provide basic information on electroanalytical analysis methods, working electrodes, techniques, and industrial, pharmaceutical, and environmental applications. An attempt was made to select some easily accessible publications that describe some advances in methodology and applications. In this chapter, the protocol to be carried out for electroanalysis is discussed. Some techniques in electroanalysis are used for both qualitative and quantitative analysis [6–9].

doi:10.1088/978-0-7503-5377-9ch3

3.2 Background of the electrochemical techniques

The identification of materials using techniques provides effective results. Every instrument has specific features and methods. The protocol is working while choosing the techniques. The different kinds of electrochemical techniques are summarized below with brief explanations [10–14].

3.2.1 Voltammetry

Voltammetry is based on the voltage–current–time relationship that occurs in a cell of three electrodes: the working electrode, the reference electrode, and the auxiliary or counter electrode. This relationship can be explained by applying potential (E) to the working electrode and recording the resulting current I flowing through the electrochemical cell. The applied potential can be changed, or the resulting current can be measured over time (t). The potential applied to the working electrode drives the reaction; it is the parameter that causes the chemical species present in a solution to be electrolyzed (reduced or oxidized) at the electrode surface. The analyte is concentrated into or onto the surface of the working electrode during a stripping analysis. Following the preconcentration step, an electrochemical measurement of the preconcentrated analyte or stripping from the electrode surface is performed using a potential scan. Voltammetry involves a three-electrode setup and it consists of reference electrode, indicator electrode and counter electrode. The current that flows between the electrodes is measured. A specific voltage is applied to a working electrode as a function of time and the current produced is measured. In voltammetric methods diffusion, convection and electrostatic migration will take place in the solution. This technique is helpful in bioanalysis to a great extent. Minimal electrolysis will occur in normal pulse voltammetry when the electrode is initially held at a voltage. The current is recorded at a point within that pulse. This process is usually repeated, with increasing potential applied in each pulse [15–22] (figure 3.1).

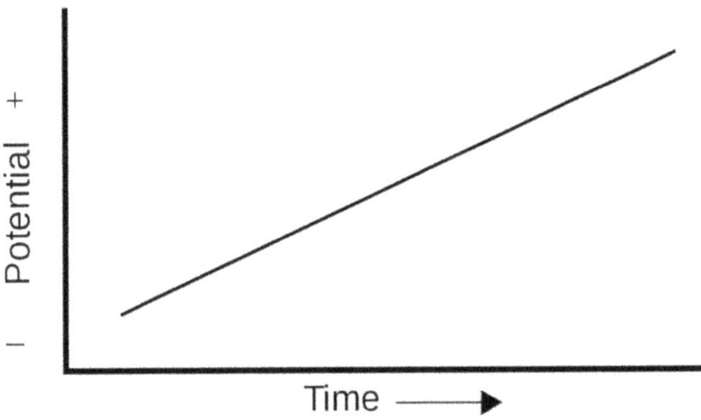

Figure 3.1. Stripping voltammetry techniques (as modern electrochemical methods).

3.2.2 Anodic stripping voltammetry (ASV)

The most common type of stripping voltammetry is anodic stripping voltammetry (ASV). The implementation of a sufficiently negative potential to initiate reduction or oxidation results in the deposition of metal ions into a mercury electrode as an amalgam or into another solid electrode as a metal. ASV is widely used for testing the quality of drinking water, surface water, and sewage or waste water, which is water that leaves the treatment plant before being discharged into surface water [23, 24] (figure 3.2).

3.2.3 Cathodic stripping analysis (CSV)

Cathodic stripping voltammetry (CSV) is a technique for analyzing low levels of analytes (primarily trace metals and sulfur-containing organic compounds) in aqueous solutions that is based on the measurement of a reductive current response as a function of a potential scan toward more negative. It is similar to the trace analysis method ASV, except that the potential is held at an oxidizing potential during the plating step, and the oxidized species are stripped from the electrode by negatively sweeping the potential [25–27] (figure 3.3).

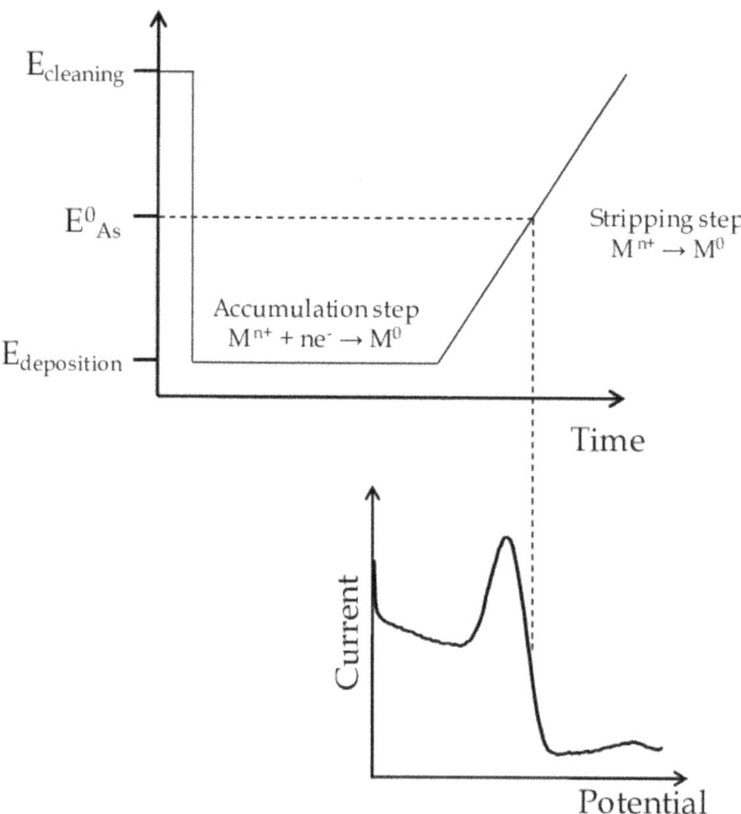

Figure 3.2. Anodic stripping voltammetry.

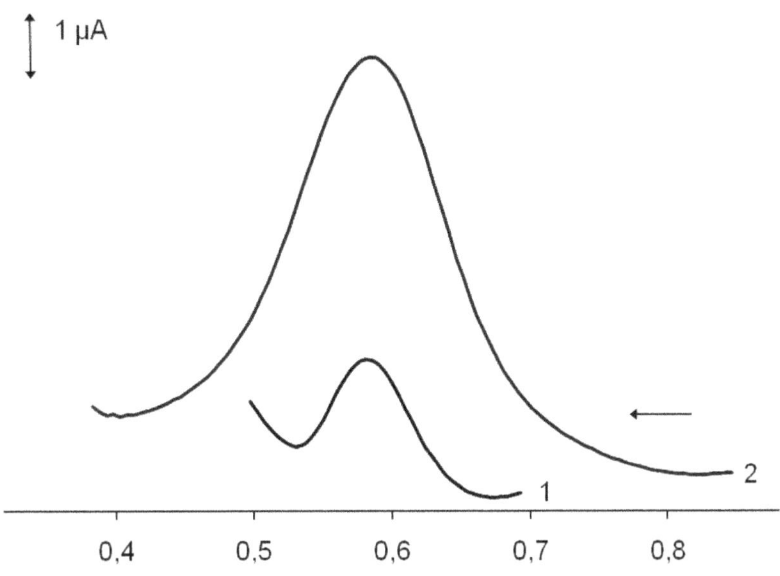

Figure 3.3. Cathodic stripping analysis.

3.2.4 Cyclic voltammetry

Cyclic voltammetry (CV) has emerged as a prominent and frequently used technique in many areas of electroanalytical chemistry. It is rarely used for quantification, but it is widely used for studying oxidation and reduction as well as learning about the redox process. CV is a rapid voltage scan technique in which the voltage scan direction is reversed. The resulting current is measured while the applied potential at the working electrode is applied in both forward and reverse directions. Normally, the forward and reverse scan rates are the same. CV can operate in single or multicycle modes (figure 3.4).

3.2.5 Conductometry

The conductance of a solution is measured using conductometry. The chemical reaction is monitored using the conductivity of a solution. The number of ions present in an analyte make a significant impact on the value of conductance. Small and highly charged ions will conduct more in an analyte. The electric current will increase if the conductance of a solution increases. The conductance of a solution depends on the distance and the surface area of the electrode. Acid base titrations and precipitation titrations are carried out using conductometry. This technique is used in environmental analysis. A potentiometer is an instrument in which potential between the two electrodes is measured. The indicator and reference electrode are used to measure the potential. The reference electrode is separated by using a salt bridge. The indicator electrode is in direct contact with the analyte solution [28] (figure 3.5).

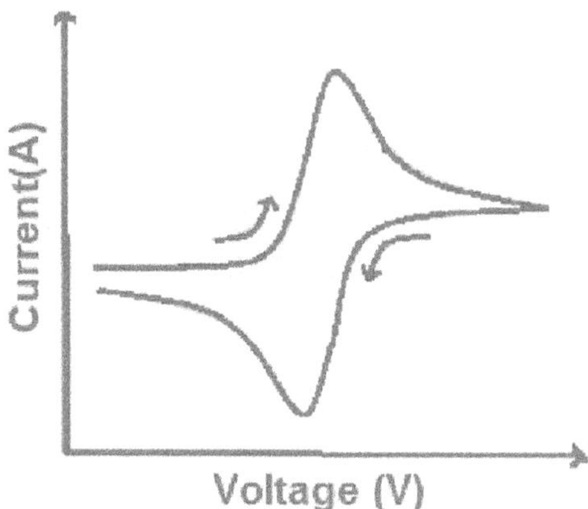

Figure 3.4. Conductometry.

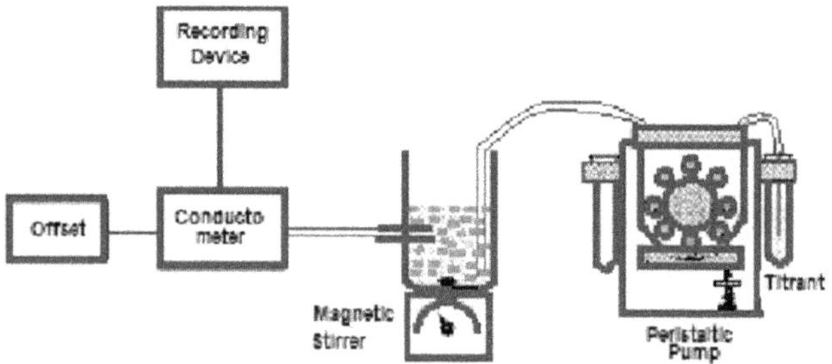

Figure 3.5. Conductometry titration with instrument setup.

3.2.6 Coulometry

Coulometry is a technique in which the electricity required for the completion of a chemical reaction is determined. In potentiostatic coulometry, the working electrode is kept at a constant potential. The electroactive species gets completely oxidized. In this method, a constant current system is used to determine the concentration. Coulometry is based on Faraday's law [28, 29]. Potentiostatic and amperostatic coulometry involve electric potential and current to be kept constant, respectively. For potentiostatic coulometry, the working current must be kept at a constant potential. When the volume of solution decreases, the rate of reactions also varies [30–33]. In coulometric titration, the current is equal to the titrant. The technique is time-consuming (figure 3.6).

3.2.7 Amperometry

Amperometry is a technique in which the current is measured in an analyte. In amperometric titration, the concentration of the analyte is determined from the curve. At the dropping mercury electrode, the current is measured at a constant

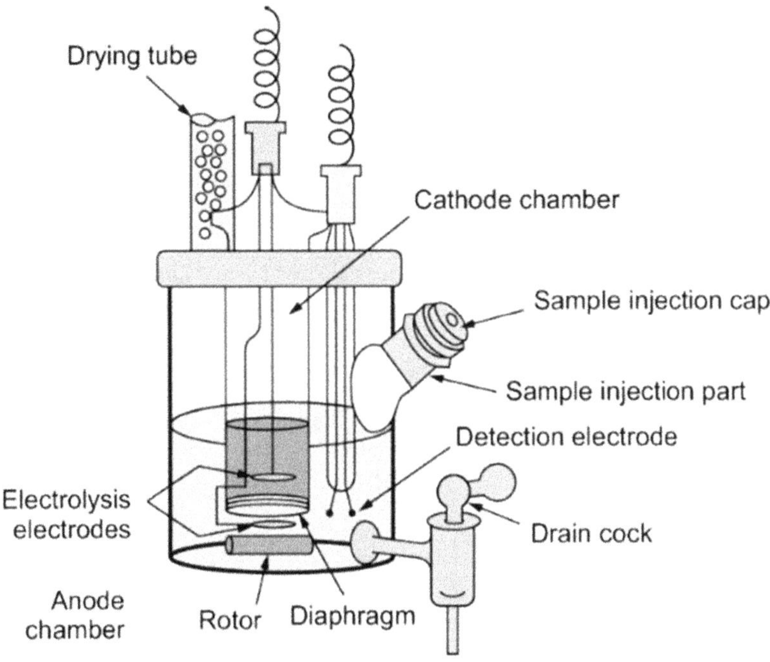

Figure 3.6. Coulometry.

voltage [34]. By the electrogravimetry method, the concentration of analyte is determined. In this method, the weight of working electrode is measured before and after the electrolysis. This technique is a rapid and sensitive technique [35, 36] (figure 3.7).

3.2.8 Fluorometry

Fluorometry is superior to spectrophotometry in terms of sensitivity and specificity. Automated immunoassays frequently use fluorometry. Although it is roughly 1000 times more sensitive than equivalent absorbance spectrophotometry, a significant issue is background interference brought on by the fluorescence of native serum. By carefully designing the filters used for spectral separation, choosing a fluorophor whose emission spectrum differs from that of the interfering chemicals, or using time- or phase-resolved fluorometry, one might reduce this interference. A species that has received exciting radiation from an external source and then released it into the environment is said to be fluorescing. The concentration of the excited species is directly inversely correlated with the intensity of the emitted (fluorescent) light [37] (figure 3.8).

3.2.9 Potentiometry

Potentiometry is an analytical method that is employed to determine the concentration of a substance in a solvent. An elevated voltmeter is used to determine the

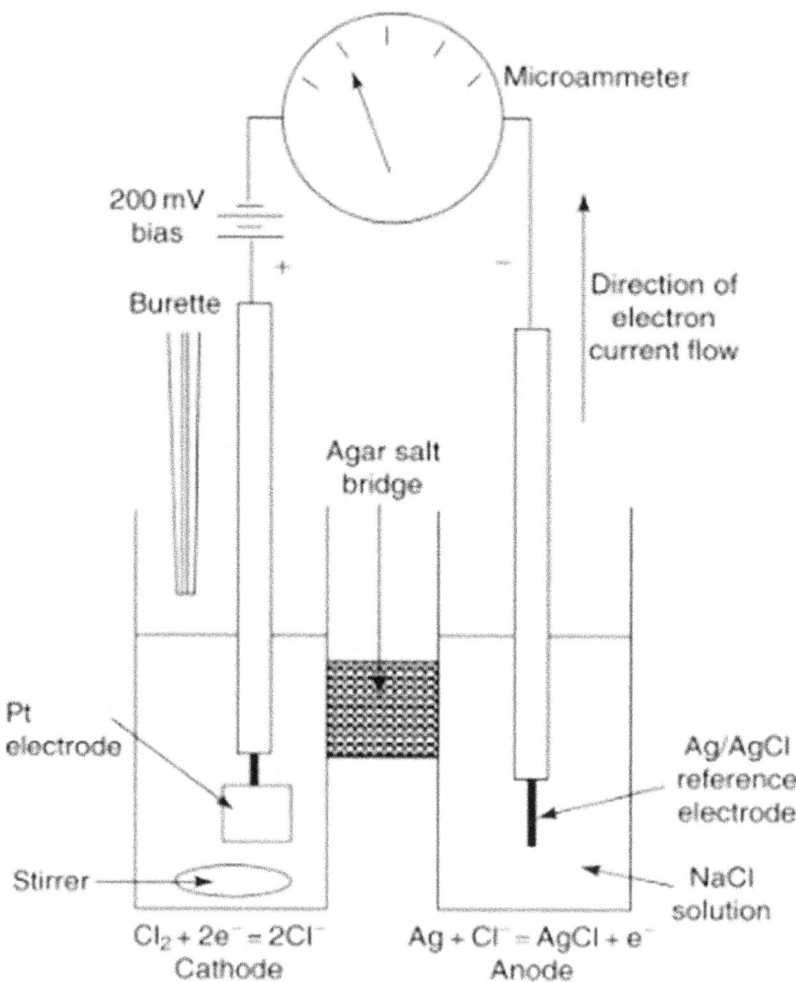

Figure 3.7. Amperometry.

potential among two electrodes in this method. The use of a high-impedance voltmeter ensures that there is no current flow. According to potentiometry principles, the change in the potential difference between two electrodes of a cell is measured. A change in the concentration of ions determines the analyte concentration. Potentiometric titration is a laboratory method for determining analyte concentration. It is employed in the analysis of acids. A chemical indicator is not used in this method. The electric potential across the substance is measured instead (figure 3.9).

3.2.10 Chromatography

Chromatography is a crucial biophysical method that makes it possible to separate, distinguish, and purify a mixture's constituent parts for qualitative and quantitative

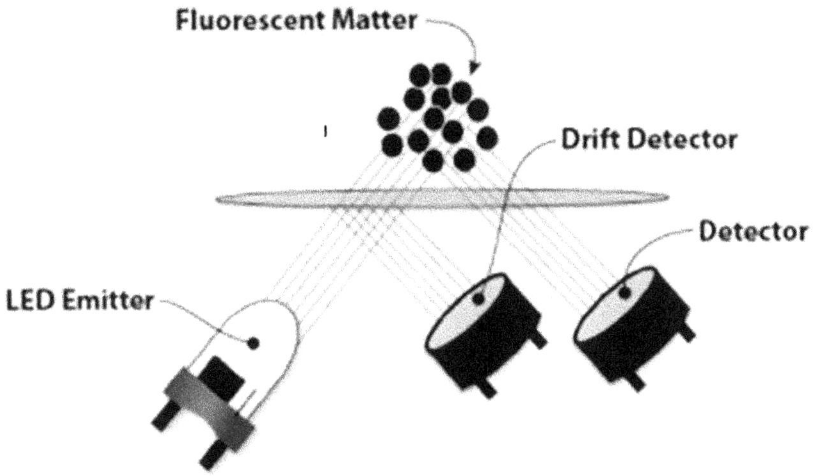

Figure 3.8. Fluorometry technique.

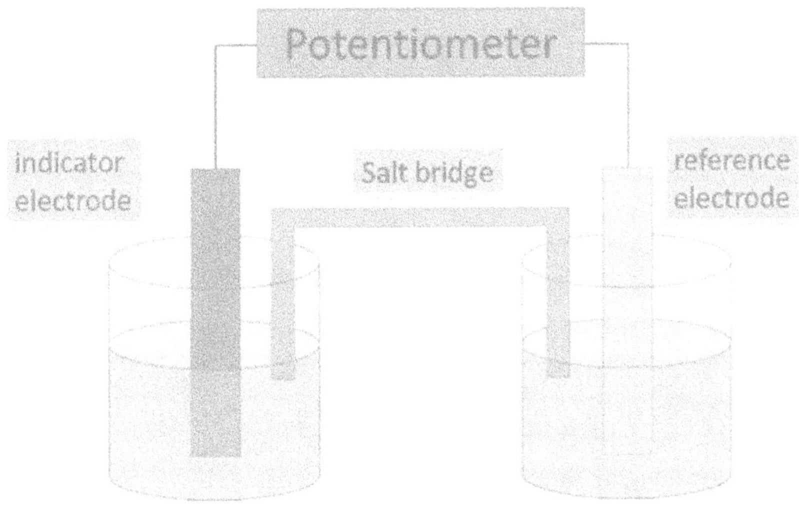

Figure 3.9. Potentiometry.

study. Based on traits like size and structure, overall charge, the presence of hydrophobic groups on the surface, and binding ability to the stationary phase, proteins can be purified. Ion exchange, surface adsorption, partition, and size exclusion are the mechanisms used in four molecular characteristics-based separation approaches. Other chromatography methods, such as column, thin layer, and paper chromatography, are based on the stationary bed. One of the most popular techniques for purifying proteins is column chromatography [38]. The measurement of how light interacts with materials is the subject of spectroscopic analysis. A material can either emit light because it has absorbed some light and re-emitted it,

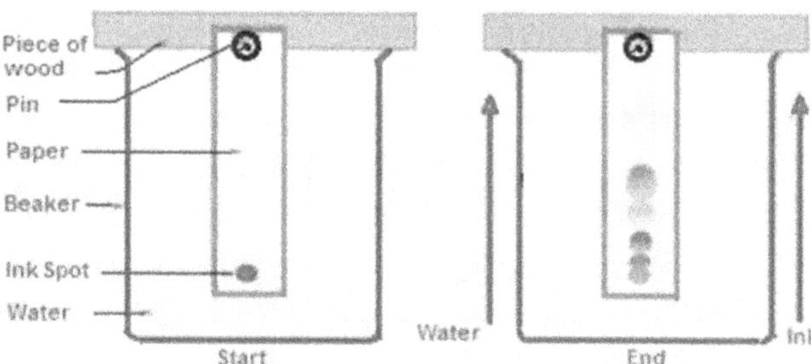

Figure 3.10. Chromatography.

because it has gained energy in another way (such as electroluminescence), or because it emits light due to its temperature. Light can be reflected, transmitted, scattered, or absorbed (incandescence). A source of heat radiation is used as a standard in contactless photothermal temperature measurement. The source consists of a heat-radiating target coupled to a ceramic heating element that has been electrically heated and is set up in a guard. The guard is given a hood to shield the target from the light reflected from the object being inspected, whose temperature is being gauged (figure 3.10).

3.3 Types of other techniques

- Adsorptive stripping voltammetry;
- Potentiometric stripping analysis (PSA);
- Normal pulse voltammetry;
- Differential pulse voltammetry (DPV);
- Square wave voltammetry (SWV).

3.4 Protocol for the selection of electrodes

The working electrode for stripping analysis must be stationary, have a favorable analyte redox behavior, a reproducible area, and a low background current over a wide range of potential [26]. The most commonly used electrodes that meet these requirements are hanging dropping mercury electrodes (HDMEs) and mercury film electrodes (MFEs).

Mercury electrodes' limited anodic potential has made them unsuitable for monitoring oxidizable substances. As a result, solid electrodes with extended anodic potential windows have been developed. Attracting the analytical interest of many, there are numerous types of solid electrodes that are used. Gold, platinum, glassy carbon electrode, and carbon paste, carbon fiber electrode, epoxy-bonded graphite electrode, and graphite electrode are examples of working electrodes. In contrast to mercury, solid electrodes have a heterogeneous surface in relation to the electrode chemical [33–44]. Activity surface heterogeneity causes deviations from expected behavior surfaces that are uniform. Working electrodes that have been used may be unaffected by being used in a specific field. The characteristics of the chosen working electrode will be improved through modification. The main idea behind the

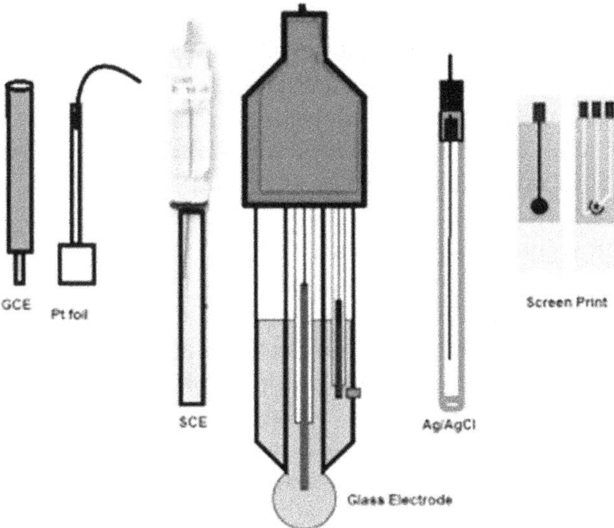

Figure 3.11. Electrodes.

modification is to incorporate a reagent on the electrode surface or into the matrix of the chosen electrode [45–51] (figure 3.11).

3.5 Protocols for electroanalysis

- Goal of the research;
- Concept of design/planning;
- Criteria of sample selection;
- Inclusion/exclusion of criteria;
- Required strategies;
- Sample size;
- Intervention of results;
- Research outcome procedures;
- Safety;
- Quality control and trail controlling;
- Result analysis;
- Discussion.

3.6 Benefits of electroanalysis

- Voltammetric methods are widely used and have made significant contributions to trace element evolution of species because they combine accuracy and sensitivity prerequisites.
- The sensitivity is sufficient and can be increased further by modifying classical voltammetric techniques (modified microelectrodes and ultramicroelectrodes) that significantly improve the method's sensitivity and selectivity.

- Voltammetry coupled with different separation methods such as HPLC, flow Injection and capillary electrophoresis, enhance the analytical properties for complex mixtures in different compounds.
- Turbid and colored solutions, which are a problem with other methods, can be easily analyzed. The separation of the excipients, in pharmaceutical analysis, is in many cases not necessary and this simplifies the preparation of samples.
- Only small volumes of samples are necessary.
- Advanced voltammetry serves as a useful tool for research scientists to studt the oxidation and reduction processes in media, adsorption processes on surfaces, and electron transfer mechanisms at chemically altered semiconductor surfaces. Electroanalytical stripping procedures have been developed for measuring down to sub-μg l^{-1} level.
- Also, these techniques combine low maintenance costs with high sensitivity and selectivity that allows the determination of low levels of analytes without prior treatments of the samples.
- These techniques have been developed for various cations, anions and organic molecules.
- Elelctroanalytical techniques (specially stripping analysis) are well known as excellent procedures for the determination of trace chemical species.
- The developed stripping voltammetric methods are simple, time-saving, selective and more sensitive for the simultaneous determination of trace substances.
- Electroanalytical methods, especially square wave voltammetry, are a very sensitive and rapid analytical method due to their high scan rate in all cases where the reacting species is accumulated by adsorption on the electrode surface.
- The short analysis time in these methods makes them very attractive for routine determination of the analytes in different samples.

3.7 Conclusion

For the first time, a powerful electrochemical strategy has been used to build effective electrode materials with an optimized role in solving in terms of high signal-to-noise ratio, rapid response, and low detection limit.

References

[1] Skoog D A, West D M and James Holler F 1995 *Fundamentals of Analytical Chemistry* 7th edn (Fort Worth, TX: Harcourt Brace College Publishers)

[2] Kissinger P and Heineman W R 1996 *Laboratory Techniques in Electroanalytical Chemistry* 2nd edn revised and expanded (Boca Raton, FL: CRC Press)

[3] Bard A J and Faulkner L R 2000 *Electrochemical Methods: Fundamentals and Applications* 2nd edn (New York: Wiley)

[4] Zoski C G 2007 *Handbook of Electrochemistry* (Amsterdam: Elsevier Science)

[5] Majeed S, Raza Naqvi S T, Najam-ul-Haq M and Ashiq M 2022 Electrochemical techniques in biosciences: conductometry, coulometry, voltammetry and electrochemical sensors *Analytical Techniques in Biosciences: From Basics to Applications* (Elsevier) ch 10

[6] Zima J, Svancara I, Barek J and Vyt ras K 2009 Recent advances in electroanalysis of organic compounds at carbon paste electrodes *Crit. Rev. Anal. Chem.* **39** 204e227

[7] Martín A, Hernandez-Ferrer J, Vazquez L, Martínez M-T and Escarpa A 2014 Controlled chemistry of tailored graphene nanoribbons for electrochemistry: a rational approach to optimizing molecule detection *RSC Adv.* **4** 132e139

[8] Salavagione H J, Díez-Pascual A M, Lazaro E, Vera S and G omez-Fatou M A 2014 Chemical sensors based on polymer composites with carbon nanotubes and graphene: the role of the polymer *J. Mater. Chem.* A **2** 14289–328

[9] Balogun Y A and Buchanan R C 2010 Enhanced percolative properties from partial solubility dispersion of filler phase in conducting polymer composites (CPCs) *Compos. Sci. Technol.* **70** 892e900

[10] Kaya T, Liu G, Ho J, Yelamarthi K, Miller K, Edwards J and Stannard A 2019 Wearable sweat sensors: background and current trends *Electroanalysis* **31** 411–21

[11] Compton R G, Foord J S and Marken F 2003 Electroanalysis at diamond-like and doped-diamond electrodes *Electroanalysis* **15** 1349–63

[12] Švancara I, Prior C, Hočevar S B and Wang J 2010 A decade with bismuth-based electrodes in electroanalysis *Electroanalysis* **22** 1405–20

[13] Yang N, Swain G M and Jiang X 2016 Nanocarbon electrochemistry and electroanalysis: current status and future perspectives *Electroanalysis* **28** 27–34

[14] Malode S J, Keerthi P K, Shetti N P and Kulkarni R M 2020 Electroanalysis of carbendazim using MWCNT/Ca-ZnO modified electrode *Electroanalysis* **32** 1590–9

[15] Settle F (ed) 1997 *Handbook of Instrumental Techniques for Analytical Chemistry* 1st edn (Englewood Cliffs, NJ: Prentice-Hall)

[16] Compton R G and Banks C E 2018 *Understanding Voltammetry* (Singapore: World Scientific)

[17] Bond A M, Duffy N W, Guo S X, Zhang J and Elton D 2005 Changing the look of voltammetry *Anal. Chem.* **77** 186–A

[18] Mabbott G A 1983 An introduction to cyclic voltammetry *J. Chem. Educ.* **60** 697

[19] Kissinger P T and Heineman W R 1983 Cyclic voltammetry *J. Chem. Educ.* **60** 702

[20] Batchelor-McAuley C, Kätelhön E, Barnes E O, Compton R G, Laborda E and Molina A 2015 Recent advances in voltammetry *ChemistryOpen* **4** 224–60

[21] Evans D H, O'Connell K M, Petersen R A and Kelly M J 1983 Cyclic voltammetry *J. Chem. Educ.* **60** 290

[22] Van Benschoten J J, Lewis J Y, Heineman W R, Roston D A and Kissinger P T 1983 Cyclic voltammetry experiment *J. Chem. Educ.* **60** 772

[23] Copeland T R and Skogerboe R K 1974 Anodic stripping voltammetry *Anal. Chem.* **46** 1257A–68a

[24] Batley G E and Florence T M 1974 An evaluation and comparison of some techniques of anodic stripping voltammetry *J. Electroanal. Chem. Interfacial Electrochem.* **55** 23–43

[25] Lucia M, Campos A and Van den Berg C M 1994 Determination of copper complexation in sea water by cathodic stripping voltammetry and ligand competition with salicylaldoxime *Anal. Chim. Acta* **284** 481–96

[26] Obata H and van den Berg C M 2001 Determination of picomolar levels of iron in seawater using catalytic cathodic stripping voltammetry *Anal. Chem.* **73** 2522–8

[27] Greulach U and Henze G 1995 Analysis of arsenic (V) by cathodic stripping voltammetry *Anal. Chim. Acta* **306** 217–23

[28] Scholz F (ed) 2010 *Electroanalytical Methods* **vol 1** (Berlin: Springer)

[29] DeFord D D and Bowers R C 1958 Electroanalysis and coulometric analysis *Anal. Chem.* **30** 613–9

[30] DeFord D D 1960 Electroanalysis and coulometric analysis *Anal. Chem.* **32** 31–7

[31] de Agostini A 2002 *Coulometric Titration* (Switzerland: Mettler-Toledo GmbH) p 3

[32] Burns J C, Stevens D A and Dahn J R 2015 In-situ detection of lithium plating using high precision coulometry *J. Electrochem. Soc.* **162** A959

[33] Johnson C and LaCourse W R 1990 *Anal. Chem.* **62** 589A–97 A

[34] Bhatt V 2016 *Essentials of Coordination Chemistry* (Academic)

[35] Hillman A R 2005 *Encyclopedia of Analytical Science* 2nd edn (Elsevier)

[36] Jill Venton B and DiScenza D J 2020 Differential pulse voltammetry *Electrochemistry for Bioanalysis* (Elsevier)

[37] Rifai N 2018 Automation in the clinical laboratory *Textbook of Clinical Chemistry and Molecular Diagnostics* (Elsevier)

[38] Coskun O 2016 Separation techniques: chromatography *North. Clin. Istanb* **3** 156–60

[39] Hareesha N, Manjunatha J G, Amrutha B M, Pushpanjali P A, Charithra M M and Prinith Subbaiah N 2021 Electrochemical analysis of indigo carmine in food and water samples using a poly (glutamic acid) layered multi-walled carbon nanotube paste electrode *J. Electron. Mater.* **50** 1230–8

[40] Tigari G and Manjunatha J G 2020 Optimized voltammetric experiment for the determination of phloroglucinol at surfactant modified carbon nanotube paste electrode *Instrum. Exp. Tech.* **63** 750–7

[41] Manjunatha J G 2020 Poly (adenine) modified graphene-based voltammetric sensor for the electrochemical determination of catechol, hydroquinone and resorcinol *Open Chem. Eng. J.* **14** 52–62

[42] Manjunatha Charithra M and Manjunatha J G 2020 Electrochemical sensing of paracetamol using electropolymerised and sodium lauryl sulfate modified carbon nanotube paste electrode *ChemistrySelect* **5** 9323–9

[43] Pushpanjali P A and Manjunatha J G 2020 Development of polymer modified electrochemical sensor for the determination of alizarin carmine in the presence of tartrazine *Electroanalysis* **32** 2474–80

[44] Tigari G and Manjunatha J G 2020 Poly (glutamine) film-coated carbon nanotube paste electrode for the determination of curcumin with vanillin: an electroanalytical approach *Monatsh. Chem.* **151** 1681–8

[45] Manjunatha J G, Raril C, Hareesha N, Charithra M M, Pushpanjali P A, Tigari G, Ravishankar D K, Mallappaji S C and Gowda J 2020 Electrochemical fabrication of poly (niacin) modified graphite paste electrode and its application for the detection of riboflavin *Open Chem. Eng. J.* **14** 90–98

[46] Manjunatha J G and Hussain C M 2022 *Carbon Nanomaterials-Based Sensors: Emerging Research Trends in Devices and Applications* (Elsevier)

[47] Raril C, Manjunatha J G, Ravishankar D K, Fattepur S, Siddaraju G and Nanjundaswamy L 2020 Validated electrochemical method for simultaneous resolution of tyrosine, uric acid, and ascorbic acid at polymer modified nano-composite paste electrode *Surf. Eng. Appl. Electrochem.* **56** 415–26

[48] Prinith N S, Manjunatha J G and Hareesha N 2021 Electrochemical validation of L-tyrosine with dopamine using composite surfactant modified carbon nanotube electrode *J. Iran. Chem. Soc.* **18** 3493–503
[49] Pushpanjali P A, Manjunatha J G and Srinivas M T 2020 Highly sensitive platform utilizing poly (l-methionine) layered carbon nanotube paste sensor for the determination of voltaren *FlatChem.* **24** 100207
[50] Raril C and Manjunatha J G 2020 Fabrication of novel polymer-modified graphene-based electrochemical sensor for the determination of mercury and lead ions in water and biological samples *J. Anal. Sci. Technol.* **11** 1–10
[51] Manjunatha J G 2020 A surfactant enhanced graphene paste electrode as an effective electrochemical sensor for the sensitive and simultaneous determination of catechol and resorcinol *Chem. Data Coll.* **25** 100331

IOP Publishing

Real-Time Applications of Advanced Electrochemical Sensing Devices

Jamballi G Manjunatha

Chapter 4

Electrochemical sensing devices for determination of hormones

Arezou Taghvimi, Samin Hamidi, Sevinc Kurbanoglu and Yousef Javadzadeh

Electrochemical sensors are categorized as superior determination methods for hormones because of their high sensitivity, simple operation, and portability. The sample pretreatment step is omitted in electrochemical methods, which is an important property of electrochemical techniques, unlike chromatographic ones. Hormones are biologically important substances that are secreted from endocrine glands. They are transported by the circulatory system to the target tissues and are presented in low amounts, emphasizing the application of a sensitive and selective determination method. Despite poor reproducibility and the hard regeneration step of electrochemical sensors, they show superior sensitivity and selectivity in clinical determination processes of hormones. Amperometry, impedimetric, potentiometry, and conductometry are various electrochemical methods used for hormone detection in the literature. In this chapter, different electrochemical sensing devices were reviewed for the determination of various hormones, which have a crucial role in body function. Changing the level of these hormones might be a sign of diseases that emphasizes the importance of a sensitive, accurate, and selective determination method in different biological mediums. Therefore, the introduction or development of a new and sensitive determination method is important in clinical and forensic laboratories.

4.1 Hormones in a general view

Hormones are chemical substances, that control important activities of the body such as inhibition and stimulation of growth, immune system, regulation of metabolism, apoptosis induction or suppression, and management of the body for the new phase of life. Therefore, hormone determination is considered important in clinical diagnosis methods acting as a sign of a metabolic disorder or progress of

doi:10.1088/978-0-7503-5377-9ch4

diseases [1]. Moreover, the determination of hormones is an indispensable part of forensic and sport medicine laboratories to control doping and nutrition disorders. Low levels of hormones in the body demand a sensitive and selective method for determination and quantification in the diagnosis of diseases [2]. Various methods such as singleplex enzyme-linked immunosorbent assays, high-performance liquid chromatography/mass spectrometry (HPLC–MS), high-performance liquid chromatography (HPLC), and gas chromatography/mass spectrometry (GC–MS) have been introduced for hormone determination. Electrochemical methods have been known as noticeable determination methods in comparison to others due to fast responsibility, low cost, easy operation, and high selectivity [3]. Electrochemical sensors/biosensors are designed to transduce a chemical/biochemical event to an electrical signal. In this regard, immobilized enzymes or an antibody are used as a bio-recognition agent in electrochemical hormones biosensors structure. Electrical signals are changed when a chemical reaction between immobilized molecules and target compounds happens [4]. Hence, various electrochemical methods have been applied for transducing the electrical signals as follows: the first one is the amperometric method in which a change in the current of the system happens; potentiometric ones detect that an alteration in the potential is noticed; next, impedimetric methods are when the impedance of the media between electrodes is considerably altered; and finally, there are conductometric methods in which the conductive properties of a system are changed [5].

Hormones determination is an important step in controlling different diseases and disorders. They are secreted from the glands and are released in the blood to regulate body function and physiology. Different organs in the body, namely the hypothalamus, anterior pituitary gland, posterior pituitary gland, adrenal cortex, testes, ovaries, thyroid glands, parathyroid glands, and pancreas secret related hormones that are transported to the target cell to manage target tissue functions. Controlling the optimum level of hormones even by technicians or patients needs rapid, sensitive, portable, and affordable methods. Electrochemical sensors/biosensors inherit the potential of accuracy, selectivity, sensitivity, minimum sample treatment process, simplicity, miniaturization ability, and low-cost characteristics in hormone determination.

4.2 Electrochemical sensing devices for hormones determination

Among different types of biosensors (piezoelectric, electrochemical, and optical), electrochemical sensors are the most applicable types used in different laboratories [4, 6]. Most of them are immunosensors and aptasensors with different approaches such as sandwich-type configurations, label-based or label-free, direct, and competitive [7]. Amperometric biosensors are the most commonly used method in the electrochemical sensing of hormones due to the presence of mature technologies, simplicity, and low costs among electro-sensing methods. In this method, the potential is fixed between two electrodes, and the current is measured when the conversion of the electroactive spices happens. Moreover, voltammetry is a method in which the current is measured along with controlled variations of potential [5, 8].

Screen-printed electrodes (SPEs) [9], carbon-based electrodes [10], and molecularly imprinted polymers (MIPs) [11] have been successfully used in the designing of amperometric biosensors. Potentiometric biosensors measure the potential difference between working and reference electrodes by a voltammeter when there is no significant current flowing through them. The potential difference happens when an electrochemical redox reaction occurs in the system [12]. Electrochemical impedance spectroscopy (EIS) is a sensitive method in which interfacial properties changes are studied when a bio-recognition event occurs at the surfaces of electrodes [2, 13]. Briefly, impedimetric biosensors are sensitive sensing devices for interfacial region changes. Conductometric biosensors work based on the changes in the conductivity or resistivity of the solution ions or electrons produced during the electrochemical reaction. Conductometric methods show low sensitivity because of possible events like double layer charge, Faradaic effects, and concentration polarization [3] (figure 4.1).

In addition, point-of-care testing (POCT) methods are getting popular because of the practical limitations of laboratory methods such as limitations in equipment, costs, laborious analytical procedures, and expert staff. It is important to consider that the results of POCT methods should be comparable with laboratory methods in terms of accuracy and precision [14]. Remarkably, great effort has been made to introduce novel POCT methods in hormone determination such as lab-on-a-chip (LOC) technologies to perform on-site and be available in pocket size [15]. Microfluidic devices and microfabrication technologies, and paper-based electrochemical biosensors (microfluidic analytical devices) are other examples of LOC systems developed for the POCT method [7, 16]. Levels of hormones, different metabolic products, and cancer markers are diagnosable by microfluidic devices. The effort for fabrication of an LOC system in the miniaturization of devices are considered to have multiplex operations, a small amount of reagents and samples, disposability, portability, low power consumption, and low cost. Consumption of low volumes of samples in the LOC system is highly considerable in biological analysis in which very small samples are available. LOC methods have been applied in the detection of proteins, DNA, hormones, pathologic organism, and also cell culture up to now [17]. However, the fabrication of LOC systems is not simple and

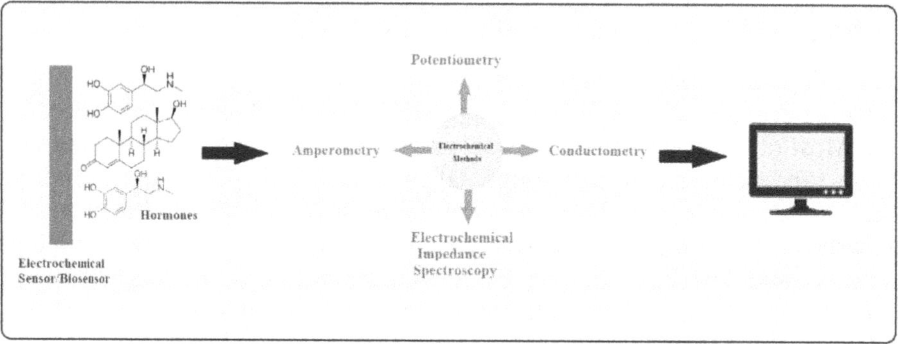

Figure 4.1. Electrochemical sensors for hormone detection.

photolithography techniques are applied in construction conditions. Moreover, low-temperature co-fired ceramics (LTCC) are needed which are produced at high temperatures [18]. Recently, 3D printers have been used to construct microfluidic devices. Unfortunately, they cannot print directly and external electrodes are needed in 3D printers [19]. In addition, it seems that the application of curable materials in microfluidic systems is complex due to obtaining acceptable adhesion between substrates [20]. However, the construction of microfluidic systems upon inexpensive materials leads to fragile microchannels that obtain semi-quantitative detection of analytes [21]. Contamination of the system, unreproducible analysis steps, reassembling of different analysis steps, and cleaning of microchannels are some problems of microfluidic systems that make reuse of them a challenging issue. Despite the present problems, research about the construction of novel flexible, disposable, portable, large-scale production processes, simple fabrication, and application of microfluidic systems, is in progress in R&D laboratories.

Electrochemical sensors are independent integrated devices providing quantitative or semi-quantitative information of analyte by using a receptor that is contacted to an electrochemical transduction element [10, 22]. Some electrochemical techniques like amperometric, potentiometric, conductometric, and impedimetric along with LOC methods have been effectively used for the determination of various analytes. These methods have been efficiently used for hormone determination as well. In this chapter, it is our aim to overview some selected current electrochemical bio/sensors that have been applied for hormone determination.

4.3 Applications of electrochemical sensors in hormones determination

4.3.1 Adrenaline

Adrenaline (AD) (also known as epinephrine) is categorized as catecholamines which act as a neurotransmitter and a significant member in central and peripheral neuroendocrine responses to different types of mental and physical stress [23]. AD is found at a nanomolar level in biological fluids such as serum, which emphasizes the presence of a sensitive determination method [24]. In a research work, a 2-Hydroxybenzimidazole modified carbon paste electrode (MCPE) was applied for efficient determination of AD in the presence of uric acid (UA), individually. In this method, the bare carbon paste electrode was modified by an optimum amount of 2-Hydroxybenzimidazole and was applied as a working electrode for the determination of AD and UC in voltammetry. A comparison between bare and MCPE, exhibited high electrochemical behavior with limits of detection (LOD) of 3.0 mM, LOQ 12.0 mM for AD and LOD of 5.1 mM, LOQ 17.8 mM for UA, respectively, with two identical different oxidation peaks [25]. (figure 4.2).

In another research, an unmodified boron-doped diamond film (BDDF) electrode was applied for the determination of AD in human urine samples by the square-wave voltammetric method. BDDFs are known as carbon-based materials with unique properties of wide potential window (up to 4 V), stable background current, and good mechanical robustness. The morphological and crystallographical

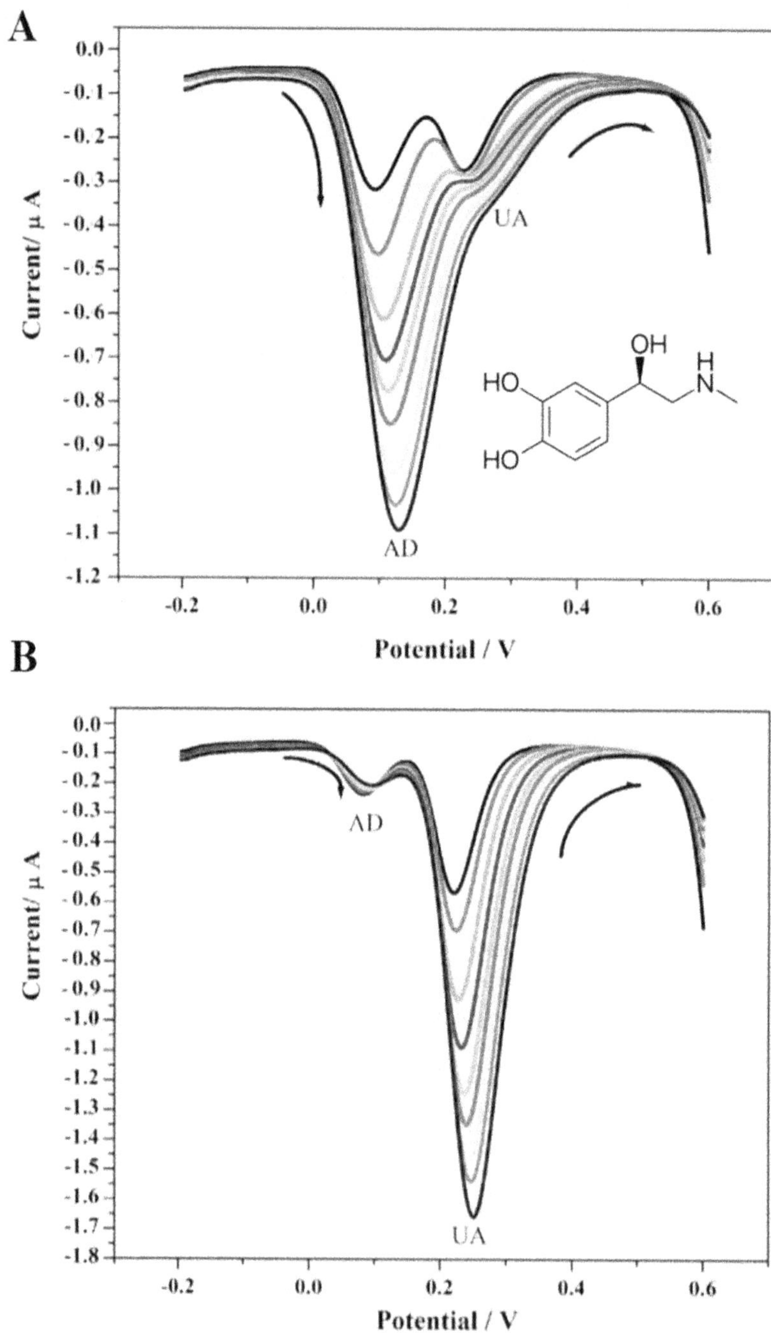

Figure 4.2. (a) Differential pulse voltammograms for interference study of AD. Inset: Chemical structure of AD. (b) Differential pulse voltammograms for interference study of uric acid. Reprinted from Materials Science for Energy Technologies [25].

Table 4.1. Different electrochemical methods for AD determination.

Electrochemical sensing device	Method	Medium	LOD	References
MCPE	CV	PBS	3 mM	[25]
BDDF	SWV	Urine	0.21 μM	[26]
rGO-MIP-GCE	CV	Urine and AD ampoules	3 nM	[27]

Abbreviations: BDDF: boron-doped diamond film; CV: cyclic voltammetry; GCE: glassy carbon electrode; MCPE: modified carbon paste electrode; MIP: molecularly imprinted polymers; PBS: phosphate buffer solution; rGO: reduced graphene oxide; SWV: square wave voltammetry.

properties of BDDF, causes BDDF to be different from other carbon-based materials. In this regard, BDDF was applied as a sensitive sensor of AD in urine without any chemical modification or pretreatment. The LOD of the method was calculated at 0.21 μM [26]. In addition, molecularly imprinted polymers (MIPs), have been efficiently used for AD-imprinted film electropolymerization, over reduced graphene oxide (rGO)-modified glassy carbon electrode (GCE) by cyclic voltammetry. The fabricated electrochemical sensor shows high selectivity owing to specific imprinted cavities for AD with a wide linear concentration range of AD 0.015 and 40 mM with LOD of 3 nM. The developed sensor showed a satisfactory result in the analysis of urine and AD in ampoule samples [27]. Some selected studies about AD detection using electrochemical methods are shown in table 4.1.

4.3.2 Human chorionic gonadotropin

Human chorionic gonadotropin (hCG) is identified as an important biomarker in the diagnosis of diseases related to the seminal system such as orchic teratoma, trophoblastic cancer, and also hydatid pregnancy. Many immunoassay methods for the determination of hCG are reported in references on ELISA [28], radioimmunoassay [29], and fluoroimmunoassay [30]. Electrochemical methods overcome the practical problems of immunoassay methods and provide rapid, simple operation, low-cost and selective methods. In a research, a novel electrochemical method had been reported for the determination of hCG through the fabrication of modified GCEs with carbon nano-onions (CNOs) gold nanoparticles (AuNPs), and polyethylene glycol (PEG) nanocomposite. The nanocomposite was deposited on the surface of the GCE by self-assembling of nanocomposite (same ratio of polyethylene, AuNPs, and commercial CNOs) via chemisorption. The CNOs/AuNPs/PEG nanocomposite showed a low LOD of 0.1 fg ml^{-1} with a detection range of 0.1 fg ml^{-1}–1 ng ml^{-1} in urine using the square wave voltammetry (SWV) technique [31]. (figure 4.3).

In other research, silver nanoparticles (AgNPs) were applied as the redox reporters and an hCG-specific binding peptide as the receptor for the determination of hCG. Peptide-induced AgNPs assembly was achieved on the electrode surface that was modified with the same sequence of peptide (recognition element) used in the AgNPs aggregation. The method showed a LOD of 0.4 mIU ml^{-1} with the linear-sweep voltammetry (LSV) method. The method was applied in human urine

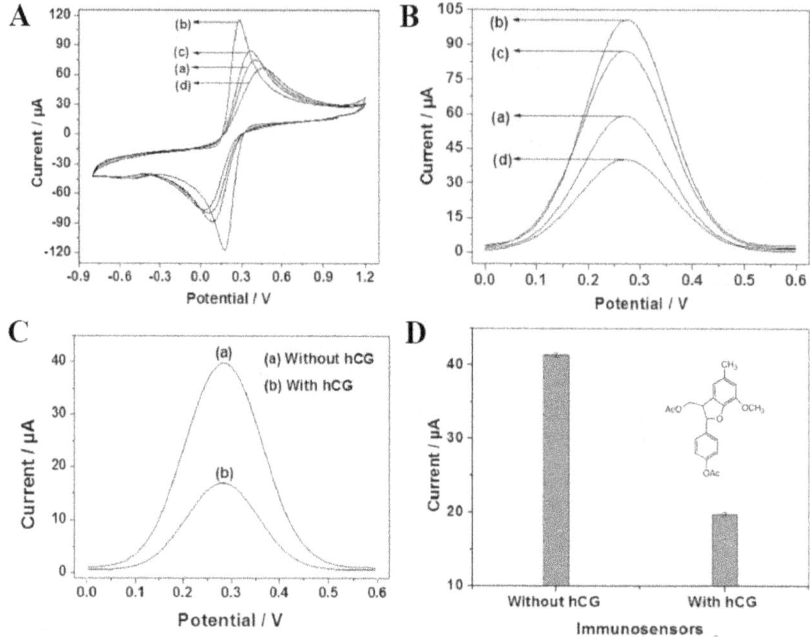

Figure 4.3. Assessment of the step-wise fabrication of the hCG-immunosensors and electrochemical signal: (A) CV curve of (a) bare-GCE, (b) CNOs/AuNPs/PEG/GCE, (c) anti-hCG/CNOs/AuNPs/PEG/GCE and (d) BSA/anti-hCG/CNOs/AuNPs/PEG/GCE. (B) SWV curve of (a) bare-GCE, (b) CNOs/AuNPs/PEG/GCE, (c) anti-hCG/CNOs/AuNPs/PEG/GCE and (d) BSA/anti-hCG/CNOs/AuNPs/PEG/GCE. (C) SWV curve of immunosensor (a) without hCG and (b) with hCG. (D) Bar diagram of the SWV peak current without hCG and with hCG. Inset: chemical structure of human chorionic gonadotropin. Reprinted from [31], copyright (2019) with permission from Elsevier.

and serum with well-amenable results [32]. Moreover, an enhanced cathodic immunosensor supported by photoanode was considered for the detection of hCG. Briefly, the immunosensor consisted of two cathodes and an anode fabricated by TiO_2:N nanotubes which were assembled by $AgInS_2$ quantum dots (QDs) to acquire the TiO_2:N/$AgInS_2$ photoanode and CNT/Pt cathodic matrix prepared by modifying CNTs and PtNPs on an indium-tin-oxide (ITO) electrode as a photocathode. The system was used to capture the hCG antibody. TiO_2:N/$AgInS_2$ photoanode acts as the signal-converting element to produce a prominent current signal, whereas immune recognition events happen on the sensing cathode to evidently change the initial current signal from the steric hindrance effect. The developed immunosensor showed sensitive detection of hCG antigen [33]. In addition, a modified carbon nanotube screen printed electrode (CNT-SPE) was used as an electrochemical device for the sensitive detection of hCG in urine samples. For this purpose, the hydroxylated surface of the CNT electrode was aminated (–NH_2) to oriental binding of hCG antibody (Ab). The SPE-CNT-NH_2-Ab showed good responses in the concentration range from 0.01×10^{-9} to 100×10^{-9} g cm^{-3} by EIS [34]. In other research, a label-free amperometric immunosensor

Table 4.2. Different electrochemical methods for human chorionic gonadotropin determination.

LOD	Medium	Method	Electrochemical sensing device	References
CNOs-AuNPs-PEG-GCE	SWV	Urine	0.1 fg ml^{-1}–1 ng ml^{-1}	[31]
Peptide-MCH-Au–AgNPs	LSV	Urine and serum	0.4 mIU ml^{-1}	[32]
Photoanode-supported cathodic immunosensor	PEC	PBS	—	[33]
CNT-NH_2-Ab-SPE	EIS	Urine	—	[34]
NPG-GSs-anti-hCG-GCE	Voltammetry	PBS	0.034 ng ml^{-1}	[35]
AuNPs-Cys-AuNPs-Anti-hCG-GE	DPV	PBS	0.3 pg ml^{-1}	[36]
nanogold-CHIT-anti-hCG-GCE	Amperometry	PBS and serum	0.1 mIU ml^{-1}	[37]

Abbreviations: Ab: antibody, AgNPs: silver nanoparticles, Anti-hCG: anti-human chorionic gonadotropin, anti-hCG: anti-human chorionic gonadotropin, AuNPs: gold nanoparticles, AuNPs: gold nanoparticles, CHIT: chitosan, CNOs: carbon nano-onions, CNT-NH_2: carbon nanotube-aminated, Cys: cysteamine, DPV: differential pulse voltammetry, EIS: electrochemical impedance spectroscopy, GCE: glassy carbon electrode, GE: gold electrode, hCG: human chorionic gonadotropin, nanogold: nanogold, PEC: photoelectrochemical, PEG: polyethylene glycol, peptide-MCH: peptide-6-mercapto-1-hexanol, SPE: screen printed electrode.

was fabricated for hCG assay by CV. For this purpose, the polished surface of GCEs was modified by graphene sheets (GSs) and nanoporous gold particles (NPG) by dropping each solution separately. After, 5 μl of hCG antibody (anti-hCG) solution was dropped on the surface of the electrode and anti-hCG/NPG/GS electrode was dried after 1 h. Hydroquinone (HQ) redox species was used as an indicator. The amperometric response variation to the concentration of hCG, the target antigen, was evaluated. The immunosensor showed good accuracy and sensitive response to hCG in the range of 0.5–40.00 ng ml^{-1} with an LOD of 0.034 ng ml^{-1} [35]. Some selected studies about human chorionic gonadotropin hormone detection using electrochemical methods are shown in table 4.2.

4.3.3 Cortisol

Cortisol is secreted by the hypothalamic–pituitary–adrenal system and is commonly named as 'stress hormone'. Cortisol maintains many vital activities of the body like blood pressure, immune responses, glucose level, carbohydrate metabolism, and central nervous system activation [38]. Knowing the importance of cortisol, the researchers were encouraged to introduce novel determination methods. Recently, a nanoporous polyimide membrane was selected as a substrate of a novel flexible cortisol sensor in human sweat. The gold particle was deposited by cryo-electron beam deposition to achieve a film thickness of 150 nm, as electrodes on the substrate surface. After incubation of the cortisol antibody, the sensor was applied for the determination of cortisol. Moreover, a machine learning (ML) algorithm was also developed to analyze sensor data in different variations of cortisol levels over time [39]. Hence, electrochemical aptamer-based sensors, are developed for biomolecular recognition combining the advantages of synthetic bioreceptors with electrochemical methods. Aptasensors use electrochemical redox reporters added to the bulk solution or covalently conjugated to the aptamer probe. Here, a novel aptamer-based sensor was designed in which the aptamer is conjugated with methylene blue that acts as a redox reporter to probe binned cortisol quantitatively on the sensor. The cortisol-specific aptamers were chemically modified with amine and thiol functional groups to provide redox reporter conjugation and attachment of aptamer to a gold electrode, respectively. SWV was used to determine cortisol in the blood serum in the concentration range of 0.05–100 ng ml^{-1} [40].

Recently, a wireless, flexible, and battery-free integrated patch is developed for real-time on-body sweat cortisol detection. The patch is equipped with a near-field communication (NFC) module that provides wireless power harvesting and data interaction with an NFC-enabled smartphone, provides battery-free patches and realizes epidermal on-body testing. The patch detects cortisol in the sweat of the skin by DPV [41]. Hence, a novel immunoelectrode was fabricated for the determination of cortisol in saliva. NiO thin film was prepared on the surface of ITO glass by RF magnetron sputtering. Then, the cortisol antibody was covalently attached to the surface of NiO/ITO by hydroxylation and silanization processes. CV and DPV were used to study the immunoelectrode's electrochemical behavior. The immunosensor showed a low LOD of 0.32 pg ml^{-1} in saliva samples [42] (figure 4.4).

It seems that, among different electrochemical sensing devices, sensitive field-effect transistors (FETs) show a growing trend in POC diagnosis due to low cost, low sample volume, miniaturization potential, and fast response. Recently, a FET was

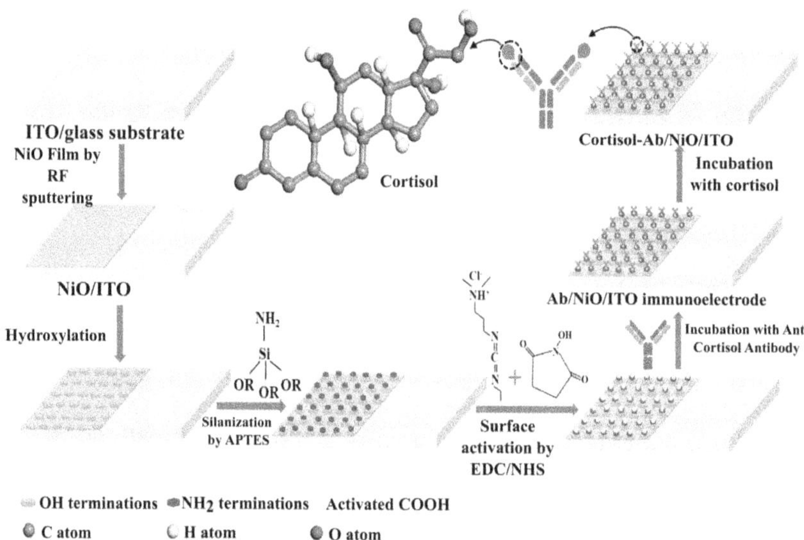

Figure 4.4. Schematic for fabrication of the immunoelectrode (Ab/NiO/ITO) and interaction of cortisol with the antibody. Reprinted from [42], copyright (2019) with permission from Elsevier.

Table 4.3. Different electrochemical methods for cortisol determination.

Electrochemical sensing device	Method	Medium	LOD	References
Nanopourse polymer deposited gold film	EIS	Human sweat	—	[39]
MB-aptamer modified gold electrode	Chronoamperometry	Blood serum	—	[40]
Electrochemical patch	DPV	Sweat	—	[41]
Ab-NiO-ITO immunoelectrode	CV and DPV	Saliva	0.32 pg ml^{-1}	[42]
ImmunoFET	EIS/potentiometry	Saliva	0.005 ± 0.002 ng ml^{-1}	[43]
Au-MIP	CV	Saliva	~200 fM	[44]
Anti-C_{ab}/ZnONRs/Au and Anti-C_{ab}/ZnO-NFs/Au	CV	Saliva	1 pM	[45]
Antibody-d-BSA-rGO	EIS	Saliva	—	[46]
Anti-C_{ab}/AuNPs/MWCNTs/PDMS	CV	Sweat	0.3 fg ml^{-1}	[47]

Abbreviations: antibody/d-BSA: antibody-denatured-bovine serum albumin, Anti-C_{ab}: anti-cortisol, Au: gold electrode, CV: cyclic voltammetry, DPV: differential pulse voltammetry, EIS: electrochemical impedance spectroscopy, MB: methylene blue, MIP: molecularly imprinted polymers, rGO: reduced graphene oxide, ZnO-NFs: zinc oxide-nanoflakes, ZnONRs: zinc oxide-nanorods.

developed for the detection of cortisol in saliva for heart failure (HF) monitoring. In this work, The device contained four silicon nitride (Si_3N_4) ISFETs, which were functionalized using 11-triethoxysilyl undecanal (TESUD) by a vapor-phase method in a saturated medium to obtain our capacitive ImmunoFET containing anti-cortisol monoclonal antibody (mAb) bound onto the functionalized ISFETs. The device was studied by the EIS method and showed a low LOD of 0.005 ± 0.002 ng ml^{-1} in saliva in the presence of other HF biomarkers [43]. In addition, a new Au@MIP sensor was designed for cortisol detection by CV and a simple antibody-free method. For this purpose, the gold layer was deposited in a GCE and then the Au@MIP layer was electrodeposited on the gold-GCE surface in the presence of 0.1 M, pH 4 acetate buffer solution containing a mixture of o-phenylenediamine (o-PD), cortisol templates, and $HAuCl_4$. The sensor was able to detect the trace amounts of cortisol by measuring the current change of the redox-active probes in response to the binding of target cortisol to the imprinted sites in the polymer. The sensor showed an LOD of ~200 fM in saliva [44]. Some selected studies about cortisol hormone detection using electrochemical methods are shown in table 4.3.

4.3.4 17β-Estradiol

Estrogens consist of estrone (E1), 17-β-estradiol (E2), estriol (E3), and 17-α-ethinylestradiol (EE2) and diethylstilbesterol (DES) are steroid hormones that are important substances responsible for various body activities. Among the above list,

17-β-estradiol is the main female sex hormone responsible for the regulation of estrous and menstrual female reproductive cycles. In addition, 17-β-estradiol plays an important role in the growth and also milk yield promotion of animals. On the other hand, the presence of estrogens in the aquatic environment causes contamination and makes it a global problem. Therefore, the introduction of a novel and sensitive analytical determination method is very important [48]. In this regard, a novel carbon dots-polyaniline-GCE electrochemical sensor (CDs-PANI/GCE) was designed to determine 17-β-estradiol from water and serum samples. CDs were synthesized by pyrolysis of iota-carrageenan and H_2O_2 (30% v/v) at 220 °C for 12 h in a Teflon-line autoclave. PANI was commercially provided for the research. The sensor was fabricated by drop-casting of CDs/PANI composite on GCE using poly (vinylidene fluoride) as a binder. The modified electrode showed an LOD of 43 nM [49] (figure 4.5).

A sensitive aptasensor of 17-β-estradiol was fabricated on disposable laser-scribed graphene electrodes (LSGE) as portable strips. At first, the aptasensor was designed on GCE and then the same method was applied to LSGEs and the results were compared. The surface of the GCE was modified in two sections as follows: the GCE was electrodeposited by β-cyclodextrin (β-CD) and then the surface was able to attach adamantane (ADA) labeled amino-modified aptamer (AF1) through guest–host interactions (AF1-ADA), which forms the first fragment of the aptasensor. Then, AF1-ADA incubates with a solution containing a thiol-modified aptamer (AF2) and self-assembles onto AuNPs containing oligonucleotide (ON1), which makes the second aptasensor fragment. Finally, the aptasensor is called AF1-ADA/ON1/AF2-Au. The electrochemical behavior of the designed aptasensor was studied in the presence of 17-β-estradiol by DPV. The method showed satisfactory results and is a potent method for the low-cost and portable detection of 17-β-estradiol [50]. In another research, an SPE was applied for the simultaneous detection of emerging pollutants (EPs) i.e. hydroquinone (HQ), paracetamol (PARA), and 17-β-estradiol (E2) in tap water by DPV. The method successfully determined EPs with low detection limits of 185, 218, and 888 nmol l^{-1}, respectively. The results of the electrochemical sensor are comparable with high HPLC as a gold analysis method.

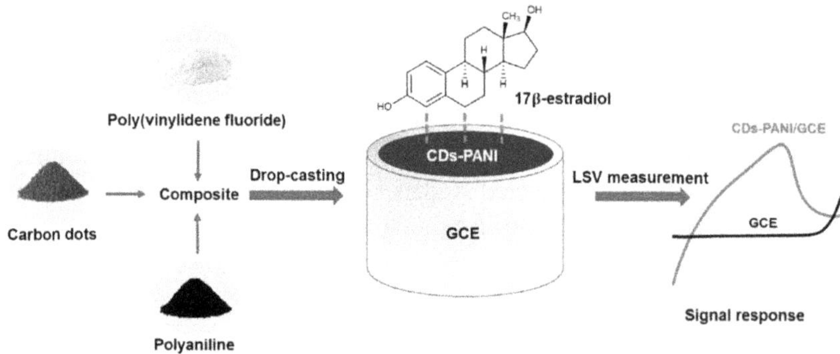

Figure 4.5. Schematic representation of the electrochemical detection of 17-β-estradiol using the CDs-PANI/GCE. Reprinted from [49], copyright (2021) with permission from Elsevier.

The simplicity, low cost, and stable analysis results present this method as a good candidate for water quality control in the supply system and the environment [51].

In other research, rGO and a metal complex porphyrin were introduced as a sensitive electrochemical method for the determination of 17-β-estradiol in river water without any sample pretreatment by DPV. In this experiment, metal complex porphyrins (MPs) have been widely used in electrochemical sensor designing due to the catalytic ability of MPs toward a wide range of redox materials. Central metal atoms in MPs play an important role in their catalytic activity of them. MPs are able to provide π–π stacking interaction with rGO to produce an MP/rGO nanocomposite. In this regard, the surface of GCE was modified by MP/rGO. Cu(II)-meso-tetra(thien-2-yl) porphyrin supported over an rGO surface and the fabricated electrochemical sensor applied for identification of 17-β-estradiol in river water. The method resulted in a low LOD of 5.3 nM without any purification step [52]. In interesting research, zinc oxide (ZnO), in nanostructure was grown on a Ag electrode surface by the aqueous chemical growth (ACG) method and resulted in ZnO nanorods (ZnONRs) nanostructure. For this purpose, The Ag electrode was immersed in zinc nitrate hexahydrate ($Zn\ (NO_3)_2{\cdot}6H_2O$) and hexamethylenetetramine ($C_6H_{12}N_4$) solutions that were tilted against the wall of the beaker. The set was put into the oven at low temperature repeated times to obtain an aligned ZnO nanostructure. 17-β-estradiol was covalently immobilized on the Ag–ZnONRs surface and studied by EIS. The method was applied in water samples with satisfactory recovery results, reproducible responses, and low LOD of 0.01 pg ml^{-1} [53].

In another work, the surface of GCE was coated with the one-pot synthesis of an rGO/rhodium nanoparticles (rGO/Rh) nanocomposite. After, the laccase enzyme biosensor (Lac/rGO-RhNP/GCE) was fabricated by dropping enzyme solution and glutaraldehyde for 1 h at 4 °C. The designed biosensor was applied for the determination of 17-β-estradiol by DPV in synthetic and real urine samples with a low LOD of 0.54 pM and high selectivity [54]. Some selected studies about 17-β-estradiol detection using electrochemical methods are shown in table 4.4.

Table 4.4. Different electrochemical methods for 17-β-estradiol determination.

Electrochemical sensing device	Method	Medium	LOD	References
CDs-PANI/GCE	LSV	Water/serum	43 nM	[49]
GCE	DPV	PBS	0.7 fM	[50]
LSGE			63.1 fM	
SPE	DPV	Tap water	888 nM	[51]
MP/rGO/GCE	DPV	River water	5.3 nmol l^{-1}	[52]
Ag–ZnONRs	EIS	Water	0.01 pg ml^{-1}	[53]
Lac/rGO-RhNP/GCE	DPV	Synthetic and real urine	0.54 pM	[54]
rGO–DHP/GCE	CV/EIS	Urine	7.7×10^{-8} M	[55]

Abbreviations: DHP: dihexadecylphosphate, DPV: differential pulse voltammetry, GCE: glassy carbon electrode, Lac: laccase enzyme, LSGE: laser-scribed graphene electrodes, LSV: linear-sweep voltammetry, MP: metal complex porphyrins, rGO: reduced graphene oxide, RhNP: rhodium nanoparticles.

4.3.5 Progesterone

Progesterone is categorized as a lipophilic steroid hormone that is responsible for pregnancy management and also menstruation in animals. Irregular levels of progesterone bring up different side effects like headache, urinary infections in women, and breast tenderness. Progesterone enters the environment after certain absorption in the body, so the determination of progesterone is an indispensable part of both clinical and environmental analysis procedures [56].

In this regard, different sensors and biosensors are developed. In a research, graphene quantum dots (GQDs) were synthesized according to the bottom-up method by pyrolysis of citric acid at 200 °C for 30–35 min. Then, the product was mixed drop-wise with NaOH under continuous stirring for some time. After neutralizing, the GQDs aqueous solution was stored at 4 °C. After, the surface of the GCE was coated with the GQDs and left to dry and finally it was activated by N-hydroxysuccinimide (NHS) and 1-ethyl-3-(3dimethylaminopropyl) carbodiimide (EDC). The modified electrode was incubated with progesterone antibody solution for 1 h, and BSA was applied as a blocking agent for nonspecific adsorption. The biosensor was applied for progesterone detection by CV in PBS. The immunosensor showed rapid and simple detection methods of progesterone for biological applications and pharmaceutical formulations [57]. In other research, a hydrogel-conjugated aptamer was used as a great platform for the construction of progesterone biosensors due to the hydrogel′s great mechanical and structural properties. The hydrophilic nature of hydrogels acts as a blocking membrane and leads to minimizing the nonspecific binding in affinity biosensors. In this regard, a combination of natural chitosan (CS) and a synthetic polymer network of hydroxyethyl cellulose (HEC) has been synthesized by blending a mixture of natural CS and synthetic HEC in the ratio of 1:1. Hence, the testosterone-specific aptamer was modified with thiol (-SH) group at 5′ end. Then, the AuNCs solution was added to the aptamer solution and mixed in the dark room. Finally, the aptamer functionalized AuNCs were blended with hybrid hydrogels for 1 h using sonication to form AuNC-aptamer-hydrogel conjugate (CS-g-HEC conjugated AuNCs-aptamer). The LOD of the modified method was calculated as 1 ng ml^{-1} by SWV in diluted blood samples [58].

In other research, a label-free aptasensor of progesterone was introduced by modifying the surface of a screen printed carbon electrode (SPCE) by aptamer/GQDs–NiO–AuNFs/f-MWCNTs/SPCE layer (aptamer-GQDs-NiO-Au hybrid nanofibers-functionlized multi-walled carbon nanotube-SPCE) for electrochemical sensing of progesterone in human serum samples and pharmaceutical formulations by EIS method. For this purpose, the GQDs were prepared by pyrolysis of citric acid at 200 °C for 30 min. The product was neutralized by NaOH and the aqueous solution of GQDs was obtained. Moreover, the NiO–AuNFs were prepared by electrospinning of polyvinyl pyrrolidone (PVP), $Ni(NO_3)_2.6H_2O$, and $HAuCl_4.3H_2O$. Therewith, GQDs–NiO–AuNFs nanocomposite was synthesized by dispersing NiO–AuNFs in GQDs suspension and stirring for 24 h. The product was dried and used for the rest of the experiment. GQDs–NiO–AuNFs/f-MWCNTs were prepared by ultrasonication of GQDs–NiO–AuNFs and MWCNTs in DMF, then 2 μl of the suspension was coated on SPCE surface and dried in air at room temperature. Then, the aptamer was immobilized on the modified electrode by immersing in PBS containing EDC and NHS to activate the terminal COOH groups for 1 h. GQDs–NiO–AuNFs/f-MWCNTs nanocomposite through covalent amide bonds between the amino groups of the aptamer and the carboxyl groups of the nanocomposite. The sensor showed an LOD of 1.86 pM in the wide dynamic concentration range of 0.01–1000 nM [56].

Hence, another progesterone sensor poly (3,4-ethylenedioxythiophene) and zirconium oxide NPs nanocomposite glassy carbon electrode (PEDOT/ZrO_2-NPs/GCE) was designed by modifying the GCE surface by poly (3,4-ethylenedioxythiophene) and zirconium oxide NPs (PEDOT/ZrO_2-NPs). The nanocomposite film was electrodeposited on the GCE surface by a saturated solution of $ZrCl_4$ and 3,4-ethylenedioxythiophene (EDOT) in presence of KCl and cycled by CV with potential scanning between −1.1 and 1.15 V with a sweep rate of 0.03 V s^{-1}. The sensor was applied in the determination of progesterone in biological fluids and pharmaceutical products with an LOD of 0.32 nM. The method presented a green synthesis process with excellent electrocatalytic activity toward the oxidation of progesterone [59] (figure 4.6).

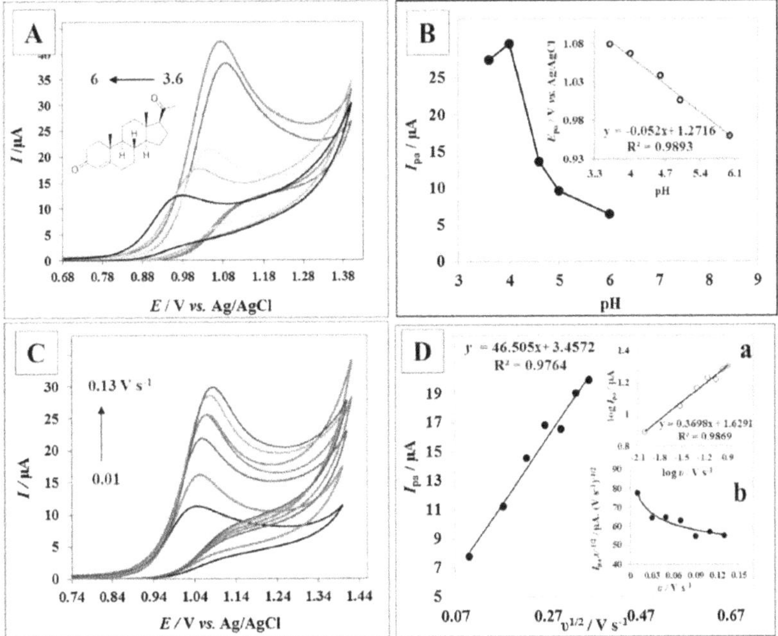

Figure 4.6. (A) CVs of PEDOT/ZrO_2-NPs/GCE in 0.1 M CBS containing 500 μM progesterone at different pHs over the range of 3.6–6.0 (scan rate: 0.1 V s^{-1}). Inset: chemical structure of progesterone. (B) Variations of redox peak current (I_{pa}) with respect to the pH of the electrolyte. Inset: linear relationship of anodic peak potential (E_p) versus pH. (C) CVs of 100 μM progesterone in 0.1 M CBS (pH 4) on the surface of PEDOT/ZrO_2-NPs/GCE at different scan rates. (D) Linear relationship of anodic peak current (I_{pa}) versus square root of scan rate ($\upsilon^{1/2}$). Inset a: relationship between the logarithmic peak current and the logarithmic scan rate for the anodic peak. Inset b: plot of scan rate normalized anodic peak current versus scan rate. Reprinted from [59], copyright (2019) with permission from Elsevier.

One more case was introduced by Gevaerd *et al* as a nonenzymatic electrochemical sensor for progesterone detection by modifying the GCE surface with imidazole-functionalized graphene oxide (GO–IMZ) as an artificial enzymatic active site. Firstly, GO–IMZ was synthesized by mixing GO suspension provided by the Hummers′ method with EDC and NHS under an ice bath, stirring vigorously for 2 h. Then, 1-(3-aminopropyl) imidazole (API), was added to the mixture and stirred overnight. The resulting product was washed and dried at 50 °C. Next, the electrode was modified by dropping 2.0 μl of GO–IMZ in methanol suspension on the GCE surface. The LOD of the sensor was 68 nmol l^{-1} and applied in the analysis of commercial pharmaceutical samples with satisfactory results [60]. Some selected studies about progesterone detection using electrochemical methods are shown in table 4.5.

Table 4.5. Different electrochemical methods for progesterone determination.

Electrochemical sensing device	Method	Medium	LOD	References
GQDs-GCE	CV	PBS	—	[57]
CS-g-HEC-AuNCs-aptamer	SWV	Diluted blood	1 ng ml^{-1}	[58]
Aptamer/GQDs–NiO–AuNFs/f-MWCNTs/SPCE	EIS	Serum	1.86 pM	[56]
PEDOT/ZrO_2-NPs/GCE	CV	Serum, plasma, and pharmaceutical products	0.32 nM	[59]
GO–IMZ-GCE	SWV/CV	Commercial pharmaceutical samples	68 nM	[60]
HRP-P4-(P4)-anti-P4-Protein-G-MBs/SPCE	Amperometry	Saliva	6.3 pg ml^{-1}	[61]
AuNP/AMBI/rGO/SPCE	EIS/CV	Calf serum and milk	0.28×10^{-9} M	[62]
SnNRs/GCE	DPV-amperometry	Commercial pharmaceutical formulations	0.12 μM	[63]
Fe_3O_4@GQD/f–MWCNTs/GCE	CV/LSV/DPV	Serum, pharmaceutical products	2.18 nM and 16.84 μA μM^{-1}	[64]
MIP/Pd/CFP electrode	CV/EIS/DPV	Pharmaceutical progesterone injections	0.05 nM	[65]
CPT-BDD	CV/SW-AdSV	Serum and drug samples	3.77 μg l^{-1}	[66]

Abbreviations: AMBI: 5-Amino-2-mercaptobenzimidazole, anti-P4-/progesterone antibody-progesterone standard, AuNCs: Au nanocubes, AuNFs: Au nanofibers, AuNPs: Au nanoparticles, CFP: carbon fiber paper, CPT-BDD: cathodically pretreated-boron-doped diamond, CS: chitosan, Fe_3O_4: nanoparticles, *f*-MWCNTs: functionalized multi-walled carbon nanotubes, GCE: glassy carbon electrode, g-HEC: hydroxyethyl cellulose hydrogel, GO–IMZ: imidazole-functionalized graphene oxide, GQDs: graphene quantum dots, HRP-P4-(P4): peroxidase-labeled progesterone, MIP: molecularly imprinted polymer, PEDOT: poly (3,4-ethylenedioxythiophene), protein-G-MBs: protein G functionalized-magnetic microbeads, rGO: reduced graphene oxide, SPCE: screen printed carbon electrode, SW-AdSV: square wave-adsorptive stripping voltammetry, ZrO_2-NPs: zirconium oxide nanoparticles.

4.3.6 Testosterone

Testosterone is known as a steroid hormone and is categorized in the androgen group that is secreted in the testes of males and the ovaries of females. However, small amounts of testosterone are secreted by the adrenal glands. Testosterone is an important male sex hormone responsible for health, protein synthesis, and human physical performance. It is mainly secreted in the testes of males and the ovaries of females and also small amounts are secreted by the adrenal glands [67]. At the same time, testosterone an anabolic androgenic steroid (AAS) has been abused by athletes as a doping substance to increase muscle mass. Therefore, the World Anti-Doping Agency (WADA) prohibited the use of testosterone to protect athletes from possible side effects [68]. Therefore, testosterone determination is highly important in clinical and forensic laboratories.

In this regard, the GCE surface was modified by electrodeposition of cobalt oxide by immersing the electrode in $Co(NO_3)_2{\cdot}6H_2O$ and Na_2SO_4 solution. The potential was adjusted in 50 mV s^{-1} between −1.0 and 1.0 V versus SCE. The developed sensor showed a reproducible oxidation response in the presence of testosterone. The modified electrode was introduced as a simple, low-cost, and stable sensor for the determination of testosterone with an LOD of 0.16 μM in 0.10 M NaOH at pH 12.5 by CV [69]. Herein, a novel testosterone immunosensor was designed by a combination of gold nanowires and polyvinyl butyral (PVB) for the determination of testosterone in serum samples by CV. Gold nanowires were electrodeposited by chronoamperometry method of ultrathin Au film in the presence of $HAuCl_4$ solution containing $HClO_4$. The polished platinum electrode was immersed in the solution containing standard testosterone antibody (T-Ab) and gold nanowires. Next, PVB ethanol solution was added to the beaker quickly and the electrode stayed there for 10 min. After, the electrode was washed and dried and used for the rest of the experiment. The fabricated T-Ab/Au nanowires/PVB/Pt showed LOD of 0.1 ng ml^{-1} with high sensitivity, long-term stability, and good reproducibility [70]. In other research, the surface of GCE was modified with GO–MIP for selective determination of testosterone in human serum by EIS. The fabrication of the sensor was as follows: 5 μl of GO was dropped on the surface of GCE at room temperature. Then, electropolymerization was carried out by immersing GCE into an acetate buffer solution containing a monomer of (o-phenylenediamine) and template (testosterone). After electropolymerization, the template was removed by washing the modified electrode in stirring ethanol and used for the rest of the experiment. The sensor showed an LOD of 0.4 fM (4.0×10^{-16} M) with long-term stability in room temperature for the detection of testosterone [71] (figure 4.7).

In one paper, the surface of GCE was modified with rGO and applied for the determination of testosterone in the presence of cationic surfactant cetyltrimethylammonium bromide (CTAB) in human urine, plasma, and commercial drug samples by DPV. The electrode was prepared by electrodeposition of rGO on the surface of GCE. The method showed a low LOD of 0.1 nM suggesting CTAB increases the electron transfer rate between solid rGO/GCE surface and electroactive

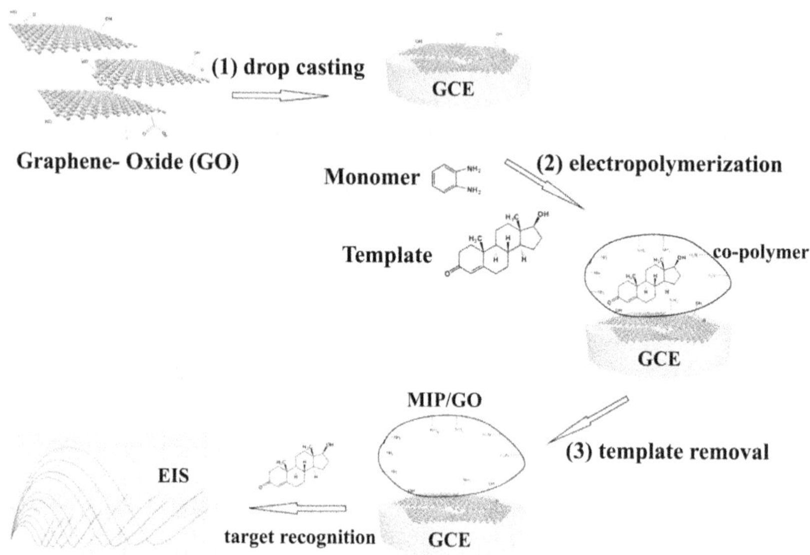

Figure 4.7. The process for the preparation of the MIP/GO composite. (1) Drop-casting graphene oxide sheets on the GCE surface. (2) Electropolymerization of the MIP layer on the surface of the GO-modified electrode. (3) Template removal and recognition of the target testosterone in different concentrations. Reprinted from [71], copyright (2017) with permission from Elsevier.

Table 4.6. Different electrochemical methods for testosterone determination.

Electrochemical sensing device	Method	Medium	LOD	References
CoO-GCE	CV	NaOH at pH 12.5	0.16 μM	[69]
T-Ab/Au nanowires /PVB/Pt	CV	Serum	0.1 ng ml^{-1}	[70]
GO–MIP-GCE	EIS	Serum	0.4 fM	[71]
rGO-GCE	DPV	Urine, serum, and commercial drug samples	0.1 nM	[72]

Abbreviations: CV: cyclic voltammetry, DPV: differential pulse voltammetry, EIS: electrochemical impedance spectroscopy, GCE: glassy carbon electrode, MIP: molecularly imprinted polymer, PVB: polyvinyl butyral, rGO: reduced graphene oxide, T-Ab: testosterone antibody.

testosterone [72]. Some selected studies about testosterone detection using electrochemical methods are shown in table 4.6.

4.3.7 Epinephrine

Epinephrine (EP) a hormone and a natural catecholamine neurotransmitter in the central nervous system (CNS) exists in the form of organic cations at the level of nM which controls vital pharmacological and physiological functions of the body.

Therefore, EP determination is important in clinical laboratories. Variation in EP level causes various diseases such as Alzheimer's, Parkinson's, hypertension, and multiple sclerosis [73].

In research, the surface of the GCE was modified by a novel nanocomposite of ordered mesoporous carbon/nickel oxide (OMC–NiO) synthesized through the hard-templating method. The nanocomposite possesses mesostructure with NiO nanocrystals embedded in the wall of the OMC. Mesoporous silica template (SBA-15), sucrose, sulfuric acid, and water were dried at 100 °C and subsequently at 160 °C. The product of SBA-15 silica was moderately polymerized and carbonized sucrose while impregnated with absolute ethanol containing nickel acetate tetrahydrate. To provide complete polymerized and carbonized sucrose inside the pores of the silica template, the dried powder was heated again at 100 °C and 160 °C after adding the aqueous solution containing sucrose and sulfuric acid. The product was calcinated at 900 °C under an Ar atmosphere to carbonize the sucrose and decompose nickel acetate into NiO. The SBA-15 silica template was removed by washing and the product was dried for the rest of the experiment. The synthesized OMC–NiO nanocomposite was dispersed in nafion ethanol and 6 μl of the solution was cast on the GCE surface, the modified sensor was applied after drying. The sensor was studied by CV and showed selective responses toward EP in the presence of UA with LOD of 8.5×10^{-8} M in spiked serum and EP injection [74] (figure 4.8).

Hence, another electrochemical sensor for the determination of EP was designed by preparing an MCPE. For this purpose, GO, 2-(5-ethyl-2,4-dihydroxyphenyl)-5,7-dimethyl-4H-pyrido[2,3-d][1,3]thiazine-4-one (EDDPT) as modifiers, graphite powder, and paraffin were mixed in optimum ratio and inserted in the bottom of a glass tube and electrical contact was provided by pushing a piece of copper wire down the glass tube. The EDDPT/GO/CPE sensor was studied by CV and DPV exhibiting excellent independent measurements for the simultaneous presence of three analytes (EP, dopamine (DA), and acetaminophen (AC)) without any interference with LOD of 0.651 M. The method had been used for the determination of EP in drug samples and human serum samples [75].

In one paper, highly ordered nanoporous thin Au films were deposited on the surface of anodic aluminum oxide (AAO) templates by sputter deposition of Au thin layer on the top side of AAO. AAO was prepared by anodizing a high-purity Al foil. The sensor was studied by different electrochemical methods of CV, LSV, and DVP. The ordered nanoporous Au electrode (ONP–Au) showed an LOD of 2.42×10^{-6} M and was applied for the determination of EP in injection ampoules with satisfactory results [76]. Some selected studies about epinephrine detection using electrochemical methods are shown in table 4.7.

4.3.8 Dopamine

Dopamine is known as another catecholamine neurotransmitter in the brain responsible for regulating many physiological functions of the CNS controlling people's daily behavior [82]. Moreover, dopamine is applied in the treatment of Parkinson's disease and dopamine-responsive dystonia. In addition, abnormal levels

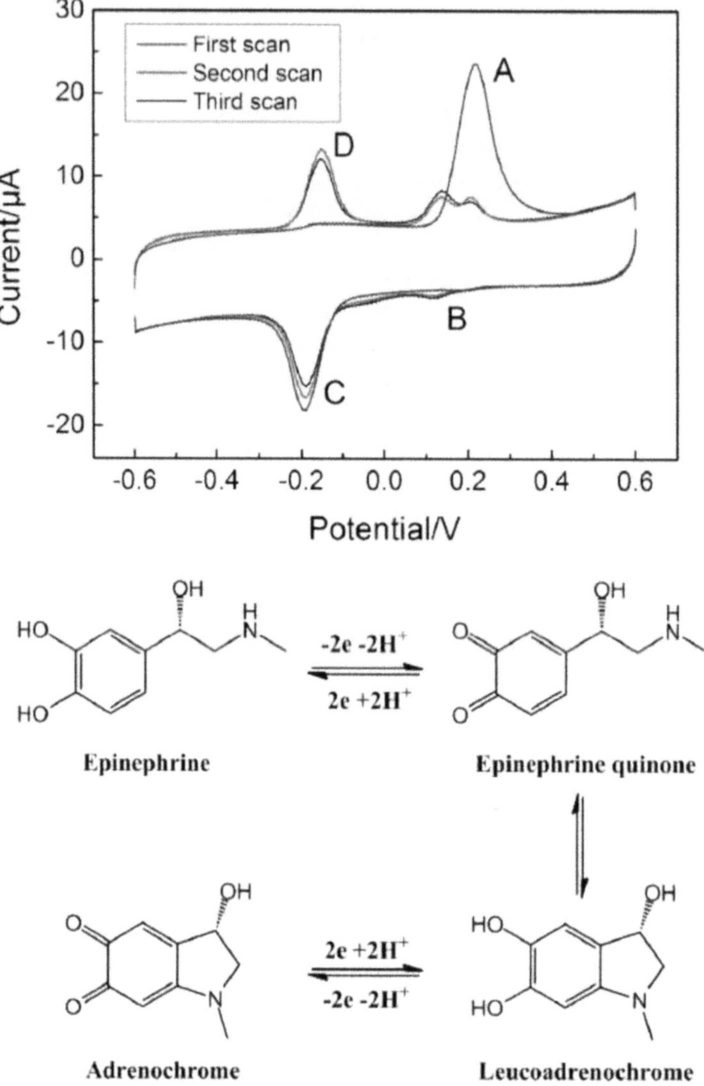

Figure 4.8. Successive cyclic voltammograms of 2.0×10^{-5} M EP on the OMC–NiO/GCE in 0.1 M PBS (pH = 7.0), scan rate: 100 mV s^{-1}. Peaks correspond to the oxidation of epinephrine to the open-chain quinone (peak A), the reduction of this quinone (peak B), the reduction of adrenochrome to leucoadrenochrome (peak C), and the oxidation of leucoadrenochrome to adrenochrome (peak D). Redox mechanism of EP on the OMC–NiO/GCE. Reprinted from [74], copyright (2021) with permission from Elsevier.

of dopamine cause diseases like attention deficit hyperactivity disorder syndrome, Alzheimer's and cardiovascular system diseases [83]. Therefore, detection levels of dopamine are important in laboratories.

In one article, a novel carbon black loaded triphenylamine based covalent organic framework (CB-loaded TPA-COF) composite was cast on the GCE surface to

Table 4.7. Different electrochemical methods for epinephrine determination.

Electrochemical sensing device	Method	Medium	LOD	References
OMC–NiO-GCE	CV	Human serum-epinephrine injections	8.5×10^{-8} M	[74]
EDDPT/GO/CPE	CV-DPV	Drug sample-human serum samples	0.651 M	[75]
ONP–Au	CV-LSV-DPV	Injection ampoules	2.42×10^{-6} M	[76]
AuNPs-TGA-CS-MWCNTs-GCE	Amperometry, DPV	Human blood serum	60 nM	[77]
Ag ions-irradiated MWCNT-ITO	SWV-CV	Human urine-blood	2 nM	[78]
NiO-rGO-GCE	CV	Human serum	10 μM	[78]
ZnO-MWCNTs-GCE	CV-DPV-EIS	Human blood serum-adrenaline titrate injection samples	0.016 μM	[79]
Ty-MWCNTs-GCE	CV-EIS	Human serum	0.51 μM	[80]
Au–Ag electrodes	CV-LSV-DPV and chronoamperometry	PBS	5.05×10^{-6} M	[81]

Abbreviations: CPE: carbon paste electrode, CS: chitosan, CV: cyclic voltammetry, DPV: differential pulse voltammetry, EDDPT: 2-(5-Ethyl-2,4-dihydroxyphenyl)-5,7-dimethyl-4H-pyrido[2,3-d][1, 3]thiazine-4-one, EIS: electrochemical impedance spectroscopy, GCE: glassy carbon electrode, GCE: glassy carbon electrode, GO: graphene oxide, ITO: indium tin oxide, MWCNTs: multi-walled carbon nanotubes, NiO: nickel oxide, OMC: ordered mesoporous carbon, ONP-Au: nanoporous Au electrode, rGO: reduced graphene oxide, TGA: thioglycolic acid, Ty: Tyrosinase enzyme.

electrochemically determine dopamine in medical injections, studied by CV and DPV. In this research, TPA-COF was synthesized by sonication of the mixture of tri (4-aminophenyl) amine (TPA–NH_2), tri (4-formylphenyl) amine (TPA-CHO), acetic acid in 1,4-dioxane/mesitylene followed by heating at 120 °C for 96 h. The yellow-to-orange precipitate was washed and used after drying. Next, CB-loaded TPA-COF was prepared with a one-pot solvothermal synthesis of different ratios of CB and TPA-COF. After, the CB-loaded TPA-COF was prepared by casting 8 μl of 1 mg of CB-doped TPA-COF, dispersed in water, ethanol, and Nafion solution on the GCE surface, and used after drying. The sensor showed satisfying stability high selectivity and good repeatability with an LOD of 0.17 μM. The experiment showed that satisfactory results of the sensor might be related to the effective host–guest interaction between dopamine molecules and the COF framework [84].

In other research, the surface of GCE was modified by Li_2TiO_3-multi-walled carbon nanotube nanocomposite (LTO-MWCNT) in order to determine dopamine in pharmaceutical formulations. Numerous electroactive sites and high surface area of the LTO-MWCNT nanocomposite are a result of well-wrapped MWCNTs in crystallized LTO cubical-shaped NPs. Acidic-treated MWCNT, anatase TiO_2, and $LiOH{\cdot}H_2O$ were stirred vigorously in ethanol and then transferred into a Teflon autoclave and heated at 180 °C for 24 h. Then, after washing and drying, the final product was annealed at 800 °C for 3 h in a horizontal tube furnace under an air atmosphere to obtain LTO-MWCNT nanocomposite. The sensor was fabricated by casting 5 μl of ethanolic suspension of LTO-MWCNT on the GCE surface. The modified electrode was studied by CV and EIS and showed an LOD of 5.7 $\mu mol\ l^{-1}$. LTO-MWCNT/GCE sensor presented good accuracy and precision with satisfactory recoveries of 98% [85] (figure 4.9).

In one research work, a new monomer for molecular imprinting sensor fabrication called thia-bilane structure (S-BIL) was synthesized and used as a molecularly imprinted polymer monomer applied in the synthesis of S-BIL MIP deposited on pencil graphite electrode (PeGE). The monomer was obtained from the addition reaction of tripyrrane 1 to nitrovinyl thiophene 2 in the presence of molecular iodine. Then after, dopamine was inserted as a template, into the electropolymerization process on the PeGE surface. After the elution step, the sensor was used for the determination of dopamine by CV and EIS. The electrode showed good selectivity for dopamine in the presence of biological interferences such as (ascorbic acid, urea, glucose, and tryptophan) [86]. Recently, a new dopamine biosensor was designed on GCE by amperometric deposition of poly(thiophene-3-boronic acid) (PT3BA) in the presence of glutaraldehyde and tyrosinase enzyme (PPO) on AuNPs, providing AuNPs-PT3BA-PPO-GCE sensor. Au film was deposited on the GCE surface by immersing the electrode in KNO_3 and $HAuCl_4{\cdot}3H_2O$. Deposition of PT3BA at 1.3 V and incorporation of Tyrosinase enzyme were carried out at the same time. The AuNPs-PT3BA-PPO biosensor was studied by CV and DPV with reproducible responses and LOD of 2×10^{-8} M for the determination of trace amounts of dopamine[87]. Some selected studies about dopamine detection using electrochemical methods are shown in table 4.8.

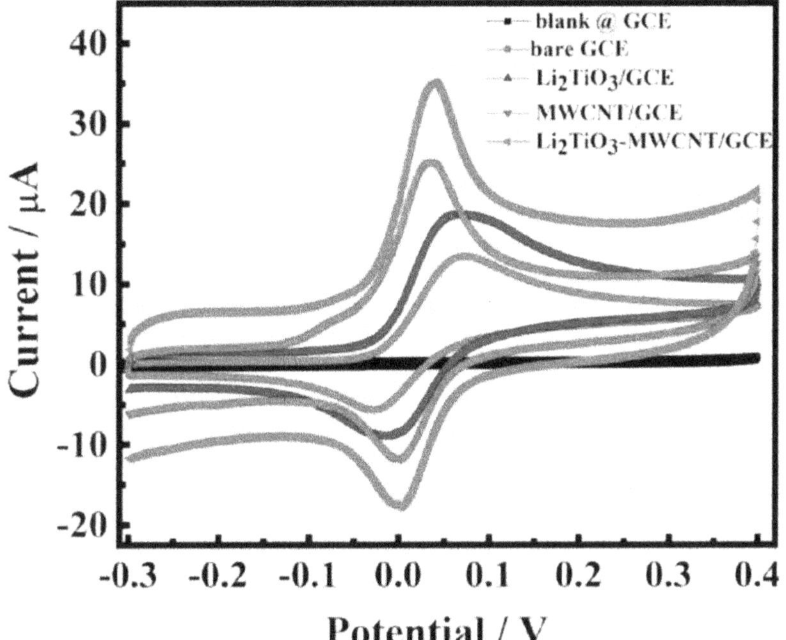

Figure 4.9. Cyclic voltammograms of 1 mM $[Fe(CN)_6]^{3-/4-}$ obtained in 0.1 M KCl solution (a) at blank GCE, bare GCE, Li_2TiO_3/GCE, MWCNT/GCE, and Li_2TiO_3-MWCNT/GCE with the scan rate of 100 mV s^{-1}. Reprinted from [85], copyright (2022) with permission from Elsevier.

4.3.9 Insulin

Insulin is known as a polypeptide hormone secreted from β-cells of the pancreas, responsible for controlling the concentration level of glucose in the blood. Diabetes is a chronic disease in which the body is not capable of producing or properly using insulin. Therefore, glucose cannot enter into cells and builds up in the blood, causing kidney disease, blindness, nerve damage, and heart disease [93]. Thus, the determination of insulin is an important and urgent issue in different clinical laboratories. In one article, the application of transition metals in the combination of MWCNTs was studied for modification of SPCE in order to determine insulin in PBS by CV. The presence of transition metals leads to an increase in the active surface area of the electrode and develops the analytical characteristics of the sensor. Therefore, two modified CoNPs-chitosan-MWCNTs-SPCE and CuNPs-chitosan-MWCNTs/SPCE were fabricated. After activation of MWCNT in an acidic medium, it was dispersed in PBS and 1 μl of chitosan, and then 10 μl of the dispersion was dropped on the SPCE surface and let to dry. Then after, CuNPs were deposited on the chitosan-MWCNTs-SPCE by immersing the electrode in $CuSO_4$ and Na_2SO_4 solution following potential application. Similarly, CoNPs were deposited on the chitosan-MWCNTs-SPCE surface by immersing the electrode in $CoCl_2$ in PBS and applying potential. The comparison between two designed modified electrodes exhibited that CoNPs-chitosan-MWCNTs-SPCE resulted in better stability and analytical

Table 4.8. Different electrochemical methods for dopamine determination.

Electrochemical sensing device	Method	Medium	LOD	References
CB-loaded TPA-COF-GCE	CV-DPV	Medical injections	0.17 μM	[84]
LTO-MWCNT/GCE	CV-EIS	Pharmaceutical formulations	5.7 μM	[85]
pS-BIL-MIP-PeGE	CV-EIS	—	20 nM	[86]
AuNPs-PT3BA-PPO-GCE	CV-DPV	—	2×10^{-8} M	[87]
PMEL-PASP- CPE	CV-DPV	PBS	0.78 nM	[88]
TFPB-TAPB-COF-Ox-MWCNT-GCE	CV-DPV-EIS	Dopamine injection-human urine	0.073 μM	[89]
Ag-ZIF-67p-GCE	CV-DPV	Tablets-dopamine hydrochloride injection	0.2 μM	[90]
AgNPs-MCPE	DPV	PBS	0.085 μM	[91]
NCDs-GCE	DPV	Human serum-urine	1.2×10^{-9} M	[92]

Abbreviations: 67p: 67 nanopinnas, AgNPs: silver nanoparticles, CB: carbon black, COF: covalent organic framework, CPE: carbon paste electrode, CV: cyclic voltammetry, DPV: differential pulse voltammetry, EIS: electrochemical impedance spectroscopy, GCE: glassy carbon electrode, MCPE: modified carbon paste electrode, MIP: molecularly imprinted polymer, NCDs: N-doped carbon dots, Ox-MWCNT:-oxidized multi-walled carbon nanotube, PASP: Poly aspartic acid, PeGE: pencil graphite electrode, PMEL: Poly melamine, pS-BIL: poly- thia-bilane, PT3BA: poly(thiophene-3-boronic acid), TFPB-TAPB: 1, 3, 5-tris(4-formylphenyl)benzene-1, 3, 5-tris(4-aminophenyl) benzene, TPA: triphenylamine, ZIF: zeolite imidazole framework.

characteristics toward insulin [94]. In another article, functionalized mesoporous silica thin film (MSTF) was deposited on the GCE surface by the electrochemically assisted self-assembly method. Then, two stranded oligonucleotides decorated the surface of GCE to cap the pore which prevents diffusion of the redox probe of (Fe $(CN)_6^{3-/4-}$) toward the GCE surface. When the insulin was added to the medium, the aptamer DNA strand shows a superior attraction toward insulin leading to dehybridization from the complementary DNA strand to open the nanopores. This task provides electrochemical probe-free diffusion along with film towards the electrode surface. The variance in the insulin peak by DPV provides insulin detection in PBS. The method provides sensitive responses with LOD of 3.0 nM [95] (figure 4.10).

In other research work, a new insulin determination method was introduced. The surface of the PeGE was modified with poly-orthophenylene diamine AuNPs—single-stranded DNA aptamer (PODA-GNPs-ssDNA-PeGE). The conductive polymer was coated on the surface of PeGE. Then after, the surface of the electrode was coated by AuNPs and subsequently, ssDNA was used to produce an aptasensor. Insulin interacted with the single-stranded ssDNA to form a G-quadruplex structure. The charge transfer resistance of the aptasensor was studied by EIS in plasma. A wide dynamic range of 1.0–1000.0 nmol l^{-1} with an LOD of 0.27 nmol l^{-1} was calculated

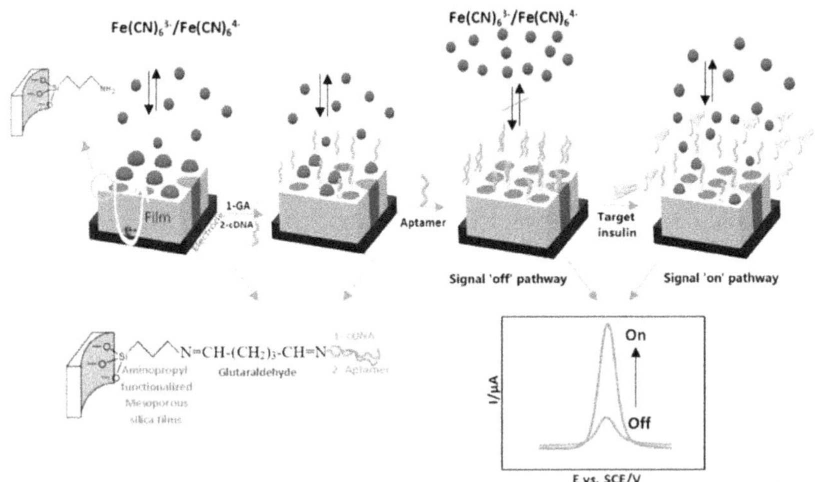

Figure 4.10. Schematic representing the configuration of MSTF applied for label-free electrochemical aptasensor utilizing an aptamer-grafted film operating as a regulating gate for the diffusion of the redox couple $Fe(CN)_6^{3-}/Fe(CN)_6^{4-}$ upon dehybridization from cDNA. Reprinted from [95], copyright (2021) with permission from Elsevier.

Table 4.9. Different electrochemical methods for insulin determination.

Electrochemical sensing device	Method	Medium	LOD	References
CuNPs-chitosan-MWCNTs-SPCE	CV	PBS	1.11 μM	[94]
CoNPs-chitosan-MWCNTs-SPCE			0.025 μM	
Aptamer-cDNA-gated mesoporous silica nanoparticles-GCE	DPV	PBS-serum	3 nM	[95]
PODA-AuNPs-ssDNA-PeGE	EIS	Serum	0.27 nmol l^{-1}	[96]
DNA2Fc@AuNPs-mDNA/MB-IBA-Au	EIS-CV	Serum	0.1 pM	[97]
EMMIPs-GCE	CV-DPV	Serum	3 pM	[98]

Abbreviations: AuNPs: gold nanoparticles, CoNPs: cobalt nanoparticles, CuNPs: copper nanoparticles, CV: cyclic voltammetry, DNA2Fc: DNA2/ferrocene, DPV: differential pulse voltammetry, EIS: electrochemical impedance spectroscopy, EMMIPs: electromagnetic molecularly imprinted polymers, GCE: glassy carbon electrode, IBA: insulin-binding aptamer, MB: methylene blue, MWCNTs: multi-walled carbon nanotubes, PeGE: pencil graphite electrode, PODA: poly-orthophenylene diamine, SPCE: screen printed carbon electrode, ssDNA: single-stranded DNA aptamer.

for this aptasensor [96]. Some selected studies about insulin detection using electrochemical methods are shown in table 4.9.

4.3.10 Parathyroid

Parathyroid (PTH) is a protein consisting of amino acids playing a vital role in controlling calcium and phosphate homeostasis. In addition, PTH is responsible for stimulating the multiple intracellular signals of cAMP, two protein kinases A and C [99]. Detection of PTH in biological fluids is helpful in early diagnosis and the prognosis of the disease in that it could be increased when prostate cancer metastases to the bone [100] or in chronic kidney diseases [101].

In a research article, a novel PTH biosensor was fabricated by modifying the surface of the gold electrode with self-assembled mercaptohexanol (SAM-mercaptohexanol) and 3-aminopropyl triethoxysilane (APTES). After, an anti-PTH solution was dropped onto the activated electrode surface to form anti-PTH–SAM-mercaptohexanol-APTES-Au. Then, a BSA solution was used to inhibit the active ends of APTES SAMs. Electrochemical CV and EIS methods had been used for studying the modified biosensor in human serum samples and artificial serum samples. The method showed good repeatability and reproducibility with the dynamic concentration range of 10–50 pg ml^{-1} PTH [102].

In a paper, molybdenum disulfide (MoS_2)-graphene composite (MG) was synthesized by self-assembly during the hydrothermal process and drop-coated on a pretreated Au electrode. Then, the PTH antibody (PTH-MAb) was immobilized on the MG-modified Au electrode to form the PTH-MAb-MG electrode. A solution containing IgG molecules bearing a separately conjugated enzyme (ALP/HRP) was used to attach the enzyme to the MG sensor. Conjugated IgG-linked enzymes to the MG electrode were carried out by the drop-casting method (AI-MG or the HI-MG sensor). Tris buffer solution containing ALP substrate was injected into the cell and electrochemical measurements were carried out. AP-MG and HP-MG sensors were fabricated in the mentioned process. The MG electrode was coated with MAb against PTH by physical absorption. PBSB solution blocked the PTH-MAb on the MG electrode. Then, the electrode was incubated with PTH antigen and washed with a rinsing buffer. Next, the MG electrode was immersed in PTH PAb solution. The sensor was applied to serum samples and studied by CV and EIS. The P-value is less than 0.01 when evaluated with a t-test using Welch's correction. This confirms that the sensor is capable of clinical applications [10]. In one research, the application of magnetic beads (MBs) was described in a novel electrochemical sensors structure to expand the detection performance of the sensor by capturing the target. The developed immunosensor was fabricated by using MBs modified by aminophenyl boronic acid (APBA) and antibody-peroxidase complex. By the addition of antibody solution on MBs, PTH is captured by antibodies on modified MBs. PEG solution was added to block nonspecific binding sites and MB-APBA/HRP-Ab/PEG was constructed. Finally, serum PTH was dropped on the MB-APBA/HRP-Ab/PEG surface. The magnetic field caused the adsorption of MB-APBA/HRP-Ab/PEG on the surface of SPE. After washing, H_2O_2 and HQ solutions were added and SWV and EIS signals were obtained. The introduced immunosensor showed a low LOD of 11.56 pg ml^{-1} by SWV and 49.30 pg ml^{-1} by EIS [103]. The provided disposable electrode is capable of analyzing PTH as a POC device (figure 4.11).

In another research paper, the surface of the Pt disk was coated with Au structure as an electrochemical transducer for the detection of PTH in blood samples. Then, Au–Pt hybrid surface was modified with 3-mercaptopropionic acid (MPA) to Au and following activation of the carboxyl groups to provide anti-PTH antibody

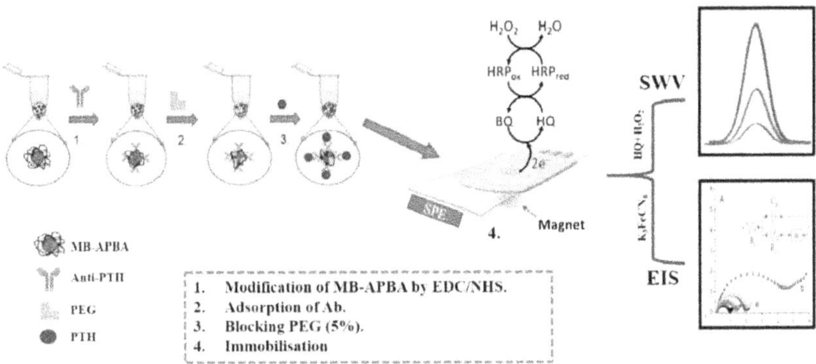

Figure 4.11. Schematic diagram showing the fabrication process and the immuno-electrochemical reaction of magneto immunosensor. Reprinted from [103], copyright (2021) with permission from Elsevier.

Table 4.10. Different electrochemical methods for PTH determination.

Electrochemical sensing device	Method	Medium	LOD	References
Anti-PTH–SAM-mercaptohexanol-APTES-Au	CV-EIS	Serum	NS	[102]
PTH-MAb-MG-Au	CV-EIS	Serum	NS	[101]
MB-APBA/HRPAb/PEG-SPE	SWV EIS	Serum	11.56 pg ml^{-1} 49.30 pg ml^{-1}	[103]
Anti-PTH-Pt-Au-hybrid	DPV EIS	Serum	0.59 pg.ml^{-1} 0.36 pg.ml^{-1}	[104]

Abbreviations: Ab/TGA: Thioglycolic acid, Ab/TSH: TSH antibody, anti-PTH: parathyroid antibody, APBA: aminophenyl boronic acid, APTES: 3-aminopropyl triethoxysilane, CILE: ionic liquid carbon paste electrode, CV: cyclic voltammetry, DPV: differential pulse voltammetry, EIS: electrochemical impedance spectroscopy, HRP: anti-human peroxidase, HRPAb: antibody-peroxidase, MB: magnetic beads, MG: molybdenum disulfide-graphene composite, PEG: polyethylene glycol, PTH: parathyroid, SAM: self-assembled, SPE: screen printed electrode.

immobilization. Casein was dropped onto the electrode surface to avoid nonspecific binding. The electrochemical measurements were carried out by EIS and DPV methods. A wide linear concentration range of 1–100 000 pg.ml^{-1} and Low LOD of 0.36 pg.ml^{-1} were obtained for the EIS method and 0.59 pg.ml^{-1} was calculated for the DPV method in serum media. The Au–Pt hybrid disk electrode was used as a POC device for highly sensitive and selective detection of PTH [104]. Some selected studies about PTH detection using electrochemical methods are shown in table 4.10.

4.3.11 Thyroid-stimulating hormone

Thyroid-stimulating hormone (TSH) is known as a glycoprotein hormone responsible for stimulating of biosynthesis and secretion of triiodothyronine (T3) and thyroxine (T4) hormones in the body, which control the metabolism of the body [105]. TSH level in the serum is a sign of thyroid abnormalities diagnosis, for example, low levels of TSH is a sign of hypopituitarism and Grave's disease and high levels of TSH is a sign of hypothyroidism and Hashimoto's thyroiditis [106]. Therefore, the determination of TSH in human serum is important for the diagnosis or prevention of different problems.

In a research article, a CPE was modified by ionic liquid (IL) and AuNPs to provide an appropriate surface for the immobilization of TSH antibody (anti-TSH) on the CPE surface. The secondary antibody, polyclonal anti-human-TSH labeled with horseradish was used for the formation of a sandwich-like structure for TSH between the anti-TSH on the CPE surface modified with AuNPs and the secondary antibody, polyclonal anti-human peroxidase (HRP-Ab). Thioglycolic acid (TGA) aqueous solution was used for the formation of a self-assembled monolayer. DPV method was used for recording the electrochemical signals. The HRP-Ab/TSH-Ab/TGA/nano-Au/CILE immunosensor showed high sensitivity and acceptable stability with an LOD of 0.02 ng ml^{-1} in the serum [107] (figure 4.12).

Other research was carried out by modifying the surface of GCE by dropping the azo compound solution (E)–5-[(4-dodecyloxyphenyl) diazenyl] isophthalic acid as

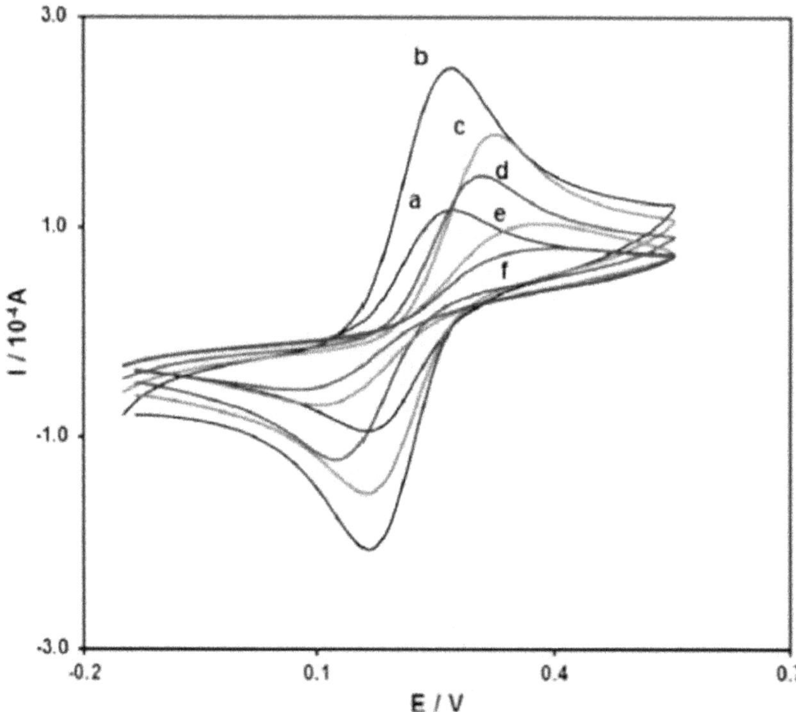

Figure 4.12. Cyclic voltammogram (CV) curves of the different electrodes in PBS (0.1 M, pH 7.0) solution containing 1 M KCl and 5 mM $Fe(CN)_6\ ^{3-/4-}$; (a) bare CILE; (b)nano-Au/CILE; (c) TGA/nano-Au/CILE; (d) Ab/TGA/nano-Au/CILE; (e)TSH/Ab/TGA/nano-Au/CILE; and (f) HRP-Ab/TSH/Ab/TGA/nano-Au/CILE. Scan rate: 0.1 V s^{-1}. Reprinted from [107], copyright (2018) with permission from Elsevier.

Table 4.11. Different electrochemical methods for thyroid-stimulating hormone determination.

Electrochemical sensing device	Method	Medium	LOD	References
HRP-Ab/TSH-Ab/TGA/nano-Au/CILE	DPV	Serum	0.02 ng ml^{-1}	[107]
gly/ab-TSH/azo-GCE	CV-SWV-EIS	PBS	0.04 μIU ml^{-1}	[105]
SPCE nanobiosensor	EIS	PBS	0.001 μIU ml^{-1}	[108]
Ab1/TSH/Ab2-HPR/loaded liposom/GCE	potentiometry	Serum	0.067 μIU ml^{-1}	[109]
AuNPs-GO/Ab-TSH/TSH/HPR-Ab-TSH/GoldMag	CV-EIS	Serum	0.005 μIU ml^{-1}	[110]

Abbreviations: azo: azo compound, CILE: ionic liquid carbon paste electrode, CV: cyclic voltammetry, DPV: differential pulse voltammetry, EIS: electrochemical impedance spectroscopy, GCE: glassy carbon electrode, Gly: glycine, GoldMag: nanogold functionalized-magnetic beads, HRP-Ab: anti-human peroxidase, TGA: thioglycolic acid nano, TSH-Ab: TSH antibody.

an electronic mediator immobilized and a medium for covalent immobilization of anti-TSH antibody (ab-TSH) on GCE surface. Immobilization of the ab-TSH was carried out by EDC/NHS through the presence of the carboxylic groups in the azo compound. Glycine (Gly) was used to block nonspecific sites on the immunosensor. In the presence of TSH antigen, the electrochemical response of the immunosensor decreased. The electrochemical behavior of the modified electrode was studied by CV, SWV, and EIS. The Gly/ab-TSH/azo-GCE showed good accurate and sensitive responses in the presence of different interferences (ascorbic acid, glucose, uric acid, and creatine) which provides availability of the immunosensor in clinical laboratories. The LOD of the method was calculated as 0.04 μIU ml^{-1} which is lower than ELISA for TSH determination [105].

In another paper, a nanobiosensor was constructed by modifying the surface of SPCE with amino-coated AuNPs coupled to an anti-TSH antibody through the formation of a peptide bond. The nanobiosensor was studied by EIS, whereas the effective resistance of the biosensor changes in the presence of TSH. Cystamine dihydrochloride provides a covalent bonding between amino groups of the surface and the carboxylic group, confirming high antibody availability. The method showed an LOD of 0.001 μIU ml^{-1} which is suitable for diagnostic purposes and makes the designed SPCE nanobiosensor a good candidate for POC and also clinical usage [108]. Some selected studies about thyroid-stimulating hormone detection using electrochemical methods are shown in table 4.11.

4.3.12 Prolactin

Prolactin (PRL) which is secreted by the actotroph cells of the anterior pituitary gland, is known as an amino acid peptide hormone regulating the metabolism, development of the pancreatic, and controlling the immune system. In addition, PRL is responsible for stimulating the secretion of milk in mammals and also the synthesis of progesterone in the corpus luteum. Therefore, the determination of PRL is of great interest in clinical laboratories.

Recently, a new immunosensor was fabricated by a GCE-based enzyme-free sandwich-type electrochemical method for the sensitive determination of PRL in serum. Multi-branched PdPt nanodendrites decorated amino-rich Fe-based metal–organic framework (PtPdNDs@NH_2-MIL53-(Fe))is used for signal amplification and AuNPs coated amino-functionalized graphene sheet (AuNPs@NH_2-GS) was used for sensing platform. After, immobilization of antibody 1(Ab1) was loaded on modified GCE by AuNPs@NH_2-GS, and the electrode was incubated with PRL. Then after, Ab2-loaded PtPdNDs@NH_2-MIL53-(Fe) was dropped on the modified electrode and after washing was used for the rest of the experiment. The modified immunosensor was studied with CV, EIS, and amperometry. The PdPtNDs@NH_2-MIL-53(Fe)-GCE showed excellent performance related to the high electrocatalytic activity of PdPtNDs@NH_2-MIL-53(Fe), as well as good conductivity and biocompatibility of AuNPs@NH_2-GS with LOD of 1.15 pg mL^{-1} [111] (figure 4.13).

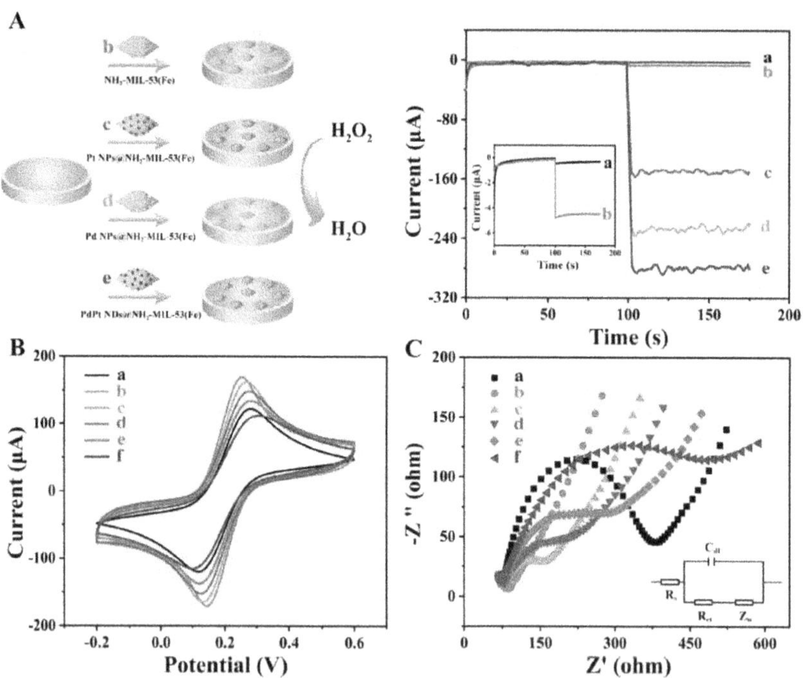

Figure 4.13. (A) Amperometric responses of the different modified electrodes in PBS (pH 7.4) containing 5 mM of H_2O_2: (a) bare GCE, (b) NH_2-MIL-53(Fe), (c) PtNPs@NH_2-MIL-53 (Fe), (d) Pd NPs@NH_2-MIL-53(Fe), (e) PdPtNDs@NH_2-MIL-53(Fe). The inset is the magnification of curves a and b. (B) CV and (C) EIS responses of the modification steps of the biosensor: (a) bare GCE, (b) AuNPs@NH_2-GS/GCE, (c) Ab1/AuNPs@NH_2-GS/GCE, (d) BSA/Ab1/AuNPs@NH_2-GS/GCE, (e) PRL/BSA/Ab1/AuNPs@NH_2-GS/GCE, (f) Ab2-label/PRL/BSA/Ab1/AuNPs@NH_2-GS/GCE. Reprinted from [111], copyright (2021) with permission from Elsevier.

Table 4.12. Different electrochemical methods for prolactin determination.

Electrochemical sensing device	Method	Medium	LOD	References
PdPtNDs@NH_2-MIL-53(Fe)-GCE	CV-EIS-amperometry	Serum	1.15 pg ml^{-1}	[111]
pPPA-CNTs-GCE	SWV	Serum-urine	3 pg ml^{-1}	[112]
nano-Au/TGA/PRL-Ab/PRL/HPR-PRL-Ab/CILE	DPV	Serum	12.5 mIU l^{-1}	[113]

Abbreviations: CNTs multi-walled carbon nano, CV: cyclic voltammetry, DPV: differential pulse voltammetry, GCE: glassy carbon electrode, NH_2-MIL-53(Fe):amino-rich Fe-based metal–organic framework, PdPtNDs: PdPt nanodendrites, pPPA: poly(pyrrolepropionic acid), SWV: square wave voltammetry.

In other research, the surface of the GCE was modified with poly (pyrrolepropionic acid)/multi-walled carbon nanotubes hybrid (pPPA/CNTs) by the simple casting of MWCN on the GCE and following electrodeposition of pPPA on the electrode surface. Anti-PRL labeled with alkaline phosphatase (AP) reacted with PRL in the sample medium providing an indirect competitive assay. The immunosensor was studied by SWV with a low LOD of 3 pg mL^{-1}. The designed pPPA-CNTs-GCE showed high sensitivity and reproducible responses in urine and serum samples [112]. Some selected studies about prolactin detection using electrochemical methods are shown in table 4.12.

4.4 Conclusion

Hormones are biologically important substances that are secreted from endocrine glands. They are transported by the circulatory system to the target tissues and are presented in low amounts, emphasizing the need for application of a sensitive and selective determination method. Despite poor reproducibility and the hard regeneration step of electrochemical sensors, they show superior sensitivity and selectivity in clinical determination processes of hormones. This chapter studied the electrochemical sensing methods for the determination of different hormones which have a crucial role in body function. Changing the level of these hormones might be a sign of diseases, emphasizing the importance of a sensitive, accurate, and selective determination method in different biological mediums. Therefore, the introduction or development of a new and sensitive determination method is important in clinical and forensic laboratories. In this regard, electrochemical sensing methods (consisting of designing sensors/biosensors) play an outstanding role due to simplicity, cost-effectiveness, and sensitivity. Moreover, the presence of electrochemical POC methods simplifies the determination of hormones in pocket size with sensitive results. Studies about POC systems are still continuing in laboratories.

References

[1] Cifrić S, Nuhić J, Osmanović D and Kišija E 2019 *Int. Conf. on Medical and Biological Engineering* 173–7

[2] Ndangili P M, Jijana A M, Baker P G and Iwuoha E I 2011 *J. Electroanal. Chem.* **653** 67–74
[3] Bahadır E B and Sezgintürk M K 2015 *Biosens. Bioelectron.* **68** 62–71
[4] Luong J H, Male K B and Glennon J D 2008 *Biotechnol. Adv.* **26** 492–500
[5] (a) Grieshaber D, MacKenzie R, Vörös J and Reimhult E 2008 *Sensors* **8** 1400–58
(b) Justino C I, Rocha-Santos T A and Duarte A C 2010 *TrAC, Trends Anal. Chem.* **29** 1172–83
[6] Yogeswaran U and Chen S-M 2008 *Sensors* **8** 290–313
[7] (a) Kudr J, Klejdus B, Adam V and Zitka O 2018 *TrAC, Trends Anal. Chem.* **98** 104–13
(b) Yáñez-Sedeño P, Campuzano S and Pingarrón J M 2016 *Sensors* **16** 1585
[8] Mohanty S P and Kougianos E 2006 *IEEE Potentials* **25** 35–40
[9] Dědík J, Janovcová M, Dejmková H, Barek J and Pecková K 2011 *Sensing in Electroanalysis* **vol 6** ed K Kalcher, R Metelka, I Švancara and K Vytřas (Univerzita Pardubice)
[10] (a) H-U, Kim H Y, Kim A, Kulkarni C, Ahn Y, Jin Y, Kim K-N, Lee M-H, Lee and Kim T 2016 *Sci. Rep.* **6** 1–9
(b) Hareesha N, Manjunatha J, Amrutha B, Pushpanjali P, Charithra M and Prinith Subbaiah N 2021 *J. Electron. Mater.* **50** 1230–8
(c) Tigari G and Manjunatha J 2020 *Instrum. Exp. Tech.* **63** 750–7
(d) Manjunatha J G 2020 *Open Chem. Eng. J.* **14** 52–62
(e) Manjunatha Charithra M and Manjunatha J G 2020 *ChemistrySelect* **5** 9323–9
(f) Tigari G and Manjunatha J 2020 *Monatsh. Chem.* **151** 1681–8
(g) Manjunatha J G 2020 *Anal. Bioanal. Electrochem.* **12** 893–903
(h) Pushpanjali P, Manjunatha J and Srinivas M 2020 *FlatChem.* **24** 100207
(i) Raril C and Manjunatha J G 2020 *J. Anal. Sci. Technol.* **11** 1–10
(j) Manjunatha J 2020 *Chem. Data Coll.* **25** 100331
[11] Pacheco J G, Rebelo P, Cagide F, Gonçalves L M, Borges F, Rodrigues J A and Delerue-Matos C 2019 *Talanta* **194** 689–96
[12] Iles R and Kallichurn H 2012 *J. Bioeng. Biomed. Sci.* **2** 4
[13] Bahadır E B and Sezgintürk M K 2016 *Artif. Cells Nanomed. Biotechnol.* **44** 248–62
[14] Faridbod F, Ganjali M R, Larijani B, Norouzi P and Hosseini M 2014 *Iran. J. Public Health* **43** 94–104
[15] Ozhikandathil J, Badilescu S and Packirisamy M 2017 *J. Neural Transm.* **124** 47–55
[16] (a) Kokkinos C, Economou A and Prodromidis M I 2016 *TrAC, Trends Anal. Chem.* **79** 88–105
(b) da Silva Neves M M P, González-García M B, Hernandez-Santos D and Fanjul-Bolado P 2018 *Curr. Opin. Electrochem.* **10** 107–11
(c) Zhu G, Yin X, Jin D, Zhang B, Gu Y and An Y 2019 *TrAC, Trends Anal. Chem.* **111** 100–17
[17] (a) Chin C D, Linder V and Sia S K 2007 *Lab Chip* **7** 41–57
(b) Becker H, Hansen-Hagge T and Gärtner C 2014 *Biological Identification* (Elsevier) pp 220–49
(c) Ozhikandathil J 2012 Microphotonics and nanoislands integrated lab-on-chips (LOCs) for the detection of growth hormones in milk *Doctoral dissertation* (Concordia University)
(d) Peterson S L, McDonald A, Gourley P L and Sasaki D Y 2005 *J. Biomed. Mater. Res.* A **72** 10–8
[18] (a) Pessoa-Neto O D, Dos Santos V B, Vicentini F C, Suarez W T, Alonso-Chamarro J, Fatibello-Filho O and Faria R C 2014 *Cent. Eur. J. Chem.* **12** 341–7

(b) Suarez W T, Pessoa-Neto O D, Dos Santos V B, de Araujo Nogueira A R, Faria R C, Fatibello-Filho O, Puyol M and Alonso J 2010 *Anal. Bioanal. Chem.* **398** 1525–33
[19] (a) Gabardo C and Soleymani L 2016 *Analyst* **141** 3511–25
(b) Ho C M B, Ng S H, Li K H H and Yoon Y-J 2015 *Lab Chip* **15** 3627–37
[20] Hamad E, Bilatto S, Adly N, Correa D, Wolfrum B, Schöning M J, Offenhäusser A and Yakushenko A 2016 *Lab Chip* **16** 70–4
[21] Martin A, Teychené S, Camy S and Aubin J 2016 *Microfluid. Nanofluid.* **20** 1–8
[22] Yohe G and Tol R S 2002 *Global Environ. Change* **12** 25–40
[23] Ribeiro J A, Fernandes P M, Pereira C M and Silva F 2016 *Talanta* **160** 653–79
[24] (a) Vedung T, Werner M, Ljung B-O, Jorfeldt L and Henriksson J 2011 *J. Hand Surg.* **36** 1974–80
(b) Pu J, Bai D, Yang X, Lu X, Xu L and Lu J 2012 *Biochem. Biophys. Res. Commun.* **428** 210–5
[25] Madhuchandra H and Swamy B K 2020 *Mater. Sci. Energy Technol.* **3** 464–71
[26] Sochr J, Švorc Ľ, Rievaj M and Bustin D 2014 *Diam. Relat. Mater.* **43** 5–11
[27] Zaidi S A 2018 *Electrochim. Acta* **274** 370–7
[28] Zhao F, Zhu W, Su J, Xue Y, Wei W and Liu S 2015 *Curr. Anal. Chem.* **11** 80–7
[29] Lund H, Paus E, Berger P, Stenman U-H, Torcellini T, Halvorsen T G and Reubsaet L 2014 *Tumor Biol.* **35** 1013–22
[30] Chu C, Li L, Li S, Li M, Ge S, Yu J, Yan M and Song X 2013 *Microchim. Acta* **180** 1509–16
[31] Rizwan M, Hazmi M, Lim S A and Ahmed M U 2019 *J. Electroanal. Chem.* **833** 462–70
[32] Xia N, Chen Z, Liu Y, Ren H and Liu L 2017 *Sens. Actuators* B **243** 784–91
[33] Lu Y, Wang H, Shi X-M, Ding C and Fan G-C 2022 *Anal. Chim. Acta* **1199** 339560
[34] Teixeira S, Conlan R S, Guy O and Sales M G F 2014 *Electrochim. Acta* **136** 323–9
[35] Li R, Wu D, Li H, Xu C, Wang H, Zhao Y, Cai Y, Wei Q and Du B 2011 *Anal. Biochem.* **414** 196–201
[36] Roushani M, Valipour A and Valipour M 2016 *Mater. Sci. Eng.* C **61** 344–50
[37] Yang G, Chang Y, Yang H, Tan L, Wu Z, Lu X and Yang Y 2009 *Anal. Chim. Acta* **644** 72–7
[38] Singh A, Kaushik A, Kumar R, Nair M and Bhansali S 2014 *Appl. Biochem. Biotechnol.* **174** 1115–26
[39] Shahub S, Upasham S, Ganguly A and Prasad S 2022 *Sens. Bio-Sens. Res.* **38** 100527
[40] Karuppaiah G, Velayutham J, Hansda S, Narayana N, Bhansali S and Manickam P 2022 *Bioelectrochemistry* **145** 108098
[41] Cheng C, Li X, Xu G, Lu Y, Low S S, Liu G, Zhu L, Li C and Liu Q 2021 *Biosens. Bioelectron.* **172** 112782
[42] Dhull N, Kaur G, Gupta V and Tomar M 2019 *Sens. Actuators* B **293** 281–8
[43] Halima H B, Bellagambi F G, Brunon F, Alcacer A, Pfeiffer N, Heuberger A, Hangouët M, Zine N, Bausells J and Errachid A 2022 *Talanta* **257** 123802
[44] Yeasmin S, Wu B, Liu Y, Ullah A and Cheng L-J 2022 *Biosens. Bioelectron.* **206** 114142
[45] Vabbina P K, Kaushik A, Pokhrel N, Bhansali S and Pala N 2015 *Biosens. Bioelectron.* **63** 124–30
[46] Kim K S, Lim S R, Kim S-E, Lee J Y, Chung C-H, Choe W-S and Yoo P J 2017 *Sens. Actuators* B **242** 1121–8

[47] Liu Q, Shi W, Tian L, Su M, Jiang M, Li J, Gu H and Yu C 2021 *Anal. Chim. Acta* **1184** 339010

[48] (a) Musa A M, Kiely J, Luxton R and Honeychurch K C 2021 *TrAC, Trends Anal. Chem.* **139** 116254

(b) Cesarino I, Cincotto F H and Machado S A 2015 *Sens. Actuators* B **210** 453–9

[49] Supchocksoonthorn P, Sinoy M C A, de Luna M D G and Paoprasert P 2021 *Talanta* **235** 122782

[50] Chang Z, Zhu B, Liu J, Zhu X, Xu M and Travas-Sejdic J 2021 *Biosens. Bioelectron.* **185** 113247

[51] Raymundo-Pereira P A, Gomes N O, Machado S A and Oliveira O N Jr 2019 *J. Electroanal. Chem.* **848** 113319

[52] Moraes F C, Rossi B, Donatoni M C, de Oliveira K T and Pereira E C 2015 *Anal. Chim. Acta* **881** 37–43

[53] Singh A C, Asif M, Bacher G, Danielsson B, Willander M and Bhand S 2019 *Biosens. Bioelectron.* **126** 15–22

[54] Povedano E, Cincotto F H, Parrado C, Díez P, Sánchez A, Canevari T C, Machado S A, Pingarrón J M and Villalonga R 2017 *Biosens. Bioelectron.* **89** 343–51

[55] Janegitz B C, dos Santos F A, Faria R C and Zucolotto V 2014 *Mater. Sci. Eng.* C **37** 14–9

[56] (a) Sitruk-Ware R 2002 *Menopause* **9** 6–15

(b) Samie H A and Arvand M 2020 *Bioelectrochemistry* **133** 107489

[57] Kumari P, Nayak M K and Kumar P 2022 *Mater. Today Proc.* **48** 583–6

[58] Velayudham J, Magudeeswaran V, Paramasivam S S, Karruppaya G and Manickam P 2021 *Mater. Lett.* **305** 130801

[59] Arvand M, Elyan S and Ardaki M S 2019 *Sens. Actuators* B **281** 157–67

[60] Gevaerd A, Blaskievicz S F, Zarbin A J, Orth E S, Bergamini M F and Marcolino-Junior L H 2018 *Biosens. Bioelectron.* **112** 108–13

[61] Serafín V, Martínez-García G, Aznar-Poveda J, Lopez-Pastor J, Garcia-Sanchez A, Garcia-Haro J, Campuzano S, Yáñez-Sedeño P and Pingarrón J 2019 *Anal. Chim. Acta* **1049** 65–73

[62] Zhao X, Zheng L, Yan Y, Cao R and Zhang J 2021 *J. Electroanal. Chem.* **882** 115023

[63] Das A and Sangaranarayanan M 2018 *Sens. Actuators* B **256** 775–89

[64] Arvand M and Hemmati S 2017 *Sens. Actuators* B **238** 346–56

[65] Cherian A R, Benny L, George A, Sirimahachai U, Varghese A and Hegde G 2022 *Electrochim. Acta* **408** 139963

[66] Uçar M and Levent A 2021 *Diam. Relat. Mater.* **117** 108459

[67] Chen H-X, Deng Q-P, Zhang L-W and Zhang X-X 2009 *Talanta* **78** 464–70

[68] Houtman C J, Sterk S S, Van de Heijning M P, Brouwer A, Stephany R W, Van der Burg B and Sonneveld E 2009 *Anal. Chim. Acta* **637** 247–58

[69] Moura S L, De Moraes R R, Dos Santos M A P, Pividori M I, Lopes J A D, de D, Moreira L, Zucolotto V and dos Santos Júnior J R 2014 *Sens. Actuators* B **202** 469–74

[70] Liang K-Z, Qi J-S, Mu W-J and Chen Z-G 2008 *J. Biochem. Biophys. Methods* **70** 1156–62

[71] Liu W, Ma Y, Sun G, Wang S, Deng J and Wei H 2017 *Biosens. Bioelectron.* **92** 305–12

[72] Levent A, Altun A, Yardım Y and Şentürk Z 2014 *Electrochim. Acta* **128** 54–60

[73] (a) Ensafi A A, Rezaei B, Zare S M and Taei M 2010 *Sens. Actuators* B **150** 321–9

(b) Reddy K K, Satyanarayana M, Goud K Y, Gobi K V and Kim H 2017 *Mater. Sci. Eng.* C **79** 93–9

[74] Yang X, Zhao P, Xie Z, Ni M, Wang C, Yang P, Xie Y and Fei J 2021 *Talanta* **233** 122545
[75] Tezerjani M D, Benvidi A, Firouzabadi A D, Mazloum-Ardakani M and Akbari A 2017 *Measurement* **101** 183–9
[76] Wierzbicka E and Sulka G D 2016 *Sens. Actuators* B **222** 270–9
[77] Dorraji P S and Jalali F 2014 *Sens. Actuators* B **200** 251–8
[78] Goyal R N and Agrawal B 2012 *Anal. Chim. Acta* **743** 33–40
[79] Shaikshavali P, Reddy T M, Gopal T V, Venkataprasad G, Kotakadi V S, Palakollu V and Karpoormath R 2020 *Colloids Surf.,* A **584** 124038
[80] Gopal P, Narasimha G and Reddy T M 2020 *Process Biochem.* **92** 476–85
[81] Wierzbicka E and Sulka G D 2016 *J. Electroanal. Chem.* **762** 43–50
[82] (a) Liu X and Herbison A E 2013 *Endocrinology* **154** 340–50
(b) Pivonello R, Ferone D, Lombardi G, Colao A, Lamberts S W and Hofland L J 2007 *Eur. J. Endocrinol.* **156** S13
(c) Prinith N S, Manjunatha J G and Hareesha N 2021 *J. Iran. Chem. Soc.* **18** 3493–503
(d) Manjunatha J, Swamy B K, Gilbert O, Mamatha G and Sherigara B 2010 *Int. J. Electrochem. Sci.* **5** 682–95
(e) Charithra M M, Manjunatha J G G and Raril C 2020 *Adv. Pharm. Bull.* **10** 247
[83] Lee W-W and Jeon B S 2014 *Curr. Neurol. Neurosci. Rep.* **14** 1–13
[84] Geng W-Y, Zhang H, Luo Y-H, Zhu X-G, Xie A-D, Wang J and Zhang D-E 2021 *Micropor. Mesopor. Mater.* **323** 111186
[85] Narayana A L, Venkataprasad G, Praveen S, Ho C W, Kim H K, Reddy T M, Julien C M and Lee C W 2022 *Sens. Actuators,* A **341** 113555
[86] Kaya H K, Cinar S, Altundal G, Bayramlı Y, Unaleroglu C and Kuralay F 2021 *Sens. Actuators* B **346** 130425
[87] Sethuraman V, Sridhar T and Sasikumar R 2021 *Mater. Lett.* **302** 130387
[88] Bonyadi S and Ghanbari K 2021 *Mater. Chem. Phys.* **267** 124683
[89] Guo H, Liu B, Pan Z, Sun L, Peng L, Chen Y, Wu N, Wang M and Yang W 2022 *Colloids Surf.,* A **648** 129316
[90] Tang J, Liu Y, Hu J, Zheng S, Wang X, Zhou H and Jin B 2020 *Microchem. J.* **155** 104759
[91] Vidya H, Swamy B K and Schell M 2016 *J. Mol. Liq.* **214** 298–305
[92] Jiang Y, Wang B, Meng F, Cheng Y and Zhu C 2015 *J. Colloid Interface Sci.* **452** 199–202
[93] (a) Ross S A, Gulve E A and Wang M 2004 *Chem. Rev.* **104** 1255–82
(b) Nagel N, Graewert M A, Gao M, Heyse W, Jeffries C M, Svergun D and Berchtold H 2019 *Biophys. chem.* **253** 106226
[94] Šišoláková I, Hovancová J, Oriňaková R, Oriňak A, Trnková L, Třísková I, Farka Z, Pastucha M and Radoňák J 2020 *J. Electroanal. Chem.* **860** 113881
[95] Asadpour F, Mazloum-Ardakani M, Hoseynidokht F and Moshtaghioun S M 2021 *Biosens. Bioelectron.* **180** 113124
[96] Ensafi A A, Khoddami E and Rezaei B 2017 *Colloids Surf.,* B **159** 47–53
[97] Zhao Y, Xu Y, Zhang M, Xiang J, Deng C and Wu H 2019 *Anal. Biochem.* **573** 30–6
[98] Zhu W, Xu L, Zhu C, Li B, Xiao H, Jiang H and Zhou X 2016 *Electrochim. Acta* **218** 91–100
[99] (a) Cohn D V and MacGregor R R 1981 *Endocr. Rev.* **2** 1–26
(b) Hunt N, Martin T, Michelangeli V and Eisman J 1976 *J. Endocrinol.* **69** 401–12
(c) Bringhurst F R, Zajac J D, Daggett A S, Skurat R N and Kronenberg H M 1989 *Mol. Endocrinol.* **3** 60–7

[100] Schwartz G G 2010 *Vitamin D Deficiency and the Epidemiology of Prostate Cancer* (Berlin: Springer) pp 797–811
[101] Levin A, Bakris G, Molitch M, Smulders M, Tian J, Williams L and Andress D 2007 *Kidney Int.* **71** 31–8
[102] Şimşek Ç S, Karaboğa M N S and Sezgintürk M K 2015 *Talanta* **144** 210–8
[103] Malla P, Liao H-P, Liu C-H and Wu W-C 2021 *J. Electroanal. Chem.* **895** 115463
[104] Yagati A K, Go A, Chavan S G, Baek C, Lee M-H and Min J 2019 *Bioelectrochemistry* **128** 165–74
[105] Smaniotto A, Mezalira D Z, Zapp E, Gallardo H and Vieira I C 2017 *Microchem. J.* **133** 510–7
[106] Szkudlinski M W, Fremont V, Ronin C and Weintraub B D 2002 *Physiol. Rev.* **82** 473–502
[107] Beitollahi H, Ivari S G and Torkzadeh-Mahani M 2018 *Biosens. Bioelectron.* **110** 97–102
[108] Saxena R and Srivastava S 2019 *Mater. Today Proc.* **18** 1351–7
[109] Cao Y, Zheng M, Cai W and Wang Z 2020 *Chin. Chem. Lett.* **31** 463–7
[110] Zhang B, Tang D, Liu B, Cui Y, Chen H and Chen G 2012 *Anal. Chim. Acta* **711** 17–23
[111] Zhang F, Huang F, Gong W, Tian F, Wu H, Ding S, Li S and Luo R 2021 *J. Electroanal. Chem.* **882** 115032
[112] Serafín V, Agüí L, Yáñez-Sedeño P and Pingarrón J 2014 *Sens. Actuators* B **195** 494–9
[113] Beitollahi H, Nekooei S and Torkzadeh-Mahani M 2018 *Talanta* **188** 701–7

IOP Publishing

Real-Time Applications of Advanced Electrochemical Sensing Devices

Jamballi G Manjunatha

Chapter 5

Polymer functionalized materials for design of electrochemical sensing devices

Murat Misir, Nida Aydogdu and Ersin Demir

Polymers have been the indispensable material of human life since they the first time were discovered. Polymers, which occupy an unimaginable place in the industry, have become one of the indispensable components of modern sensors. Polymers are promising candidates for the development of sensor analyzers due to their exceptional flexibility and lightweight nature. In addition, polymer functional materials that can show conductivity provide an alternative to the world of sensors. Therefore, numerous applications of polymer materials have been made in the production and development of electrochemical sensors. Polymer-supported sensors with these unique properties have applications in a wide range of areas, from pharmacy to medicine, from environmental analysis to food analysis. Moreover, electrochemical sensors treated with polymer-functionalized materials; enable the development of on-site, online, portable, and fast analysis devices. Moreover, polymer-based sensors have also taken a place in daily life due to their small size, high sensitivity, and suitability for routine analysis. In addition, polymers, the most remarkable industrial material of the last 50 years, have contributed greatly to the production of numerous gas sensors. In this section, polymer-functionalized materials in the development, construction, and improvement of electrochemical sensors are discussed in detail. Real electrochemical sensor application examples of polymer functional materials used in recent years were examined. The application areas of polymer-functionalized material-based sensors, which are developed for substances that must be detected quickly, reliably, and accurately, are evaluated. The recent past, current status, and future perspectives of these sensors are discussed. In addition, important validation parameters such as working range, limits of the assay, and selectivity of studies conducted in the last five years were evaluated.

doi:10.1088/978-0-7503-5377-9ch5

Finally, the general trends and challenges in the production of electrochemical sensors with polymeric functions are discussed.

5.1 Introduction

A polymer is a substance with unique properties consisting of very large molecules or macromolecules composed of subunits such as many repeating monomers. Due to their unlimited properties, they are used in a wide range of industries, such as medicine and food. Polymer expression is one of the most important chemicals of the 20th century. The first expression of polymers occurred in the 1920s, and in 1953 H Staudinger was awarded the Nobel Prize for it. However, the first plastic material for human life was based on nitrocellulose in the middle of the 18th century. From the past to the present, polymer materials have been an important component of human life. Polymers, which have an activity in almost all fields, have created a unique and privileged field of study in the scientific world. Apart from this, it also constitutes a promising and remarkable material class in multidisciplinary fields. With the discovery of conductive polymers in the 1970s, these polymer materials have greatly influenced electrochemical sensors' production, development, and miniaturization. The first conductive polymer, polyacetylene, was synthesized. For this, Shirakawa, MacDiarmid, and Heeger were awarded the 2000 Nobel Prize in Chemistry [1]. Since the 1980s, conductive polymers have become interesting and promising materials for electrochemical sensing. Especially, this material with these unique properties is used in developing miniaturized sensors for electrochemical sensors, providing on-site analysis, and is portable. Also, polymer functional materials are one of the most serious candidates for the development of user and environmentally-friendly sensors. Moreover, these polymer-supported electrochemical devices, which are easy to use and inexpensive to manufacture, are increasing rapidly as a hybrid material with polymer function. They are an indispensable part of sensors due to their high strength, electrical properties, natural structure, corrosion resistance, easy processing, and lightweight. In sensors, semiconductor and conductive polymers, whose valence band and conduction band are very close to each other, are one of the materials used directly. Although conductive polymers do not show as conductive properties as metals, they contribute to the development of sensors with super-perfect sensitivity by acting synergistically with hybrid materials. Electrochemical sensor sensing devices with conductive polymer structure show conductivity properties mainly due to the conjugated chain consisting of double bonds in the structure. The number of nanosensor sensing devices produced with conductive polymers such as polyaniline, polyglycine, and polypyrrole allows electrochemical analysis of substances. As a result, common polymer functional materials used in electrochemical sensing devices are conductive polymers, hydrogels, molecular imprinting polymers, and nanocomposites (figure 5.1).

In this section, polymer functional materials designed for electrochemical sensing devices are discussed in detail. Grouping was done according to the preferred electrochemical methods for analyte determinations. Here, we examined more than 100 articles according to the studies carried out in the last five years in terms of

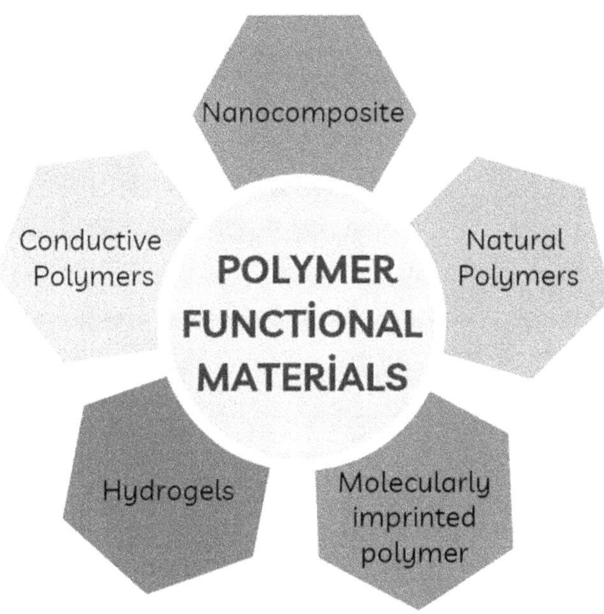

Figure 5.1. Common polymer materials used in electrochemical sensing devices.

polymer material types, working electrodes, analyte type, working range, detection limits, interference study, and analytical applications. The polymer and its derivatives used for nanosensor development were evaluated. In addition, the advantages of polymer-supported materials used for this electrochemical sensing were discussed. The type of coating of polymer functional materials on the electrode surface was also investigated. Consequently, the general trends and challenges in the production of electrochemical sensors with polymeric functional materials are discussed in this chapter.

5.2 Electrochemical sensing

Electrochemical techniques for the examination and determination of electroactive samples have attracted great interest in recent years because they are inexpensive, applicable and low detection limits can be obtained [2, 3]. Electrochemical sensing has allowed us to characterize a large number of systems, providing information such as stoichiometry of charge transfer in chemical reactions and mass transfer rate of electroactive components. Electrochemical methods are environmentally friendly considering that fewer organic solvents are used compared to other analytical methods and analysis can be made with fewer samples. In addition, they offer us many advantages as they are selective, sensitive, easy to apply, and highly reproducible. There are various subclasses such as potentiometry and voltammetry. The electrochemical methods mentioned in this review are shown in figure 5.2. Based on the literature, it is seen that methods such as amperometry, differential pulse, square wave, and cyclic voltammetry are more preferred, and methods such as polarography, and

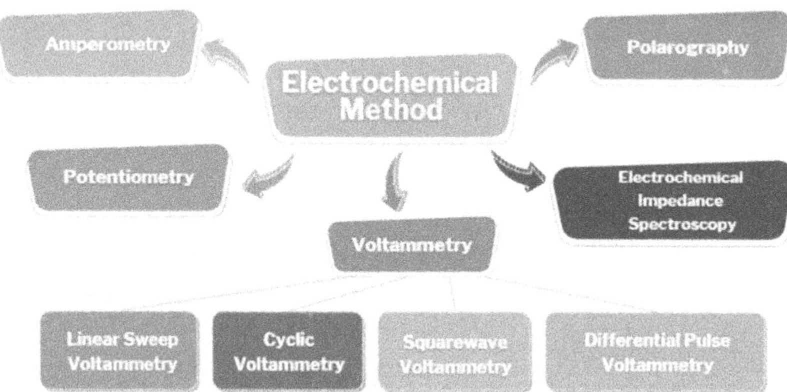

Figure 5.2. Electrochemical methods used with polymer sensors.

EIS are less preferred in electrochemical determination using polymer materials. For electrochemical sensing, various working electrodes are available such as carbon paste electrode (CPE) [4, 5], graphite paste electrode (GPE) [6, 7], indium tin oxide (ITO) electrode [8, 9], gold electrode (AuE) [10, 11], glassy carbon electrode (GCE) [12–14], and fluorine-tin oxide electrode (FTO) [15, 16] in analyte determinations. Furthermore by coating the surface of these electrodes with various materials such as nanocomposites [12, 14], polymers [17, 18], or nanoparticles [14, 19], new-generation hybrid nanosensors can be produced and more sensitive determination of electroactive materials have been used.

5.2.1 Amperometry

The amperometric technique, which constitutes an important class among electrochemical sensing tools, can be defined as a technique where the applied voltage produces an electric current proportional to the concentration of the analyte. This technique consists of two electrodes immersed in the electrochemical cell and a controlled potential is applied by a potentiostat/galvanostat instrumentation device. In some cases, it may include a triple electrode system connected with an electrochemical device. This electrochemical technique allows the determination of analytes with high accuracy and sensitivity even in complex environments to a high level of satisfaction.

In studies carried out with amperometry, natural polymers such as chitosan, especially poly(Celestine blue), polymethyl methacrylate acid (PMMA), polypyrrole (PPy), poly(methylene blue) (PMB), polydopamine (PDA), molecularly imprinted polymer (MIP), poly (3-aminobenzylamine) (PABA), polyaniline (PANI), and polythiophene functional polymers were used. The working electrode treated with polymer material, which is used directly as an indicator electrode (solid polymer electrode (SPE)), is the most serious alternative candidate to conventional electrodes. In addition, electrochemical nanosensors that can start from some conductive monomers and polymerize on the electrode surface have also been developed.

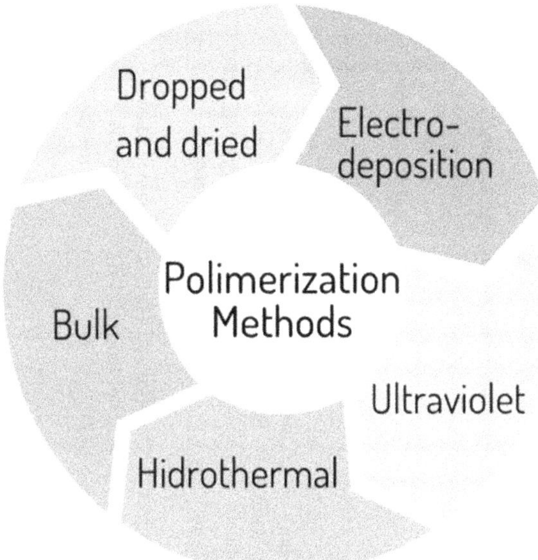

Figure 5.3. Commonly used polymerization methods.

New-generation hybrid nanosensors has been produced by coating the surface of the main electrodes such as GCE, ITO electrode, AuE, FTO, pencil graphite electrode (PGE), graphite rod electrode (GRE), and screen-printed carbon electrode (SPCE). Here, four coating techniques such as electrochemical deposition, electropolymerization, dropped and dried, and constant potential technique were used for the determination of an analyte in real samples. In addition, electrochemical tools with unique properties have been developed with two or more nanomaterials in a multifunctional structure that contains at least one functional polymer (figure 5.3).

Working wonders in the world of electrochemical sensing devices, polymers exhibit very sensitive analyses with the amperometric technique. Analytes at nM levels of sensors improved with polymers are determined qualitatively and quantitatively by amperometric technique. Although nanosensors have been developed with a wide variety of polymers, they are still of interest today. We can say that this interest has increased with the discovery of molecularly imprinted polymer (MIP). Although not as much as in other electrochemical techniques, the detection limits of analytes can be reduced to nM by using MIP sensors with the amperometric method. In addition, MIP-based electrochemical sensors show excellent hyper selectivity as they form a unique cavity according to the molecular shape of the analyte. Therefore, in recent years, scientists have adopted new-generation electrochemical sensor instruments based on MIP. Combining all these extraordinary features, the world of sensors will lead the future of flexible, portable, fast, sensitive, and highly accurate online electrochemical analysis devices. Polymer-supported sensors will become indispensable in electrochemical sensor devices such as wearable, portable, body-adhesive, and disposable types (table 5.1).

Table 5.1. Polymer functional nanomaterials in amperometric technique.

Polymer composition	Electrode	Analyte	Linear range (μM)	LOD (nM)	LOQ (nM)	Samples	Electrode preparation method	Interference	References
5-HT/CHT/GO	GCE	Serotonin	—	—	—	—	Electrochemical deposited	—	[12]
AChE/ $PNRE_{thaline}HNO_3$/ Fe_2O_3NP	GCE	Acetylcholine	2.5–60	1060		Synthetic urine	Electropolymerization	—	[22]
Au@poly(TTP)/rGO	GCE	Dopamine	0.02–232	11.5	—	Injection sample	Electropolymerization	Paracetamol, uric acid, glucose, mannose, fructose, folic acid	[20]
AuNP/P-Celestine blue	GE	H_2O_2	11.7–1290	3900	—	Milk	Electropolymerization	Ascorbic acid, uric acid, acetic acid	[23]
BPA-PMMA	SPE	Bisphenol A	0.01–8.0	166.4	—	Water, tap water, saline, mineral water and cola	Dropped and polymerization	Naphthalene, methylene blue, methyl red and phenol	[24]
CB/f-MWCNTs	GCE	Bisphenol A	0.1–130	80	—	Bottle samples	Dropped and dried at room temperature	Alcohol, phenol, antioxidant, dopamine, Ca^{2+}, K^+, Na^+, Mg^{2+}, and Zn^{2+}	[25]
Chitosan/β-Cyclodextrin/ WCNTs Composite Film	ITO	Tyrosine	1.0–100 and 100–1000	600	—	Artificial urine samples	Co-electrodeposition	—	[8]

Co/PPy	GCE	Nitrite	2–3318	350	—	Pickled juice	Constant potential technique	KCl, NaCl, $CuSO_4$, $CaCl_2$, sodium citrate, sucrose, maltose, fructose, glucose, ascorbic acid, lactic acid, acetic acid, and natural phenol rutin	[26]
Fc-PLL/Gox	GCE	Glucose	0–0.010	0.023	—	Serum and blood	Dried at room temperature	Uric acid, dopamine, and ascorbic acid	[27]
GDH/PMB/Au-MWCN	AuE microneedles	Glucose	50–5000	7000	—	Artificial interstitial fluid	Electropoly merization	L-lactate, ascorbic acid,uric acid, glucose	[28]
GDH/PMB/Au-MWCN	AuE microneedles	Lactate	10–100	3000	—	Artificial interstitial fluid	Electropolymerization	L-lactate, ascorbic acid,uric acid, glucose	[28]
Gox/ Hexacyano ferrate/CHT	SPE	Glucose	0.001–0.05	—	—	Serum and blood	Air dried	—	[29]
ITO/ZnO-PDA/GOx	GCE	Glucose	15–120	6200	—	—	Deposition	—	[9]
MAA	GCE	p-nitrophenol	10–100	—	—	—	Air dried	—	[30]
MIP/Au@CNTs	GCE	Prednisolone	1–210	3	—	Blood serum	Electropolymerization	Ascorbic acid, uric acid, acetaminophen, xanthine, chlordiaze poxide, albumin, diazepam, hypoxanthine, metolazone, glucose, napr oxen, nitrite, ephedrine, dopamin	[21]

(Continued)

Table 5.1. (*Continued*)

Polymer composition	Electrode	Analyte	Linear range (μM)	LOD (nM)	LOQ (nM)	Samples	Electrode preparation method	Interference	References
MIP/MPBA/AuNP	GCE	H_2O_2	0.6–20	160	—	—	—	—	[31]
Ox/P- Methylene blue/Au-MWCNT	Au microneedles	Lactate	2400	10–200	—	Artificial interstitial fluid, human serum	—	Ascorbic acid, uric acid, glucose	[32]
PABA/f-CNTs/Au	SPE	Glucose	560–2800	67×10^3	—	—	Drip dry	Ascorbic acid, uric acid, dopamin	[33]
PABA/poly(styrene sulfonate) (PSS) LBL	FTO	Dopamine	0.1–1.0	62.8	—	—		Uric acid, ascorbic acid, glucose	[15]
PANI/CQDs	FTO	Dopamine	10–90	101.3	—	—	—	Ascorbic acid, glucose	[34]
PANI/CuO	PGE	Glucose	—	—	—	—	Polymerization	Uric acid, ascorbic acid, fructose and sucrose	[35]
PANI-AuNPs(6 nm)-GOx/GOx	GRE	Glucose	100–16 500	70×10^3	—	Serum	Deposition	Fructose, mannose, galactose, xylose, saccharose	[36]
PMB/AuNP	GCE	AA and 5-ASA	—	400	—	Pharmaceutical samples	Polymerization	—	[19]
PMB/AuNP	GCE	AA and 5-ASA	—	64	—	Pharmaceutical samples	Polymerization	—	[19]
Polythiophene/ CuO	GCE	Hydrogen peroxide	20–3300	3860	—	—	Polymerization	—	[37]

Ppy/GOx/DGNs/Gr	GCE	Glucose	0–19.9 × 10^3	0.07 × 10^6		Human serum, saliva, wine, milk, juice	Polymerization	Uric acid, ascorbic acid, saccharose, xylose, galactose, mannose, and fructose	[38]
PPy/TA/CTAB NCs	SPCE	Dopamine	0.5–2.0	290	—	—	Polymerization	Ascorbic acid, uric acid	[39]
PtNPs/PAA	SPCE	H_2O_2	0–300	52	—	Antiseptic Hair lightener 'Liquid oxygen'	Deposition	Citric acid, glucose, mannitol, salicylic acid, ascorbic acid, dehydroascorbic acid, urea, resveratrol	[40]

In a study published by Guler *et al* in 2022, Au nanoparticles (AuNPs) and poly ([2,2′;5′,2″]-terthiophene-3′-carbaldehyde)/reduced graphene oxide (poly(TTP)/rGO) have developed a multifunctional layered as Au@poly(TTP)/rGO/GCE sensitive electrochemical sensor [20]. The sensitivities of the polymeric electrochemical sensor were investigated by techniques such as cyclic voltammetry (CV), differential pulse voltammetry (DPV), amperometry, and electrochemical impedance spectroscopy (EIS). The surface morphology of the electrochemical sensor device was illuminated by field emission scanning electron microscopy (FESEM). They determined that the linear operating range of the developed hybrid sensor is between 0.02 and 232 μM with a sensitivity of 315 $\mu A\ mM^{-1}\ cm^{-2}$ for the analysis of dopamine. They also calculated the limit of detection (LOD) value as 11.5 nM. Moreover, the electrochemical sensor with polymer function has achieved dopamine determination with excellent selectivity, fast response time and low excellent selectivity, and high recovery. They proved the accuracy and precision of the Au@poly(TTP)/rGO/GCE electrochemical next-generation sensor, which was built by performing analytical applications on injection samples. They also investigated the shelf life of the sensor and found that after 37 days there was approximately 10% lower loss of dopamine current. As a result, they developed a portable nanosensor with excellent selectivity, extraordinary sensitivity, and rapid analysis, containing NPs, polymers, and carbonaceous materials for the determination of dopamine, an important substance.

In another study (Wang, 2022), a modified GCE (MIP/Au@CNTs/GCE) containing molecularly imprinted membranes on a Au@CNTs nanocomposite was developed for the determination of prednisolone (PNS), an anti-inflammatory or immunosuppressant drug [21]. First, the Au@CNTs hybrid material was coated on the GCE surface. Then, the electropolymerization technique modified ρ-amino thiophenol and tetrabutylammonium perchlorate on Au@CNTs/GCE. Scanning electron microscopy (SEM) and x-ray powder diffraction (XRD) illuminated the surface morphology and microstructure of the developed modified electrode. Electrochemical techniques such as CV, DPV, and amperometry were used for the determination of PNS. The recommended MIP/Au@CNTs/GCE sensor linear range and LOD were found to be 1–210 μM and 0.003 μM, respectively. The validity and accuracy of the developed polymer-supported MIP/Au@CNTs/GCE and the recommended amperometric technique were performed in real blood serum samples of young athletes. Statistical evaluation was made with the ELISA method and the relative standard deviation (RSD) was found to be less than 4.15%. In addition, the selectivity of the functional polymer electrochemical sensor built for the PNS was investigated in the presence of various metabolic substances using the amperometry method in 0.1 M phosphate buffer solution (PBS) at −0.95 V. As a result, they developed a multifunctional layered nanosensor containing nanoparticles, polymers, and carbonaceous materials with excellent selectivity, exceptional sensitivity, and rapid analysis for the detection of PNS, an anti-inflammatory or immunosuppressant drug.

5.2.2 Cyclic voltammetry

The basic working principle of CV is to obtain a voltammogram by measuring the current between the indicator electrode and the counter electrode as a function of potential in two directions (anodic and cathodic scanning). This results in a triangular potential curve in an electrochemical cell consisting of a three-electron system. A reference electrode is applied with a working electrode immersed in a stationary solution, and it measures the resulting current between the working electrode and the counter electrode depending on the applied potential. Although qualitative and quantitative analysis has been made with the CV technique, its main purpose is to examine the electrochemical behavior of the analyte. In addition, this technique is applied in the kinetic model, the material transport process and the number of electrons transferred in the electrode reaction. This electrochemical technique, unlike other methods, will not be satisfactory but provides adequate detection limits. However, it is a very interesting technique because it has a very fast analysis time. Consequently, the qualitative studies carried out by CV are therefore limited by using a polymer nano-device.

Polymer-supported nanomaterials such as poly(Prussian blue), polyaniline (PANI), methacrylic acid (MAA), MIP, PDA, and polyethyleneimine (PEI) were used in electrochemical detection by CV. They have coated functional polymer materials on indicator electrode surfaces, such as carbon ionic liquid electrodes (CILEs), GCEs, AuEs, SPCEs, FTOs, PGEs, and carbon plate electrodes. These processes sometimes start from the monomer and are coated with the electro-polymerization technique on the main electrode surface, and sometimes they are modified directly on the electrode surface of the polymer material by the drop-dry technique. It is produced as a single layer containing only polymer in the production of electrochemical sensing sensors. At the same time, polymer-supported and multi-layered new-generation sensors have been developed using different nanomaterials. For this purpose, agents with different advantages such as metal nanoparticles, carbonaceous materials, and metal phthalocyanine were used. The qualitative and quantitative analyzes of very low levels of analytes as pM were performed using polymer functional materials with CV electrochemical technique. Here, the unique selection and sensitivity of the MIP come to the fore (table 5.2).

In their study published in 2019, Xing *et al* developed a GCE modified with polymer Prussian blue (PB) and carboxyl-functional multi-walled carbon nanotubes (MWNTs-COOH). They produced a simple and highly sensitive electrochemical detection tool for the determination of bisphenol B (BPB) with the developed polymer functional nanosensor [41]. They characterized the morphology of the PB/MWCNTs-COOH/GCE nanosensor using FESEM, XRD, EIS, and CV. They showed that the carbonaceous material and the polymer agent improved the bisphenol B anodic signal with their synergistic effect. For bisphenol B determination, they found the working range to be between 50 nm and 175 μM under optimum conditions with the CV technique, and the limit of detection (S/N = 3) was calculated as 0.5 nM. They also showed that poly(Prussian blue)/MWCNTs-COOH/GCE exhibited excellent stability, sensitivity, and reproducibility. Finally, bisphenol

Table 5.2. Polymer functional nanomaterials in CV technique.

Polymer composition	Electrode	Analyte	Linear range (μM)	LOD (nM)	LOQ (nM)	Samples	Electrode preparation method	Interference	References
AuNPs-MCA-rGO	CILE	Bisphenol A	0.004–18	1.1	—	PVC food package,PVC bottle, PC baby bottle, and PC water bottle	Dropped on the CILE	Hydroxyphenol, carbamazepine, diallyl bisphenol A, p-nitrophenol	[42]
Cu-PANI	GCE	Ascorbic and gallic acid	—	—	—	—	Electrodeposited	—	[43]
DNT/type-2 dengue virus/GO	AuE	2,4-dinitrotoluene		1.64×10^6	—	Aqueous solution	UV-polymerization	—	[10]
MAA	GCE	p-nitrophenol	2–400	200	700	Wastewater	Air dried	2,4-dichlorophenol, 4-bromo-2-nitrophenol and 3-NP	[30]
MAA	Carbon plate electrode	Salbutamol	4.17–4178	6.68×10^3	—	—	Polymerization	Glycine, L-glutamic acid, 4-aminophenol, bisphenol A, and L-ascorbic acid	[44]
MIP/ Polydopamine	AuE	1,3,5-trinitroperhydro-1,3,5-triazine	0.1×10^{-3}–0.01	0.1	—	Aqueous solution	Electropolymerization	TMA, IPA and 4NP	[11]
MIP-TNT	AuE	2,4,6-trinitrotoluene	5.49×10^3	2.19×10^6	—	Aqueous solution	Polymerization, dropped-dried	—	[45]
PEI-PC/DPNs/ AuNPs	SPCE	Bisphenol A	0.01–1.0	6.63	—	Tap water	Dropped and dried	—	[46]
PG-BP	GCE	Bisphenol A	0.043–55	7.8	—	Drinking water bottle, urine	Dropped and dried	Catechol, ascorbic acid, uric acid, Fe^{3+}, Ca^{2+}, Na^+, Br^{2+}, Cl^-	[47]
poly(Prussian blue) /MWCNTs-COOH	GCE	Bisphenol B	0.05–175	0.5	—	Water	Deposition	Cl–, Na^+, Pb^{2+}, SO_4^{2-}, NO_3^-	[41]

Polydopamine	AuE	2,4,6-trinitrotoluene	0.1×10^{-3}–0.01	0.050	—	Aqueous solution	Electropolymerization	TMA, IPA and 4NP	[11]
PTHBN/AuNPs/Gox	PGE	Glucose	2.975–2087.0	30.4×10^3	101.5×10^3	Dextrose solution, cherry juice	Electropolymerization	Fructose, galactose, glutamic acid, sucrose, citric acid and uric acid	[48]
RGO/MAA	Carbon plate electrode	Salbutamol	4.17–4178	3.46×10^3	—	—	Polymerization	Glycine, L-glutamic acid, 4-aminophenol, bisphenol A, and L-ascorbic acid	[44]
ZnPc/TiO_2 NRAs	FTO	Bisphenol A	0.047–52.1	8.6	—	PC baby bottle, PVC bottle, PVC food package, milk	Deposition	Methyl orange, sudan III, phenanthrene, anthracene, fluoranthene	[16]
Poly(Glutamic Acid) Layered	MWCNTPE	Indigo carmine	5.0–60.0	4200	—	Gems chocolate and water	Electrodeposition	—	[49]
Polyglycine	GPE	Hg (II)	150–1500	6600	—	Groundwater and blood sample	Electropolymerization	—	[7]
Polyglycine	GPE	Pb (II)	200–450; 500–1200	800	—	Groundwater and blood sample	Electropolymerization	—	[7]
Polyglycine	CPE	Alizarin carmine	4.0–100	980	—	Sewage water	Electropolymerization	Zn^{2+}, Na^+, Mg^{2+}, K^+, Ca^{2+}, urea, sucrose, glucose, oxalic acid	[50]
Poly (Adenine)	GPE	Catechol	2.0–150	24	—	—	Electropolymerization	—	[18]
Polyglycine	CPE	Mitoxantrone	0.04–10	320	1000		Electropolymerization		[51]
Poly (Alizarin carmine)/SLS	CPE	Paracetamol	0.4–100	60	—	Tablet formulation	Electropolymerization	Dopamine, and estriol	[52]

B analysis in real samples was successfully performed with the functional polymer electrochemical sensor. As a result, in this study, a sensitive, selective, economical, and fast electrochemical detection tool for the quantitative detection of bisphenol B containing polymer-supported and carbonaceous material has been introduced to the literature.

5.2.3 Differential pulse voltammetry

In the 1960s, efforts were made to design new voltammetric techniques similar to, but more sensitive than CV for their electrochemical behavior. By measuring desired Faradaic currents against load currents, highly sensitive pulsed voltammetric techniques were discovered. The basic working principle is based on applying a potential in the form of a pulse and measuring the Faradaic current that occurs in a small time period at the end of the applied pulse [53]. Thus, pulse voltammetric tools allow qualitative and quantitative analysis of analytes at very low concentrations such as nM or even pM. Studies using the DPV technique, which exhibits a very sensitive detection feature, are unlimited and new studies are carried out day by day. The basic research with this technique is the development, improvement, or commercial production of indicator electrodes. For this purpose, polymer materials come to the fore. Conducting polymers provide precise measurements, determining the analyte even in complex matrix environments. In addition, due to the flexible nature of polymer functional materials, they contribute greatly to the development of on-site, portable, or disposable electrochemical detection devices.

In the DPV studies, natural polymers like chitosan, nafion, and other polymers such as a MIP, co-polymer of pyrrole-histidine (PyHis), PMB, PANI, poly thionine-methylene blue (PTH-MB), PDA, poly(ionic liquid) (PIL), MAA, PDA: Polydopamine, poly3,4-ethylene dioxythiophene (PEDOT), PEI, polyphenylenediamine and PPy functional polymers were used in the analyte determination. Sensor fabrication with functional polymer materials was carried out by electropolymerization of the monomer on the main electrode surface or by coating the polymer directly on the main electrode surface by drop-drying technique. The GCE is the most preferred indicator electrode in studies carried out by DPV. In addition, there are a few studies using a PGE and SPCE treated with polymer hydride materials for the analysis of agents. In the construction of electrochemical sensors, only a polymer layer is used in some cases, and polymers containing different composite materials are sometimes used with a synergistic effect. The most important of these are carbonaceous materials and NPs that are polymer-supported in multilayer sensor construction (table 5.3).

Koyun and Sahin, in their study in 2018, developed a AuNP/PMB-modified PGE and performed their surface morphology characterization studies [54]. The working electrode (PGE) was coated by electropolymerization of methylene blue (MB) in a two-step procedure and then modified with AuNPs. For the characterization of GNP/PMB/PGE, EIS, CV, x-ray photoelectron spectroscopy (XPS), Fourier transform Infrared spectroscopy (FTIR), and SEM techniques were used. They then evaluated the analytical performance for nitrite determination with a polymer-

Table 5.3. Polymer functional nanomaterials in differential pulse voltammetric technique.

Polymer composition	Electrode	Analyte	Linear range (μM)	LOD (nM)	LOQ (nM)	Samples	Electrode preparation method	Interference	References
(1S,2S)-PSDO@MIP	GCE	Pseudoephedrine	1×10^{-9}–1×10^{-8}	2.98×10^{-7}	—	Tablet, serum	Electropolymerization	Ascorbic acid, dopamine, uric acid, paracetamol, Na^+, SO_4^{2-}	[56]
[Poly (Py-co-PyHis)] @MIP	GCE	Teriflunomide	0.1×10^{-6}–1.0×10^{-6}	11.38×10^{-6}	—	Serum, tablet	Electropolymerization	K^+, Cl^-, Na^+, SO_4^{2-}, dopamine, ascorbic acid, uric acid, paracetamol	[57]
5-HT/CHT/GO	GCE	Serotonin	0.005–10.0	1.6	—	Human serum, artificial saliva, artificial tears	Electrochemical deposited	Noradrenalin, dopamine, glucose	[12]
AuNP/PMB	PGE	Nitrite	5–5000	314	—	Sausage, mineral water	Electrodeposition and electropoly merization	Glucose, uric acid, ascorbic acid, dopamine, KCl, NaCl	[54]
AuNPs/PTH-MB	GE	Human serum albumin	10^{-10}–10^{-4} ppt	3.0×10^{-11} ppt	—	Urine	Electropolymerization	L-Gly, L-Glu, L-Cys, L-Try, L-His, DA, AA, Hemoglobin and BSA	[58]
AuNPs@PANi/	GSPE	Dopamine	1–100	800	—	Human serum	Electropolymerization	Serotonin, uric acid	[59]
AuPd NPs/PDA/ MWCNTs-CS-IL	GCE	Cholestanol	1×10^{-6}–50×10^{-6}	0.2×10^{-3}	—	—	Dropped and electropoly merization	—	[60]
CMOF-MIPIL	GCE	Bisphenol A	0.005–5.0	4.0		Water samples, plastic bottle and fresh liquid milk	Dropped onto electrode surface	PN, ET, 2-NP, 4-NP, HQ, and HBPA	[61]
Fe_3O_4@Au-Cys/ PANi/Gr	SPE	Dopamine	20–1000	2190	—	Pharmaceutical and urine	Dried in a desiccator	KCl, $NaNO_3$, NaCl, glucose, Urea	[62]

(*Continued*)

Table 5.3. (*Continued*)

Polymer composition	Electrode	Analyte	Linear range (μM)	LOD (nM)	LOQ (nM)	Samples	Electrode preparation method	Interference	References
Fe_3O_4@ Au-Cys/PANi/	Graphite SPE	Uric acid	20–1000	1800	—	Pharmaceutical and urine	Dried in a desiccator	KCl, $NaNO_3$, NaCl, glucose, Urea	[62]
Fe_3O_4@SiO_2@IIP	GCE	Lead (II)	1.21–975	609	—	Tap water, rain water, river water, and and fruit juice	Polymerization	Ca^{2+}, Mg^{2+}, Zn^{2+}, Ni^{2+}, Cr^{3+}, Ag^+, Hg^{2+}, Na^+, K^+, Cd^{2+} and Cu^{2+}	[63]
L-His-MWCNTs @PDMS-5/MIP	GCE	Tetracycline	1×10^{-5}–1×10^{-4}	2.642×10^{-3}	—	Capsule, human serum, tap water	Hydrothermal polymerization	Dopamin, ascorbic acid, uric acid, Na^+, SO_4^{2-}, K^+, NO_3^-, Mg^{2+}, Cl^-	[64]
MAA	CPE	Thiuram	20×10^{-3}–125×10^{-3}	1.7×10^{-6}	—	Pesticide	Polymerization (MIP)	—	[65]
MIP(TEOS: L-tryptophan)	GCE	Bisphenol F	1×10^{-9}–10×10^{-9}	0.291×10^{-6}	0.971×10^{-6}	Serum, plastic bottled water	Electropolymerization	Dopamine, acetic acid, Na^+, Cl^-, K^+, SO_4^{2-}	[66]
mMIPs/rGO-ZIF-8	GCE	Catechin	0.01×10^{-3}–10.0	0.003		Green tea	Dried at room temperature	NaCl, KCl, $CaCl_2$, $Al(NO_3)_3$, $FeSO_4$ and isoquercitrin	[67]
MoS_2–PANi/rGO	GCE	Ascorbic acid	0.0–8000	22.2×10^3	—	Human serum and urine	Dried in a vacuum oven at 60 °C for 12 h	$ZnSO_4$, glucose, $NaNO_3$, citric acid, glycine, and the mixture	[68]
MoS_2–PANi/rGO	GCE	Dopamine	5.0–500	700	—	Human serum and urine	Dried in a vacuum oven at 60 °C for 12 h	$ZnSO_4$, glucose, $NaNO_3$, citric acid, glycine, and the mixture	[68]
MoS_2–PANi/rGO	GCE	Uric acid	1.0–500	360	—	Human serum and urine	Dried in a vacuum oven at 60 °C for 12 h	$ZnSO_4$, glucose, $NaNO_3$, citric acid, glycine, and the mixture	[68]
MWCNT/IIP-MBT-AA	CPE	Methylmercury	2.59–2.82	2495	—	Natural water	—	Fe^{3+}, Hg^{2+}, Cd^{2+}, Pb^{2+}	[69]

MWCNT-COOH /PTH/PtNP	GCE	Myricetin/Rutin	0.01–15	3.8/1.7	—	—	Electrodeposition	L-cysteine, ascorbic acid, citric acid, glucose	[70]
MWCNTs/MIP	CPE	Metribuzin	0.001–100	0.5	1.66	Tomatoes and potatoes samples	Bulk polymerization	Chloropyrifos, dimethoate, glyphosate, and thiamethoxam	[71]
Ni^{2+}@PNR HN/ MWCNT/Nafion	GCE	Glucose	0.02–1.0	6.5	—	Human serum	Dropped and dried at room temperature	Ascorbic acid, dopamine, uric acid	[72]
p(ANI-o-PD)@MIP	GCÊ	Leflunomide	1.0×10^{-9}–10×10^{-9}	291×10^{-3}	969×10^{-3}	Serum, tablet	Electropolymerization	K^+, Cl^-, Na^+, SO_4^{2-}, dopamine, ascorbic acid	[73]
P(HEMA-MAAsp) @MIP	GCE	Somatostatin	1.0×10^{-8}–1.0×10^{-7}	0.175×10^{-6}	0.584×10^{-6}	Serum, tablet	Electropolymerization	K^+, Cl^-, Na^+, SO_4^{2-}, dopamine, ascorbic acid	[74]
PANi-H_2SO_4@Au	GCE	Dopamine	10 100; 7–100	67005250	—	—	Drop casting on the electrode surface	Ascorbic acid and uric acid	[75]
PBCB/Fe2O3	GCE	Epinephrine	0.05–15	310	—	Adrenaline	Electropolymerization	Ascorbic Acid	[76]
PEDOT/ CNTs/Au NTs	PDMS electrode	Urea	1000–0.1×10^6	1×10^5	—	Artificial sweat serum	Electropolymerization	Glucose, lactate, riboflavin, UA, sodium chloride, potassium chloride, calcium chloride, and ammonia	[77]
PEI-rGO-Au-NCs	SPCE	β-lactoglobulin	1×10^{-6}–0.1 ppm	1×10^{-6} ppm	—	Milk	placed in an oven at 60 °C for 20 min	—	[78]
Poly[(Cys)VIMBF_4]/ AuNPs	GCE	Alpha-fetoprotein	0.03×10^{-3}–5×10^{-3} ppm	2×10^{-6} ppm	—	—	Electrochemical deposition and polymerization	AA, L-Cys, Gly, L-His, HSA, NSE, PSA, IgG	[79]

(*Continued*)

Table 5.3. (*Continued*)

Polymer composition	Electrode	Analyte	Linear range (μM)	LOD (nM)	LOQ (nM)	Samples	Electrode preparation method	Interference	References
Polyaniline/PTH/ AuNP@ZIF-67/	GCE	Tyrosine	0.01–4.0	0.79	—	Human serum	Deposition and electropoly merization	Ascorbic acid, glucose, lysine, alanine, glycine, cysteine, dopamine, tryptophan, phenylalanine, histidine, tyrosine	[13]
Polyphenylenediamine	GCE	Entacapone	0.1–5.0	50	—	Human serum	Electropolymerization	L-dopa, C-dopa	[55]
PPy/TA/CTAB NCs	SPCE	Dopamine	10–50	6390	—	—	Polymerization	—	[39]
Pyrrole/ AuNPs@rGO	GCE	Amyloid A	0.01×10^{-3}– 0.2 ppm	5×10^{-6} ppm	—	Milk	Electropolymerization	—	[80]
Poly(L-methionine)	CPE	Voltaren	2–50	100	—	Pharmaceutical formulation	Electropolymerization	Zn^{2+}, Sr^{2+}, Mg^{2+}, K^{+}, Cu^{2+}, Ca^{2+}, Al^{3+}, urea, sucrose, oxalic acid, glucose	[81]
Poly(glutamine)	CPE	Curcumin	0.4–6; 6–10	27.9	—	Food supplement	Electropolymerization	Vaniline	[82]
Poly(Tyrosine)	GPE	Catechol	2.0–10; 15–50	300	1000	Tap water	Electropolymerization		[83]

supported electrochemical nanosensor. Since they found the highest peak signal at pH 5.0, they chose it as the optimum supporting electrolyte and obtained a linear operating range of 5–5000 μM at this pH and using the GNP/PMB/PGE sensor. They calculated the LOD value as 0.314 μM. Finally, they have successfully applied analytical applications to real samples such as commercial sausage and mineral water to prove the accuracy and precision of the polymer functional electrode as GNP/PMB/PGE.

In a study published in 2021 by Radi and Abd-Ellatief, they developed an electro-polymerized polyphenylenediamine (Po-PD)-based MIP electrochemical sensor on a GCE surface for the determination of entacapone (ETC) [55]. For this, the electro-polymerization technique coated the o-phenylenediamine monomer (o-PD) directly on the GCE surface. Steps such as electropolymerization of the analyte, mold removal, and bonding were elucidated by CV and EIS, and the electrochemical behavior of ferricyanide and ferrocyanide. They obtained a linear operating range of 0.1–5.0 μM against ETC with a MIP-based electrochemical sensor. They found the LOD value of 50 nM using the polymer-supported sensor with DPV. They examined the selectivity of the polymer sensor they developed in the presence of levodopa and carbidopa and showed that it did not affect the ETC analysis. Finally, they have successfully applied analytics to detect ETC in human serum samples added using the MIP sensor.

As a result, highly sensitive, selective, and reproducible analyses were performed with electrochemical tools with functional polymer material with DPV. Single or multilayer sensors consisting of monomers and polymers allow the detection of analytes at nM or even pM levels. In addition, since MIP sensors are produced according to the specific chemical form of the analyte, these nanosensors show superior selectivity. Consequently, the functional polymer makes a great contribution to the development of flexible, portable, and disposable electrochemical devices with unique properties.

5.2.4 Square wave voltammetry

In the 1960s, the concept of pulse voltammetry conducted with a range of scanning rates emerged, which could be an exciting direction for the further advancement of voltammetry. Square wave voltammetry (SWV) based on Faradaic current has been developed. To further advance the SWV technique, a series of new SWV variants developed in recent years have been produced. Contrary to the CV technique, SWV is one of the most advanced impact techniques and is applicable in cases where it is relatively more complex. Therefore, it is among the most common electrochemical techniques. SWV is a candidate as a very good tool in electroanalysis, although the technique is more intuitively complex [84]. Moreover, SWV is a fast, sensitive, and highly selective electroanalytical method. Therefore, SWV and derivative tools allow qualitative and quantitative analysis of analytes at very low concentrations such as nM or even pM. The studies carried out with the SWV, which exhibits a very sensitive detection feature, are endless and new studies have been carried out every day. For this purpose, polymer materials are the first to come to mind in electrochemical tools. In particular, conductive polymers provide precise measurements by detecting the analyte even in complex matrix environments. In addition, due to the flexible nature of polymer functional materials, it makes a positive contribution to the development of on-site, portable, or disposable electrochemical sensing devices.

In the study carried out with SWV and its derivatives, poly(brilliant blue), MIP, poly(phenylenediamine) (P(o-PD)), PANI, poly(amido amine) dendrimer (PAMAM), poly (neutral red) (PNR), polyamic acid (PAA), and PDA functional polymer materials were used. These studies were performed using main indicator electrodes such as boron-doped diamond electrodes (BBDEs), GCEs, PGEs, saturated calomel electrodes, SPCEs, and carbon black electrodes. Sensor production with functional polymer materials, electrodeposition on the main electrode surface and drop-drying technique were the most applied. During this process, the monomer is coated on the main electrode surface by electropolymerization, or conductive polymers are coated directly on the main electrode surface by the drop-drying technique. In the construction of electrochemical sensors, nanosensors have been produced, sometimes coated with a single layer of polymer, and sometimes by combining different composite materials with a synergistic effect. With SWV, it allows the determination of analytes at nM levels with polymer-supported nanosensors. In addition, these nanomaterials allow selective, fast, and sensitive analysis even in a complex matrix environment (table 5.4).

Afzali *et al* in 2019 developed a polymer-supported nanosensor for the determination of the cancer inhibitor as imiquimod (IMQ) [85]. For this, they fabricated a multilayer electrochemical sensor with nanomaterials containing ionic liquid-based MIP and AuNP/GO on the GCE surface. The morphology of the polymer-supported MIP/Au/GO/GCE sensor was characterized by FTIR and SEM. They optimized some parameters in their proposed SWV technique and determined a linear operating range of 0.02–20.0 μM in 0.1 M PBS at pH 7.0 under optimum conditions. The limit of quantification (LOQ) and detection (LOD) values were determined at 0.02 and 0.006 μM, respectively. They achieved low RSD values and good repeatability results with the MIP/Au/GO/GCE sensor. As a result, they proved that the developed polymer-supported electrochemical sensor device can be successfully applied for IMQ detection in real samples.

In another study with SWV, Louw *et al* (2020) developed a polymer-supported nanosensor for the measurement of norfloxacin used as an antibiotic in environmental samples [86]. For this, they fabricated a modified SPCE containing PAA semiconductor polymer and cobalt nanoparticles (CoNPs). PAA, a semiconductor polymer, is coated on the SPCE surface by electrodeposition technique. They elucidated the morphology of the polymer functional composite electrode they developed with FTIR, transmission electron microscopy, CV, and small-angle x-ray scattering (SAXS) techniques. Using the recommended CoNP/PAA/SPCE sensor with SWV, they obtained a linear operating range of 1.1–10 μM according to the 0.75 anodic signal of norfloxacin. They also calculated the LOD value as 0.979 μM. Reliable, selective, and reproducible determination of norfloxacin was performed with a polymer-supported electrochemical sensor. Moreover, an alternative, cost-effective and routine monitoring for low-level analyte determinations has yielded promising results.

Table 5.4. Polymer functional nanomaterials in SWV technique.

Polymer composition	Electrode	Analyte	Linear range (μM)	LOD (nM)	LOQ (nM)	Samples	Electrode preparation method	Interference	References
AgNP–rGO/P-Brilliant blue	PGE	Rosuvastatin	0.005–0.5	2.17	—	Rosuvastatin tablets, human plasma	Electrodeposition	Ascorbic acid, citric acid, starch, glucose, sucrose, uric acid and magnesium stearate	[87]
Au-Cu@BSA-GNRs	GCE	Bisphenol A	0.01–2;2.0–70	4.0		Tap water, bottled water, baby bottle, food storage container	Dried at ambient temperature	Na^+, K^+, Pb^{2+}, Cd^{2+}, Fe^{3+}, Cl^-, $H_2PO_4^-$, HPO_4^{2-}, SO_4^{2-}, NO_3^-, SCN^-, bisphenol S, 4-nitrophenol, catechol, hydroquinone, and phenol	[88]
CPT	BDDE	Bisphenol A	0.43–219	131	—	Tap water, mineral water, lake water	Electropolymerization	—	[89]
CPT	BDDE	Bisphenol S	0.79–319	239	—	Tap water, mineral water, lake water-	Electropolymerization	—	[89]
MIP/[APMIm]Br/BN-HPC	GCE	Citrinin	1×10^{-6}–10×10^{-3} ppm	1×10^{-7} ppm	—	Red yeast rice, Rice, wheat	Dropped on the electrode surface	Aflatoxin B1, deoxynivalenol, zearalenone, patulin, lovastatin	[90]
MIP/Au/GO	GCE	Imiquimod	0.02–20.0	6.0	—	Placebo pharma ceutical sample and IMQ cream	dried in an oven at 60 °C for 24 h	Quinolone, imidazole, adenine, guanine, uric acid	[85]
MIP/BOMC-IL-AuNPs	GCE	Zearalenone	1.5×10^{-6}–3.14×10^{-3}	3.14×10^{-4}	—	Corn, rice, beer	Electropolymerization	—	[91]
MIP/PtPd-NPC	GCE	Patulin	6.48×10^{-5}–0.06	0.048	—	Juice	Electropolymerization	5-hydroxymethyl furfural, aflatoxin B1, citrinin, alternariol, K^+, Mg^{2+}, Na^+, Cl^-, SO_4^{2-}, NO_3^-	[92]
o-PD	GE	Glucose	0.001 25–3.20	1.25	—	Human serum	Electropolymerization	Sucrose, dopamine, starch, and bovine serum albumin	[93]

(*Continued*)

Table 5.4. (*Continued*)

Polymer composition	Electrode	Analyte	Linear range (μM)	LOD (nM)	LOQ (nM)	Samples	Electrode preparation method	Interference	References
P(Cz-co-ANI)	GCE	2,4,6-trinitrotoluene	0.44–4.4	110	365	Soil	Electropolymerization and deposition	Analgesic drug, acetylsalicylic acid, sweetener, and sugar	[94]
P(Cz-co-ANI)	GCE	2,4-dinitrotoluene	0.54–5.4	164	549	—	Electropolymerization and deposition	Analgesic drug, acetylsalicylic acid, sweetener, and sugar	[94]
P(Cz-co-ANI)	GCE	1,3,5-trinitro-1,3,5-triazacy clohexane	0.28–5.74	57.4	189.5	Soil	Electropolymerization and deposition	Analgesic drug, acetylsalicylic acid, sweetener, and sugar	[94]
P(Cz-co-ANI)	GCE	Octahydro-1,3,5,7-tetranitro-1,3,5,7-tetrazocine	0.16–3.37	33.76	111.4	—	Electropolymerization and deposition	Analgesic drug, acetylsalicylic acid, sweetener, and sugar	[94]
PAMAM/AgNP/MWCNT/PNR	Saturated calomel electrode	Paracetamol	0.16–2000	53	—	Human urine, pharmaceuticals	Electropolymerization	L-Tryptophan, riboflavin, epinephrine	[95]
Poly(CTAB)/MWCNTs	PGE	Bisphenol A	0.002–0.808	0.134		Drinking water bottleBaby bottleBaby teether	Electropolymerization	Na^+, K^+, Ca^{2+}, Mg^{2+}, Fe^{3+}, Cu^{2+}, PO_4^{3-}, Cl^-, NO_3^-, phenol, para-nitrophenol, 2,4 dinitrophenol	[96]
Polyamic acid /CoNP	SPCE	Norfloxacin	1100–1000	0.98×10^6		—	Electrodeposition	—	[86]
SH-β-CD/NPGL	GE	Bisphenol A	0.3–100	60	—	Tissue paper	Dried at room temperature	NH_4^+, K^+, Na^+, NO_3^-, Cl^-, SO_4^{2-}	[97]
β-CD/MWCNTs	GCE	Bisphenol S	0.5–60	50		Drinking water, tap water	Dropped	Ca^{2+}, Mg^{2+}, SO_4^{2-}, PO_4^{3-}, Cl^-, Fe^{3+}, Al^{3+}, ascorbic acid, dopamine, uric acid	[98]
μPAD	Carbon black electrode	Bisphenol A	0.1–0.9; 1–20	30	—	River and drinking water	Dropped	2,4 D, paraoxon, atrazine, lead, cadmium	[99]

5.2.5 Other detection tools

Linear sweep voltammetry (LSV), electrochemical impedance spectroscopy, polarography, and potentiometry techniques are rarely used for the determination of analytes using polymer-supported nanomaterials. Among these techniques, LSV and pM levels allowed analyte determinations. In addition, among these techniques, polyvinyl chloride (PVC) is the most used in the electrochemical potentiometry sensing technique [100]. Sensor applications were made by using functional polymer materials such as beta-cyclodextrin (βCD), poly(L-serine), MIP, PANI, PDA, PEDOT, poly(styrene sulfonate) (PSS), poly(o-phenylenediamine), PAA, MAA, and PPy. These studies were performed using main indicator electrodes such as SPEs, SPCEs, GCEs, polyethylene terephthalate (PET) electrode, dropping mercury electrodes (DMEs), pencil core electrodes (PCEs), PVC membrane electrodes, carbon paste electrodes (CPEs), ion-selective electrodes (ISEs). As a result, very sensitive, selective, reproducible, and reliable studies have been carried out with polymers and other electrochemical techniques, which have attracted attention in recent years (table 5.5).

5.3 Conclusion

Polymers have formed an important class of materials from the first moment they entered human life. Polymers, which have an important place in the plastics industry, have given promising results in the production of modern sensors. Polymers are the most serious candidates for the development of electrochemical sensor analyzers because of their exceptional flexibility and lightweight nature. In addition, polymer functional materials that can show conductivity are an alternative to the world of sensors. Therefore, a large number of polymer material applications have been made in the production and development of electrochemical sensors. Therefore, in the production and development of electrochemical sensors, beta-cyclodextrin (βCD), poly(L-serine), MIP, PANI, PDA, PEDOT, many polymer materials such as PSS, poly(o-phenylenediamine), PAA, MAA, and PPy have been used. In addition, monolayer or multilayer nanomaterials containing polymers have been produced. The production methods of polymer-supported sensing devices are electrodeposition, drop-drying, annealing, and potential application. Polymer-supported sensing devices with these unique features are applied in a wide range of areas from pharmacy to medicine, from environmental analysis to food analysis. These materials greatly contribute to the development of online, portable, and fast analyzers. As a result, rapid, reliable, and accurate detection of polymer functional materials used in recent years in real electrochemical sensor applications is provided.

Table 5.5. Polymer functional nanomaterials in other electrochemical techniques.

Polymer composition	Electrode	Analyte	Method	Linear range (μM)	LOD (nM)	LOQ (nM)	Samples	Electrode preparation method	Interference	References
GO-MWCNT-βCD	SPE	Bisphenol A	LSV	0.05–5.0; 5.0–30.0	6.0	—	Tap water, bottled water, lake water	Dropped	Na^+, K^+, Ca^{2+}, Pb^{2+}, Cl^-, SO_4^{2-}, NO_3^-, ascorbic acid, dopamine, acetaminophen	[101]
MWCNTs-β-CD	SPCE	Bisphenol A	LSV	0.125–2.0; 2.0–30	13.76	—	Lake water, tap water	Dropped	Acetaminophen, bisphenol S, NaCl, KCl	[102]
Poly(L-serine)	GCE	Naproxen	LSV	4.3–65	690	—	Water	Electropoly merization	Caffeine, acetaminophen	[103]
ssDNA-MB/Aptamer	Au/PET electrode	Bisphenol A	LSV	4.4×10^{-6}–0.440	1.75×10^{-3}	—	Tap water, PC drinking bottle	Dried with nitrogen	Phenol, 4-nitrophenpl, hydroquinone, 4,4′-bisphenol, bisphenol B, and 6F-bisphenol	[104]
Poly(niacin)	GPE	Riboflavin	LSV	5–65	782	2760.8	Multivitamin tablet	Electropoly merization	Ascorbic acid, dopamine	[17]
AuPd NPs/PDA/MWCNTs-CS-IL	GCE	Cholestanol	EIS	0.1×10^{-6}–60×10^{-6}	0.05×10^{-3}	—	Protein of the serum sample	Dropped and electropoly merization	Cholesterol, stigmasterol, estradiol, vitamin D3, testosterone, ascorbic acid, uric acid, progesterone, glucose, estrone and dopamine	[60]
Graphene-PEDOT: PSS	Whatman paper No. 3	Carcinoembryonic antigens	EIS	0.77–14 ppm	450 ppm	—	Human serum	Spinning coating and dip coating	—	[105]
MIP/Fe_3O_4-NCs-APTES	GCE	Bisphenol A	EIS	8.76×10^{-3}–21.9	0.876	—	Seawater, fish	Dried under nitrogen atmosphere	—	[106]

MIP/ MWCNTs	GCE	Trinitroperhydro-1,3,5-triazine	EIS	0.001–1.0; 0.1–10	4.5×10^{-6} mg l^{-1}	—	Tap water samples	Electropoly merization	Na^+, K^+, Ca^{2+}, Mg^{2+}, Fe^{3+}, Cl^-, SO_4^{2-}, NO_3^-	[107]
MIP-AF	SPE	Tryptophan	EIS	1×10^{-5}–2×10^{-3};4×10^{-3}–80	8×10^{-3}		Human HT29 cancer cell culture media, milk	Polymerization	—	[108]
poly (o-phenyle nediamine)	SPGE	Mycotoxin Zearalenone	EIS	7.75×10^{-3}–0.62	0.62	—	Cornflakes	Electropoly merization	Nivalenol, ochratoxin A, fumonisin B1, fumonisin B2, zearalenone, deoxynivalenol	[109]
Poly o-phenylene diamine	SPGE	Deoxynivalenol	EIS	0.01–1.68	1.68×10^{-3}		Cornflakes	Electropoly merization	Nivalenol, ochratoxin A, fumonisin B1, fumonisin B2, zearalenone	[110]
Polyamic acid /CoNP	SPCE	Norfloxacin	EIS	100–500	0.228×10^6	—	—	Electro deposition	—	[86]
MAA	DME	Galegine	Polarography	7.86–786	3.2×10^4	9.7×10^4	Extracted from Galega officinalis	Polymerization	—	[111]
Acrylamide	PCE	4,4′–DIADPE	Potentiometry	0.1–500	46.3	—	Wastewater	Electropoly merization	—	[112]
MIP/MAA	PVC membrane electrode	Clarithromycin	Potentiometry	1–5000	800		Pharmaceutical formulation	—	Na^+, K^+, Mg^{2+}, Ca^{2+}, NH_4^+, CO_3^{2-}, HPO_4^{2-}, $H_2PO_4^-$, erythromycin, glucose	[113]
MWCNTs/ TiO2/ Polyaniline	CPE	Mirtazapine	Potentiometry	0.09–1×10^4	32.03	—	Pure form, pharmaceutical tablets, spiked biological samples	Polymerization	NaCl, $MgCl_2$, Talc, Urea, $FeSO_4$, $FeCl_3$, KCl, $CuCl_2$, $Pb(CH_3COO)_2$, ascorbic acid, uric acid, alanine, leucine, melitracen HCL, olanzapine, flupentixol diHCl, gabapentin, starch, glucose, lactose	[114]

(Continued)

Table 5.5. (*Continued*)

Polymer composition	Electrode	Analyte	Method	Linear range (μM)	LOD (nM)	LOQ (nM)	Samples	Electrode preparation method	Interference	References
Ni(II)-PVC/GO/SPE	ISE	Ni(II)	Potentiometry	100–0.3	200	—	Tap water, well water, and mineral water	Dried at room temperature	Na^+, K^+, Cu^{2+}, Ca^{2+}, Mg^{2+}, Fe^{3+}, Zn^{2+}	[115]
Ni^{2+}/itaconic acid	ISE	Ni (II)	Potentiometry	10–1 × 10^5	5000	—	—	Polymerization	Ag^+, K^+, Cd^{2+}, Cu^{2+}, Co^{2+}, Zn^{2+}, Pb^{2+}, Mg^{2+}	[116]
PPy: LAC	ISE	Lactate	Potentiometry	100–10 000	81 × 10^3	—	Human sweat, human tear, blood	Galvano static method	Glucose, ascorbic acid, urea, chloride	[117]
PVC membrane/PPy	GE	Fluoxetine	Potentiometry	1.0–1000	6300	—	Pharmaceutical sample	Polymerization	Na^+, K^+, Mg^{2+},Ca^{2+}, NH_4^+, Cl^-,NO_3^-, lactose, glucose	[118]
PVC/Gr	PME	Cefotaxime	Potentiometry	0.3–1 × 10^4	90	—	Pharmaceutical preparation	Dried at room temperature	—	[119]
SB-Salpr	CPE	Al(III)	Potentiometry	1.0–1 × 10^4	210	—	Titration of ethylene diaminetetraacetic acid	—	Ce^{3+}, Cd^{2+}, La^{3+}, Zn^{2+}, Cr^{3+}, Cu^{2+}, Fe^{3+}, Co^{2+}, Mn^{2+}, Ni^{2+}, Mg^{2+}, Ag^+	[120]

Abbreviations

(Cys) $VIMBF_4$	1-[3-(N-cystamine)propyl]-3-vinylimidazolium tetrafluoroborate
4,4′-DIADPE	4,4′-diaminodiphenyl ether
5-ASA	5-aminosalicylic acid
AA	ascorbic acid
ABS	Acetate buffer solution
[APMIm]Br	1-aminopropyl-3-methylimidazolium bromide
ATP	aminothiophenol
β-CD	β-cyclodextrins
BBDE	Boron-doped diamond electrode
BN-HPC	Ionic liquid decorated boron and nitrogen co-doped hierarchical porous carbon
BSA	bovine serum albumin
CHT	Chitosan
CMOF-MIPIL	Conductive metal organic framework and molecularly imprinted poly (ionic liquid)
CILE	Carbon ionic liquid electrode
Co	Cobalt
DGNs	Dendritic Gold Nanostructures
DGr	Dopamine@graphene
EGCG	Epigallocatechin gallate
EIS	Electrochemical impedance spectroscopy
Fc	Ferrocene
GE	gold electrode
GO	Graphene oxide
GOx	Glucose oxidase
GRE	Graphite rod electrode
IIP	Ion imprinted polymer
LBL	Layer-by-layer
LSV	linear sweep voltammetry
MAA	Methacrylic acid
mMIP	magnetic molecularly imprinted polymers
o-PD	o-phenylenediamine
PABA	poly(3-aminobenzylamine)
PANI	polyaniline
PBS	poly(sodium 4styrenesulfonate)
PCE	Pencil core electrode
PDA	Polydopamine
PEDOT	Poly 3,4-ethylenedioxythiophene
PEI	Polyethyleneimine
PET	Polyethylene terephthalate electrode
PLL	Poly-L-lysine
PMB	Poly(methylene blue)
PME	Potentiometric electrodes
PMMA	Polymethyl methacrylate
PPy	Polypyrrole
PTH-MB	polythionine-methylene blue

SB-Salpr	*N*,*N*′-bis(salicylidene)-1,3-propanediamine
SPCE	screen-printed carbon electrode
SPE	Solid polymer electrolyte
SPGE	Screen-printed gold electrode
ZIF-8	Zeolitic Imidazolate Frameworks-8

References

[1] Chandrasekhar P 1999 Basics of conducting polymers (CPs) *Conducting Polymers, Fundamentals and Applications* (Springer) pp 3–22

[2] Prinith N S, Manjunatha J G and Hareesha N 2021 Electrochemical validation of L-tyrosine with dopamine using composite surfactant modified carbon nanotube electrode *J. Iran. Chem. Soc.* **18** 3493–503

[3] Manjunatha J G 2020 A surfactant enhanced graphene paste electrode as an effective electrochemical sensor for the sensitive and simultaneous determination of catechol and resorcinol *Chem. Data Collect.* **25** 100331

[4] Tigari G and Manjunatha J G 2020 Optimized voltammetric experiment for the determination of phloroglucinol at surfactant modified carbon nanotube paste electrode *Instrum. Exp. Tech.* **63** 750–7

[5] Manjunatha J G, Kumara Swamy B E, Gilbert O, Mamatha G P and Sherigara B S 2010 Sensitive voltammetric determination of dopamine at salicylic acid and TX-100, SDS, CTAB modified carbon paste electrode *Int. J. Electrochem. Sci.* **5** 682–95

[6] Charithra M M, Manjunatha J G G and Raril C 2020 Surfactant modified graphite paste electrode as an electrochemical sensor for the enhanced voltammetric detection of estriol with dopamine and uric acid *Adv. Pharm. Bull.* **10** 247–53

[7] Raril C and Manjunatha J G 2020 Fabrication of novel polymer-modified graphene-based electrochemical sensor for the determination of mercury and lead ions in water and biological samples *J. Anal. Sci. Technol.* **11** 1–10

[8] Bai X, Wu Y, Deng L, Gong L, Xu T, Song W *et al* 2022 Imprinted electrochemical sensor of tyrosine based on chitosan β-cyclodextrin/multi-walled carbon nanotubes composite film *Curr. Anal. Chem.* **18** 495–503

[9] Fedorenko V, Damberga D, Grundsteins K, Ramanavicius A, Ramanavicius S, Coy E *et al* 2021 Application of polydopamine functionalized zinc oxide for glucose biosensor design *Polymers (Basel)* **13** 2918

[10] Tancharoen C, Sukjee W, Yenchitsomanus P, Panya A, Lieberzeit P A and Sangma C 2021 Selectivity enhancement of MIP-composite sensor for explosive detection using DNT-dengue virus template: a co-imprinting approach *Mater. Lett.* **285** 129201

[11] Leibl N, Duma L, Gonzato C and Haupt K 2020 Polydopamine-based molecularly imprinted thin films for electro-chemical sensing of nitro-explosives in aqueous solutions *Bioelectrochemistry* **135** 107541

[12] Tertis M, Sîrbu P, Suciu M, Bogdan D, Pana O, Cristea C *et al* 2022 An innovative sensor based on chitosan and graphene oxide for selective and highly-sensitive detection of serotonin *ChemElectroChem.* **9** e202101328

[13] Chen B, Zhang Y, Lin L, Chen H and Zhao M 2020 Au nanoparticles @metal organic framework/polythionine loaded with molecularly imprinted polymer sensor: Preparation, characterization, and electrochemical detection of tyrosine *J. Electroanal. Chem.* **863** 114052

[14] Aydogdu N, Ozcelikay G and Ozkan S A 2022 Rapid and sensitive electrochemical assay of cefditorena with MWCNT/Chitosan NCs/Fe_2O_3 as a nanosensor *Micromachines* **13** 1348
[15] Panapimonlawat T, Phanichphant S and Sriwichai S 2021 Electrochemical dopamine biosensor based on poly(3-aminobenzylamine) layer-by-layer self-assembled multilayer thin film *Polymers (Basel)* **13** 1488
[16] Fan Z, Fan L, Shuang S and Dong C 2018 Highly sensitive photoelectrochemical sensing of bisphenol A based on zinc phthalocyanine/TiO_2 nanorod arrays *Talanta* **189** 16–23
[17] Manjunatha J G, Raril C, Hareesha N, Charithra M M, Pushpanjali P A, Tigari G *et al* 2021 Electrochemical fabrication of poly (niacin) modified graphite paste electrode and its application for the detection of riboflavin *Open Chem. Eng. J.* **14** 90–8
[18] Manjunatha J G 2020 Poly (adenine) modified graphene-based voltammetric sensor for the electrochemical determination of catechol, hydroquinone and resorcinol *Open Chem. Eng. J.* **14** 52–62
[19] Abad-Gil L and Brett C M A 2022 Poly(methylene blue)-ternary deep eutectic solvent/Au nanoparticle modified electrodes as novel electrochemical sensors: optimization, characterization and application *Electrochim. Acta* **434** 141295
[20] Guler M, Kavak E and Kivrak A 2022 Electrochemical dopamine sensor based on gold nanoparticles electrodeposited on a polymer/reduced graphene oxide-modified glassy carbon electrode *Anal. Lett.* **55** 1131–48
[21] Wang W 2022 Electrochemical sensor based on molecularly imprinted membranes at Au@CNTs nanocomposite-modified electrode for determination of prednisolone as a doping agent in sport *Int. J. Electrochem. Sci.* **17** 220222
[22] da Silva W and Brett C M A 2020 Novel biosensor for acetylcholine based on acetylcholinesterase/poly(neutral red)—deep eutectic solvent/Fe_2O_3 nanoparticle modified electrode *J. Electroanal. Chem.* **872** 114050
[23] Sangeetha N S and Narayanan S S 2019 Amperometric H_2O_2 sensor based on gold nanoparticles/poly (celestine blue) nanohybrid film *SN Appl. Sci.* **1** 732
[24] Leepheng P, Limthin D, Onlaor K, Tunhoo B, Phromyothin D and Thiwawong T 2021 Modification of selective electrode based on magnetic molecularly imprinted polymer for bisphenol A determination *Jpn. J. Appl. Phys.* **60** SCCJ03
[25] Thamilselvan A, Rajagopal V and Suryanarayanan V 2019 Highly sensitive and selective amperometric determination of BPA on carbon black/f-MWCNT composite modified GCE *J. Alloys Compd.* **786** 698–706
[26] Lü H, Wang H, Yang L, Zhou Y, Xu L, Hui N *et al* 2022 A sensitive electrochemical sensor based on metal cobalt wrapped conducting polymer polypyrrole nanocone arrays for the assay of nitrite *Microchim. Acta* **189** 26
[27] Estrada-Osorio D V, Escalona-Villalpando R A, Gutiérrez A, Arriaga L G and Ledesma-García J 2022 Poly-L-lysine-modified with ferrocene to obtain a redox polymer for mediated glucose biosensor application *Bioelectrochemistry* **146** 108147
[28] Bollella P, Sharma S, Cass A E G and Antiochia R 2019 Minimally-invasive microneedle-based biosensor array for simultaneous lactate and glucose monitoring in artificial interstitial fluid *Electroanalysis* **31** 374–82
[29] Nikitina V N, Karastsialiova A R and Karyakin A A 2023 Glucose test strips with the largest linear range made via single step modification by glucose oxidase-hexacyanoferrate-chitosan mixture *Biosens. Bioelectron.* **220** 114851

[30] Ata S, Feroz M, Bibi I, Mohsin I, Alwadai N and Iqbal M 2022 Investigation of electrochemical reduction and monitoring of p-nitrophenol on imprinted polymer modified electrode *Synth. Met.* **287** 117083

[31] Yang S, Bai C, Teng Y, Zhang J, Peng J, Fang Z *et al* 2019 Study of horseradish peroxidase and hydrogen peroxide bi-analyte sensor with boronate affinity-based molecularly imprinted film *Can. J. Chem.* **97** 833–9

[32] Bollella P, Sharma S, Cass A E G and Antiochia R 2019 Microneedle-based biosensor for minimally-invasive lactate detection *Biosens. Bioelectron.* **123** 152–9

[33] Sriwichai S and Phanichphant S 2022 Fabrication and characterization of electrospun poly (3-aminobenzylamine)/ functionalized multi-walled carbon nanotubes composite film for electrochemical glucose biosensor *Express Polym. Lett.* **16** 439–50

[34] Ratlam C, Phanichphant S and Sriwichai S 2020 Development of dopamine biosensor based on polyaniline/carbon quantum dots composite *J. Polym. Res.* **27** 183

[35] Jose S, Das S, Vakamalla T R and Sen D 2022 Electrochemical glucose sensing using molecularly imprinted polyaniline–copper oxide coated electrode *Surf. Eng. Appl. Electrochem.* **58** 260–8

[36] German N, Ramanaviciene A and Ramanavicius A 2021 Dispersed conducting polymer nanocomposites with glucose oxidase and gold nanoparticles for the design of enzymatic glucose biosensors *Polymers (Basel)* **13** 2173

[37] Rashed M A, Ahmed J, Faisal M, Alsareii S A, Jalalah M, Tirth V *et al* 2022 Surface modification of CuO nanoparticles with conducting polythiophene as a non-enzymatic amperometric sensor for sensitive and selective determination of hydrogen peroxide *Surf. Interfaces* **31** 101998

[38] German N, Popov A, Ramanavicius A and Ramanaviciene A 2022 Development and practical application of glucose biosensor based on dendritic gold nanostructures modified by conducting polymers *Biosensors* **12** 641

[39] Abdi M M, Azli N F W M, Chaibakhsh N, Lim H N, Tahir P M, Karimi G *et al* 2021 Nonenzymatic dopamine biosensor based on tannin nanocomposite *J. Polym. Sci.* **59** 428–38

[40] Jiménez-Pérez R, González-Rodríguez J, González-Sánchez M-I, Gómez-Monedero B and Valero E 2019 Highly sensitive H2O2 sensor based on poly(azure A)-platinum nanoparticles deposited on activated screen printed carbon electrodes *Sensors Actuators* B **298** 126878

[41] Xing Y, Wu G, Ma Y, Yu Y, Yuan X and Zhu X 2019 Electrochemical detection of bisphenol B based on poly(Prussian blue)/carboxylated multiwalled carbon nanotubes composite modified electrode *Measurement* **148** 106940

[42] Jalilian R, Ezzatzadeh E and Taheri A 2021 A novel self-assembled gold nanoparticles-molecularly imprinted modified carbon ionic liquid electrode with high sensitivity and selectivity for the rapid determination of bisphenol A leached from plastic containers *J. Environ. Chem. Eng.* **9** 105513

[43] Moreno M T, Mellado J M R and Medina A 2022 Rapid electrochemical determination of antioxidant capacity using glassy carbon electrodes modified with copper and polyaniline. Application to ascorbic and gallic acids *Biointerface Res. Appl. Chem.* **13** 23

[44] Limthin D, Leepheng P, Onlaor K, Tunhoo B, Klamchuen A, Thiwawong T *et al* 2022 Enhancing the sensitivity and selectivity of salbutamol detection using reduced graphene oxide combined with molecularly imprinted polymers (RGO/MIP) *Jpn. J. Appl. Phys.* **61** 1033

[45] Puttasakul T, Tancharoen C, Sukjee W, Pintavirooj C and Sangma C 2021 Detection of 2,4,6-Trinitrotoluene by MIP-composite based electrochemical sensor *9th Int. Electr. Eng. Congr.* (Piscataway, NJ: IEEE) vol 2021 pp 559–62

[46] Shim K, Kim J, Shahabuddin M, Yamauchi Y, Hossain M S A and Kim J H 2018 Efficient wide range electrochemical bisphenol-A sensor by self-supported dendritic platinum nanoparticles on screen-printed carbon electrode *Sens. Actuators* B **255** 2800–8

[47] Cai J, Sun B, Li W, Gou X, Gou Y, Li D *et al* 2019 Novel nanomaterial of porous graphene functionalized black phosphorus as electrochemical sensor platform for bisphenol A detection *J. Electroanal. Chem.* **835** 1–9

[48] Tan B and Baycan F 2023 Fabricating a new immobilization matrix based on a conjugated polymer and application as a glucose biosensor *J. Appl. Polym. Sci.* **140** e53268

[49] Hareesha N, Manjunatha J G, Amrutha B M, Pushpanjali P A, Charithra M M and Prinith Subbaiah N 2021 Electrochemical analysis of indigo carmine in food and water samples using a poly(glutamic acid) layered multi-walled carbon nanotube paste electrode *J. Electron. Mater.* **50** 1230–8

[50] Pushpanjali P A and Manjunatha J G 2020 Development of polymer modified electrochemical sensor for the determination of alizarin carmine in the presence of tartrazine *Electroanalysis* **32** 2474–80

[51] Manjunatha J G 2018 Highly sensitive polymer based sensor for determination of the drug mitoxantrone *J. Surf. Sci. Technol.* **34** 74–80

[52] Manjunatha Charithra M and Manjunatha J G 2020 Electrochemical sensing of paracetamol using electropolymerised and sodium lauryl sulfate modified carbon nanotube paste electrode *ChemistrySelect* **5** 9323–9

[53] Gulaboski R 2022 Future of voltammetry *Maced. J. Chem. Chem. Eng.* **41** 151–62

[54] Koyun O and Sahin Y 2018 Voltammetric determination of nitrite with gold nanoparticles/poly(methylene blue)-modified pencil graphite electrode: application in food and water samples *Ionics (Kiel)* **24** 3187–97

[55] Radi A and Ragaa Abd-Ellatief M 2021 Molecularly imprinted poly-o-phenylenediamine electrochemical sensor for entacapone *Electroanalysis* **33** 1578–84

[56] Karadurmus L, Ozcelikay G, Armutcu C and Ozkan S A 2022 Electrochemical chiral sensor based on molecularly imprinted polymer for determination of (1S,2S)-pseudoephedrine in dosage forms and biological sample *Microchem. J.* **181** 107820

[57] Çorman M E, Cetinkaya A, Armutcu C, Bellur Atici E, Uzun L and Ozkan S A 2022 A sensitive and selective electrochemical sensor based on molecularly imprinted polymer for the assay of teriflunomide *Talanta* **249** 123689

[58] Zhang G, Yu Y, Guo M, Lin B and Zhang L 2019 A sensitive determination of albumin in urine by molecularly imprinted electrochemical biosensor based on dual-signal strategy *Sens. Actuators* B **288** 564–70

[59] Selvolini G, Lazzarini C and Marrazza G 2019 Electrochemical nanocomposite single-use sensor for dopamine detection *Sensors* **19** 3097

[60] Jalalvand A R, Zangeneh M M, Jalili F, Soleimani S and Díaz-Cruz J M 2020 An elegant technology for ultrasensitive impedimetric and voltammetric determination of cholestanol based on a novel molecularly imprinted electrochemical sensor *Chem. Phys. Lipids* **229** 104895

[61] Lei X, Deng Z, Zeng Y, Huang S, Yang Y, Wang H *et al* 2021 A novel composite of conductive metal organic framework and molecularly imprinted poly (ionic liquid) for highly sensitive electrochemical detection of bisphenol A *Sens. Actuators* B **339** 129885

[62] Nontawong N 2018 Fabrication of a three-dimensional electrochemical paper-based device (3D-ePAD) for individual and simultaneous detection of ascorbic acid, dopamine and uric acid *Int. J. Electrochem. Sci.* 6940–57

[63] Dahaghin Z, Kilmartin P A and Mousavi H Z 2020 Novel ion imprinted polymer electrochemical sensor for the selective detection of lead(II) *Food Chem.* **303** 125374

[64] Sulym I, Cetinkaya A, Yence M, Çorman M E, Uzun L and Ozkan S A 2022 Novel electrochemical sensor based on molecularly imprinted polymer combined with L-His-MWCNTs@PDMS-5 nanocomposite for selective and sensitive assay of tetracycline *Electrochim. Acta* **430** 141102

[65] Soysal M and Karagözler A E 2018 Moleküler Baskılı Polimerler ile Modifiye Edilmiş Karbon Pasta Elektrotlarla Thiuramın Voltametrik Tayini *Eur. J. Sci. Technol.* 323–33

[66] Kaya S I, Corman M E, Uzun L and Ozkan S A 2022 Simple preparation of surface molecularly imprinted polymer based on silica particles for trace level assay of bisphenol F *Anal. Bioanal. Chem.* **414** 5793–803

[67] Fu Y, You Z, Xiao A and Liu L 2021 Magnetic molecularly imprinting polymers, reduced graphene oxide, and zeolitic imidazolate frameworks modified electrochemical sensor for the selective and sensitive detection of catechin *Microchim. Acta* **188** 71

[68] Li S, Ma Y, Liu Y, Xin G, Wang M, Zhang Z *et al* 2019 Electrochemical sensor based on a three dimensional nanostructured MoS_2 nanosphere-PANI/reduced graphene oxide composite for simultaneous detection of ascorbic acid, dopamine, and uric acid *RSC Adv.* **9** 2997–3003

[69] Mesa R, Khan S, Sotomayor M D P T and Picasso G 2022 Using carbon paste electrode modified with ion imprinted polymer and MWCNT for electrochemical quantification of methylmercury in natural water samples *Biosensors* **12** 376

[70] Liu C, Huang J and Wang L 2018 Electrochemical synthesis of a nanocomposite consisting of carboxy-modified multi-walled carbon nanotubes, polythionine and platinum nanoparticles for simultaneous voltammetric determination of myricetin and rutin *Microchim. Acta* **185** 414

[71] Atef Abdel Fatah M, Abd El-Moghny M G, El-Deab M S and Mohamed El Nashar R 2023 Application of molecularly imprinted electrochemical sensor for trace analysis of Metribuzin herbicide in food samples *Food Chem.* **404** 134708

[72] Chang Z 2018 Study of the enzyme-free glucose biosensor based on Ni^{2+}@ poly (neutral red) hybrid nanocomposites (Ni^{2+}@PNR HN)/MWCNTs/Nafion modified electrode *Int. J. Electrochem. Sci.* **13** 1754–72

[73] Cetinkaya A, Kaya S I, Çorman M E, Karakaya M, Bellur Atici E and Ozkan S A 2022 A highly sensitive and selective electrochemical sensor based on computer-aided design of molecularly imprinted polymer for the determination of leflunomide *Microchem. J.* **179** 107496

[74] Ozkan E, Çorman M E, Nemutlu E, Ozkan S A and Kır S 2022 Development of an electrochemical sensor based on porous molecularly imprinted polymer via photopolymerization for detection of somatostatin in pharmaceuticals and human serum *J. Electroanal. Chem.* **919** 116554

[75] Mahalakshmi S and Sridevi V 2019 Conducting, crystalline and electroactive polyaniline-Au nanocomposites through combined acid and oxidative doping pathways for biosensing applications: detection of dopamine *Mater. Chem. Phys.* **235** 121728
[76] Tomé L I N and Brett C M A 2019 Polymer/iron oxide nanoparticle modified glassy carbon electrodes for the enhanced detection of epinephrine *Electroanalysis* **31** 704–10
[77] Liu Y-L, Liu R, Qin Y, Qiu Q-F, Chen Z, Cheng S-B *et al* 2018 Flexible electrochemical urea sensor based on surface molecularly imprinted nanotubes for detection of human sweat *Anal. Chem.* **90** 13081–7
[78] Wang B, Hong J, Liu C, Zhu L and Jiang L 2021 An electrochemical molecularly imprinted polymer sensor for rapid β-lactoglobulin detection *Sensors* **21** 8240
[79] Wu Y, Wang Y, Wang X, Wang C, Li C and Wang Z 2019 Electrochemical sensing of α-fetoprotein based on molecularly imprinted polymerized ionic liquid film on a gold nanoparticle modified electrode surface *Sensors* **19** 3218
[80] Zhang Z, Chen S, Ren J, Han F, Yu X, Tang F *et al* 2020 Facile construction of a molecularly imprinted polymer–based electrochemical sensor for the detection of milk amyloid A *Microchim. Acta* **187** 642
[81] Pushpanjali P A, Manjunatha J G and Srinivas M T 2020 Highly sensitive platform utilizing poly(l-acmethionine) layered carbon nanotube paste sensor for the determination of voltaren *FlatChem.* **24** 2452–627
[82] Tigari G and Manjunatha J G 2020 Poly(glutamine) film-coated carbon nanotube paste electrode for the determination of curcumin with vanillin: an electroanalytical approach *Monatsh. Chem.* **151** 1681–8
[83] Manjunatha J G 2020 A promising enhanced polymer modified voltammetric sensor for the quantification of catechol and phloroglucinol *Anal. Bioanal. Electrochem.* **12** 893–903
[84] Demir E, İnam O and İnam R 2018 Determination of ophthalmic drug proparacaine using multi-walled carbon nanotube paste electrode by square wave stripping voltammetry *Anal. Sci.* **34** 771–6
[85] Afzali M, Mostafavi A and Shamspur T 2019 Developing a novel sensor based on ionic liquid molecularly imprinted polymer/gold nanoparticles/graphene oxide for the selective determination of an anti-cancer drug imiquimod *Biosens. Bioelectron.* **143** 111620
[86] Louw C J, Hamnca S and Baker P G L 2020 Voltammetric and impedimetric detection of norfloxacin at Co nanoparticle modified polymer composite electrodes *Electroanalysis* **32** 3170–9
[87] El-Zahry M R and Ali M F B 2019 Enhancement effect of reduced graphene oxide and silver nanocomposite supported on poly brilliant blue platform for ultra-trace voltammetric analysis of rosuvastatin in tablets and human plasma *RSC Adv.* **9** 7136–46
[88] Mahmoudi E, Hajian A, Rezaei M, Afkhami A, Amine A and Bagheri H 2019 A novel platform based on graphene nanoribbons/protein capped Au–Cu bimetallic nanoclusters: application to the sensitive electrochemical determination of bisphenol A *Microchem. J.* **145** 242–51
[89] Freitas J M, Wachter N and Rocha-Filho R C 2020 Determination of bisphenol S, simultaneously to bisphenol A in different water matrices or solely in electrolyzed solutions, using a cathodically pretreated boron-doped diamond electrode *Talanta* **217** 121041
[90] Hu X, Liu Y, Xia Y, Zhao F and Zeng B 2021 A novel ratiometric electrochemical sensor for the selective detection of citrinin based on molecularly imprinted poly(thionine) on ionic

liquid decorated boron and nitrogen co-doped hierarchical porous carbon *Food Chem.* **363** 130385
[91] Hu X, Wang C, Zhang M, Zhao F and Zeng B 2020 Ionic liquid assisted molecular self-assemble and molecular imprinting on gold nanoparticles decorated boron-doped ordered mesoporous carbon for the detection of zearalenone *Talanta* **217** 121032
[92] Hu X, Xia Y, Liu Y, Zhao F and Zeng B 2021 Determination of patulin using dual-dummy templates imprinted electrochemical sensor with PtPd decorated N-doped porous carbon for amplification *Microchim. Acta* **188** 148
[93] Sehit E, Drzazgowska J, Buchenau D, Yesildag C, Lensen M and Altintas Z 2020 Ultrasensitive nonenzymatic electrochemical glucose sensor based on gold nanoparticles and molecularly imprinted polymers *Biosens. Bioelectron.* **165** 112432
[94] Sağlam Ş, Üzer A, Erçağ E and Apak R 2018 Electrochemical determination of TNT, DNT, RDX, and HMX with gold nanoparticles/poly(carbazole-aniline) film-modified glassy carbon sensor electrodes imprinted for molecular recognition of nitroaromatics and nitramines *Anal. Chem.* **90** 7364–70
[95] Devi C L and Narayanan S S 2019 Poly (amido amine) dendrimer/silver nanoparticles/multi-walled carbon nanotubes/poly (neutral red)-modified electrode for electrochemical determination of paracetamol *Ionics (Kiel)* **25** 2323–35
[96] Bolat G, Yaman Y T and Abaci S 2018 Highly sensitive electrochemical assay for bisphenol A detection based on poly (CTAB)/MWCNTs modified pencil graphite electrodes *Sens. Actuators* B **255** 140–8
[97] Zhang R, Zhang Y, Deng X, Sun S and Li Y 2018 A novel dual-signal electrochemical sensor for bisphenol A determination by coupling nanoporous gold leaf and self-assembled cyclodextrin *Electrochim. Acta* **271** 417–24
[98] Filik H, Avan A A and Yetimoğlu E K 2019 Multiwalled carbon nanotubes β-cyclodextrin modified electrode for electrochemical determination of bisphenol S in water samples *Russ. J. Electrochem.* **55** 70–7
[99] Jemmeli D, Marcoccio E, Moscone D, Dridi C and Arduini F 2020 Highly sensitive paper-based electrochemical sensor for reagent free detection of bisphenol A *Talanta* **216** 120924
[100] Isildak I, Attar A, Demir E, Kemer B and Aboul-Enein H Y 2018 A novel all solid-state contact PVC-membrane beryllium-selective electrode based on 4-hydroxybenzo-15-crown-5 ether ionophore *Curr. Anal. Chem.* **14**
[101] Alam A U and Deen M J 2020 Bisphenol A electrochemical sensor using graphene oxide and β-cyclodextrin-functionalized multi-walled carbon nanotubes *Anal. Chem.* **92** 5532–9
[102] Ali M Y, Alam A U and Howlader M M R 2020 Fabrication of highly sensitive bisphenol A electrochemical sensor amplified with chemically modified multiwall carbon nanotubes and β-cyclodextrin *Sens. Actuators* B **320** 128319
[103] Hung C-M, Huang C P, Chen S-K, Chen C-W and Dong C-D 2020 Electrochemical analysis of naproxen in water using poly(l-serine)-modified glassy carbon electrode *Chemosphere* **254** 126686
[104] Yu Z, Luan Y, Li H, Wang W, Wang X and Zhang Q 2019 A disposable electrochemical aptasensor using single-stranded DNA–methylene blue complex as signal-amplification platform for sensitive sensing of bisphenol A *Sens. Actuators* B **284** 73–80
[105] Yen Y-K, Chao C-H and Yeh Y-S 2020 A graphene-PEDOT:PSS modified paper-based aptasensor for electrochemical impedance spectroscopy detection of tumor marker *Sensors* **20** 1372

[106] Zhang R-R, Zhan J, Xu J-J, Chai J-Y, Zhang Z-M, Sun A-L *et al* 2020 Application of a novel electrochemiluminescence sensor based on magnetic glassy carbon electrode modified with molecularly imprinted polymers for sensitive monitoring of bisphenol A in seawater and fish samples *Sens. Actuators* B **317** 128237

[107] Alizadeh T, Atashi F and Ganjali M 2019 Molecularly imprinted polymer nano-sphere/multi-walled carbon nanotube coated glassy carbon electrode as an ultra-sensitive voltammetric sensor for picomolar level determination of RDX *Talanta* **194** 415–21

[108] Alam I, Lertanantawong B, Sutthibutpong T, Punnakitikashem P and Asanithi P 2022 Molecularly imprinted polymer-amyloid fibril-based electrochemical biosensor for ultrasensitive detection of tryptophan *Biosensors* **12** 291

[109] Radi A, Eissa A and Wahdan T 2020 Molecularly imprinted impedimetric sensor for determination of mycotoxin zearalenone *Electroanalysis* **32** 1788–94

[110] Radi A-E, Eissa A and Wahdan T 2021 Impedimetric sensor for deoxynivalenol based on electropolymerised molecularly imprinted polymer on the surface of screen-printed gold electrode *Int. J. Environ. Anal. Chem.* **101** 2586–97

[111] Azimi M, Ahmadi Golsefidi M, Varasteh Moradi A, Ebadii M and Zafar Mehrabian R 2020 A novel method for extraction of galegine by molecularly imprinted polymer (MIP) technique reinforced with graphene oxide and its evaluation using polarography *J. Anal. Methods Chem.* **2020** 1–9

[112] Ma M, Zhang Y and Liu J 2022 Adsorption of 4,4′-diaminodiphenyl ether on molecularly imprinted polymer and its application in an interfacial potentiometry with double poles sensor *Chem. Pap.* **76** 1691–705

[113] Mahmoudi S, Rashedi H and Faridbod F 2018 A molecularly imprinted polymer (MIP)-based biomimetic potentiometric sensing device for the analysis of clarithromycin *Anal. Bioanal. Electrochem.* **10** 1654–67

[114] Abdallah N A 2022 Application of titanium oxide decorated multi-walled carbon nanotubes/polyaniline as a transducer polymer for the potentiometric determination of mirtazapine *ChemistrySelect* **7** e202202985

[115] Darroudi A 2023 Electrophoretic deposition of graphene oxide on screen-imprinted carbon electrode and its modification using Ni2+-imprinted polymer as ionophere by a potentiometric sensor for determination of nickel ions *Anal. Bioanal. Chem. Res.* **10** 63–70

[116] Hamidi N, Alizadeh T and Madani M 2018 A novel potentiometric Ni2+-sensor based on a Ni^{2+} ion-imprinted polymer *Anal. Bioanal. Electrochem.* **10** 281–91

[117] Mengarda P, Dias F A L, Peixoto J V C, Osiecki R, Bergamini M F and Marcolino-Junior L H 2019 Determination of lactate levels in biological fluids using a disposable ion-selective potentiometric sensor based on polypyrrole films *Sensors Actuators* B **296** 126663

[118] Madani M 2019 Poly(Pyrrole) conducting polymer solid-state sensor for potentiometric determination of fluoxetine *Anal. Bioanal. Electrochem.* **11** 647–56

[119] Majdi M, Mizani F and Mohammad-Khah A 2022 Quantitative monitoring of cefotaxim ions by a new potentiometric sensor based on molecularly imprinted polymer *Anal. Bioanal. Electrochem.* **14** 100–15

[120] Khoshnood R S, Akbari S and Chenarbou T M 2022 Determination of Aluminium by potentiometry using carbon paste ion selective electrode based on N,N′-bis(salicylidene)-1,3-propanediamine ligand *J. Anal. Chem.* **77** 1057–61

Chapter 6

Metal organic framework based electrochemical sensing devices

Monireh Zamani-Kalajahi and Sevinc Kurbanoglu

Metal–organic frameworks (MOFs) refer to a group of porous materials that consist of metal-containing nodes and organic linkers. Because of unique features such as high surface area, porosity, and structural and functional tenability, these emerging materials are extensively applied as electrode-modifying materials individually or in combination with other electroactive materials. Thus electrochemical sensing fields based on MOFs as novel electrochemical sensing platforms are growing rapidly and gaining popularity. Electrochemical sensors are considered sensitive, selective, rapid, precise, and simple analytical devices for analyzing various targets. This chapter highlights a comprehensive overview of various recently developed advanced electrochemical sensing applications of MOF-based materials. Moreover, the fabrication of different electrochemical sensor systems using various electrodes and their modification procedures utilizing MOFs including pristine MOFs modified MOFs, MOF-based composites, and MOF-derived materials are summarized. This overview is hoped to provide a new and creative perspective for future studies in this rapidly expanding field.

6.1 Introduction

Analytical field researchers are always looking for new analytical strategies that are fast, easy, inexpensive, environmentally friendly, and yet reliable. According to the characteristics and concentration range of the analyte, the nature of the sample matrix, the number of samples to be analyzed, the time required, and the available facilities, a suitable analytical technique is selected. Among other analytical methods, electrochemical detection methods have attracted much attention from many researchers due to their unique advantages. In recent decades, ultrasensitive analysis has become an essential demand [1]. Therefore, many researchers have

doi:10.1088/978-0-7503-5377-9ch6

focused on improving strategies for enhancing electrochemical response for example modification with various functional materials. To date, many materials with micro/nanostructures and exclusive properties have been used for electrochemical sensing platforms. Among these materials, MOFs can be mentioned, having attracted fast and growing attention from many researchers in the field of electrochemical sensors due to their diverse properties including large surface area, adjustable physicochemical properties, superior catalytic activity, and high porosity [2, 3]. This chapter highlights the recent advances during the period of 2018–22 related to the electrochemical sensors based on MOFs for various applications. Also, the reliable and effective approaches to improve the stability, selectivity, and sensitivity of the MOF-based electrochemical sensors are also demonstrated.

6.1.1 Electrochemical sensors

Electrochemical sensors, the largest and the oldest category of chemical sensors, are devices that use an electrode as a transducer element in the presence of an analyte and give information about the composition of a system. An electrochemical sensing device converts the chemical energy of the selective interaction between the sensing surface and the analyte into a useful analytical signal which can further be used for the qualitative and quantitative measurement of a variety of important chemical species in different areas, ranging from food technology to clinical diagnostics. Different families of electrochemical sensors can be recognized depending on their functions. They include amperometric (changes in electric current resulting from oxidation or reduction of the analyte is measured); potentiometric (change of membrane potential is measured when no current is passed); conductometric (change of conductance of analyte is measured); impedimetric (change of impedance is monitored); and voltammetric (change in the potential of the working electrode is measured). Compared to other analytical methods, electrochemical sensing techniques are now attracted great interest because of their analytically significant features including cost-effective sensing ability, simplicity of the instrumentation and procedures, high sensitivity, multiple analytes sensing at the same time, and the element speciation ability etc [4–6]. Today, most electrochemical sensing techniques are based on the use of electrodes modified by different (nano) materials, since they present improved electrocatalytic performance. Electrochemical sensors based on diverse materials (e.g., MOFs, graphene, carbon nanotubes (CNTs), molecularly imprinted polymers (MIPs), metals, and metal oxides) are currently in wide use for the analysis of various chemicals such as glucose, proteins, antibiotics, pesticides, heavy metals, hormones, etc. In the following sections, we summarize recent developments in MOF-based electrochemical sensors.

6.2 Metal–organic frameworks (MOFs)

MOFs are crystalline porous coordination polymers that are constructed by a combination of metal species and organic linkers and characterized as 1D, 2D, or 3D unlimited networks [7]. These newly emerged porous materials owing to their diverse properties including large surface area (typically 1000–10 000 $m^2\ g^{-1}$), numerous micropores, flexible, modifiable structure, tunable pore sizes (3–9.8 nm), rich active sites, and controllable morphologies [8, 9], compared with traditional porous materials such as activated carbons and zeolites, have been extensively utilized in energy storage [10], catalysis [11], chemical sensors [12], separation [13] and others in various forms including pristine MOFs, MOF composites, and MOF derivatives. The usual MOFs include University of Oslo (UiO), Porous Coordination Network (PCN), zeolitic imidazolate framework (ZIF), coordination pillared-layer (CPL), isoreticular metal–organic framework (IRMOF), Materials of Institute Lavoisier (MIL), and so on [14].

6.2.1 Structures of MOFs

As can be seen in figures 6.1 and 6.2, MOFs are created by the incorporation of metals or metal clusters with organic linkers which are mostly nitrogen-containing ligands or carboxylic acids [8, 15]. Therefore, MOF structures are affected by the geometry of metal clusters and also the size and shape of organic linkers. Accordingly, by properly choosing metals and ligands, numerous MOFs with desired structure, pore size, and functionality for special applications can be obtained. By investigating the relationship between the property and structure of MOFs, it is possible to achieve newly designed MOFs that can be used for desired purposes. Pore surfaces are appropriate sites for introducing functional materials into MOF structures. And this post-synthetic approach will be valuable for incompatible functional groups with MOF synthesis.

The porous structure of MOFs makes it possible for guest species with different favorable characteristics such as various nanomaterials, polymers, and biochemicals to be placed inside these pores to considerably expand the applications of MOFs [16–19].

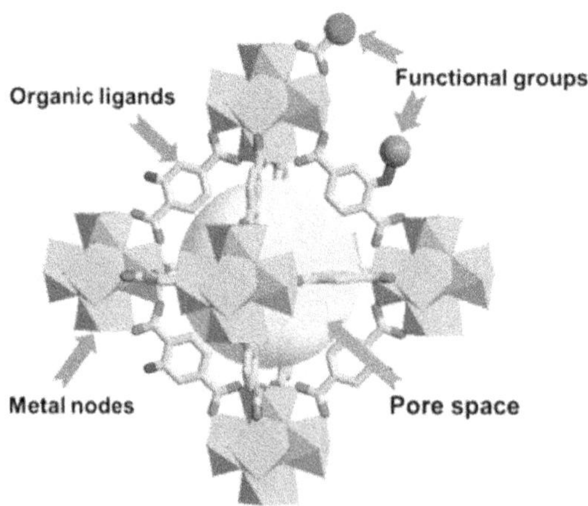

Figure 6.1. Schematic structure of functional MOF. Reprinted from [8], copyright (2019) with permission from Elsevier.

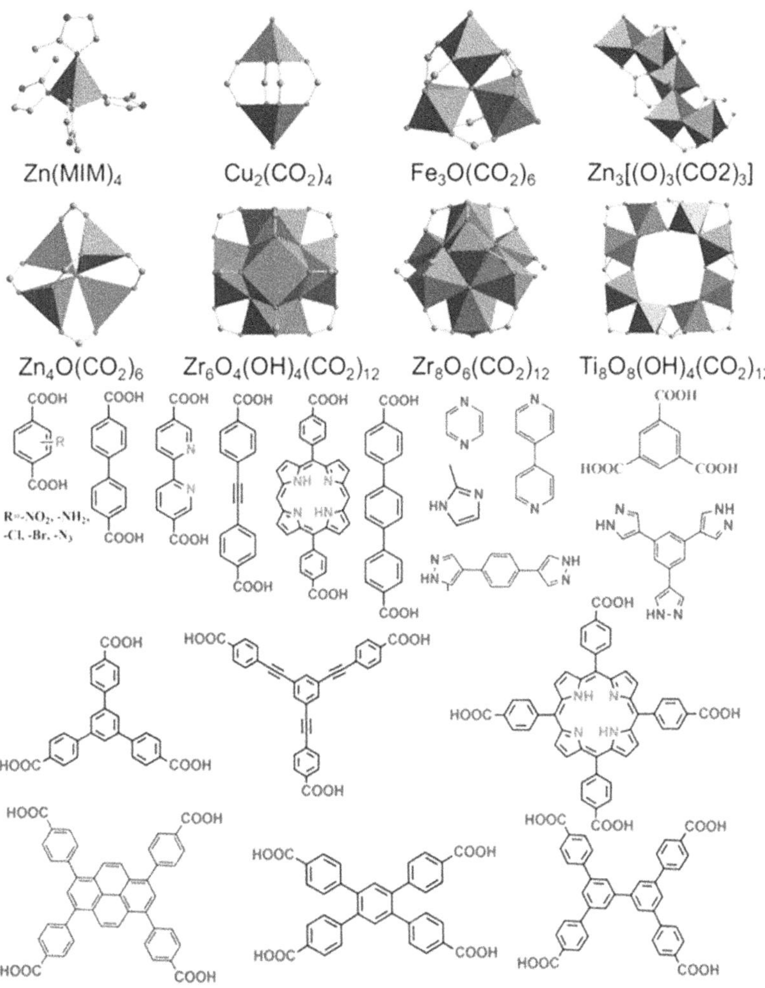

Figure 6.2. Some metals or metal clusters and organic linkers for MOFs. Reprinted from [8], copyright (2019) with permission from Elsevier.

6.2.2 Stability of MOFs

For various applications of MOFs, stability, mainly chemical, mechanical, and thermal stability, is an essential requirement. The chemical stability of MOFs, especially, is defined as their resistance to exposure to different chemicals, such as solvents, moisture, acids, and bases. Unfortunately, most MOFs suffer from the drawback of relatively poor stability, which would hamper their practical applications [8, 20]. In the recent past, great endeavors have been made for the development of strategies to create more robust MOFs and many remarkable advances have been attained [21–25]. The stability of MOFs in water and acidic or basic media is important for the development of stable, reproducible, and sensitive electrochemical sensors and biosensors. Some of the strategies to overcome poor water and acid/base stability are summarized in table 6.1.

Table 6.1. Some approaches to improve MOF stability.

Approaches to improve water stability	References
Integration of medium to long alkyl groups within IRMOF-3	[26]
Converting the organic ligands on the MOF surface into an amorphous carbon layer via heating under N_2 while maintaining the structure and pore size distribution	[27]
Decorating their surface with hydrophobic and permeable polydimethylsiloxane (PDMS) thin layers via vapor deposition technique without changing the surface area of the original MOFs	[28]
Incorporating MOFs into polymers	[29]
Modifiing the inside channels of MOF-5 via *in situ* polymerization of aromatic acetylenes.	[30]
Doping metal ions (Cu^{2+}, Cd^{2+}, or Fe^{2+}) into a gyroidal MOF, STU-1A, resulting in heterometallic MOFs	[31]
Approaches to improve acid/base stability	**References**
Employed Fe-TCPP) TCPP=tetrakis(4carboxyphenyl)porphyrin) as a heme-like ligand and Zr6 clusters as nodes to afford a highly stable, mesoporous MOF, PCN-222(Fe), which can remain stable in 8 M HCl for 24 h	[32]
Metals like lanthanides, cobalt, nickel, manganese, and cadmium are also good for constructing stable MOFs	[33, 34]

The thermal stability of MOFs is also important for applications requiring high temperatures and determination through the number and strength of the metal–linker bond. For example, MOFs with high thermal stability are often fabricated using high-valence metal ions such as Ti^{4+}, Zr^{4+}, Al^{3+}, and Ln^{3+} [8]. Owing to the practical properties of MOFs such as wonderful porosity and big surface area, they have inherent poor mechanical stability. It was indicated that the functional groups on the organic linkers could improve the mechanical stability [35].

6.3 MOF-based materials for electrochemical sensing applications

Owing to the recent expansion of applications of MOF-based materials in various fields of science, we are witnessing their rapidly growing and important role in the sensing of various analytes in various fields of research, including industry, food, environment, and biomedicine [36–43]. The role of MOFs in electrochemical sensing can be attributed to signal amplification through their inherent catalytic activity as nanozymes that originates from their active metal nodes or linkers [44]. Detection sensitivity of electrochemical sensing methods can be improved by MOF-based materials through concentrating of analytes near the electrode surface, resulting in response signal amplification, owing to their high surface area and porosity. Studies on the electrochemical sensing performance of MOF-based materials are currently underway and their applications have been reviewed extensively in recent years [2, 3, 9, 45–55]. These sensors have been employed for the detection of heavy metal ions from groundwater, and various types of organic pollutants from the water bodies. Moreover, electrochemical sensing of antibiotics, phenolic compounds, and pesticides has been investigated [49, 56]. Before 2018, the use of MOFs for the modification of electrode surfaces to detect biomolecules was not studied extensively [57]. Thus, few reports were available on the investigation of the electrochemical behaviors of MOFs until 2018 [58–63]. Considering these recent review papers, this chapter is mainly focused on the most newly developed electrochemical sensing applications of the various MOF-based materials and the practical strategies to improve the performance of MOF-based electrochemical sensors. In this way, application of pristine and modified MOFs, their composites and derivatives-based electrochemical sensors will be discussed in the next sections.

6.3.1 Pristine MOFs as electrochemical sensing platforms

MOFs are considered promising materials in electrochemical sensing due to the presence of metal ions inserted into the MOFs, which allows catalysis of the oxidation/reduction of various compounds, leading to the production of electrochemical responses. Several conventional redox-active MOFs, such as Cu-MOF, Cr-MOF, Co-MOF, and Ni-MOF, have been designed as modifying materials for powerful electrocatalytic electrodes for non-enzymatic detection purposes. The electrochemical response signals can be effectively amplified through these MOFs due to their ultra-high porosities, large surface areas, and desirable electrocatalytic activities [2, 9]. Considering these attractive and important features, the direct use of some pristine MOFs with excellent electrocatalytic properties has resulted in the detection of different analytes ranging from heavy metals and various pollutants to biologically important compounds (table 6.2). In one research, a Cr-MOF (MIL-53-Cr^{III})-modified glassy carbon electrode (GCE) was used for non-enzymatic detection of H_2O_2 [57]. MIL-53-Cr^{III} was prepared through a solvothermal method with high purity and crystallinity, which presented excellent chemical and electrochemical stability in an alkaline medium. The +3 oxidation state of Cr in MIL-53-Cr^{III} could be easily reduced to a +2 oxidation state in an alkaline medium and mediated the direct reduction of H_2O_2 (figure 6.3). The MIL-53-Cr^{III}/GCE sensor exhibited superior electrocatalytic activity for the reduction of H_2O_2 with a wide dynamic range, low detection limit, good sensitivity, and excellent long-term stability. Ling *et al* [64] reported electrochemical sensing of some biologically important species such as glucose, glycine, L-tryptophan (L-Trp), and ascorbic acid (AA) through surface modification of flexible sensors with conductive MOFs (Co-MOF and Cu-MOF) with high porosity, large surface area, and tunable catalysis capability to obtain highly specific and sensitive implantable electrochemical detection. The obtained sensors provided high sensitivity, selectivity, reversibility, and long life sensing capabilities of biologically active chemicals in organs, tissues and interstitial fluids even in very complicated media and severe deformations with the ability to continuously monitor for 20 days because of the minimal usage of biochemical targets.

Table 6.2. Recent applications of pristine MOF-based materials as electrochemical sensing platforms.

MOF-based material	Electrode	Analyte	Method	Matrix	Linear range	Limit of detection	RSD %	References
Cr-MOF (MIL-53-Cr^{III})	MIL-53-Cr^{III}/GCE	H_2O_2	DPV	Human serum	25–500 μM	3.52 μM	4.2	[57]
Cu-MOF	Cu-MOF/SPE	H_2O_2 AA L-His	CV	Live/dead cell	10–1000, 1400–6800 10–2400 μM —	4.1 μM 2.94 5.3	—	[65]
Zn-MOF	Zn-MOF/AuE	L-Cys L-Met L-Cys–Cys	CV	—	200–600 μM	3.12 μM 1.93 2.05	—	[66]
Co-MOF	Co-MOF@Nafion/PGE	H_2O_2	CV	Lens solutions	1000–10 000 μM	—	6.1	[69]
Ni-MOF	Ni-MOF-PVP/GCE	Nitrobenzene	CV	Lake and tap water	0.2–1000 μM	0.097 μM	—	[72]
Bi/MIL-101(Cr)	Bi/MIL-101(Cr)/CCE	Cd(II) Pb(II)	DPASV	Lake and groundwater	0.1–90 ng ml^{-1} 0.1–90 ng ml^{-1}	0.06 ng ml^{-1} 0.07 ng ml^{-1}	1.7 0.88	[102]
Ln-MOF(ZJU-27)	ZJU-27/GCE	Cd(II) Pb(II)	SWASV	Drink and lake water	0.1–1.0 μM 0.5–2 μM	1.66 nM 1.10 nM	7.3	[103]

Abbreviations: GCE: glassy carbon electrode; SPE: Screen-printed electrode; DPV: differential pulse voltammetry; CV: cyclic voltammetry; AuE: gold electrode; AA: ascorbic acid; L-His: L-Histine; L-Cys: L-cysteine; L-Met: L-methionine; L-Cys–Cys: L-cystine; PGE: pyrolytic graphite electrode; PVP: polyvinylpyrrolidone.

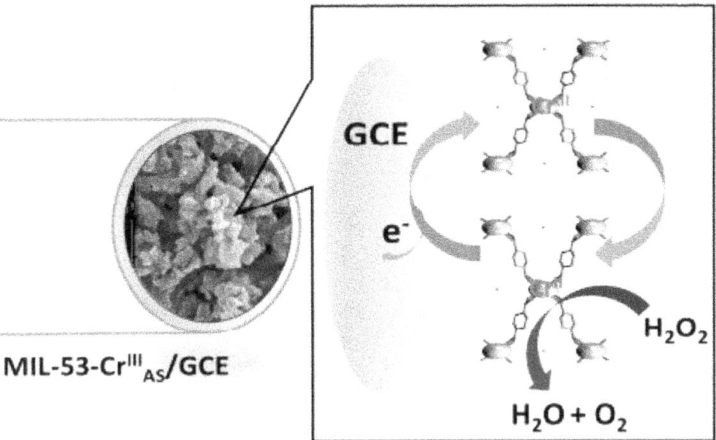

Figure 6.3. A possible mechanism for the direct reduction of H_2O_2 at MIL-53-$Cr^{III}{}_{AS}$/GCE. Reprinted from [57], copyright (2018) with permission from Elsevier.

In another work, Ling *et al* [65], synthesized a Cu-MOF with 2D layered topology and integrated it with flexible electrochemical sensors. The construction of Cu-MOF was accomplished by the coordination of Cu^{2+} ions and the oxygen atoms in carboxylic groups. Therefore, a layered structure was obtained such that these layers were connected with hydrogen bonds. The obtained Cu-MOF sensor was utilized for the electrochemical detection of some important biomolecules such as L-Histidine, H_2O_2, and ascorbic acid. The selected sensor provided a linear range comparable to other sensors which offer similar sensitivity. Moreover, the cytotoxicity study results showed that the proposed sensor indicated its prospective usage in *in vivo* sensing, because of providing good biocompatibility. Wu *et al* [66], used 2,5-thiophene dicarboxylic acid as the main ligand for the synthesis of new 2D Zn-MOF. Since, thiophene and its derivatives have been extensively applied as an electrode modifier material resulting in superior electrocatalytic performance [67, 68]. The air and solvent stability and conductivity of the main ligand were improved by 2,2′-dipyridyl as an auxiliary ligand. It was utilized as a 'lock' to stabilize the framework by chelating the Zn^{2+} metal. As a result, the Zn-MOF could be directly applied to modify the Au electrode (AuE) surface without any additional post-procedure. The excellent electrochemical activity of the as-prepared Zn-MOF/AuE was confirmed by electrochemical biosensing of some S-containing amino acids such as L-cystine, L-methionine, and L-cysteine.

Most recently, Portorreal-Bottier and co-workers [69] fabricated a 2D-Co-MOF with a nitrogen coordination environment based on layered double nanosheets with enzyme-like activity for modification of graphite electrode surface and development of an electrochemical sensor for the rapid H_2O_2 detection through the catalytic oxidation of H_2O_2 at neutral pH, with wide linear range, high stability, sensitivity, and selectivity. (figure 6.4)

It has been illustrated that the materials with nano-sized structures possess many advantages such as increased external surface areas, resulting in enhancing the

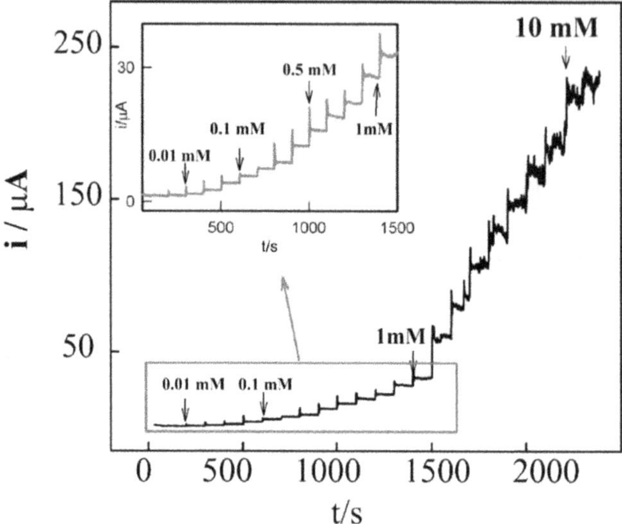

Figure 6.4. Chronoamperometric curve recorded at 0.90 V and 750 rpm after successive additions of hydrogen peroxide to the cell solution, within the hydrogen peroxide concentration range of 0–12 mM, of a pyrolytic graphite electrode coated with 2D-Co-MOF@Nafion composite with 1.20×10^{-9} mol of electroactive cobalt centers. Reprinted from [69], copyright (2022) with permission from Elsevier.

electroactive surface area and improving catalytic reactions. Compared to traditional micrometer-size MOFs, nano-MOFs have enhanced the number of available active sites and facilitated mass diffusion of analytes toward the active sites, resulting in improved electrocatalytic activity and sensing sensitivity by remarkably reducing the limit of detections [9, 70, 71]. In this regard, Arul *et al* [72] modified a GCE through the preparation of polyvinylpyrrolidone-capped Ni-MOF (Ni-MOF-PVP), by a simple *in situ* solvothermal process. The prepared Ni-MOF exhibited a spherical shape with an average particle size of 322 nm that after adding the PVP to Ni-MOF was reduced to 80 nm, according to transmission electron microscopy (TEM) images. The Ni-MOF-PVP-modified GCE was used for highly sensitive and selective electrochemical sensing of nitrobenzene in lake and tap water samples. Despite all these effective approaches, by utilizing some special methods for modification of MOFs their catalytic performance, and as a result, their capability for electrochemical detection applications can be effectually upgraded. These approaches include: (i) MOFs modification by introducing an appropriate functional group into their structure through various practical methods in order to enhance their catalytic performances [73]; (ii) the use of catalytically active species such as electroactive molecules, enzymes, and metal nanomaterials in combination with MOFs can improve their catalytic efficiency; (iii) fabrication of MOF-based composites using highly conductive and mechanically stable materials such as graphene oxides, carbon-based materials, enzymes, antibodies, and aptamers in order to enhance electrochemical activities of MOFs and introduce new properties and functions [74, 75]; (iv) construction of MOF derivatives, which include both inherent properties of MOFs and new features of high conductivity and stability [76]. These strategies for improving the electrochemical performance of MOFs will be described in the following. Recent applications of pristine MOF-based materials as electrochemical sensing platforms are tabulated in table 6.2.

6.3.2 Functionalized MOFs as electrochemical sensing platforms

Due to the low electrical conductivity and redox activity of MOFs, the sensibility of the MOF-based electrochemical sensors and their applications are limited. So in order to extend their applicability, one of the accessible and flexible ways is chemical modifications of MOFs through loading the desirable functional groups/electroactive molecules into them. This approach is easily possible through the characteristics including having a large specific surface area of MOFs and the presence of a large number of pores with different sizes in them.

Generally, modification of MOFs is carried out by functionalization of their metal nodes and/or organic linkers [73, 77]. Functionalization of MOFs as sensing platforms is an excellent strategy to upgrade the sensitivity and selectivity of electrochemical sensors and biosensors by means of the possibility of hydrogen bonding between analyte molecules and MOFs and elevating electron transfer between MOFs and analytes [78, 79]. Previously reported works present wide applications of functionalized MOFs in the electrochemical sensing of various analytes. For example, Rezki and co-workers [80] synthesized amine-functionalized Cu-MOF nanospheres through a facile solvothermal route and successfully utilized them for the electrochemical assay of hepatitis B infection biomarker, i.e., hepatitis B surface antigen (HBsAg). Covalent interaction between the amino-functionalized ligand in Cu-MOF and the carboxyl group of the antibodies provided immobilization of the bioreceptor/antibody to Cu-MOF. Therefore, the amine-functionalized Cu-MOF in addition to providing immobilization of the antibody onto the electrode surface also acts as an electroactive material that produces an electrochemical signal. The HBsAg detection using Cu-MOF was assayed using the differential pulse voltammetry (DPV) method and by proportionally decreasing the response with the increasing of the analyte concentration. This procedure allowed sensitive detection of HBsAg within a wide linear range in real human serum samples using amine-functionalized Cu-MOF as an easy-to-use and low-cost immunosensor. Gao *et al* [81] reported the first example of the electrochemical sensing application of sulfo-functionalized MOFs (sulfo-MIL-101). The sensing electrode was obtained by modifying a graphite paste electrode (GPE) with sulfo-MIL-101, and used for the electrochemical detection of dopamine. The performance of sulfo-MIL-101-GPE, MIL-101-GPE, and bare GPE were investigated for the dopamine voltammetric responses. The sulfo-MIL-101-GPE exhibited amplified voltammetric signals and superior sensitivity to dopamine, due to the catalytic activity of sulfo-modified MOFs and of the electrical conductivity of graphite. The efficient interactions between the sulfo group and dopamine through hydrogen bonding and electrostatic interactions simplified the electron transfer phenomenon. Cassani and co-workers [82] introduced a new Cu(II)-MOF based on a propargyl carbamate-functionalized isophthalate ligand as a platform for electrochemical sensing of nitrite. Their synthesis strategy involved the use of a new organic linker 5-(2-{[(prop-2-yn-1-yloxy)carbonyl]amino}ethoxy)isophthalic acid (named Cu-YBDC). The GCE was modified by Cu-YBDC in order to achieve GC/Cu-YBDC modified electrode. Finally, a GC/Au/Cu-YBDC electrode was obtained by introducing the modified Cu-MOF in $HAuCl_4$ solution followed by electrochemical reduction of Au(III) absorbed on the propargyl group. The effective electrochemical performance of GC/Au/Cu-YBDC was confirmed by detection of nitrite in drinking water with a wide linear range, ultrafast response time (<2 s), and a low detection limit. Su *et al* [83] fabricated toluidine blue (TB)-functionalized MOF thin films through simple *in situ* electrochemical reduction approach. The synthesized TB@MOFs presented excellent stability and electrochemical performance depending on the TB concentrations captured in the MOFs. So a sensitive label-free immunosensor was developed based on TB@MOFs detecting platform for the electrochemical sensing of Indole-3-acetic acid (IAA, a plant hormone) in actual plant samples with good selectivity and sensitivity and wide linear range (figure 6.5).

The incorporation of MOFs with high-affinity aptamers has led to the development of advanced detection strategies for diagnostics of clinical and food/water safety [79]. When MOFs are functionalized with aptamers, they may be used to construct high-performance signal probes, especially constructing an efficient electrochemical aptasensor for various applications, such as environmental monitoring, food safety examination, and biomolecule analysis [84]. Some examples of this research area are summarized in table 6.3. Moreover, additional analytical figures of merit of each reported research and more examples of functionalized MOF applications are presented in table 6.3.

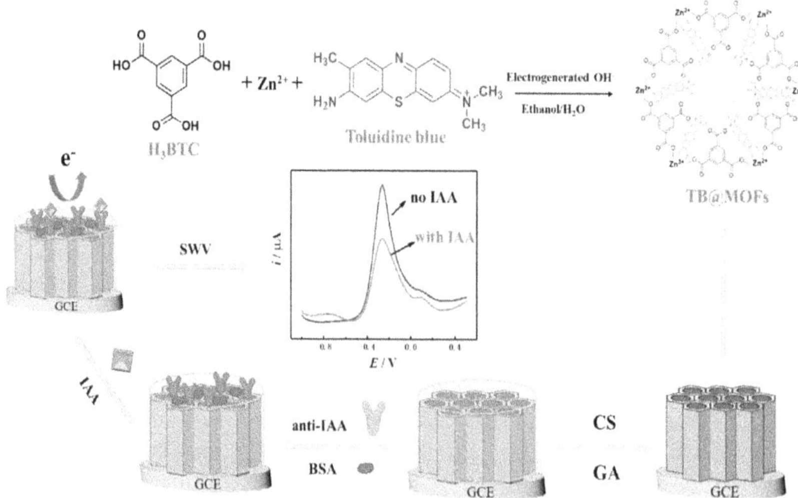

Figure 6.5. Schematic diagram of the preparation of TB modified MOF modified electrode (TB@MOFs/GCE) and immunosensor fabrication Reprinted from [83], copyright (2021) with permission from Elsevier. SWV: square wave voltammetry; CS: chitosan; GA: glutaraldehyde; BSA: bovine serum protein.

Table 6.3. Recent applications of functionalized-MOF-based electrochemical sensing platforms.

MOF-based material	Electrode	Analyte	Method	Matrix	Linear range	Limit of detection	References
NH_2-functionalized Cu-MOF	Anti-HBsAg/Cu-NH_2BDC/GCE	HbsAg	DPV	Human serum	1–500 ng ml^{-1}	0.73 ng ml^{-1}	[80]
Sulfo-functionalized MOF	Sulfo-MIL-101/GPE	Dopamine	CV	Medical injections	0.07–100 μM	0.043 μM	[81]
Cu(II)-MOF based on a propargyl carbamate-functionalized isophthalate ligand	Cu-YBDC/Au/GC	Nitrite	CV	Drinking waters	20–160, 160–120 000, 1200–8000 μM	5.0 μM	[82]
TB-functionalized MOF	TB@MOFs/GCE	IAA	SWV	Seeds	0.0025–5 ng ml^{-1}	0.0014 ng ml^{-1}	[83]
Invertase enzyme modified MOF	Invertase/CP/Au/Cu-MOF/AuE	Dam MTase	DPV	Clinical samples	0.002–1 U ml^{-1}	0.001U ml^{-1}	[104]
NH_2-Ni-MOF	MB–DNA/anti-CRP/AuNS/NH_2-Ni-MOF/AuE	CRP	SWV	Human serum	0.0001–100 ng ml^{-1}	29×10^{-6} ng ml^{-1}	[105]
Fc-functionalized ZIF-67	ZIF-67@Fc/AMNFs	HERS2	CV	Human serum	0.5–1000 pg ml^{-1}	155 fg ml^{-1}	[106]
Sulfur-containing nanocapsule-based MOF	Co-TMC4R-BDC/GCE	Cu2+ Pb2+ Cd2+ $Hg2^+$	CV	Aqueous solution	0.05–12.0 μM 0.05–13.0 μM 0.1–17.0 μM 0.75–18.0 μM	13 μM 11 μM 26 μM 18 μM	[107]

Abbreviations: GCE: glassy carbon electrode; SPE: Screen-printed electrode; DPV: differential pulse voltammetry; CV: cyclic voltammetry; AuE: gold electrode; PVP: polyvinylpyrrolidone; NH_2BDC: 2-amino terephetalic acid; HBsAg: Hepatitis B surface antigen; GPE: graphite paste electrode; IAA: Indole-3-acetic acid; Cu-YBDC: a terminal alkyne function branching out from the Cu-MOF; TB: toluidine blue; MTase: DNA methyltransferase; CP: Capture probe: 5′-AGAAGTCTTAe(CH2) 6eNH2–3; CRP: C-reactive protein; MB: methylene blue; CCE: carbon cloth electrode; DPASV: Differential pulse anodic stripping voltammetry, FC: Ferrocene; AMNFs: antimonate nano flakes; HERS2: human epidermal growth factor receptor-2; Co-TMC4R-BDC: sulfur-containing nanocapsule-based electrochemical metal–organic framework (MOF) sensor $[Co_2(TMC4R)(1,4\text{-}BDC)_2(\mu 2\text{-}H_2O)]\cdot 3DMF\cdot CH_3OH\cdot 5CH_3CN\cdot H_2O$. (TMC4R = tetra(4-mercaptopyridine)calix[4]resorcinarene, 1,4-BDC = 1,4-benzenedicarboxylic acid.

6.3.3 MOF-based composites as electrochemical sensing platforms

In comparison with single materials, MOF composites present higher stability, catalytic performance, greater specific surface area and porosity, and other functions. Highly conductive nanomaterials like metallic nanoparticles, carbon nanomaterials, and so on could form composites with promoted features compared to individual materials. Noble-metal nanoparticles, such as Pd, Au, and Pt despite having unique properties, including good biocompatibility and superior electrocatalytic performance, are easily aggregated. To overcome this disadvantage, MOFs despite having large surface area and adjustable porosity are considered as promising candidates to provide an appropriate platform for the insertion of noble-metal nanoparticles to produce noble-metal nanoparticles/MOFs composite. Therefore, the obtained MOF-based composite, in addition, to prohibiting noble-metal nanoparticle aggregation, can facilitate the availability of the pores for both product and reactants. Newly, many researchers have focused on the use of the noble-metal nanoparticles/MOFs composite as a substrate for electrochemical biosensing applications [78]. For example, in the research reported by Wang and co-workers [78], a sensitive and easy electrochemical biosensor was fabricated for the sensing of telomerase (a critical biomarker for early cancer diagnostics). This strategy involved anchoring Pd nanoparticles on amine-functionalized Cr-based MOFs (Pd/MIL101-NH_2) as a redox mediator and capturing DNA (cDNA) followed by investigation of its electrocatalysis activity as a signal probe for detection of telomerase in cancer cells (figure 6.6). Remarkably, the method presented an enzyme-free and PCR-free electrochemical detection approach, which had no need of extra separation stages.

In another work, the research group of Wang *et al* [85] synthesized Cu-MOFs/ordered mesoporous carbon (OMC) composite through a fast and easy strategy, for the first time in 2018. The obtained Cu-MOFs/OMC nanohybrid exhibited novel electrocatalytic activity characteristics, applied as an electrochemical sensing platform for the detection of hydrazine in environmental samples.

Gao *et al* [86], synthesized ERG-UiO-67-bpy composite as an effective sensing platform for the electrochemical detection of norepinephrine. In the proposed strategy the GCE was modified with Zr-MOF with 2,2′-bipyridyl-5,5′-dicarboxylate (UiO-67-bpy) and graphene oxide that electrochemically reduced to graphene (ERG), subsequently. Both UiO-67-bpy and ERG exhibited simultaneous electrocatalytic efficacy for norepinephrine oxidation, resulting in plenty of increased voltammetric response. The authors suggested that the possibility of forming a hydrogen bond between norepinephrine and the bipyridyl group of UiO-67-bpy, promoted electron transfer on the ERG-UiO-67-bpy/GCE surface. Consequently, the developed electrochemical sensor showed good sensitivity, stability, reproducibility, and wide linear range.

Xu *et al* [87] used Ni-MOF@ CNTs composite for the electrochemical detection of bisphenol A (BPA). Due to the large specific surface area of Ni-MOF and the high conductivity of CNTs, their combination as a new electrochemical sensing platform led to improved performance as electrochemical sensors compared to each

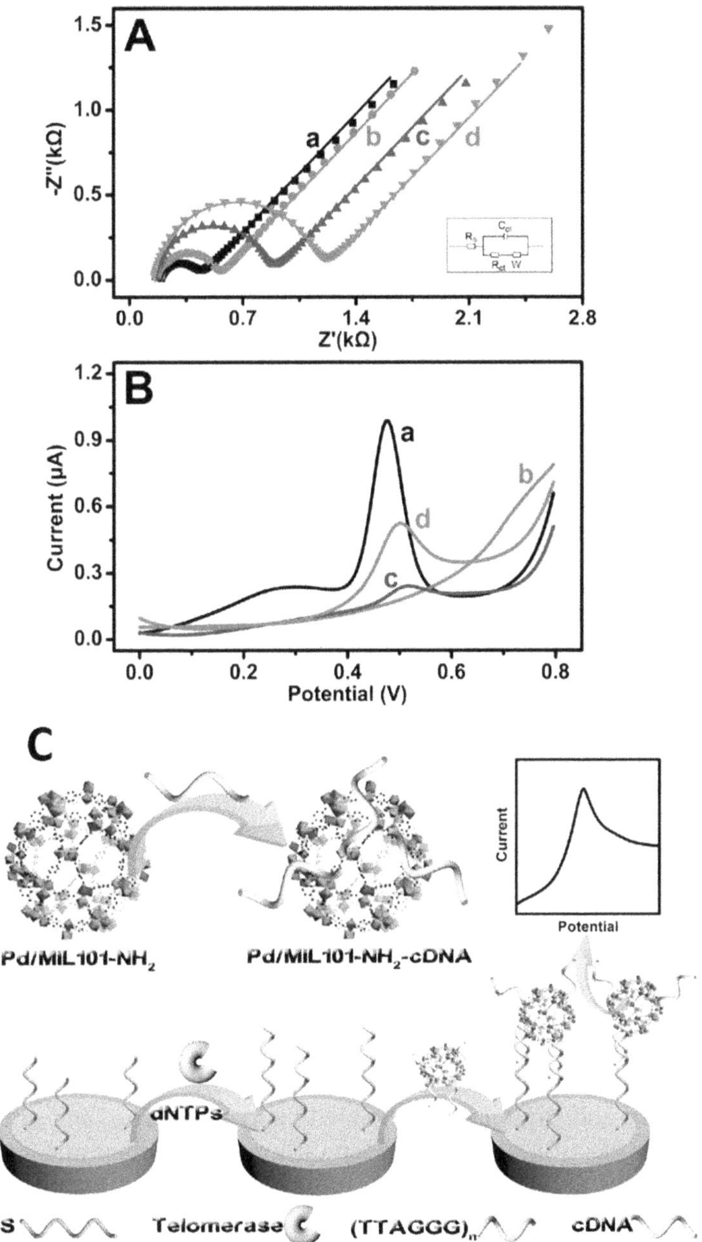

Figure 6.6. (A) Nyquist plot of EIS for bare GE (a), TS/GE (b), (extract+TS)/GE (c), extract+TS/Pd/MIL101-NH_2-cDNA)/GE (d), respectively. (B) DPV responses of bare GE (a), TS/GE (b), (heated extract+TS/Pd/MIL101-NH_2-cDNA)/GE (c), and extract+TS/Pd/MIL101-NH2-cDNA)/GE (d), respectively. (C) Schematic sensing principle of the electrochemical detection of telomerase activity. Reprinted from [78], copyright (2019) with permission from Elsevier.

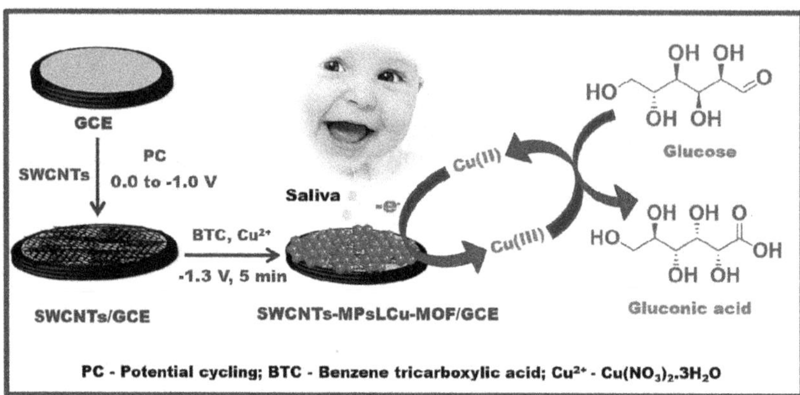

Figure 6.7. Schematic illustration of Cu-MOF deposited on SWCNTs/GC electrode and application for non-enzymatic ultrasensitive determination of glucose in human saliva. Reprinted from [88], copyright (2020) with permission from Elsevier.

one alone. Arul *et al* [88] prepared single-walled carbon nanotubes (SWCNTs)/Cu-MOF composite through electrodeposition on GCE and applied the prepared modified electrode for glucose detection in saliva samples (figure 6.7).

Recently, Du *et al* [89] synthesized gold nanoparticles (AuNPs)/Ce-MOFs nanocomposites, then coupled them with 3D network nanocomposites of electrochemical-reduced graphene oxide (ERGO)/CNTs. The electrochemical response was enhanced through the simultaneous amplification effect of AuNPs, Ce-MOFs, and ERGO/CNTs. The designed sensor provided low detection limits and wide linear ranges for the simultaneous determination of catechol and hydroquinone.

The combination of biomolecules such as enzymes, nucleic acids, and antibodies with MOFs has improved their electrocatalytic activities, resulting in enhancing their sensitivities, selectivity, and linear concentration range as electrochemical sensors [37]. Biomolecules can effectively be immobilized in the porous structure of MOFs with large specific surface areas [90, 91]. For example, in the research reported by Wang *et al* [92], the successful immobilization of xanthine oxidase (XOD) onto a Cu-MOF film was conducted. Subsequently, the applicability of the as-prepared biocomposite was demonstrated as a sensitive electrochemical biosensor for hypoxanthine and xanthine detection with a low detection limit and wide linear range. Promoted electron transfer between the electrode surface and XOD occurred by using Cu-MOF as an effective substrate for entrapping and biocatalysing of XOD enzyme. In 2022, Rasheed *et al* [75] and Lavanya *et al* [93] published comprehensive review papers on MOF-based composites for electrochemical sensing applications. Table 6.4 represents a summary of more electrochemical sensing applications of MOF composites.

6.3.4 MOF derivatives as electrochemical sensing platforms

MOFs can be applied as self-templates or precursors for the preparation of porous carbon materials, metal oxides, metal nanoparticles, and carbon/metal composites

Table 6.4. Recent applications of MOF-based composite materials as electrochemical sensing platforms.

MOF-based material	Electrode	Analyte	Method	Matrix	Linear range	Limit of detection	References
Pd/MIL101-NH_2-cDNA	Pd/MIL101-NH_2-cDNA/AuE	Telomerase	DPV	Human serum	5×10^2–1.62×10^7 HeLa cells/ml	11.25 HeLa cells/ml	[78]
Cu-MOFs/OMC	Cu-MOFs/OMC/GCE	Hydrazine	CV	Environmental samples	0.5–711 μM	0.35 μM	[85]
ERG-UiO-67-bpy	ERG-UiO-67-bpy/GCE	Norepinephrine Uric acid	DPV	Human urine	0.2–100 μM 0.6–100 μM	0.026 μM 0.025 μM	[86]
Ni-MOF@CNTs	Ni-MOF@CNTs/GCE	BPA	DPV	Receipt, movie tickets and water buckets	0.001–1 μM	0.35 nM	[87]
Ni-MOF/AuNPs/CNTs/PVA	Ni-MOF/AuNPs/CNTs/PVA film electrode	HIV DNA	CV	Human serum	0.01–1 μM	0.13 nM	[108]
SWCNTs-MPsLCu-MOF	SWCNTs-MPsLCu-MOF/GCE	Glucose	AMP	Human saliva	0.02–80 μM	1.72 nM	[88]
Sn-MOF@CNT	Sn-MOF@CNT/AuE	H_2O_2	CV	Water samples	0.22–500 μM	4.7×10^{-3} μM	[109]
ERGO/CNTs/AuNPs@Ce-MOF	RGO/CNTs/AuNPs@Ce-MOFs/GCE	CC HQ	CV	Human serum	3–190 μM 3–200 μM	0.73 μM 0.44 μM	[89]
Cu@UiO-67-BPY/GO	Cu@ UiO-67-BPY/ERG	Nitrite Histamine	DPV	Tap and river water	10–6000 μM 0–100 μM	1.2 μM 0.595 μM	[110]
Cu-MOF/SNDGr	Cu-MOF/SNDGr/PGE	STLHC	DPV	Tablet and human serum	0.05–2.67 μM	0.038 μM	[111]

g-C_3N_4/Cu-DTO MOF	g-C_3N_4/Cu-DTO MOF/GCE	BPSIP	DPV	River water sample	0.04–0.1, 0.1–1.1 μM	0.02 μM	[112]
XOD@Cu-MOF/SA	XOD@Cu-MOF/SA/GCE	Hypoxanthine Xanthine	DPV	Fish samples	0.01–10 μM 0.01–10 μM	0.0023 μM 0.0064 μM	[92]
sDNA-Cu-MOF-AuNPs	sDNA-Cu-MOF/GO@Au/GCE	Hg^{2+}	DPV	Diary products	0.10 aM–100 nM	0.001 aM	[113]
Fe_3O_4@UiO-66/Cu@Au	Fe_3O_4@UiO-66/Cu@Au/SPGE	Cardiac troponin I	DPV	Human serum	0.05–100 ng ml^{-1}	16 pg ml^{-1}	[114]

Abbreviations: GCE: glassy carbon electrode; SPE: Screen-printed electrode; DPV: differential pulse voltammetry; CV: cyclic voltammetry; AuE: gold electrode; OMC: ordered mesoporous carbon; ERG: electrochemically reduced to grapheme; bpy: 2,2′-bipyridyl-5,5′-dicarboxylate; CNTs: carbon nanotubes; BPA: bisphenol A; SWCNTs: single-walled carbon nanotubes; AMP: Amperometry; MPsL: microparticles like, CC: catechol; HQ: hydroquinone; SA: sodium alginate; PGE: pencil graphite electrode; SNDGr: sulfur and nitrogen co-doped graphene; STLHC: sertraline hydrochloride; BPSIP: 4-(4-isopropoxy-benzenesulfonyl)-phenol; DTO: dithiooxamide; SPGE: screen-printed gold electrode.

[47, 94, 95]. Unique properties such as great surface area and high porosity of these MOF-derived materials are related to pristine MOFs. Moreover, the obtained MOF-derived carbons/metals possess superiorities such as high stability and conductivity. Integrating these properties together has led to providing opportunities for use in the electrochemical sensing field [96, 97]. Figure 6.8 shows some available synthesis routes for MOF-derived materials as electrochemical sensors.

As an example of the electrochemical sensing application of MOF-derived materials, the research recently done by Rezapasand and co-workers can be mentioned [98]. They prepared MOFs-derived Zn–Ni–P through a simple phosphorization of Zn–Ni–MOF precursor in N_2 atmosphere. They obtained Zn–Ni–P electrode materials exhibited superior electrocatalytic activity toward isoprenaline (IPN) oxidation (figure 6.9). Cao *et al* [99], fabricated the MOF-derived Co_3O_4@C/graphene nanoplates (GNP) composite using a two-steps procedure involving a solvothermal process for Co-MOF precursor and GNPs synthesis and the next step involving calcination treatment under N_2 atmosphere. The prepared Co_3O_4@C/GNP indicated high sensitivity for electrochemical sensing of some hazardous photographic developing agents including catechol, hydroquinone, and metal with good selectivity and stability, and excellent reproducibility. This applicability rose from important advantages such as a large surface area and a fast electron transfer rate of GNPs and the strong electrical catalytic ability of the Co_3O_4@C hybrid. In another work, Saravanakumar *et al* [100], for the first time, developed a new electrochemical sensor based on carbonized Cu-4,4′-bipyridine-trimesic interlinked MOF (CBT-MOF) and applied it for the effective electrochemical detection of caffeine. A simple solvothermal method was used to preparation of CBT-MOF. And then carbonization was performed at various temperatures. The best electrocatalytic

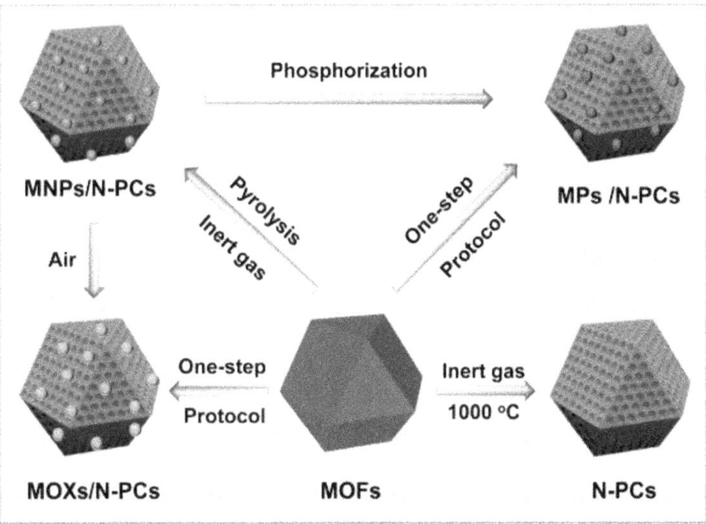

Figure 6.8. Synthesis routes for MOF-derived materials as the electrochemical sensor platforms. Reprinted from [9], copyright (2020) with permission from Elsevier.

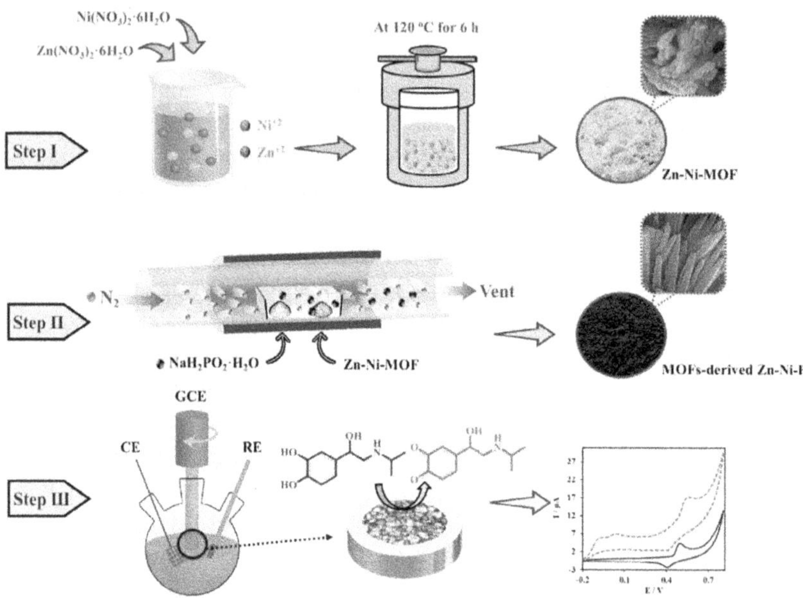

Figure 6.9. Illustration of MOFs-derived Zn–Ni–P preparation and its electrocatalytic activity. Reprinted from [98], copyright (2022) with permission from Elsevier.

activity was obtained with carbonization at 500 °C, indicating enhanced conductivity of the CuO nanoparticles and subsequently enhanced electron transfer kinetics between the CuO nanoparticles and caffeine at this temperature. More applications in this field have been summarized in a review paper reported by Goncalves and co-workers [101]. The most recently published research is given in table 6.5.

6.4 Conclusion

The development of highly selective and ultrasensitive analytical sensors for the determination of various analytes is of great importance. This is especially due to the new demands and needs that have arisen in medical diagnoses, environmental analysis, food analysis, and monitoring the production of some products. Compared to other sensing methods, electrochemical methods have been promising in the recent two decades because they are inexpensive, simple, reliable, good sensitive, and able to analyze element speciation. According to recent demands, most electrochemical techniques have applied modified electrodes because of their improved electrocatalytic performance. MOFs, as emerging porous crystalline materials, have attracted growing attention on account of their potential applications in sensing fields. The utilization of MOFs, especially as electrochemical sensing platforms is one of their most significant and potential applications. Due to their unique characteristics, including larger surface area, high porosity, adjustable pore size, chemical functionality, and good electrochemical activity, MOFs and their derivatives can be employed as promising platforms for signal amplification in electrochemical sensing. Accordingly, MOFs can present enzyme-like catalytic activity for

Table 6.5. Recent applications of MOF-derivatives-based electrochemical sensing platforms.

MOF-based material	Electrode	Analyte	Method	Matrix	Linear range	Limit of detection	References
Zn–Ni–P	Zn–Ni–P/GCE	IPN	DPV		0.2–5000 μM, 5000–14 000 μM	0.06 μM	[98]
Co_3O_4@C/GNP	Co_3O_4@C/GNP/GCE	MT HQ CC	DPV	Water samples	0.01–80.0 μM	5.1 nM 14.7 nM 169 nM	[99]
CuO NPs	CuO NPs/GCE	Caffeine	AMP	Coffee samples	1.0–10.0 μM	0.019 μM	[100]
N-doped carbon coated CoNi alloy	N-doped carbon coated CoNi alloy/GCE	4-AP ACOP	CV	Human urine and serum	0.05–60, 60–250 μM 0.05–40, 40–150 μM	5.2 nM 3.8 nM	[115]
CN-NiO	CN-NiO/GCE	Glucose	AMP	Stock solution	0.5–3000 μM	0.5 μM	[116]
Ti_3C_2/N-PC	Ti_3C_2/N-PC/GCE	HQ CC	DPV	Industrial wastewate	0.5–150 μM	4.8 nM 3.1 nM	[117]
Co_3O_4/NCNTs	Co_3O_4/NCNTs/GCE	H_2O_2 Glucose	AMP	Stock solution	5.0–11 000 μM	1.0 μM 5.0 μM	[118]
NiO@ZnO sphere	NiO@ZnO/GCE	Isoniazid	DPV	Tablets and mouse serum	0.8–800 μM	0.25 μM	[119]
CuOx@mC	CuOx@mC/GCE	Glyphosate	DPV	Fruits samples	1.0×10^9–1.0×10^2 μM	7.69×10^{-10} μM	[120]

Abbreviations: GCE: glassy carbon electrode; SPE: Screen-printed electrode; DPV: differential pulse voltammetry; CV: cyclic voltammetry; AMP: Amperometry; IPN: Isoprenaline; GNP: graphene nanoplates; 4-AP: 4-aminophenol; ACOP: acetaminophen; CN-NiO: carbon and nitrogen atomic doped NiO nanocomposite; N-PC: nitrogen-doped porous carbon; NCNTs: N-doped carbon nanotubes; mC: mesoporous carbon.

a range of substrates, which allows them to be used as nanozymes for the fabrication of various types of sensors. In short, the electrochemical sensing capability of MOFs comes from the following features: (i) owing to the large surface areas, tunable porosity, and cavity structures MOFs can be used to coat materials for electrodes; (ii) The highly porous structure, as well as large surface area of MOFs, provides the concentration of analytes and their mass transfer resulting in amplifying signal response; (iii) the specific size and shape of the existing cavities in MOFs make the MOF-based electrochemical sensors more selective towards specific analytes. With the achievements of sensitivity enhancing strategies, including surface modification of MOFs, and incorporation of various kinds of (nano)materials such as metal nanoparticles, graphene, and carbon nanotubes, MOFs-based electrochemical sensors present very promising applications. The studies reported in this chapter demonstrated clearly the applicability of this sensing strategy for the detection of analytes in complex matrices such as biological fluids, pharmaceutical drugs, and environmental and food samples. This overview may provide innovative concepts for future research on MOF-based materials in this rapidly extending field.

References

[1] Mahato K and Wang J 2021 Electrochemical sensors: from the bench to the skin *Sens. Actuators* B **344** 130178

[2] Liu S *et al* 2020 Metal-organic frameworks and their derivatives as signal amplification elements for electrochemical sensing *Coord. Chem. Rev.* **424** 213520

[3] Chuang C H and Kung C W 2020 Metal–organic frameworks toward electrochemical sensors: challenges and opportunities *Electroanalysis* **32** 1885–95

[4] Shukla P *et al* 2022 Metal-organic-frames (MOFs) based electrochemical sensors for sensing heavy metal contaminated liquid effluents: a review *Nanoarchitectonics* **3** 46–60

[5] Zhao F *et al* 2019 Metal-organic frameworks-based electrochemical sensors and biosensors *Int. J. Electrochem. Sci.* **14** 5287–304

[6] Azzouz A *et al* 2019 Nanomaterial-based electrochemical sensors for the detection of neurochemicals in biological matrices *TrAC, Trends Anal. Chem.* **110** 15–34

[7] Wang Y *et al* 2014 Construction of an electrochemical sensor based on amino-functionalized metal-organic frameworks for differential pulse anodic stripping voltammetric determination of lead *Talanta* **129** 100–5

[8] Jiao L *et al* 2019 Metal–organic frameworks: structures and functional applications *Mater. Today* **27** 43–68

[9] Liu C-S, Li J and Pang H 2020 Metal-organic framework-based materials as an emerging platform for advanced electrochemical sensing *Coord. Chem. Rev.* **410** 213222

[10] Jiang Q *et al* 2022 Recent progresses of metal-organic framework-based materials in electrochemical energy storage *Mater. Today Sustain.* **19** 100174

[11] Wei Y-S *et al* 2020 Metal–organic framework-based catalysts with single metal sites *Chem. Rev.* **120** 12089–174

[12] Yi F-Y *et al* 2016 Metal-organic framework-based chemical sensors *ChemPlusChem.* **81** 675–90

[13] Zhao Y *et al* 2021 Metal-organic framework based membranes for selective separation of target ions *J. Membr. Sci.* **634** 119407

[14] Zhang X *et al* 2021 Carbon-based MOF derivatives: emerging efficient electromagnetic wave absorption agents *Nanomicro Lett.* **13** 1–31

[15] Guillerm V *et al* 2014 A supermolecular building approach for the design and construction of metal–organic frameworks *Chem. Soc. Rev.* **43** 6141–72

[16] Lu W *et al* 2014 Tuning the structure and function of metal–organic frameworks via linker design *Chem. Soc. Rev.* **43** 5561–93

[17] Lyndon R *et al* 2020 Tuning the structures of metal–organic frameworks via a mixed-linker strategy for ethylene/ethane kinetic separation *Chem. Mater.* **32** 3715–22

[18] Zhang J-P *et al* 2018 Controlling flexibility of metal–organic frameworks *Natl. Sci. Rev.* **5** 907–19

[19] Cai G *et al* 2021 Metal–organic framework-based hierarchically porous materials: synthesis and applications *Chem. Rev.* **121** 12278–326

[20] Ding M, Cai X and Jiang H-L 2019 Improving MOF stability: approaches and applications *Chem. Sci.* **10** 10209–30

[21] Bosch M, Zhang M and Zhou H-C 2014 Increasing the stability of metal-organic frameworks *Adv. Chem.* **2014** 1155

[22] Wang Z *et al* 2021 The controllable synthesis of urchin-shaped hierarchical superstructure MOFs with high catalytic activity and stability *Chem. Commun.* **57** 8758–61

[23] Lv X-L *et al* 2019 Ligand rigidification for enhancing the stability of metal–organic frameworks *J. Am. Chem. Soc.* **141** 10283–93

[24] Healy C *et al* 2020 The thermal stability of metal-organic frameworks *Coord. Chem. Rev.* **419** 213388

[25] Ding M and Jiang H-L 2021 Improving water stability of metal–organic frameworks by a general surface hydrophobic polymerization *CCS Chem.* **3** 2740–8

[26] Nguyen J G and Cohen S M 2010 Moisture-resistant and superhydrophobic metal–organic frameworks obtained via postsynthetic modification *J. Am. Chem. Soc.* **132** 4560–1

[27] Yang S J and Park C R 2012 Preparation of highly moisture-resistant black-colored metal organic frameworks *Adv. Mater.* **24** 4010–3

[28] Zhang W *et al* 2014 A facile and general coating approach to moisture/water-resistant metal–organic frameworks with intact porosity *J. Am. Chem. Soc.* **136** 16978–81

[29] Shih Y H *et al* 2017 A simple approach to enhance the water stability of a metal-organic framework *Chem. Eur. J.* **23** 42–6

[30] Ding N *et al* 2016 Partitioning MOF-5 into confined and hydrophobic compartments for carbon capture under humid conditions *J. Am. Chem. Soc.* **138** 10100–3

[31] Zhu X-W, Zhou X-P and Li D 2016 Exceptionally water stable heterometallic gyroidal MOFs: tuning the porosity and hydrophobicity by doping metal ions *Chem. Commun.* **52** 6513–6

[32] Feng D *et al* 2012 Zirconium-metalloporphyrin PCN-222: mesoporous metal–organic frameworks with ultrahigh stability as biomimetic catalysts *Angew. Chem. Int. Ed.* **51** 10307–10

[33] Lu X-F *et al* 2016 An alkaline-stable, metal hydroxide mimicking metal–organic framework for efficient electrocatalytic oxygen evolution *J. Am. Chem. Soc.* **138** 8336–9

[34] Dong J *et al* 2015 Ultrastrong alkali-resisting lanthanide-zeolites assembled by [Ln60] nanocages *J. Am. Chem. Soc.* **137** 15988–91

[35] Moosavi S M *et al* 2018 Improving the mechanical stability of metal–organic frameworks using chemical caryatids *ACS Cent. Sci.* **4** 832–9

[36] Wang H, Lustig W P and Li J 2018 Sensing and capture of toxic and hazardous gases and vapors by metal–organic frameworks *Chem. Soc. Rev.* **47** 4729–56

[37] Qiu Q *et al* 2019 Recent advances in the rational synthesis and sensing applications of metal-organic framework biocomposites *Coord. Chem. Rev.* **387** 60–78

[38] Li X *et al* 2019 Water contaminant elimination based on metal–organic frameworks and perspective on their industrial applications *ACS Sustain. Chem. Eng.* **7** 4548–63

[39] Cheng W *et al* 2021 Applications of metal-organic framework (MOF)-based sensors for food safety: enhancing mechanisms and recent advances *Trends Food Sci. Technol.* **112** 268–82

[40] Yang G L *et al* 2021 Applications of MOFs as luminescent sensors for environmental pollutants *Small* **17** 2005327

[41] Fang X, Zong B and Mao S 2018 Metal–organic framework-based sensors for environmental contaminant sensing *Nanomicro Lett.* **10** 1–19

[42] Chen W and Wu C 2018 Synthesis, functionalization, and applications of metal–organic frameworks in biomedicine *Dalton Trans.* **47** 2114–33

[43] Al Sharabati M, Sabouni R and Husseini G A 2022 Biomedical applications of metal–organic frameworks for disease diagnosis and drug delivery: a review *Nanomaterials* **12** 277

[44] Xu W *et al* 2022 Engineering metal-organic framework-based nanozymes for enhanced biosensing *Curr. Anal. Chem.* **18** 739–52

[45] Liu L *et al* 2018 The applications of metal–organic frameworks in electrochemical sensors *ChemElectroChem.* **5** 6–19

[46] Tajik S *et al* 2021 Performance of metal–organic frameworks in the electrochemical sensing of environmental pollutants *J. Mater. Chem.* A **9** 8195–220

[47] Gao L-L and Gao E-Q 2021 Metal–organic frameworks for electrochemical sensors of neurotransmitters *Coord. Chem. Rev.* **434** 213784

[48] Kempahanumakkagari S *et al* 2018 Metal–organic framework composites as electrocatalysts for electrochemical sensing applications *Coord. Chem. Rev.* **357** 105–29

[49] Kajal N *et al* 2022 Metal organic frameworks for electrochemical sensor applications: a review *Environ. Res.* **204** 112320

[50] Çorman M *et al* 2022 Metal-organic frameworks as an alternative smart sensing platform for designing molecularly imprinted electrochemical sensors *TrAC, Trends Anal. Chem.* **150** 116573

[51] Garg N, Deep A and Sharma A L 2022 Recent trends and advances in porous metal-organic framework nanostructures for the electrochemical and optical sensing of heavy metals in water *Crit. Rev. Anal. Chem.* 1–25

[52] Lin Y, Huang Y and Chen X 2022 Recent advances in metal-organic frameworks for biomacromolecule sensing *Chemosensors* **10** 412

[53] Abdelkareem M A *et al* 2022 High-performance effective metal-organic frameworks for electrochemical applications *J. Sci.: Adv. Mater. Devices* **7** 100465

[54] Dourandish Z *et al* 2022 A comprehensive review of metal–organic framework: synthesis, characterization, and investigation of their application in electrochemical biosensors for biomedical analysis *Sensors* **22** 2238

[55] Ma T *et al* 2020 Application of MOF-based materials in electrochemical sensing *Dalton Trans.* **49** 17121–9

[56] Devaraj M *et al* 2021 Metal organic framework based nanomaterials for electrochemical sensing of toxic heavy metal ions: progress and their prospects *J. Electrochem. Soc.* **168** 037513

[57] Lopa N S *et al* 2018 A base-stable metal-organic framework for sensitive and non-enzymatic electrochemical detection of hydrogen peroxide *Electrochim. Acta* **274** 49–56

[58] Domenech A *et al* 2007 Electrochemistry of metal–organic frameworks: a description from the voltammetry of microparticles approach *J. Phys. Chem.* C **111** 13701–11

[59] Li C *et al* 2017 Boosted sensor performance by surface modification of bifunctional rht-type metal–organic framework with nanosized electrochemically reduced graphene oxide *ACS Appl. Mater. Interfaces* **9** 2984–94

[60] Yang J, Zhao F and Zeng B 2015 One-step synthesis of a copper-based metal–organic framework–graphene nanocomposite with enhanced electrocatalytic activity *RSC Adv.* **5** 22060–5

[61] Shu Y *et al* 2017 Ni and NiO nanoparticles decorated metal–organic framework nanosheets: facile synthesis and high-performance nonenzymatic glucose detection in human serum *ACS Appl. Mater. Interfaces* **9** 22342–9

[62] Hosseini H *et al* 2013 A novel electrochemical sensor based on metal-organic framework for electro-catalytic oxidation of L-cysteine *Biosens. Bioelectron.* **42** 426–9

[63] Peng Z *et al* 2016 A novel electrochemical sensor of tryptophan based on silver nanoparticles/metal–organic framework composite modified glassy carbon electrode *RSC Adv.* **6** 13742–8

[64] Ling W *et al* 2018 Materials and techniques for implantable nutrient sensing using flexible sensors integrated with metal–organic frameworks *Adv. Mater.* **30** 1800917

[65] Ling W *et al* 2019 A novel Cu-metal-organic framework with two-dimensional layered topology for electrochemical detection using flexible sensors *Nanotechnology* **30** 424002

[66] Wu X-Q *et al* 2019 Design of a Zn-MOF biosensor via a ligand 'lock' for the recognition and distinction of S-containing amino acids *Chem. Commun.* **55** 4059–62

[67] Zejli H, Goud K Y and Marty J L 2018 Label free aptasensor for ochratoxin A detection using polythiophene-3-carboxylic acid *Talanta* **185** 513–9

[68] Mahakul P C *et al* 2018 Investigation of optical and electrical properties of MWCNT/rGO/poly (3-hexylthiophene) ternary composites *J. Mater. Sci.* **53** 8151–60

[69] Portorreal-Bottier A *et al* 2022 Enzyme like activity of cobalt-MOF nanosheets for hydrogen peroxide electrochemical sensing *Sens. Actuators* B **368** 132129

[70] Majewski M *et al* 2018 NanoMOFs: little crystallites for substantial applications *J. Mater. Chem.* A **6** 7338–0

[71] Wang S *et al* 2022 Synthesis strategies and electrochemical research progress of nano/microscale metal–organic frameworks *Small Sci.* **2** 2200042

[72] Arul P and John S A 2018 Size controlled synthesis of Ni-MOF using polyvinylpyrrolidone: new electrode material for the trace level determination of nitrobenzene *J. Electroanal. Chem.* **829** 168–76

[73] Zhang X *et al* 2019 Recent progress in the design fabrication of metal-organic frameworks-based nanozymes and their applications to sensing and cancer therapy *Biosens. Bioelectron.* **137** 178–98

[74] Tong P *et al* 2020 Research progress on metal-organic framework composites in chemical sensors *Crit. Rev. Anal. Chem.* **50** 376–92

[75] Rasheed T and Rizwan K 2022 Metal-organic frameworks based hybrid nanocomposites as state-of-the-art analytical tools for electrochemical sensing applications *Biosens. Bioelectron.* **199** 113867

[76] Zhou J *et al* 2022 Electrochemical determination of levofloxacin with a Cu–metal–organic framework derivative electrode *J. Mater. Sci., Mater. Electron.* **33** 9941–50
[77] Razavi S A A and Morsali A 2019 Linker functionalized metal-organic frameworks *Coord. Chem. Rev.* **399** 213023
[78] Wang L *et al* 2019 Fabrication of amine-functionalized metal-organic frameworks with embedded palladium nanoparticles for highly sensitive electrochemical detection of telomerase activity *Sens. Actuators* B **278** 133–9
[79] Gupta R *et al* 2022 Recent progress in aptamer-functionalized metal-organic frameworks-based optical and electrochemical sensors for detection of mycotoxins *Crit. Rev. Anal. Chem.* 1–22
[80] Rezki M *et al* 2021 Amine-functionalized Cu-MOF nanospheres towards label-free hepatitis B surface antigen electrochemical immunosensors *J. Mater. Chem.* B **9** 5711–21
[81] Gao L-L *et al* 2019 Graphite paste electrodes modified with a sulfo-functionalized metal-organic framework (type MIL-101) for voltammetric sensing of dopamine *Microchim. Acta* **186** 1–9
[82] Cassani M C *et al* 2021 A Cu (II)-MOF based on a propargyl carbamate-functionalized isophthalate ligand as nitrite electrochemical sensor *Sensors* **21** 4922
[83] Su Z *et al* 2021 Electrochemically-assisted deposition of toluidine blue-functionalized metal-organic framework films for electrochemical immunosensing of Indole-3-acetic acid *J. Electroanal. Chem.* **880** 114855
[84] Karimzadeh Z *et al* 2022 Aptamer-functionalized metal organic frameworks as an emerging nanoprobe in the food safety field: promising development opportunities and translational challenges *TrAC, Trends Anal. Chem.* **152** 116622
[85] Wang L *et al* 2018 Facile synthesis of metal-organic frameworks/ordered mesoporous carbon composites with enhanced electrocatalytic ability for hydrazine *J. Colloid Interface Sci.* **512** 127–33
[86] Gao L-L *et al* 2019 Synergetic effects between a bipyridyl-functionalized metal-organic framework and graphene for sensitive electrochemical detection of norepinephrine *J. Electrochem. Soc.* **166** B328
[87] Xu C *et al* 2020 Unique 3D heterostructures assembled by quasi-2D Ni-MOF and CNTs for ultrasensitive electrochemical sensing of bisphenol A *Sens. Actuators* B **310** 127885
[88] Arul P *et al* 2020 Tunable electrochemical synthesis of 3D nucleated microparticles like Cu-BTC MOF-carbon nanotubes composite: enzyme free ultrasensitive determination of glucose in a complex biological fluid *Electrochim. Acta* **354** 136673
[89] Du Y *et al* 2022 Signal synergistic amplification strategy based on functionalized CeMOFs for highly sensitive electrochemical detection of phenolic isomers *Microchem. J.* **177** 107285
[90] An H *et al* 2019 Incorporation of biomolecules in metal-organic frameworks for advanced applications *Coord. Chem. Rev.* **384** 90–106
[91] Fu X, Ding B and D'Alessandro D 2023 Fabrication strategies for metal-organic framework electrochemical biosensors and their applications *Coord. Chem. Rev.* **475** 214814
[92] Wang Z *et al* 2019 Electrochemical biosensing of chilled seafood freshness by xanthine oxidase immobilized on copper-based metal–organic framework nanofiber film *Food Anal. Methods* **12** 1715–24
[93] Lavanya J *et al* 2022 Metal-organic frameworks composites for electrochemical detection of heavy metal ions in aqueous medium *J. Electrochem. Soc.* **169** 047525

[94] Wu H B and Lou X W 2017 Metal-organic frameworks and their derived materials for electrochemical energy storage and conversion: promises and challenges *Sci. Adv.* **3** eaap9252
[95] Chen T *et al* 2020 Recent progress on metal–organic framework-derived materials for sodium-ion battery anodes *Inorg. Chem. Front.* **7** 567–82
[96] Xu Q *et al* 2019 MOF-derived N-doped nanoporous carbon framework embedded with Pt NPs for sensitive monitoring of endogenous dopamine release *J. Electroanal. Chem.* **839** 247–55
[97] Muthurasu A and Kim H Y 2019 Fabrication of hierarchically structured MOF-Co_3O_4 on well-aligned CuO nanowire with an enhanced electrocatalytic property *Electroanalysis* **31** 966–74
[98] Rezapasand S *et al* 2022 Metal-organic frameworks-derived Zn–Ni–P nanostructures as high performance electrode materials for electrochemical sensing *J. Electroanal. Chem.* 116441
[99] Cao M *et al* 2022 Robust and selective electrochemical sensing of hazardous photographic developing agents using a MOF-derived 3D porous flower-like Co_3O_4@ C/graphene nanoplate composite *Electrochim. Acta* **409** 139967
[100] Saravanakumar V *et al* 2022 Cu-MOF derived CuO nanoparticle decorated amorphous carbon as an electrochemical platform for the sensing of caffeine in real samples *J. Taiwan Inst. Chem. Eng.* **133** 104248
[101] Goncalves J M *et al* 2021 Recent trends and perspectives in electrochemical sensors based on MOF-derived materials *J. Mater. Chem.* C **9** 8718–45
[102] Shi E *et al* 2019 The incorporation of bismuth (III) into metal-organic frameworks for electrochemical detection of trace cadmium (II) and lead (II) *Microchim. Acta* **186** 1–11
[103] Ye W *et al* 2020 Electrochemical detection of trace heavy metal ions using a Ln-MOF modified glass carbon electrode *J. Solid State Chem.* **281** 121032
[104] Chen Y *et al* 2019 A dual-response biosensor for electrochemical and glucometer detection of DNA methyltransferase activity based on functionalized metal-organic framework amplification *Biosens. Bioelectron.* **134** 117–22
[105] Wang Z *et al* 2018 NH_2-Ni-MOF electrocatalysts with tunable size/morphology for ultrasensitive C-reactive protein detection via an aptamer binding induced DNA walker–antibody sandwich assay *J. Mater. Chem.* B **6** 2426–31
[106] Xu Y *et al* 2022 An ultra-sensitive dual-signal ratiometric electrochemical aptasensor based on functionalized MOFs for detection of HER2 *Bioelectrochemistry* **148** 108272
[107] Wang F-F *et al* 2022 A sulfur-containing capsule-based metal-organic electrochemical sensor for super-sensitive capture and detection of multiple heavy-metal ions *Chem. Eng. J.* **438** 135639
[108] Lu Q *et al* 2021 Flexible paper-based Ni-MOF composite/AuNPs/CNTs film electrode for HIV DNA detection *Biosens. Bioelectron.* **184** 113229
[109] Rani S *et al* 2020 Sn-MOF@ CNT nanocomposite: an efficient electrochemical sensor for detection of hydrogen peroxide *Environ. Res.* **191** 110005
[110] Zhang H-J *et al* 2022 A novel copper-functionalized MOF modified composite electrode for high-efficiency detection of nitrite and histamine *J. Electrochem. Soc.* **169** 077511
[111] Habibi B *et al* 2021 Direct electrochemical synthesis of the copper based metal-organic framework on/in the heteroatoms doped graphene/pencil graphite electrode: highly sensitive and selective electrochemical sensor for sertraline hydrochloride *J. Electroanal. Chem.* **888** 115210

[112] Singh A K *et al* 2021 Development of g-C3N4/Cu-DTO MOF nanocomposite based electrochemical sensor towards sensitive determination of an endocrine disruptor BPSIP *J. Electroanal. Chem.* **887** 115170
[113] Zhang X *et al* 2020 Electrochemical DNA sensor for inorganic mercury (II) ion at attomolar level in dairy product using Cu (II)-anchored metal-organic framework as mimetic catalyst *Chem. Eng. J.* **383** 123182
[114] Sun D *et al* 2019 Electrochemical dual-aptamer-based biosensor for nonenzymatic detection of cardiac troponin I by nanohybrid electrocatalysts labeling combined with DNA nanotetrahedron structure *Biosens. Bioelectron.* **134** 49–56
[115] Niu X, Bo X and Guo L 2021 Ultrasensitive simultaneous voltammetric determination of 4-aminophenol and acetaminophen based on bimetallic MOF-derived nitrogen-doped carbon coated CoNi alloy *Anal. Chim. Acta* **1145** 37–45
[116] Jia S, Wang Q and Wang S 2021 Ni-MOF/PANI-derived CN-doped NiO nanocomposites for high sensitive nonenzymic electrochemical detection *J. Inorg. Organomet. Polym. Mater.* **31** 865–74
[117] Huang R *et al* 2020 A strategy for effective electrochemical detection of hydroquinone and catechol: decoration of alkalization-intercalated Ti_3C_2 with MOF-derived N-doped porous carbon *Sens. Actuators* B **320** 128386
[118] Qin Y *et al* 2020 MOF derived Co3O4/N-doped carbon nanotubes hybrids as efficient catalysts for sensitive detection of H_2O_2 and glucose *Chin. Chem. Lett.* **31** 774–8
[119] Wang J *et al* 2020 An electrochemical sensor based on MOF-derived NiO@ ZnO hollow microspheres for isoniazid determination *Microchim. Acta* **187** 1–8
[120] Gu C *et al* 2020 Ultrasensitive non-enzymatic pesticide electrochemical sensor based on HKUST-1-derived copper oxide@ mesoporous carbon composite *Sens. Actuators* B **305** 127478

IOP Publishing

Real-Time Applications of Advanced Electrochemical Sensing Devices

Jamballi G Manjunatha

Chapter 7

Electrochemical detection of toxic metal ions in food products

Milan Z Momčilović

Food contamination by metals is one of the main topics in food safety issues. Heavy metals present in foodstuff are always a huge problem due to their toxic effects on living beings. In this chapter, only the levels of lead, cadmium, mercury and copper in different types of solid and liquid food products will be addressed. For determination of heavy metal content in the samples of some vegetables, fruit, cereals, seafood, and alcoholic drinks, the role of electrochemical techniques mostly based on anodic stripping voltammetry will be illustrated through many useful examples. Rapid, accurate, sensitive, and selective electroanalytical systems to measure the undesirable metal levels in food are herein represented as a counter-weight to robust and expensive instrumental techniques.

7.1 Introduction

Metals are naturally occurring elements which can be found in a variety of food products in an excessive extent. Contamination of foods by metals has numerous sources, mostly of natural or anthropogenic origin. Human activities which involve industrial, manufacturing, and agricultural processes related to industrial plants, mining, ore smelting, landfill and treatment of municipal waste, pesticide and fertilizer use, are sometimes responsible for redistribution and concentrating metals in areas that are not naturally metals-enriched. From these areas, metals can enter the food supply through the air, water, and soil.

High quantities of heavy metals present in groundwater used for irrigation during food production may be a significant source of food pollution. Another example of direct source of pollution is combustion fumes coming from the petrochemical complex which cause metal deposition on the soil and plants [1]. Pollution of the soil from which foods are produced could be a source of metals in livestock feeding or

doi:10.1088/978-0-7503-5377-9ch7

eventually the human diet which stands at the end of the food chain. Naturally occurring metals in sea water, especially mercury, accumulate in fish tissue as they feed and absorb through the gills as they swim. Rice can absorb arsenic from soil treated with pesticides or from irrigation water. Exposure of humans to metals comes from different foods they eat and has cumulative effect. Hence, by combining the foods, harmful metals from individual food sources can reach a level of concern.

Metal contaminants can occur at every stage of food production. Mechanisms of food contamination, types of metals, and their final content in the food products depend on many factors. Some of the metals, such as sodium, potassium, magnesium, calcium, zinc, cobalt, copper, iron, manganese, and molybdenum, are essential elements for humans and human body must have appropriate amounts of them [2]. In the other hand, at certain concentrations they may be toxic in some of their forms. However, there are metals which are toxic even in tiny amounts and can be a major health problem. The intake of heavy metals is especially harmful because they are not chemically or biologically degradable and they often tend to accumulate in the tissue of the living organisms.

Heavy metals such as lead, cadmium, arsenic, chromium, and mercury when present in foodstuffs are a potential hazard. Contamination by mercury and cadmium is an urgent health issue since these metals are stable in contaminated sites and express the complex mechanism of biological toxicity. Once absorbed, they are accumulated in the body and can lead to the emergence of certain diseases [3]. Metallic arsenic is an ecological toxin naturally present in all soil types while food grown on soil may contain arsenic absorbed from it. The same applies to lead, which interferes with different physiological pathways in plants and unlike other metals it has no biological role. It adversely affects the photosynthetic and chlorophyll metabolism of the plant and is retained in the plant structure. If the plant is edible, this might be one of the sources of lead in food along with pollution by air and water during cultivation [4].

Toxic effects of some heavy metals are very well known. For example, lead can harm the liver, cause anemia, pulmonary and cardiovascular dysfunction, and reduced pulmonary function [5]. Cadmium can cause degenerative bone disease, kidney dysfunction, liver and lung damage, metabolic syndromes associated with Zn and Cu, and cancer [5]. Arsenic is responsible for cardiovascular dysfunction, skin and hair conditions and liver damage while exposure to mercury through food leads to hepatotoxicity and dysfunction of the renal and central nervous system. All these elements also cause severe gastrointestinal disorders [5]. One of the most widely studied mechanisms of action of toxic metals is oxidative damage due to direct generation of free radical species and depletion of antioxidant reserve [6].

Reference values for heavy metals in foods are precisely defined by health agencies. Daily maximum safe exposure levels for heavy metals are named differently by the different agencies as oral reference dose, provisional total daily intake, or minimal risk level. Those values, in one term called reference values, are for cadmium 1 $\mu g\ kg^{-1}\ day^{-1}$ (EPA 1989), for lead 0.26 $\mu g\ kg^{-1}\ day^{-1}$ (young children) and 0.16 $\mu g\ kg^{-1}\ day^{-1}$ (older children and adults) (FDA 2018), for methylmercury 0.1 $\mu g\ kg^{-1}\ day^{-1}$ (EPA 2001), etc [7].

Considering the detrimental impact of heavy metals on the human body, it is necessary to monitor their presence in food. Many instrumental techniques have been established in common food analysis for the determination of heavy metals. The typical techniques include inductively coupled plasma mass spectrometry (ICP-MS) [8], inductively coupled plasma atomic emission spectroscopy (ICP-AES) [9], x-ray fluorescence spectrometry (XRF) [10], and atomic absorption spectrometry (AAS) [11]. However, since these techniques are too time-consuming and involve expensive and sophisticated equipment with excessive cost of maintenance, their applications are complicated and impractical. Having in mind this problem, there is a constant need for quick and simple determination techniques which could be employed in monitoring of certain heavy metals in all types of food samples.

In this chapter, various recently developed electrochemical platforms will be presented with the focus in determination of lead, cadmium, mercury, and copper in the samples of food such as rice, eggs, tea, wheat, corn, soybean, banana, apple, lettuce, orange, and some vegetables. Due to its nature, liquid food is usually more prone to contamination during raw material collection, processing, and packaging. Hence, presence of some of the toxic metal in milk, juice, wine, and beer will be addressed as well.

7.2 Electrochemical approach in heavy metal detection

Stripping voltammetry is an electroanalytical technique which encompasses a wide variety of options based upon the metal deposition and dissolution phenomena occurring on the electrode surface. Despite long-term intensive research of these techniques, there is no general principle for metal stripping on a solid electrode which has been established. Although homogenous liquid amalgams of mercury were initially used as ideal electrodes, other options are nowadays employed in anodic stripping voltammetry (ASV). Various less toxic solid electrodes have been designed with the tendency to attain as many of the favorable attributes of mercury drop electrode as possible, such as retarded hydrogen evolution reaction (HER), low background currents, reproducible surface, and narrow stripping peaks [12]. As striking examples, metallic electrodes based on silver, gold, platinum, iridium, and bismuth should be kept in mind.

The bismuth film electrodes (BiFEs) give narrow, well resolved, reproducible stripping peaks, due to the ability of bismuth to form solid metal alloys with a range of metals [12]. The Bi film is typically generated by electrodepositing or co-depositing Bi with the analyte ions onto a carbon support electrode. The reason why Bi films are preferred over bulk Bi electrodes is because the film deposition/co-deposition process can be regarded as an easy, reliable way of obtaining a reproducible surface [12].

Carbon materials such as carbon nanotubes [13], carbon microspheres [14], dopped ordered mesoporous carbons [15], are along with graphene and graphite the most frequently considered in the field of commercial electrode modification for ASV since they are more inert than metals, have low background currents, and these materials can lower the values of limits of detection (LODs) [12]. multi-walled carbon nanotubes (MWCNTs) are also popular in fabrication of sensors for pesticide detection as in the case of, for instance, oxyfluorfen [13] or clomazone [16]. In combination with carbons, biological molecules such as DNA, enzymes and bacteria are also increasingly being used as modifiers due to their high specificity.

Carbon paste electrodes as electrochemical sensors are often used to determine concentrations of some compounds which might be already labeled as benchmark compounds in electroanalytical science, such as dopamine [17], epinephrine [18], vanillin [19], curcumin [20], catechol [21], phloroglucinol [22], hydroquinone [23], resorcinol [23], etc. Also, one should mention boron doped diamond (BDD) electrode as a specific type of carbon electrode with immense potential in many applications.

Screen-printed electrodes are also very popular in ASV. In this type of electrode, the reference and counter electrodes are printed alongside the working electrode, producing a compact sensor system. They are usually made from materials such as various carbons or metal oxides printed as ink and integrated into a support. Depending on the application, their surface can be further functionalized, for instance, through the addition of metal nanoparticles or metal complexing agents on the surface of the working electrode.

For solutions containing a mixture of metals, to ensure deposition of all metals, typically the potential must be made more negative than the most negative $E^{\circ\prime}$ metal. For solid electrodes, it is important to understand the voltammetric behavior of the metals of interest on the solid electrode of choice before choosing the deposition potential (E_{dep}). Furthermore, the overpotential of deposition ($E_{dep}-E^{\circ\prime}$) is also likely to strongly influence the morphology of the metal deposits formed on the surface [12]. Composition of the solution matrix and the knowledge of which metals are likely to exist in which form is of paramount importance for electrochemical analysis.

All food samples have a complex composition. Before the voltammetric analysis, certain pretreatment steps should be performed to make the metal ions 'more available' to the electrode. So, thermal, dry-ashing, microwave and ultrasound methods are the most often used in food sample pretreatment prior to any analysis while every method has its own adapted procedure.

7.3 Novel sensors applied in anodic stripping voltammetry

ASV is a two-step process which involves initial cathodic reduction of metal ion to elemental form on the electrode surface (known also as electrodeposition or pre-concentration step) at a potential more negative than the formal potential of the M^{n+}/M redox couple ($E^{\circ\prime}$) for some time, followed by anodic oxidation or dissolution of the previously deposited metal back to the form of a metal ion (stripping step). The rate of deposition is increased in order to improve the LOD. Oxidative stripping peaks are investigated in current-potential curves and correlated to metal concentrations. The potential at which the stripping peak occurs and the area under that peak depend on the chemical identity of the species and amount of metal deposited, respectively, which is the very basis of the ASV method. Different metal concentrations give different peak surface areas and different current intensities, which are used for calibration plot [12].

The simplest method of performing the stripping step in ASV is by linearly sweeping the current in the anodic direction (LSASV). However, in order to further increase sensitivity of the detection, more complex potential waveforms can be adopted as in case of square wave (SWASV) and differential pulse anodic stripping voltammetry (DPASV). The latter is often employed in metal detection. Briefly, in DPSAV, a small, increased voltage pulse is superimposed over the voltage ramp while the current is sampled just before the pulse is applied and just before the pulse ends. The difference between the two recorded currents is plotted versus the base potential. In this way, enhancement of sensitivity is achieved in contrast to normal pulse voltammetry.

7.3.1 Sensors based on GCE

The fact that graphene is a strong, stiff, and exceptionally light carbon-based material with specific electronic and physical properties due to its high conductivity is crucial in enhancing the sensitivity and electron transport properties of the modified glassy carbon electrode (GCE). Tris (2,2′-bipyridyl)ruthenium(II)/graphene/Nafion® modified GCEs can be used as the working electrode in differential pulse voltammetry (DPV) for the determination of lead, cadmium, and copper ions in mussel and oyster samples [24]. As explained by Palisoc *et al*, tris (2,2′-bipyridyl) ruthenium(II) was applied as a redox mediator while Nafion® was used as a polymer known for its anti-fouling properties in the modifying electrodes. Along with DPV, cyclic voltammetry (CV) was used to investigate the reversibility and stability of the modified electrodes. In this paper, oyster and mussel samples were harvested and analyzed during the months of April, June, and August. For pretreatment, they were subjected to white ashing and acid digestion.

DPV is a stripping technique which consists of three steps, already mentioned: accumulation or pre-concentration, deposition or quiet time equilibration, and stripping of the anaylte. In the present case, DPV was applied with the following parameters: a pulse amplitude of 0.0025 mV, interval time of 0.50 ms, pulse time of 0.25 ms, and step size of 0.002 mV. The accumulation time and deposition time were both held constant at 30 s. An anodic current peak at 1.0 V noticed in CV is the reduction potential of $[Ru\ (bpy)_3]^{2+}$. It was observed that at this potential the highest peak was obtained with the combination of 4 mg of $[Ru\ (bpy)_3]^{2+}$ and 3 mg of graphene. Atomic absorption spectroscopy (AAS) was performed for validation of the results obtained from voltammetry measurements.

The employed tris (2,2′-bipyridyl)ruthenium(II)/graphene/Nafion® modified GCE produced calibration curves with linearity from 48 to 745 ppb for lead, 49–618 ppb for cadmium, and 28–472 ppb for copper. The LODs of the best electrode for lead, cadmium, and copper ions were determined to be 48, 49, and 28 ppb, respectively. The electrode exerted fine behavior in determination of these three heavy metals by DPV in the seafood samples. Along with their presence in almost all samples, other metals such as zinc, tin, mercury, and iron can be also detected by DPV.

Direct application of electroanalytical techniques where expensive equipment is not available is reported in the case of lead, copper and cadmium determination in edible plant samples collected from the government area of Zamfara State, Nigeria [25]. Metals were determined in the samples of guinea corn, maize seeds, and millet as the main food in this region. The study was motivated by the high mortality of local women and children near the territory where processing activities of the lead ores were done. Since this could promote metal contamination of the soil, water and vegetables, there was an urgent need for investigating the sort and levels of the pollutants which may constitute health hazards. Prior to electrochemical measurements all samples need to be pretreated. Plant samples were air-dried in a dust-free environment for three weeks and then ground in a blender. The digestion of samples was carried out by sequential treatment with 65% HNO_3 and 70% $HClO_4$ at elevated temperatures until a clear solution was obtained. After cooling, probes were filtered and filled with deionized water. Linear sweep ASV technique was used with the glassy carbon as working, Ag/AgCl as reference and platinum as auxiliary electrode to determine concentrations of heavy metals. Voltammetric peaks for lead, copper and cadmium noticed at −495, −19.4 and −675 mV, respectively, were used for construction of the corresponding calibration plots. The concentrations of lead, copper and cadmium in the food samples ranged between 5.70–79.91, 11.17–41.21 and 0.00–5.74 mg kg^{-1}, respectively, which is above the FAO/WHO limits.

Magnetic nanoparticles can contribute to handy modification of working electrodes. In a work of Wu *et al*, Fe_3O_4 nanoparticles, Fe_3O_4/MWCNTs and Fe_3O_4/fluorinated MWCNTs nanocomposites-modified electrodes were constructed and used for the simultaneous determination of ions of cadmium, lead, copper, and mercury in soybean by the aid of SWASV [26]. Inductively coupled plasma mass spectroscopy and atomic fluorescence spectrometry (AFS) were employed as background techniques used for the purpose of validation. SWASV was performed in a 0.1 M acetate buffer solution at pH 5.0. The deposition potential of −1.2 V for 180 s was applied to adsorb metal ions under stirring. Then, the SWASV responses were recorded between −1.0 and 0.6 V with a step potential of 5 mV, amplitude of 0.02 V, and frequency of 25 Hz. A desorption potential of 1.0 V for 210 s was performed to remove the residual metals.

The sensitivity of Fe_3O_4/F-MWCNTs was higher than that of Fe_3O_4/MWCNTs or Fe_3O_4. This Fe_3O_4/F-MWCNTs sensor exhibited excellent performance with regards to its selectivity, stability, and analysis reproducibility. It showed the sensitivity of 108.79, 125.91, 160.85, and 312.65 $\mu A\ mM^{-1}\ cm^{-2}$ toward cadmium, lead, copper, and mercury ions, respectively, which was obviously higher than that of Fe_3O_4/MWCNTs and Fe_3O_4. Additionally, the Fe_3O_4/F-MWCNTs sensor exhibited the wider linear detection ranges of 0.5–30 μM for cadmium, lead, and copper, and 0.5–20 μM for mercury ions. The LODs of the Fe_3O_4/F-MWCNTs sensor were 0.05, 0.08, 0.02, and 0.05 nM (signal-to-noise ratio (SNR) of 3) for cadmium, lead, copper, and mercury ions, respectively.

A novel B and N co-doped porous carbon material (named BCN) derived from a metal-organic framework could be successfully used for cadmium and lead ions determination in vegetable samples named *Beta vulgaris var.* cicla L [27]. It was shown that the surface structure and N content of C ZIF-8 were optimized by combining B and N and thereby become responsible for introducing more effective binding sites and accelerating the mass-transfer rate of target ions. After optimizing the parameters, the sensor represented as BCN-Nafion/GCE could accurately distinguish Cd^{2+} in the linear range from 1 to 150 $\mu g\ l^{-1}$ and Pb^{2+} in the linear range from 2 to 150 $\mu g\ l^{-1}$, with LODs of 0.41 and 0.93 $\mu g\ l^{-1}$, respectively. Also, the sensor exhibited superior repeatability, stability, and specificity during the performance testing with a recovery rate as high as 101.56%–104.81%.

7.3.2 Sensors based on CPE

Activated carbons (ACs) are amorphous carbon-based materials commonly prepared from different carbon precursors by chemical or physical activation. In production of ACs, great attention is dedicated to revalorization of biomass wastes and the reduction of the carbon footprint. Commercially, ACs are mostly produced from coal, wood, peat, and coconut shells [28].

There is an interesting idea and skillful application of ACs in electrochemical heavy metal detection. As reported by Palisoc *et al*, multilayer graphene paste electrode (MGPE) modified with AC from coconut husk can be employed to simultaneously detect lead (Pb^{2+}) and cadmium (Cd^{2+}) ions via ASV in the vegetable samples including sweet potato tops, corchorus, broccoli, and Chinese cabbage [29]. The coconut shells were chopped, dried, and carbonized in a protective atmosphere of nitrogen. The obtained char was mixed with zinc chloride, dried again and then activated at 850 °C in the same condition as carbonization, washed with deionized water and neutralized with 0.1 M hydrochloric acid.

Unmodified MGPE was prepared by mixing graphene nanopowder with mineral oil while the modified electrodes were prepared by substituting the corresponding amount of graphene nanopowder with the prepared coconut husk activated carbon of different proportions (from 1% to 9% w/w). Pretreatment of the vegetable samples included chopping into small pieces and drying them in a furnace. Then, the known mass of the sample was carbonized on a hot plate and transferred to the furnace to dry it out. The sample ashes were dissolved in nitric acid and put back on the hot plate for the nitric acid to evaporate. The sample was then transferred to a salt solution for ASV analysis while AAS was used for verifying the obtained results. The optimized parameters in ASV analysis were established as follows: accumulation time of 75 s, deposition time of 90 s, and −0.95 V as deposition potential. The limit of detection for lead and cadmium ions were 43 and 56 ppb, respectively. The limit of quantification for lead and cadmium ions were 132 and 169 ppb, respectively. Finally, it was shown that the heavy metal concentrations in the food samples were above the maximum limit as dictated by the World Health Organization (WHO).

Simple, accurate, selective, and highly sensitive simultaneous determination of cadmium, copper and mercury by using SWASV in tuna fish, shrimp, rice, and tobacco samples is possible by carbon paste electrode technique involving silica nanoparticles modified by a newly synthesized Schiff base ligands (L-MSNPs/CPE) as a modifier [30]. Due to interactions of these ligands with metal ions, a newly synthesized Schiff base contributed to selectivity and sensitivity of the method. The Schiff base mentioned is actually N,N-bis(3-(2-thenylidenimino)propyl)piperazine (BTPP). This compound is capable of forming complexes with target ions which is indirectly responsible for low detection limits of targeted heavy metal ions in the voltammetric assays. Food samples were digested by the aid of hydrogen peroxide, nitric acid and potassium persulfate. The pH of the working solution was 2 as adjusted by Britton–Robinson buffer. Optimized conditions for SWASV were as follows: deposition potential of −1.1 V versus Ag/AgCl; deposition time of 60 s; resting time of 10 s; SW frequency of 25 Hz; pulse amplitude of 0.15 V; and dc voltage step height of 4.4 mV. The attained detection limits were 0.3, 0.1 and 0.05 ng ml^{-1} for the determination of Cd^{2+}, Cu^{2+} and Hg^{2+}, respectively. Interference study revealed that ions of Ca^{2+}, Co^{2+}, Cr^{3+}, Cr(VI), Al^{3+}, Fe^{3+}, Pb^{2+}, Mn^{2+}, Ni^{2+}, Zn^{2+}, Sr^{2+}, Mg^{2+}, Li^{+}, Na^{+}, $SO_4{}^{2-}$, Cl^{-}, $NO_3{}^{2-}$, $NO_2{}^{-}$, $S_2{}^{-}$ and $ClO_4{}^{-}$, at 10 000 ng ml^{-1} level, have no significant influence on the signals of 500 ng ml^{-1} each of targeted heavy metal ions with deviations below 5%. So, L-MSNPs/CPE has shown fast response, durability and high selectivity for the simultaneous determination of targeted species.

As reported by Ping *et al*, bismuth oxide nanoparticles and the ionic liquid *n*-octylpyridinium hexafluorophosphate can be used to fabricate a novel carbon composite electrode for simultaneous determination of cadmium and lead ions in the samples of milk by the means of SWASV [31]. It is stated that this electrode combines the unique advantages of nanomaterials and ionic liquid with the low cost, easy preparation route, and great anti-fouling capacity improved by the ionic liquid. The electrochemical principle is based on the following. The reduction of bismuth oxide to bismuth is performed at −1.2 V (versus Ag/AgCl) for 300 s in 0.1 M KOH solution. SWASV is used in the buffer solution and the sample solution. Pre-concentration step is performed at −1.2 V for 180 s with an equilibration period of 10 s, and a square wave stripping scan from −1.2 to −0.3 V. The parameters for the square wave measurement are square wave amplitude of 20 mV; potential step of 5 mV; and frequency of 20 Hz. Before each measurement, a precondition step at potential of −0.3 V is applied for 30 s. Under optimized conditions, the linear range of the applied electrode ranged from 3 to 30 μg l^{-1} for both metal ions with a detection limit of 0.15 μg l^{-1} for cadmium and 0.21 μg l^{-1} for lead. In conclusion, the proposed method appeared to be sensitive, reliable, and effective for the determination of trace levels of cadmium and lead in milk samples.

A variant where CPE is used as a rotating electrode brings some more advantages. Bismuth film electrodes can be combined with a rotating disk electrode (RDE) as reported in case of lead determination in beer samples [32]. In this work, lead and bismuth were simultaneously deposited by reduction at −1 V on a rotating carbon paste disk electrode, equilibrated for 15 s, and then the pre-concentrated metals were oxidized by scanning the potential of the electrode from −1 to −0.4 V using a square-wave waveform. The electrode showed good linearity from 5 to 85 μg l^{-1} of lead and detection limit of 0.27 μg l^{-1}. Beer samples were digested with dry-ashing and wet-ashing procedures. ICP-AES was applied for the validation of the method.

7.3.3 Sensors based on SPCE

Determination of cadmium and lead ions in the samples of rice, wheat, sorghum, and corn can be performed by DPASV with great sensitivity using an amino-functionalized multilayer titanium carbide (NH_2–$Ti_3C_2T_x$) as a modifier on screen-printed carbon electrode (SPCE) [33]. The carbide was prepared by grafting (3-amino-propyl) triethylsilane (APTES) onto the surface of $Ti_3C_2T_x$. As confirmed by the density functional theory, electron-rich amino groups in functionalized carbide can coordinate metal ions and promote their efficient accumulation on the electrode. As this material has unique multilayer structure, large active surface area, strong adsorption capacity and excellent electrical conductivity, it exerts superior electrochemical behavior in determination of cadmium and lead with good linear relationships in a range from 10 to 500 $\mu g\ l^{-1}$. DPASV technique was performed in 0.2 M acetate buffer solution (ABS). The heavy metal ions were first pre-enriched under a deposition voltage of -1.6 V with stirring for 200 s, and then the system was balanced for 10 s. The stripping voltammetry was recorded in the potential range from -1.0 to -0.4 V. The voltammetric parameters included the following: amplitude of 25 mV; potential increase of 5 mV; frequency of 25 Hz; and quiet time of 2 s. The pretreatment of all food samples was carried out by wet-heat digestion method. Samples were weighed, soaked in water, ground, and a mixture of nitric and perchloric acid was added into supernatant which was then heated at 37 °C for 8 h digestion. Then, the pH value of the digestive liquid was adjusted to pH 4.5 and filled to defined volume with 0.2 M ABS (pH = 4.5).

There is a useful mechanistic interpretation of electrode reactions crucial for sensitivity and selectivity of cadmium and lead determination in this case. The suggested scenario relies on the mechanism of adsorption, on one side, and mechanism of determination, on the other. The first mechanism includes adsorption of metal ions by electrostatic interactions and surface complexation. Coordination interaction between –NH_2 groups and metal ions can lead to efficient adsorption of metal ions. In the present case, –NH_2 groups are introduced onto the surface of $Ti_3C_2T_x$ through APTES graft. The coordination interaction, along with deformation charge density, and partial density of states (PDOS) of the bonding atoms before and after Cd^{2+} binding is presented in figure 7.1. As can be observed, the lengths of coordination bonds between N and Cd^{2+} are 2.98 and 3.05 Å, showing a typical coordination bond feature while the lengths of coordination bonds between N and Pb^{2+} are 2.81 and 2.84 Å. Figure 7.1(c) shows that Pb^{2+}-APTES compound presents a larger deformation charge density, indicating the electrons are easier to transfer between Pb^{2+} and APTES. Thus, more Pb^{2+} can be enriched and reacted, resulting in a larger current response. The overall electrochemical determination of metal ions using ASV involves three processes: preliminary accumulation, electrochemical reduction, and anodic striping. Ions are first adsorbed on the surface of NH_2–$Ti_3C_2T_x$/SPCE by condensation, and then the adsorbed ions are selectively electrodeposited onto the electrode surface under -1.4 V for 200 s. Finally, the reduced elemental form of metals is selectively oxidated to generate oxidation current where –NH_2 groups in functionalized $Ti_3C_2T_x$ is responsible for efficient complexation of metal ions.

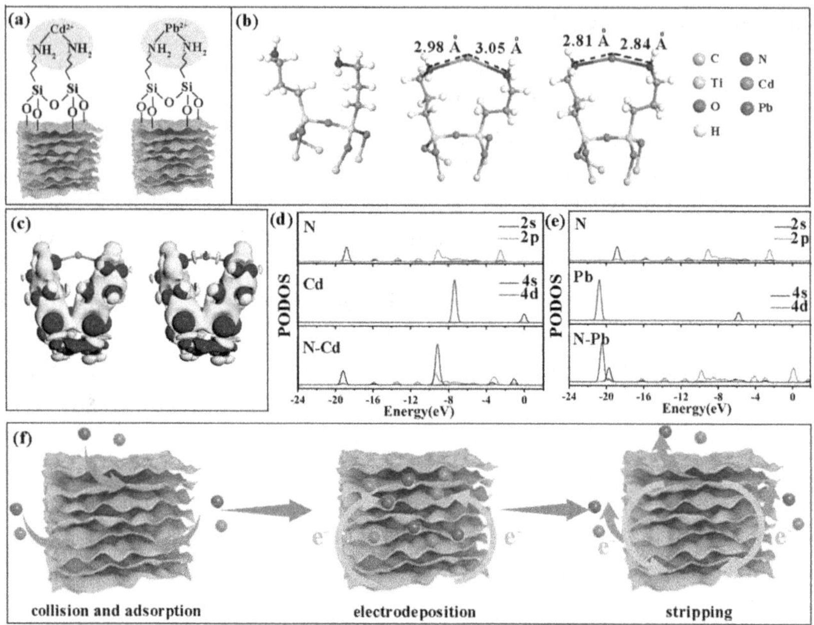

Figure 7.1. (a) Schematic diagram of coordination interaction. (b) The optimal configuration of APTES condensation compound and its complexes. (c) Deformation charge density. Blue and yellow cloud density represent electron enrichment and depletion, respectively. (d) PDOS of the bonding atoms before and after Cd^{2+} binding. (e) PDOS of the bonding atoms before and after Pb^{2+} binding. (f) Schematic diagram of Cd^{2+} and Pb^{2+} determination on $NH_2–Ti_3C_2T_x$. Reprinted from [33], copyright (2023) with permission from Elsevier.

The electrochemical performances of all three electrodes used in this study are given in figure 7.2. Precisely, figures 7.2(a) and (b) display CV and electrochemical impedance spectroscopy (EIS) for the fero-feri system. It is seen that after modification with $Ti_3C_2T_x$, the peak currents for SPCE increase, which is attributed to its specific surface area and remarkable charge mobility. Further modification to $NH_2–Ti_3C_2T_x$ leads to even higher peak currents. In figures 7.2(c) and (d) the well-defined stripping currents for both Cd^{2+} and Pb^{2+} can be noticed for all electrodes while $NH_2–Ti_3C_2T_x$/SPCE shows the highest response.

The achieved detection limits for Cd^{2+} and Pb^{2+} determination by $NH_2–Ti_3C_2T_x$/SPCE were 0.41 and 0.31 $\mu g\ l^{-1}$, respectively. Their recoveries in food samples were determined in the ranges from 90.70% to 105.53% for cadmium and from 91.18% to 102.68% for lead. The results were verified by ICP-MS

A disposable electrochemical sensor for determination of cadmium and lead content in honey and milk samples can be performed as a screen-printed electrode (SPE) modified with single-walled carbon nanohorns (SWCNHs) and electroplated bismuth film [34]. Carbon-based modifying material used here was firstly discovered by Iijima's group dealing with laser ablating the pure graphite without metal catalyst [35]. In contrast to graphene and CNTs, which commonly contain metal residuals, SWCNHs have no residual metal impurities responsible for potential interference prone to obstruct practical electrochemical application [36]. Also, the SWCNHs

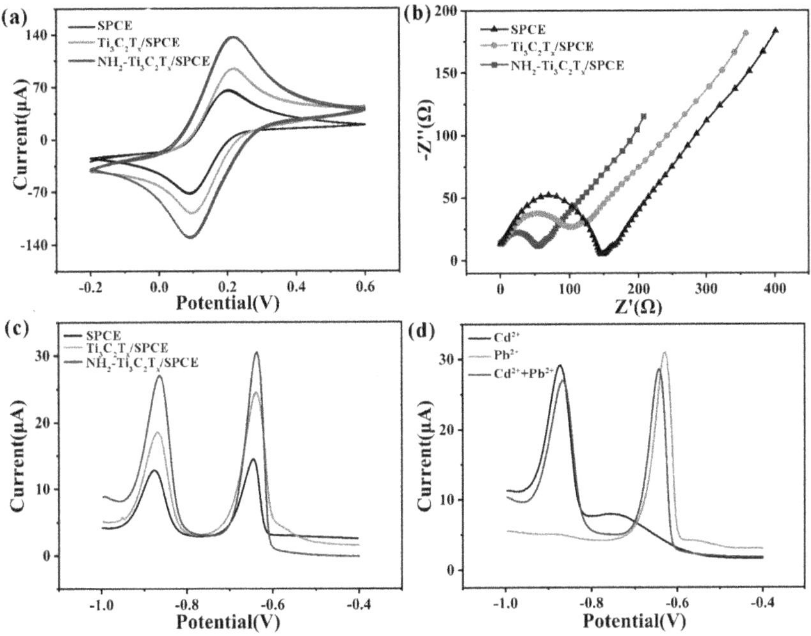

Figure 7.2. (a) Cyclic voltammograms and (b) EIS plots obtained at different electrode in 0.1 M KCl containing 5 mM $[Fe(CN)_6]^{3-/4-}$. (c) DPASV plots obtained at different electrodes toward 300 μg L^{-1} Cd^{2+} and Pb^{2+}. (d) DPASV plots obtained at NH_2–$Ti_3C_2T_x$/SPCE under the condition that these two ions exist individually and simultaneously. Reprinted from [33]. copyright (2022) with permission from Elsevier.

possess unique internal and interstitial nanopore structure, excellent biocompatibility, and high conductivity [37]. Such features are shown to be very important in the field of electroanalytical chemistry.

In the mentioned case, by using an *in situ* approach, a bismuth film (BiF) is electroplated on the surface of SPE/SWCNHs resulting in excellent electrochemical performance in simultaneous detecting of Cd^{2+} and Pb^{2+} by SWASV. The commercially fabricated SPE with a working area of 0.07 cm^2 is used with prior sonication in deionized water and drying in nitrogen stream. SWCNHs employed as modifying agent are dispersed in DMF and sonicated for 1 h to obtain a homogeneous suspension after which a small drop is placed on the surface of the working SPE and spontaneously dried at room temperature. The stripping method is always a more demanding option, but it has its advantages for sure. Precondition (clean step) at +0.2 V which should last for 50 s is done followed by a time-controlled electrochemical deposition performed at −1.2 V for 150 s, and by an equilibration time continued for 10 s, and by a positive-applied potential stripping varied from −1.2 to −0.2 V. Here, 0.1 M acetate buffer solution at pH 4.5 shows the most optimum conditions. The parameters of SWASV are as follows: frequency, 20 Hz; potential step, 5 mV; square wave amplitude, 25 mV. During the electrodeposition and precondition steps test solutions should be stirred by a magnetic stirrer. BiF electroplating procedure implies Bi deposition from standard detection solutions or real-sample extractions containing bismuth nitrate. Since surface-active compounds were reported to foul the bismuth film electrodes by causing passivation of active sites on the electrode surface, investigating this influence certainly deserves attention. For this purpose, three kinds of surfactants, namely: Triton *X*-100, sodium dodecyl sulfate (SDS), and hexadecyltrimethylammonium bromide (CTAB) were tested and they generated huge impact on the analytical performance of target metal ions. In addition, it is expedient to investigate the interferences when other metal ions are present in the matrix since they might be co-deposited with analyte ions. However, in the present case, the results demonstrated that the tested metal ions (divalent ions of Ca, Zn, Ni, and Mg) show no obvious effect on the response of Cd^{2+} and Pb^{2+} determination. Range of linearity for determination of both heavy metal ions by SPE/SWCNHs is from 1 to 60 $\mu g\ l^{-1}$. The detection limit for determination of Cd^{2+} and Pb^{2+} is 0.2 and 0.4 $\mu g\ l^{-1}$, respectively. In testing reproducibility of the same examined electrode, negligible changes after 30 successive operations are noticed.

Graphene is a rather new carbon-based nanomaterial which consists of sheets of sp^2 hybridized carbon atoms bonded with π-electron clouds and all arranged into a honeycomb structure [38]. It definitely deserves attention in electrochemical field due to its large specific surface, high electron conductivity, and excellent biocompatibility [39]. Recently, new electrochemical route for preparing electrochemically reduced graphene oxide (ErGO) has been introduced [40]. This advantageous, rapid, and economic option provides material which could form a stable film on the electrode surface. Since graphene is hardly soluble in most of the common solvents, with ERGNO there is no need for fabrication of film-based electrodes which is often a demanding and time-consuming step.

ErGO can be used for preparing a novel disposable electrode for selective sensing of cadmium and lead ions by SWASV in milk sample extracts [41]. Film of ErGO is prepared by on-step electrodeposition of the exfoliated graphene oxide onto the surface of the SPE and further *in situ* plating with bismuth film. Exfoliation of graphite oxide to graphene oxide is achieved by ultrasonication of graphite oxide dispersion by the aid of a supersonic cleaner. Chemically reduced graphene oxide (CrGO) is made by the chemical reduction of graphene oxide with hydrazine and then dispersed in dimethylformamide.

For the electrochemical analyses, the fabrication of an SPE was done on a semi-automatic screen printer with 150 μm screen mesh. The preparation of SPE/ErGO was executed in a one-step electrodeposition with a working potential of 0.8 V (versus Ag/AgCl) applied for 600 s in a nitrogen purged GO suspension magnetically agitated. For the sake of comparison, the SPE modified with CRGNO film (SPE/CrGO) was made by drop-casting of CrGO-DMF suspension on the SPE surface and dried at room temperature in the air.

Determination of lead ions was done in eight different types of retail packaged milk purchased from a local supermarket in China. Pretreatment of milk samples was performed by ultrasonication-assisted acid digestion method. The measurements in SWASV contained an initial precondition step, a time-controlled electrochemical deposition, an equilibration period, and a positive-applied potential square wave stripping scan. The parameters of SWASV mode were as follows: $E_{cond} = 0$ V, $t_{cond} = 60$ s; $E_{dep} = 1.2$ V, $t_{dep} = 150$ s; $t_{eq} = 10$ s; $E_{begin} = 1.2$ V, $E_{end} = 0.3$ V, $E_{step} = 5$ mV, $E_{ampl} = 25$ mV, $f = 25$ Hz. A magnetic stirrer was used to stir the test solutions during the electrodeposition and precondition steps. The ions of heavy metals have different electrochemical behaviors in different electrolytes. This is the reason why preliminary analyses should involve testing of electrolytes, such as HCl, HNO_3, $HClO_4$, Na_2SO_4, acetate buffer solution, and phosphate buffer solution. Acetate buffer solution has been shown to be the best choice for Cd^{2+} and Pb^{2+} determination in the present case due to the well-defined peaks with largest peak current. The interferences tests from other metal ions evidenced that they have no considerable influence on the stripping response of cadmium and lead. In analyzing interference effects, small concentration of ferrocyanide can be added to analytical solution to mask the copper ions which may be the problem in some cases. Results show that the stripping signals for the analyte ions could be recovered by this action. On the other hand, surfactants such as CTAB, SDS and Triton *X*-100 exhibited profound influence on the analytical performance of target metal ions, especially for cadmium. The linear range of the sensor based on ErGO and bismuth film extends from 1 to 60 $\mu g\, l^{-1}$ for both heavy metal ions with a deposition potential of +1.2 V and a deposition time of 150 s. The detection limits for the determination of cadmium and lead ions were 0.5 and 0.8 $\mu g\, l^{-1}$, respectively.

The rGO can be electrochemically deposited on the surface of bare SPCE to improve its electrochemical properties and used for fast, sensitive, and low-cost determination of lead in real juice, preserved eggs and tea samples by SWV [42]. Usually, the disposable sensor is commercial platform of miniaturized three-electrode system including silver/silver chloride pseudo reference electrode and two carbon electrodes acting as working and counter electrode. In the described case, the diameter of the working electrode is 3 mm. For modifying the SPCE, GO powder is dispersed in phosphate buffer at alkaline value and ultrasonicated to obtain a uniform suspension. This suspension (80 μl) is then dropped on the surface of the working electrode integrated in SPE. In order to electrochemically deposit rGO onto this electrode, the system is scanned in cyclic voltammetry from −1.4 to 0 V for 10 cycles with a scan rate of 50 mV s^{-1}. After careful washing with double-distilled water, and drying in air, this disposable sensor is ready for use. The accumulation potential and accumulation time are factors of paramount importance in SWASV. The deposition potential and deposition time should be respectively optimized by testing of the standard solution of lead. Deposition potential is changed, and the resulting peak current is monitored. The highest current response determines which potential should be applied. In the present case, the accumulation potential of −1.2 V is applied to the rGO-SPCE for 420 s. After the pre-concentration step, the voltammograms are recorded by applying square-wave potential scan to positive direction with amplitude of 30 mV, step potential of 2 mV, and frequency of 20 Hz. The calibration curve for determination of lead ions by the described rGO-SPCE gives linear response in the range from 5 to 200 ppb. A detection limit is obtained with the calculation based on SNR equal to 3, and this limit is 1 ppb which is rather low while the detection range can be rated as wide. In an interference test, a competitive adsorption behavior of cadmium ions is responsible for obvious reducing of the stripping peak current for lead ions. However, this is the case only at concentrations of cadmium above 100 ppm. Since this sensor displays no selectivity toward cupric ion (Cu^{2+}) in the solution, interference test shows that lead can be effectively determined in the solution containing copper at concentrations lower than 100 ppb.

Heavy metals residues in cereals can bioaccumulate in the human body, which makes the monitoring of frequently consumed cereals especially important. For this purpose, a sensor for trace detection of cadmium ions in rice samples was developed by direct screen-printing graphene pastes on the surface of SPCE [43]. The electrochemical set-up consisted of the carbon or graphene working electrode, carbon counter electrode, and Ag/AgCl reference electrode. Since the graphene electrode can provide good conductivity, it does not require bismuth film pre-concentration. As a working electrolyte, basic buffer at pH 8.2 was selected. The rice samples were digested with concentrated nitric acid and pure water as a pretreatment procedure prior to electrochemical analysis. Graphene slurry was dispersed in ultrapure water and centrifuged in order to obtain graphene paste. For DPV, the electrodes were initially pre-concentrated under stirring at a potential of −1.0, −1.1, −1.2, −1.3, −1.4, and −1.5 V for 200, 300, 400, 450, 500, or 550 s, respectively. After a rest for 10 s, the electrical currents were measured by DPV from −1.3 to −0.5 V and recorded. The screen graphene-printed electrodes were cleaned for reuse. The Cd^{2+} was first reduced to Cd on the surface of the used sensor at negative potential. Then, the stripping curve of Cd oxidized to Cd^{2+} was recorded. The stripping peak intensity of Cd was affected by the detection conditions. In this analysis recoveries ranged between 82% and 102%. The detection limit of this proposed method is 10^{-7} mol l^{-1}. The obtained results from portable potentiostat were similar to the results obtained by inductively coupled plasma mass spectrometry.

Nafion® is sulfonated tetrafluoroethylene-based fluoropolymer-copolymer which is frequently used in electroanalytical analyses due to its binding and ion conducting features helpful in providing a suitable interface between the Nafion membrane and electrode materials. Keawkim *et al* showed that the cation exchange property of Nafion can be combined with the selective complexing ability of a crown ether (dibenzo-24-crown-8 (D24C8)) in order to fabricate a highly sensitive *in situ* plated Bi film D24C8/Nafion modified SPCE (Bi-D24C8/Nafion SPCE) [44]. This sensor was effectively applied for the simultaneous quantification of trace amounts of cadmium and lead ions in rice and rice products samples by automated sequential injection analysis (SIA)-ASV with remarkably improved sensitivity and linearity in the range from 0.5 to 60 μg l^{-1} for both metals. LODs were determined to be 0.11 μg l^{-1} for Pb^{2+} and 0.27 μg l^{-1} for Cd^{2+}, respectively.

7.3.4 Sensors based on graphite electrodes

Graphite as an allotrope of carbon can be visualized as a bunch of stacked graphene layers in a honeycomb lattice structure. Each of the layers has a free valence electron that enables bonding with other graphene layers and other elements through the Van der Waals interaction. Three main types of graphite are natural flake, kish, and synthetic graphite. Natural flake graphite is sourced by mining and generally is impure and has poor crystallinity in terms of rotational stacking faults. Kish graphite, which contains iron impurities is a byproduct in making steel. Highly ordered pyrolytic graphite is a synthetic form that has excellent Bernal stacking of the graphene layers. Graphite has many applications. It found its beneficial role in production of pens and batteries, dry lubricants, refractory materials, moderating nuclear reactors, etc. For scientists, probably one of the most attractive application of graphite is related to its exfoliation and producing graphene.

Graphite extracted from waste zinc-carbon battery can be used in forming electrochemical sensor for determination of trace amounts of cadmium and lead in the herbal food supplements [45]. In order to increase the electrode's sensitivity, a graphite electrode was modified with bismuth nanoparticles (BiNPs), MWCNT and Nafion via the drop-coating method. ASV was applied as the working technique. BiNPs are frequently used for modifying a working electrode's surface due to their broad electrochemical window and low toxicity. MWCNTs can enhance an electrode's electrical conductivity, decrease the probability of surface fouling, and increase the rate of electrochemical reactions. Nafion also can contribute to anti-fouling capacity and high permeability to cations while it remains chemically inert. The idea in the present work is that a synergistic effect of these materials should greatly increase the selectivity and sensitivity of the working graphite electrode. The procedure for graphite extraction from the waste zinc-carbon batteries and its processing is as follows. After disassembling the battery, a graphite rod is cleaned by submerging it in propanol and sonicating. Cleaned graphite is dried in a furnace and the tip of the electrode is polished with silicon-carbide sandpaper (different grades) and alumina slurry (different granulations) on a glass slide. After further cleaning by sonicating in ethanol and later in deionized water it is then wrapped with Teflon tape in order to insulate the lateral exposure of the rod. The fabricated graphite electrode is kept in a desiccator at room temperature until use.

It was established that the optimum mixture of modifiers which would give the highest anodic current peaks for cadmium and lead, contained 1 mg of BiNPs, 1 mg of MWCNTs and 5 ml Nafion-ethanol. This mixture was sonicated in ethanol solution, and it was dropped at the graphite electrode with the use of a micropipette. Prior to measurements in herbal samples, optimization procedure in ASV was done. It was concluded that the initial potential of −0.86 V should be used, the optimal deposition time was determined to be 105 s, and optimum scan rate was 0.10 V s^{-1}. Modification of the electrode surface by the applied modifiers remarkably enhanced the voltammetric response of the determined heavy metals. The calibration curves of the bare graphite electrode exhibited linearity from 100 to 1000 ppb, while the modified electrode showed linearity from 5 to 1000 ppb. Low detection limits were achieved (115.37 ppb for Cd^{2+} and 112.1 ppb for Pb^{2+} at the bare electrode; 1.06 ppb for Cd^{2+} and 0.72 ppb for Pb^{2+} at the modified electrode).

7.3.5 Polarography in food analysis

An optimized SWASV procedure for the determination of cadmium, lead and copper ions in white wine samples using a rotating thin mercury-film electrode (TMFE) can be used as a very sensitive system [46]. UV photo-oxidative digestion of the wine sample was applied prior to voltammetric analysis. Measurements for cadmium were carried out at deposition potential of −950 mV versus Ag/AgCl/3 M KCl with deposition time of 15 min. Simultaneously, separate measurements were done for lead and cadmium at deposition potential of −750 mV with deposition time of 30 s. The optimized voltammetric parameters determined with evaluating the SNR were as follows: E_{SW} 20 mV, f 100 Hz, E_{step} 8 mV, t_{step} 100 ms, t_{wait} 60 ms, t_{delay} 2 ms, and t_{meas} 3 ms. It was revealed that the electrochemical behavior was reversible bielectronic for both metals, and kinetically controlled monoelectronic for Cu. Techniques of DPASV, AAS, and ICP-MS were applied for the sake of comparison of results and testing recovery. A good agreement was found with the applied SWASV. This method has shown the linearity of the response up to −4 g l^{-1} for cadmium and lead, and −15 g l^{-1} for Cu. The detection limits in determination of cadmium, lead and copper ions by this method were determined to be 7, 1.2 and 6.6 ng l^{-1}, respectively, which is remarkable.

Another case for determination the concentration of lead ions in beer samples was developed by Zapata-Flores *et al* by applying differential pulse polarography (DPP) coupled to ASV [47]. Lead was deposited by reduction at −900 mV on the surface of the static mercury drop electrode (SMDE) and then re-oxidized by scanning the potential from −900 to +100 mV. The signal which belongs to lead was identified at around −450 mV. The SMDE showed an excellent linear behavior at the concentration range from 49.75 to 476.1 μg l^{-1} with the limit of detection of 26 μg l^{-1} and the limit of quantification of 87 μg l^{-1}.

As shown by Es'haghi *et al*, a novel solid phase microextraction technique using a hollow fiber-supported sol–gel combined with MWCNTs can be coupled with DPASV for determination of lead, cadmium, and copper ions in rice samples [48]. Hanging mercury drop electrode (HMDE) was used as a working electrode. In this study, ASV conditions included: deposition potential of −0.8 V, deposition time of 60 s, equilibration time of 5 s, pulse amplitude of 0.050 05 V, pulse time of 0.04 s, sweep rate of 0.0149 V S^{-1}, stirring rate of 2000 rpm, and potential scan range of −0.9 to 0.07 V. Response was linear in the range from 0.05 to 500 μg l^{-1} for cadmium and lead and from 0.01 to 100 μg l^{-1} for copper. It was shown that the metal concentrations at ppb levels can be determined easily. This method is characterized by its simplicity, rapidness, high sensitivity, truly little sample solution and no expensive and toxic organic solvents.

7.4 Integrated electrochemical platforms

Recently, Zhang *et al* presented an inexpensive and user-friendly electrochemical platform to detect cadmium, lead and mercury ions in the samples of milk, and orange and apple juices by synchronously developing smartphone connectivity, solid-state-microwave flow digestion and nano-Au-modified electrode [49].

This platform successfully saves most of manual operations in sample pretreatment and determination, due to its programmable and powerful flow digestion based on solid-state microwave technology. In this system, a smartphone is used as the information terminal for resource-limited users which can perform electrochemical analyses and share data to the cloud. Solid-state-microwave flow digestion and DPSV are automatically performed in order to simplify pretreatment and analyses steps. In establishing this system, the parameters were optimized, and results validated by ICP-MS

As represented in figure 7.3, the platform is composed of: a sensor module, a wireless potentiometric controller, a solid-state-microwave digestion device, and necessary fluidics components. The sensor module basically includes the following: a counter electrode, a pseudo reference electrode, a working electrode, and a pH electrode. Potential difference between the pH electrode and the pseudo reference electrode is deposited as a record for pH sensing. The wireless potentiometric controller is connected to a smartphone by Wi-Fi. It performs fluidics operations, digestion procedures and DPSVs according to the instructions chosen by the phone. The solid-state-microwave digestion device has a three-channel microwave source and a microwave cavity containing three separated rooms. For each channel of the microwave source, its output could be independently set to a low level. Target sample is digested in digestion loops under programmable microwave environments, cooled in a cooling loop, and then collected in a measurement cell for further determination.

The three-electrode system and the pH electrode were fabricated on a small silicon wafer with standard semiconductor techniques. During heavy metal determination, influence from Zn^{2+}, Cu^{2+}, Fe^{2+}, Fe^{3+}, Mg^{2+}, Cr^{3+}, Cr^{6+}, Sr^{2+}, Al^{3+}, Ba^{2+}, Ni^{2+} or Be^{2+} was investigated with a level of 1 mg l^{-1}. In such a matrix, the proposed sensor faced 3.5%–5.5% signal loss. Recovery tests revealed that targeted Cd^{2+}, Pb^{2+} and Hg^{2+} ions were well recovered. As determined by the instrumental method, LODs of Cd^{2+}, Pb^{2+} and Hg^{2+} were 1.1, 1.0 and 1.2 μg l^{-1}, respectively. Limits of quantification of Cd^{2+}, Pb^{2+} and Hg^{2+} were 3.3, 3.0 and 3.6 μg l^{-1}, respectively.

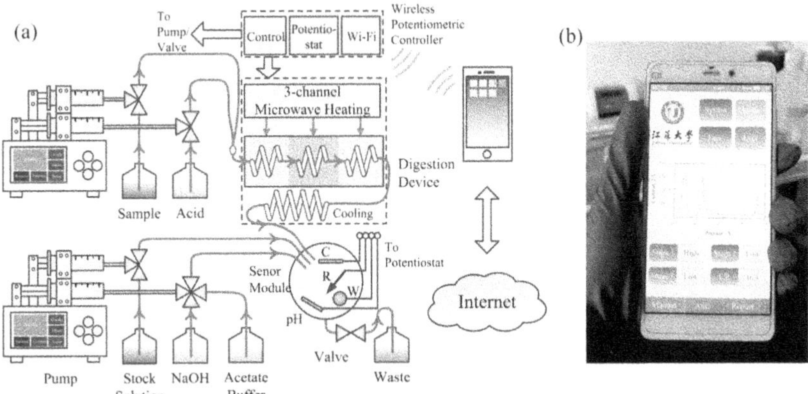

Figure 7.3. (a) Basic principle and (b) typical user interface of the smartphone-based electrochemical platform. Reprinted from [49], copyright (2020) with permission from Elsevier.

7.5 Application of nucleic acid aptamer in food safety

Aptamers are typically peptides or oligonucleotides, such as single stranded DNA or RNA, artificially synthesized *in vitro* by the method called SELEX (systematic evolution of ligands by exponential enrichment). The molecules are folded into stable spatial structures such as stem rings, hairpins, and G-quadruplex under Van der Waals' forces, hydrogen bonds, hydrophobic interactions, electrostatic interactions, and other weak interactions [50]. Such structures can form binding pockets and clefts within a well-defined 3D formation suitable for the specific recognition of target molecules including metal ions, small molecules, macromolecules, microorganisms, and cells [51]. They are characterized by the following advantages: specificity, target versatility, high binding affinity compared to small-molecule ligands, high temperature and chemical stability, and they are easy and cost-effective to prepare and modify [52].

An electrochemical sensor for detection of lead ions at the ppt level in organic and ordinary Chinese cabbage and spinach is reported by Zhang *et al* [53]. This sensor is based on the stable porphyrin-functionalized metal-organic framework (porph@MOF) as peroxidase-mimicking catalyst for signal amplification. Namely, by using the specific recognition of DNAzyme to Pb^{2+} ions, the proposed strategy allows precise determination of lead ions in the presence of other metal ions. This sensor is characterized by the property of simple operation, rapidity, high sensitivity, and selectivity, offering very handy application in determination of lead in real samples of leaf vegetables.

The principle of this sensor is given in figure 7.4. In the preparation process, Au nanoparticles modified GCE was used as sensing platform for DNA2 immobilization

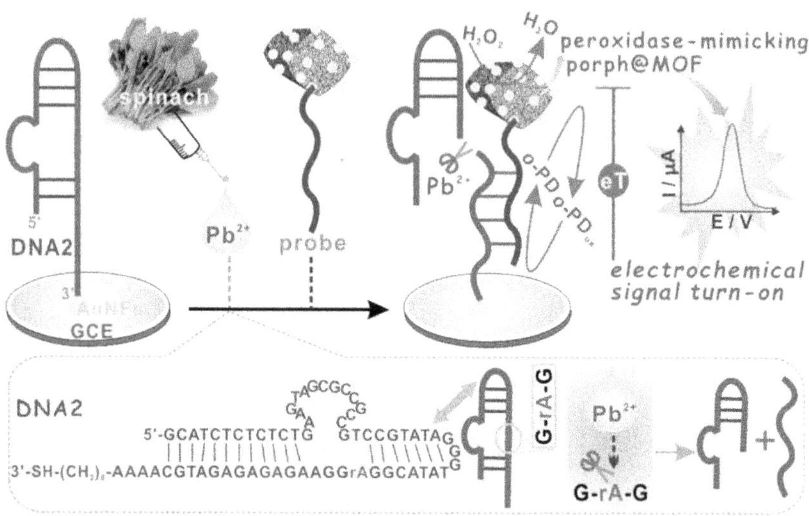

Figure 7.4. Schematic illustration of electrochemical sensing toward Pb^{2+} in leaf vegetables based on DNA2 as recognition element and porph@MOF as mimic peroxidase. Reprinted from [53], copyright (2020) with permission from Elsevier.

via Au–S bond. After the hairpin DNA2 is catalytically cleaved at the ribonucleotide (rA) site by lead ion, the short DNA2 fragment further hybridized with porph@MOFDNA1 via six A-T and six G-C base pairs. Importantly, the DNA2 cleavage and hybridization can be accomplished in one-step without laborious washing steps. This way, the operation of the process is considerably simplified. Following this, the anchored porph@MOF as a mimic peroxidase exhibits catalytic activity toward oxidation of *o*-PD by hydrogen peroxide, and then 2,2′-diaminoazobenzene (*o*-PDox) as catalytic product is electrochemically reduced to output signal. Since porph@OF is only present when DNA2 is specifically cleaved by Pb^{2+}, the electrochemical signal reflects concentration of lead ions in sample solution, offering an effective strategy for assessing the quality of vegetables polluted with lead. After sonication, a GCE was coated with AuNPs and DNA2, rinsed with tris(2-carboxyethy) phosphine hydrochloride buffer, and passivated with 6-mercapto-hexanol (MCH). Samples of cabbage and spinach were fired in a furnace at 550 °C, dissolved in 1% HNO_3 and heated on an electric stove. When nitric acid evaporated, pure water was added to the samples. CV was performed at a scan rate of 50 mV s^{-1}, and SWV was measured with a step potential of 4 mV, a frequency of 10 Hz, and amplitude of 25 mV. A good linear relationship between SWV response and logarithm of Pb^{2+} concentration was found in the range from 50 pM to 5 μM. The sensor attained a low LOD of 5 pM (which corresponds to 1 ppt by weight of the sample) and this way emphasized potential needed for routine monitoring of lead content in leaf vegetables. The interference tests which involved the ions of Cd^{2+}, Cu^{2+}, Mg^{2+}, Ni^{2+}, Zn^{2+}, Ca^{2+}, Na^{+}, and K^{+}, demonstrated that the sensor exhibited good anti-jamming ability for Pb^{2+} detection.

An electrochemical aptasensor for simultaneous detection of cadmium and lead ions in fruit and vegetable samples was reported for the first time by Yuan *et al* [54]. In this case, simultaneous detection of heavy metal ions is performed by a competitive strategy based on the highly specific affinity of aptamers to metal ions. Namely, the aptamers specifically and strongly bound to cadmium and lead ions and form folded structures, as represented in figure 7.5. Consequently, the hybridization of aptamers and their complementary sequences breaks. The complementary sequences of metal ions aptamer are immobilized on the gold electrode via an Au–S bond. The aptamers of cadmium and lead ions labelled with ferrocene and methylene blue, respectively, are also linked on the gold electrode through complementary base pairing. Without the presence of these ions, the sensor is able to provide high-intensity electrochemical signals. However, the presence of the heavy metal ions leads to their specific bounding to aptamers, destroying the rigid double-stranded structure and releasing the aptamers away from the gold electrode. With this action, the electrochemical signal of the modified electrode weakens. To measure this signal, SWV can be applied.

In the present case, pretreatment of the samples included wet digestion. The orange and lettuce samples were washed, drained, and blended. Probes were spiked with different quantities of heavy metals mixed solution and digested. Then, 1 ml 1% HNO3 and 10 mmol l^{-1} PBS buffer (pH 7.5) was added and pH was adjusted to 7.0. A graphite furnace-atomic absorption spectrophotometer (GF-AAS) was used for validation of the technique. The electrode was tested in SWV with the parameters set

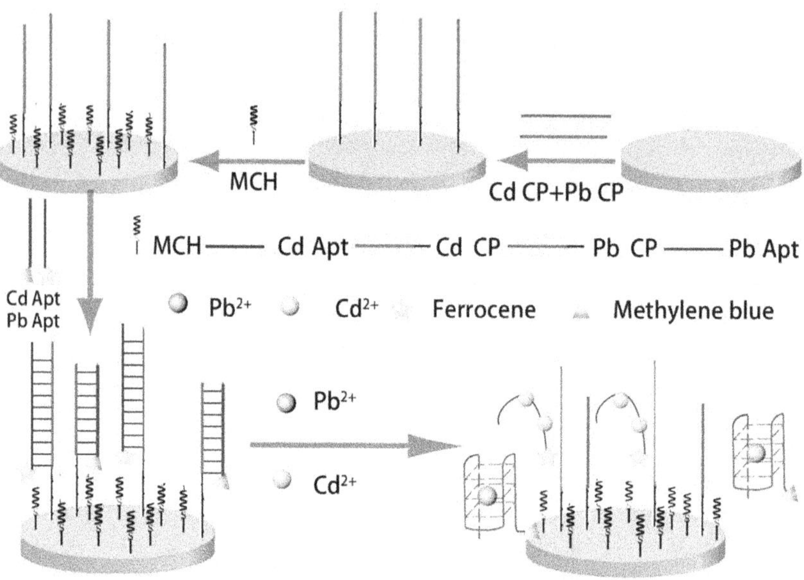

Figure 7.5. Schematic illustration of electrochemical aptasensor for detection of cadmium and lead ions. Reprinted from [54], copyright (2022) with permission from Elsevier.

as follows: potential 0.6 to −0.6 V, frequency of 25 Hz, step potential of 4 mV, and the amplitude of 25 mV. Under the optimal conditions, the electrochemical aptasensor showed a linear response for heavy metal determination in the range from 0.1 to 1000 nmol l^{-1} and the LODs of cadmium and lead ions attained 89.31 and 16.44 pmol l^{-1}, respectively. Excellent stability and reproducibility of the electrode measurements were exhibited while recoveries were determined in the range from 90.06% to 97.24%. In addition, the interference effect involving other metal cations, such as Co^{2+}, Zn^{2+}, Mn^{2+}, Ca^{2+}, Cu^{2+}, and Fe^{2+}, was investigated. It was concluded that this electrochemical aptasensor only had a significant response to targeted cadmium and lead ions which makes it highly selective in this field of application.

Although not proved for food samples, it is interesting to mention the work of Yang *et al* based on a nanorod-like nanocomposite of three-dimensional rGO and polyaniline (3DrGO@PANI) synthesized via *in situ* chemical oxidative polymerization method, which was employed as a sensitive layer of a DNA adsorbent for detecting mercury ions in aqueous solution by the means of EIS [55]. Polyaniline is recognized as a conducting polymer widely used in electrochemical applications because it is cheap, stable in the environment, and has pronounced electroactivity, and unique doping/dedoping chemistry [56]. Amino-group-rich 3D-rGO@PANI exhibits high affinity toward the immobilization of T-rich DNA strands, which prefer to bind with mercury ions to form T–Hg^{2+}–T coordination. Results implied that the proposed nanosensor could be used for highly sensitive and selective determination of heavy metal ions in various environmental detections. Two advantages of this concept stand out: (i) the presence of 3D-rGO within the nanocomposite further improves the specific surface area and electrochemical performance; and (ii) the high intensity of amino groups in PANI ensures sufficient DNA strands immobilized onto its surface, resulting in high sensitivity for heavy metal ion detection.

7.6 Conclusion

Several heavy metals when present in food can cause harmful biological and cellular effects in humans and other living beings. Their presence in food can greatly diminish the benefits of consuming healthy products such as fruit and vegetables, naturally rich in organic compounds with acknowledged antioxidant, antimicrobial, and anticancer effects. In order to improve food security, the sources of heavy metal contamination must be reduced while content of certain heavy metals in targeted food products should be monitored on a daily basis.

The revolutionary advent of practical and sensitive voltammetric techniques in analytical chemistry of metals brought this field to the stage where many original and sophisticated approaches enabled the determination of submicromolar concentration of metals in real samples. Although it is not easy to give an overview on voltammetric techniques applied in metal detection because of unique thermodynamic and kinetic processes at the electrode/electrolyte interface, examples based on ASV described in this chapter might contribute to common knowledge, motivate and instruct the researchers in new studies.

It is obvious that the metal pre-concentration step which precedes the voltammetric potential scan is of major significance in improving the sensitivity of the electrode. For modifying the surface of GCE, CPE, and SPCE, materials such as bismuth films, graphene-based materials, nanoparticles and aptamers have shown the best performance due to their specific effects in electron transfer. The other feature of these materials is based in the fact that they are able to provide the enhanced electrode response towards the targeted metal ions upon modification which is a deciding attribute for selectivity of the analysis.

Everyday application of the described voltammetric sensing in monitoring of metals in food samples is still to be evaluated. Minimization of public health challenges would certainly embrace rapid and practical electrochemical analyses. However, serial production of dimensionally stable electrodes based on nanomaterials, their commercialization, testing of equipment, definition and validation of the methods within accredited laboratories, definitely deserve a space in the years to come. Hopefully, ideas presented in this chapter might pave the way to electrodes integrated into small-scale measuring devices that will be the future of heavy metal detection in food samples, or even more.

Acknowledgments

This chapter is part of the work supported by the Ministry of Science, Technological Development and Innovations of the Republic of Serbia.

References

[1] Salvo A, La Torre G L, Mangano V, Casale K E, Bartolomeo G, Santini A, Granata T and Dugo G 2018 Toxic inorganic pollutants in foods from agricultural producing areas of Southern Italy: level and risk assessment *Ecotoxicol. Environ. Saf.* **148** 114–24

[2] Zoroddu M A, Aaseth J, Crisponi G, Medici S, Peana M and Nurchi V M 2019 The essential metals for humans: a brief overview *J. Inorg. Biochem.* **195** 120–9

[3] Chen D, Hu B and Huang C 2009 Chitosan modified ordered mesoporous silica as micro-column packing materials for on-line flow injection-inductively coupled plasma optical emission spectrometry determination of trace heavy metals in environmental water samples *Talanta* **78** 491–7

[4] Al Othman Z A 2010 Lead contamination in selected foods from Riyadh city market and estimation of the daily intake *Molecules* **15** 7482–97

[5] Munir N *et al* 2022 Heavy metal contamination of natural foods is a serious health issue: a review *Sustainability* **14** 161

[6] Ercal N, Gurer-Orhan H and Aykin-Burns N 2001 Toxic metals and oxidative stress part I: mechanisms involved in metal-induced oxidative damage *Curr. Top. Med. Chem.* **1** 529–39

[7] Wong C, Roberts S M and Saab I N 2022 Review of regulatory reference values and background levels for heavy metals in the human diet *Regul. Toxicol. Pharmacol.* **130** 105122

[8] Akoury E, Baroud C, Kantar S E, Hassan H and Karam L 2022 Determination of heavy metals contamination in thyme products by inductively coupled plasma mass spectrometry *Toxicol. Rep.* **9** 1962–7

[9] Pehlivan E, Arslan G, Gode F and Altun T 2008 Determination of some inorganic metals in edible vegetable oils by inductively coupled plasma atomic emission spectroscopy (ICP-AES) *Grasas Aceites* **59** 239–44

[10] Pytlakowska K, Kocot K, Hachuła B, Pilch M, Wrzalik R and Zubko M 2020 Determination of heavy metal ions by energy dispersive X-ray fluorescence spectrometry using reduced graphene oxide decorated with molybdenum disulfide as solid adsorbent *Spectrochim. Acta* B **167** 105846

[11] Atasoy M, Yildiz D, Kula I and Vaizogullar A I 2023 Determination and speciation of methyl mercury and total mercury in fish tissue samples by gold-coated W-coil atom trap cold vapor atomic absorption spectrometry *Food Chem.* **401** 134152

[12] Borrill A J, Reily N E and Macpherson J V 2019 Addressing the practicalities of anodic stripping voltammetry for heavy metal detection: a tutorial review *Analyst* **144** 6834

[13] Milićević J S, Ranđelović M S, Momčilović M Z, Zarubica A R, Mofarah S S, Matović B and Sorrel C C 2020 Multiwalled carbon nanotubes modified with MoO_2 nanoparticles for voltammetric determination of the pesticide oxyfluorfen *Microchim. Acta* **187** 429

[14] Ranđelović M, Momčilović M, Matović B, Babić B and Barek J 2015 Cyclic voltammetry as a tool for model testing of catalytic Pt- and Ag-doped carbon microspheres *J. Electroanal. Chem.* **757** 176–82

[15] Stojmenović M, Momčilović M, Gavrilov N, Pašti I A, Mentus S, Jokić B and Babić B 2015 Incorporation of Pt, Ru and Pt-Ru nanoparticles into ordered mesoporous carbons for efficient oxygen reduction reaction in alkaline media *Electrochim. Acta* **153** 130–9

[16] Ranđelović M S, Momčilović M Z, Milićević J S, Đurović-Pejčev R D, Mofarah S S and Sorrel C C 2019 Voltammetric sensor based on Pt nanoparticles suported MWCNT for determination of pesticide clomazone in water samples *J. Taiwan Inst. Chem. Eng.* **105** 115–23

[17] Manjunatha J G, Kumara Swamy B E, Mamatha G P, Raril C, Nanjunda Swamy L and Fattepur S 2018 Carbon paste electrode modified with boric acid and TX-100 used for electrochemical determination of dopamine *Mater. Today Proc.* **5** 22368–75

[18] Manjunatha J G, Kumara Swamy B E, Mamatha G P, Gilbert O, Chandrashekar B N and Sherigara B S 2010 Electrochemical studies of dopamine and epinephrine at a poly (tannic acid) modified carbon paste electrode: a cyclic voltammetric study *Int. J. Electrochem. Sci.* **5** 1236–45

[19] Hareesha N, Manjunatha J G, Amrutha B M, Sreeharsha N, Basheeruddin Asdaq S M and Anwer K 2021 A fast and selective electrochemical detection of vanillin in food samples on the surface of poly(glutamic acid) functionalized multiwalled carbon nanotubes and graphite composite paste sensor *Colloids Surf. A Physicochem. Eng. Asp.* **626** 127042

[20] Raril C, Manjunatha J G and Tigari G 2020 Low-cost voltammetric sensor based on an anionic surfactant modified carbon nanocomposite material for the rapid determination of curcumin in natural food supplement *Instrum. Sci. Technol.* **48** 561–82

[21] Manjunatha J G 2019 Electrochemical polymerised graphene paste electrode and application to catechol sensing *Open Chem. Eng. J.* **13** 81–7

[22] Tigaria G and Manjunatha J G 2020 Optimized voltammetric experiment for the determination of phloroglucinol at surfactant modified carbon nanotube paste electrode *Instrum. Exp. Tech.* **63** 750–7

[23] Manjunatha J G 2020 Poly (adenine) modified graphene-based voltammetric sensor for the electrochemical determination of catechol, hydroquinone and resorcinol *Open Chem. Eng. J.* **14** 52–62

[24] Palisoc S T, Uy D Y S, Natividad M T and Lopez T B G 2017 Determination of heavy metals in mussel and oyster samples with tris (2,2′-bipyridyl) ruthenium (II)/graphene/ Nafion® modified glassy carbon electrodes *Mater. Res. Express* **4** 116406

[25] Ogunlesi M, Okiei W, Adio-Adepoju A and Oluboyo M 2017 Electrochemical determination of the levels of cadmium, copper and lead in polluted soil and plant samples from mining areas in Zamfara State, Nigeria *J. Electrochem. Sci. Eng.* **7** 167–79

[26] Wu W, Jia M, Zhang Z, Chen X, Zhang Q, Zhang W, Lid P and Chen L 2019 Sensitive, selective and simultaneous electrochemical detection of multiple heavy metals in environment and food using a lowcost Fe_3O_4 nanoparticles/fluorinated multi-walled carbon nanotubes sensor *Ecotoxicol. Environ. Saf.* **175** 243–50

[27] Huang R, Lv J, Chen J, Zhu Y, Zhu J, Wågberg T and Hu G 2023 Three-dimensional porous high boron-nitrogen-doped carbon for the ultrasensitive electrochemical detection of trace heavy metals in food samples *J. Hazard. Mater.* **442** 130020

[28] Momčilović M, Purenović M, Bojić A, Zarubica A and Ranđelović M 2011 Removal of lead (II) ions from aqueous solutions by adsorption onto pine cone activated carbon *Desalination* **276** 53–9

[29] Palisoc S T, Estioko L C D and Natividad M T 2018 Voltammetric determination of lead and cadmium in vegetables by graphene paste electrode modified with activated carbon from coconut husk *Mater. Res. Express* **5** 085035

[30] Afkhami A, Soltani-Felehgari F, Madrakian T, Ghaedi H and Rezaeivala M 2013 Fabrication and application of a new modified electrochemical sensorsing nano-silica and a newly synthesized Schiff base for simultaneous determination of Cd^{2+}, Cu^{2+} and Hg^{2+} ions in water and some foodstuff samples *Anal. Chim. Acta* **771** 21–30

[31] Jianfeng P, Jian W and Yibin Y 2012 Determination of trace heavy metals in milk using an ionic liquid and bismuth oxide nanoparticles modified carbon paste electrode *Chin. Sci. Bull.* **57** 1781–7

[32] Ghanjaoui M E, Srij M, Hor M, Serdaoui F and El Rhazi M 2012 Fast procedure for lead determination in alcoholic beverages *J. Mater. Environ. Sci.* **3** 85–90

[33] Chen Y, Zhao P, Hu Z, Liang Y, Han H, Yang M, Luo X, Hou C and Huo D 2023 Amino-functionalized multilayer $Ti_3C_2T_x$ enabled electrochemical sensor for simultaneous determination of Cd^{2+} and Pb^{2+} in food samples *Food Chem.* **402** 134269

[34] Yao Y, Wu H and Ping J 2019 Simultaneous determination of Cd(II) and Pb(II) ions in honey and milk samples using a single-walled carbon nanohorns modified screen-printed electrochemical sensor *Food Chem.* **274** 8–15

[35] Iijima S, Yudasaka M, Yamada R, Bandow S, Suenaga K, Kokai F and Takahashi K 1999 Nano-aggregates of single-walled graphitic carbon nano-horns *Chem. Phys. Lett.* **309** 165–70

[36] Valentini F *et al* 2016 Sensor properties of pristine and functionalized carbon nanohorns *Electroanalysis* **28** 2489–99

[37] Liu Z, Zhang W, Qi W, Gao W, Hanif S, Saqib M and Xu G 2015 Label-free signalon ATP aptasensor based on the remarkable quenching of tris(2,2′-bipyridine)-ruthenium(II) electrochemiluminescence by single-walled carbon nanohorn *J. Chem. Soc., Chem. Commun.* **51** 4256–8

[38] Momčilović M Z, Milićević J S and Ranđelović M S 2021 Recent advances in electrochemical determination of pesticides *J. Nanosci. Nanotechnol.* **21** 5795–811

[39] Liu Y X, Dong X C and Chen P 2012 Biological and chemical sensors based on graphene materials *Chem. Soc. Rev.* **41** 2283–307

[40] Zhou M, Wang Y L, Zhai Y M, Zhai J F, Ren W, Wang F A and Dong S 2009 Controlled synthesis of large-area and patterned electrochemically reduced graphene oxide films *Chemistry* **15** 6116–20

[41] Ping J, Wang Y, Wu J and Ying Y 2014 Development of an electrochemically reduced graphene oxide modified disposable bismuth film electrode and its application for stripping analysis of heavy metals in milk *Food Chem.* **151** 65–71

[42] Jian M J, Liu Y Y, Zhang Y L, Guo X S and Cai Q 2013 Fast and sensitive detection of Pb^{2+} in foods using disposable screen-printed electrode modified by reduced graphene oxide *Sensors* **13** 13063–75

[43] Teng Y, Zhang Y, Zhou K and Yu Z 2018 Screen graphene-printed electrode for trace cadmium detection in rice samples combing with portable potentiostat *Int. J. Electrochem. Sci.* **13** 6347–57

[44] Keawkim K, Chuanuwatanakul S, Chailapakul O and Motomizu S 2013 Determination of lead and cadmium in rice samples by sequential injection/anodic stripping voltammetry using a bismuth film/crown ether/Nafion modified screen-printed carbon electrode *Food Control* **31** 14–21

[45] Palisoc P, Vitto R I M and Natividad M 2019 Determination of heavy metals in herbal food supplements using bismuth/multi-walled carbon nanotubes/nafion modified graphite electrodes sourced from waste batteries *Sci. Rep.* **9** 18491

[46] Illuminati S, Annibaldi A, Truzzi C, Finale C and Scarponi G 2013 Square-wave anodic-stripping voltammetric determination of Cd, Pb and Cu in wine: Set-up and optimization of sample pre-treatment and instrumental parameters *Electrochim. Acta* **104** 148–61

[47] Zapata-Flores E D J, Gazcón-Orta N E and Flores-Vélez L M 2016 A direct method for the determination of lead in beers by differential pulse polarography-anodic stripping voltammetry *J. Mater. Environ. Sci.* **7** 4467–70

[48] Es'haghi Z, Khalili M, Khazaeifar A and Rounaghi G H 2011 Simultaneous extraction and determination of lead, cadmium and copper in rice samples by a new pre-concentration technique: hollow fiber solid phase microextraction combined with differential pulse anodic stripping voltammetry *Electrochim. Acta* **56** 3139–46

[49] Zhang W, Liu C, Liu F, Zou X, Xu Y and Xu X 2020 A smart-phone-based electrochemical platform with programmable solidstate-microwave flow digestion for determination of heavy metals in liquid food *Food Chem.* **303** 125378

[50] Bing T, Zheng W, Zhang X, Shen L, Liu X, Wang F, Cui J, Cao Z and Shangguan D 2017 Triplex-quadruplex structural scaffold: a new binding structure of aptamer *Sci. Rep.* **7** 10

[51] Liu R, Zhang F, Sang Y, Katouzian I, Jafari S M, Wang X, Li W, Wang J and Mohammadi Z 2022 Screening, identification, and application of nucleic acid aptamers applied in food safety biosensing *Trends Food Sci. Technol.* **123** 355–75

[52] Sawan S, Errachid A, Maalouf R and Jaffrezic-Renault N 2022 Aptamers functionalized metal and metal oxide nanoparticles: recent advances in heavy metal monitoring *Trends Anal. Chem.* **157** 116748

[53] Zhang X, Huang X, Xu Y, Wang X, Guo Z, Huang X, Li Z, Shi J and Zou X 2020 Single-step electrochemical sensing of ppt-level lead in leaf vegetables based on peroxidase-mimicking metal-organic framework *Biosens. Bioelectron.* **168** 112544

[54] Yuan M, Qian S, Cao H, Yu J, Ye T, Wu X, Chen L and Xu F 2022 An ultra-sensitive electrochemical aptasensor for simultaneous quantitative detection of Pb^{2+} and Cd^{2+} in fruit and vegetable *Food Chem.* **382** 132173

[55] Yang Y, Kang M, Fang S, Wang M, He L, Zhao J, Zhang H and Zhang Z 2015 Electrochemical biosensor based on three-dimensional reduced graphene oxide and polyaniline nanocomposite for selective detection of mercury ions *Sens. Actuators* B **214** 63–9

[56] Wu Q, Xu Y, Yao Z, Liu A and Shi G 2010 Supercapacitors based on flexible graphene/polyaniline nanofiber composite films *ACS Nano* **4** 1963–70

IOP Publishing

Real-Time Applications of Advanced Electrochemical Sensing Devices

Jamballi G Manjunatha

Chapter 8

New electrochemical sensing devices for determination of fungicides and herbicides

Hulya Silah and Bengi Uslu

Environmental contamination with different pollutants, especially fungicides and herbicides, is an essential concern on a worldwide scale and the risk concerned with exposure to contaminants present in soil, air, water is acknowledged as an important threat to human health and other living things because of their bioaccumulation. To avoid these hazards, monitoring of these pollutants in environmental, industrial, and biomedical analyses, by using cost-effective and portable instruments, has become an important field of increasing interest over the past decade. Electrochemical methods have been employed in recent years for analysis of environmental, biological, and pharmaceutical compounds because of their fast response, accuracy, sensitivity, simplicity, low-cost, and high dynamic range. Currently, analyzing techniques based on electrochemical methods offer important advantages for the determination of fungicides and herbicides in the different sample matrices. This chapter highlights the progress in determination of fungicide and herbicides using electroanalytical methods and points towards the problems and required endeavors to reach electrochemical detection appropriate for on-site implementations.

8.1 Introduction

It cannot be denied that increment in globalization, population, and intense agricultural activities are accountable for an increase in environmental contamination that now creates a considerable challenge for scientists and anti-pollutionists (Durodola *et al* 2022).

Kinds of environmental contaminants are classified as (Kailas *et al* 2020):

- agricultural contaminants and nutrients;
- inorganic contaminants;
- organic contaminants;

doi:10.1088/978-0-7503-5377-9ch8

- radioactive contaminants;
- pathogens;
- suspended solids;
- thermal contamination.

Contaminants and effluents available in the environmental medium include various types of organic matter such as fungicides, herbicides, insecticides, pesticides, food additives, detergents, pharmaceuticals, dyes, plastics, fibers, volatile organic compounds, solvents, oils, and inorganic matters such as heavy metals, mineral acids, trace elements, inorganic salts, and radioactive pollutants. After entering the environmental medium, these inorganic and organic substances can be exposed to complicated environmental transformations, which may induce potential damage to the environment (Dong *et al* 2020).

Pesticides, also called plant prevention products, are utilized to control diseases, pests, and weeds. These compounds are categorized according to kind of pest, implementation technique, chemical structure, and mechanism of action. These agricultural chemicals involve fungicides, herbicides, insecticides, molluscicides, acarides, and plant growth regulators (Brycht *et al* 2021, Raluca *et al* 2022).

Considering the potential harm of these toxic fungicide and herbicide compounds to the environment, it is essential to introduce a basic, sensitive, selective, and fast method based on identifying and quantifying these effluents for analysis in environmental samples (Dong *et al* 2020).

The determination of herbicides and fungicides in various samples has already been actualized by conventional techniques involving methods based on spectrophotometry, high-performance liquid chromatography, gas chromatography, capillary electrochromatography, mass spectrometry, liquid chromatography/tandem mass spectrometry, thin-layer chromatography, atomic absorption spectroscopy, atomic emission spectroscopy, inductively coupled plasma mass spectroscopy, inductively coupled plasma optical emission spectroscopy, ultraviolet–visible spectroscopy, spectrofluorometric methods, capillary zone electrophoretic methods.

This chapter is focused on the novel progressions (2010–2023) for sensing of fungicides and herbicides in different environmental samples especially in water and vegetable matrices by electrochemical techniques. The important requirements for an ideal chemosensor involve sensitivity for target effluent analyte, selectivity, short analysis time, extended shelf life and probability of miniaturization at low-cost. In this way, we have investigated the recent studies in determining of fungicides and herbicides using electrochemical methods by focusing on the sensor architecture, electrochemical method, linearity range, sensitivity, selectivity, and application areas.

8.2 Fungicides and herbicides

To prevent the destruction of about one-third of crops in various growth and harvesting steps by pests, the usage of fungicides and herbicides is necessary in agricultural activities (Ensafi *et al* 2017). Fungal contamination induces considerable damage to harvests for human consumption every day, causing weak crops, inadequate food quality, and giant economic decrement. To prevent these troubles, the usage of fungicides has been increased over the last years (Koukouvinos *et al* 2021). Pesticides that are designed to kill fungi or their spores are referred to as fungicides (Brycht *et al.* 2021). Currently, modern fungicides do not kill fungi, they only inhibit growing for a period of days or weeks (Rouabhi 2010). Basic chemical groups of fungicides involves triazoles (i.e. propiconazole, triadimefon), pyrimidines (i.e. fenarimol), strobilurins (i.e. fluoxastrobin, trifloxystrobin), polyoxins (i.e. polyoxin D), benzimidazoles (i.e. thiophanate-methyl), dicarboxamides (i.e. vinclozolin), phenylamides (i.e. mefenoxam), carbamates÷ phosphonates, dithiocarbamates, aromatic hydrocarbons, peroxides, nitriles, phenylpyrolles, cyanoimidazole, carboxamides, biofungicides (Rouabhi 2010). Herbicides are the major class of the agricultural chemicals usaged as plant prevention agents. Especially, one group of these chemicals which submits a high selectivity for certain pests and is mostly applied in agriculture for the treatment of soil is represented by the phenylureas (Buleandra *et al* 2019).

Fungicides and herbicides are usually used in agricultural activities to augment crop yield and aid in monitoring the vectors of diverse diseases, but extreme and uncontrolled utilization can give rise to damage to the environment, such as the pollution of soil, water, food, and air accordingly causing diseases and damage to living things (Silva *et al* 2019b). Residue and degradation products of fungicides and herbicides are defined by their durability, toxicity, and bioaccumulative attitude in environment (Rahmani *et al* 2018). In particular, their permanent utilization and exposure may also have undesired effects on health aspects of living things such as infertility, carcinogenicity, neurological, respiratory, and immunological diseases. Research in recent years has represented that these pollutants have been responsible for reproductive toxicity, cytotoxicity, immunotoxicity, and genotoxicity effects. The World Health Organization (WHO) categorizes them as neurotoxic, carcinogenic, and teratogenic (Erkmen *et al* 2020, Raluca *et al* 2022).

Furthermore, several of these chemical effluents have endocrine disruptor properties, therewith breaking the natural hormonal functions. Except for their environmental and human health implications these pollutants additionally induce various cancer diseases and disruption of reproductive cardiovascular functions (Kumunda *et al* 2021).

Because most pesticides are toxic, and their residuals may induce significant health problems, their maximum residue limits (MRLs) are stated by established legislation and are subject to safety control (Brycht *et al.* 2021). Also, WHO and the European Union (EU) have produced instructions that define concentration limits of effluents in surface water, drinking water, and groundwater (Pardeshi and Dhodapkar 2022). For these reasons, it is important to improve environmental preservation through determination and monitoring.

8.3 Electrochemical determination of fungicides and herbicides and their applications

Fungicides and herbicides are ubiquitous, persistent, bioaccumulative, toxic, and exist in trace amounts in complicated environmental matrices with the different concentrations ranging from 10−9 to 10−12 M (Pardeshi and Dhodapkar 2022).

Methods and techniques of fungicides and herbicides residue detection have been given in documents of the Food and Agriculture Organization (FAO) of the United Nations (and recently in documents of the Joint FAO/WHO Meeting on Pesticide Residues) where definitions and validation data are enlisted for analytical methods for the detection of residuals of this pesticides (and their metabolites) in feed commodities, foodstuffs, animal tissues and trading products of assorted kinds (Šelešovská *et al* 2021).

In spite of the selectivity and sensitivity of chromatographic and spectroscopic detection techniques, the necessities of preconcentration procedures, troublesome sample preparation, very costly devices, and professional specialty restrict their routine and on-site implementations (Deep *et al* 2014). Also, the other deficits of these methods are that they require a large sample volume, are time-consuming, and need a large number of organic solvents with separation and extraction steps (Demir and Silah 2020).

Amongst different analytical methods for qualitative and quantitative analysis of these compounds, electroanalytical methods are one of the most encouraging procedures for their numerous superiorities, such as sensitivity, high accuracy and reliability, low cost, portability, reliability, consistency, rapid responsiveness, accuracy, their simplicity of application, and have relatively short assay time compared to other analytical techniques (Bozal-Palabiyik and Uslu 2016, Dindar *et al* 2021). The most widespread electrochemical methods in analysis of fungicides and herbicides are based on voltammetry and polarography. Various voltammetric techniques involving cyclic voltammetry (CV), differential pulse voltammetry (DPV), differential pulse anodic stripping voltammetry (DPASV), differential pulse cathodic stripping voltammetry (DPCSV), linear sweep voltammetry (LSV), and square wave voltammetry (SWV) were implemented successfully with high selectivity and high sensitivity for the detection of different inorganic and organic species (Bitew and Amare 2020).

The working electrode plays a significant part in the electrochemical reactions, and electrode species determines the selectivity, sensitivity, efficiency, and reproducibility of the developed technique (figure 8.1). The different types of conventional electrode materials such as gold, platinum, graphite, carbon paste, glassy carbon, and conductive polymers have been presented as relevant electrochemical sensors in various analyses in recent years.

Electrochemical techniques have been used for herbicides and fungicides due to their superiorities such as rapid analysis times, perfect sensitivity with very low detection limits, simultaneous detection of several analytes, etc. Selected electrochemical analyses of the last decade for the determination of various fungicides and herbicides are given in tables 8.1 and 8.2.

In some cases, traditional working electrodes used as electrochemical sensors such as mercury, GCE, platine and gold have relatively poor success, because of the deficiency of sensitivity and selectivity. Amongst the assorted strategies utilized to overcome these disadvantages and drawbacks, the utilization of materials deposited as thin films on the surface of electrodes is commonly implemented (Mekeuo *et al* 2022). Utilizing bare electrodes throughout the determination of effluents has some

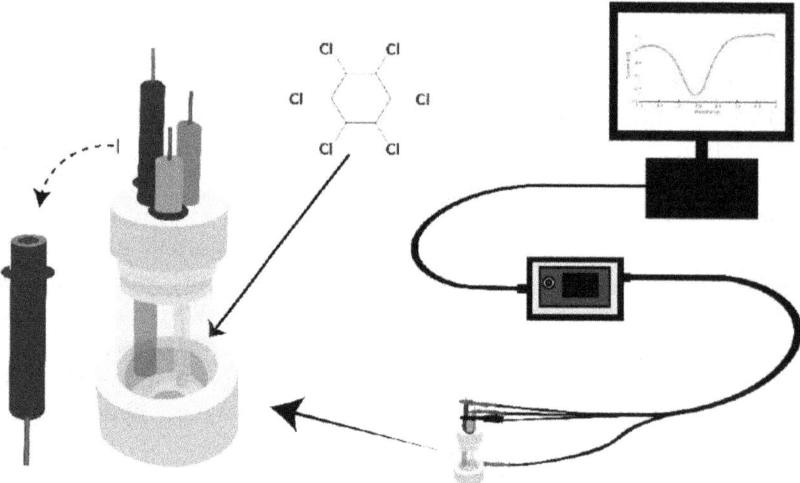

Figure 8.1. Electrochemical scheme utilized in the operation consisting of an electrochemical cell, a potentiostat connected to a computer system to operate the measurement and record data. Reproduced from Noori *et al* 2021, CC BY 4.0.

disadvantages involving high overpotential of analyte oxidation or reduction, electrode passivation, and slow direct electron transfer kinetics (Kumunda *et al* 2021). In recent years, various materials including carbonaceous materials, clay minerals, metal nanoparticles, mesoporous (organo)silica, graphene, zeolites, natural or synthetic polymers, quantum dots (graphene, carbon), etc., are widely employed as surface modifier to increase the electrical conductivity. Depending on the surface modification of the electrode, the selectivity and the sensitivity of the modified sensors also increase in determining various analytes. Otherwise, developing an electrocatalyst to enhance the formed current of an analyte with high sensitivity and low overpotential is one of the most significant aims of searchers (Lavanya *et al* 2021, Mekeuo *et al* 2022). Surfaces of electrodes, modified with various nanomaterials, can be generated by electrochemical deposition, chemical vapor deposition, chemical attachment with/without crosslinking agents, and stirring of nanoparticles with electrode materials (Ozcelikay *et al* 2018).

Multi-walled carbon nanotubes (MWCNTs) are commonly acknowledged to promote electron transfer reactions when utilized as electrode modifier material and they are an attractive group of nanomaterials because of their outstanding features such as high electrical conductivity, significant mechanical strength, high surface area, and good both chemical, physical, and electrochemical performances (Chaabani *et al* 2022). Khadem *et al* designed a highly sensitive sensor to detect the dicloran pesticide in environmental and biological samples. MIP and MWCNTs were used as modifiers in carbon paste electrode composition. The prepared electrode was utilized to detect the dicloran after the optimizing experimental parameters involving instrumental parameters of SWV such as the concentration of supporting electrolyte, pH, deposition potential amount, potential frequency, time

Table 8.1. Some reported electrochemical applications for the detection of different fungicides.

Analyte	Method	Electrode	Limit of detection	Linearity range	Interferences	Applications	References
Acibenzolar-s-methyl	SWAdSV	HMDE	3.0×10^{-9} M	1.0×10^{-8}–6.0×10^{-8} M	Thiophanate-methyl, methidathion, metam, acephate, aldicarb, cyromazine, clothianidin, dodine, methamidophos	Soil, river and tap water samples	Guziejewski *et al* (2014)
Azoxystrobin	DPV	BDDE	8.4×10^{-8} M	5.0×10^{-7}–1.55×10^{-4} M	Difenoconazole, propiconazole, cyproconazole tebuconazole, epoxiconazole, triclopyr, chlorpyriphos, imidacloprid, tribenuron, glyphosate	Model solution, river and drinking water samples	Šelešovská *et al* (2019)
Azoxystrobin	DPV	m-AgSAE	7×10^{-7} M	2.0×10^{-6}–5.0×10^{-5} M	—	River water samples	Šelešovská *et al* (2019b)
Benzovindiflupyr	SWAdSV	PGE	0.023 μM	0.10–12.5 μM	Azoxystrobin, propiconazole, prothioconazole, and picoxystrobin	Seized product samples	Carvalho *et al* (2023)
Bixafen	SWV	TRGOPE	31.5 nM	0.1–2.5 μM	Prothioconazole and fluopyram	Tap and river water samples	Brycht *et al* (2022)
Bromacil	DPV	BDDE	1.26×10^{-6} M	5.00×10^{-6}–7.50×10^{-5} M		River water samples	Brycht *et al* (2016)
Carbendazim	DPSV	CHTCN	1.21 ng cm^{-3}	25–490 ng cm^{-3}	K^+, Na^+, Ag^+, Ca^{2+}, CO_3^{2-}, Cl^-, HCO_3^-, NO_3^-, dimethoate, tebufenozide, imidacloprid, simazine)	Tap and river water samples	Kalijadis *et al* (2017)
Carbendazim	DPAdsV	TCP-CPE	3.0×10^{-7} M	5.0×10^{-7}–1.0×10^{-5} M	Linuron,SO_4^{2-}, Br^-, NO_3^-, Cl^-	River water samples	Ashrafi *et al* (2012)

Carbendazim	DPV	$ZnFe_2O_4$/SWCNTs/GCE	0.09 μM	1.0–100 μM	K^+, Na^+, NH_4^+, NO_3^-, Cl^-, HCO_3^-, $H_2PO_4^-$, CO_3^{2-}, SO_4^{2-}, Mg^{2+} ascorbic acid, anthocyanin, catechin, Pb^{2+}, Zn^{2+}, Cu^{2+}, Cd^{2+}, pyrimethanil, triadimenol, paranitrophenol, tricyclazole	Tomatoes, apple, leeks, and water samples	Dong *et al* (2017)
Carbendazim	SWV	TiO_2/CPE	1.71×10^{-8} M	1.0×10^{-7} M–4.2×10^{-4} M	$MnSO_4$, KNO_3, NaCl, $MgCl_2$, $FeSO_4$, $ZnCl_2$, 2,4-D, Amitrole, diuron	Soil and water samples	Killedar *et al* (2022)
Carbendazim	DPV	Yb_2O_3/*f*-CNF	6.0 nM	0.05–3035 μM	Amatryn, thiamethoxam, parathion, 2,4,6-trichlorophenol, carbofuran, potassium, calcium, iron, magnesium, aluminum, chlorine, copper, sulfate	Carrot, radish, lake water, and pond water samples	Krishnapandi *et al* (2023)
Carbendazim	DPV	GCE/CNT	0.08 μM	0.1–2.0 μM	—	Commercial formulation samples	Mekeuo *et al* (2022)
Carbendazim	SWV	Pd NPs /PGE	1.8×10^{-8} M	2.00×10^{-7}–1.60×10^{-6} M	Ascorbic acid, catechol, urea, NaCl, uric acid, hydroquinone, captopril, Cd^{2+}, Pd^{2+}, ranitidine	Synthetic urine and river water samples	Wong *et al* (2021)
Carbendazim	DPV	MIP/CdMoO4/g-C_3N_4–15/GCE	2.5×10^{-12} M	1.0×10^{-11}–1.0×10^{-9} M	—	Orange and apple juice samples	Yola (2022)
Carbendazim	SWV	GO/GCE	1.38×10^{-8} M	1.0×10^{-7}–2.5×10^{-4} M	Amitrole, diuron, 2, 4-D, KNO_3, $CuSO_4$, $MgCl_2$, $MnSO_4$, $ZnCl_2$,NaCl, $FeSO_4$	Soil and water samples	Ilager *et al* (2022)

(*Continued*)

Table 8.1. (*Continued*)

Analyte	Method	Electrode	Limit of detection	Linearity range	Interferences	Applications	References
Carbendazim	SWV	Zeolit modified CPE	3.51×10^{-9} M	0.10×10^{-6}–2.35×10^{-6} M	Na^+, Ca^{2+}, NH_4^+, Mg^{2+}, Cu^{2+}, Al^{3+}, OH^-, Cl^-, SO_4^{2-}, atrazine, endosulfan, cyproconazole, linuron	Grape juice and agricultural formulation samples	Maximiamo *et al* (2018)
Carbendazim	SWV	MoS_2 QDs@MWCNTs/GCE	2.6×10^{-8} M	4.0×10^{-8}–1.0×10^{-6} M	$MgCl_2$, KCl, $CaCl_2$, $Pb(NO_3)_2$, ascorbic acid, carotene	Platycodon grandiflorum and pear samples	Zhang *et al* (2021)
Chlornitrofen	SWAdSV	Hg(Ag)FE	3.00×10^{-8} M	1.0×10^{-7}–1.5×10^{-6} M	Fenfuram, bromacil, closantel, oxycarboxcine, cadmium, copper, zinc	River water samples	Brycht *et al* (2016)
Cyazofamid	SWAdSV	(Hg(Ag)FE)	1.64×10^{-6} M	4.00×10^{-6}–2.00×10^{-5} M	Nitrothalisopropyl, fenoxanil, aclonifen, clothianidin, metam-sodium trihydrate, acibenzolar-S-methyl, lead, cadmium, copper, zinc	River water and potato samples	Brycht *et al* (2014)
Cymoxanil	SWSV	HMDE	7.1 μg l^{-1}	23.6–1950 μg l^{-1}	Cyromazine, alanycarb, acifluorfen, anilazine, Na^+, Mg^{2+}, K^+, Co^{2+}, Cu^{2+}, Ni^{2+}, Zn^{2+}, Pb^{2+}, NH_4^+, Cl^-, NO_3^-, CN^-	Pharmaceutical dosage forms, river and tap water samples	Mercan and Inam (2010)
Difenoconazole	DPV	BDDE	0.049 μM	0.1–40.0 μM	Cyproconazole, epoxiconazole, propiconazole, tebuconazole, azoxystrobin, chlorpyriphos	Real pesticide preparations, water samples	Šelešovská *et al* (2021)
Dimoxystrobin	Amperometry	BDDE	0.38 μM	1–60 μM	Acetylsalicylic acid, dipyrone, paracetamol, and diclofenac	Tap, river and mineral water samples	Dornellas *et al* (2014)

Fenfuram	SWV	BDDE	6.3×10^{-6} M	2.4×10^{-5}–2.6×10^{-4} M	—	Triticale seed and river water samples	Brycht *et al* (2015a)
Fenhexamid	SWV	CPE/MWCNTs	0.52 μM	1.74–157.48 μM	—	Wine grapes and blueberries samples	Brycht *et al* (2021)
Fenoxanil	SWV	(Hg(Ag)FE)	6.5×10^{-8} M	5×10^{-7}–4×10^{-6} M	Clothianidin, cyazofamid, cyanofenphos, dinotefuran, metam-sodium trihydrate, moroxydine, nitrothalisopropyl, lead, cadmium, copper, zink	Spiked rice and river water samples	Brycht *et al* (2015c)
Fenoxanil	SWAdSV	(Hg(Ag)FE)	2.8×10^{-11} M	1×10^{-10}–9×10^{-10} M	Clothianidin, cyazofamid, cyanofenphos, dinotefuran, metam-sodium trihydrate, moroxydine, nitrothalisopropyl, lead, cadmium, copper, zink	Spiked rice and river water samples	Brycht *et al* (2015c)
Fludioxonil	SWV	Hg(Ag)FE	5.81×10^{-7} M	2.00×10^{-6}–2.25×10^{-5} M	—	River water samples	Brycht *et al* (2015b)
Glyphosate	DPV	HF-PGE/CuO/ MWCNTs–IL	1.3 nM	5.0–1.1 μM	Zn^{2+}, Ca^{2+}, Cd^{2+}, Mg^{2+}, NH_4^+, Na^+, NO_3^-, Br^-, PO_4^{3-}, SO_4^{2-}, Bialaphos, Glufosinate, Tridemorph, Cypermethrin, Chlorpyrifos, (Aminomethyl) phosphonic acid	River water and soil samples	Gholivand *et al* (2018)
Itraconazole	SWV	BDDE	1.79×10^{-8} M	7.9×10^{-8}–1.2×10^{-6} M	K^+, Na^+, Cu^{2+}, Fe^{2+}, Ca^{2+}, Cd^{2+}, Mg^{2+}, Pb^{2+}, Zn^{2+}, Cl^-, NO_3^-, SO_4^{2-}, Triton X-100,	Biala and tap water samples	Mielech-Łukasiewicz and

(*Continued*)

Table 8.1. (*Continued*)

Analyte	Method	Electrode	Limit of detection	Linearity range	Interferences	Applications	References
					tetrabutylamonium bromide, sodium dodecyl sulfate, methylparaben, triclosan, voriconazole, ketoconazole, clotrimazole		Starczewska (2019)
Kresoxim-methyl	SWV	BDDE	2.6×10^{-7} M	8.7×10^{-7} mol l^{-1}–3.4×10^{-5} M	Trifluoxystrobin, Fluoxastrobin	Grape juice samples	Dornellas *et al* (2013)
Mancozeb	SWAdSV	GCE	7.0 μM	10–90 μM	—	Commercial pesticide formulations	Lopez-Fernandez *et al* (2015)
Mancozeb	DPV	WO_3/rGO/GCE	0.0038 μM	0.05–70.0 μM	Ca^{+2}, K^+, Mg^{+2}, pentachlorophenol, methyl parathion, trichlorophenol, endosulfan, bentazone, ascorbic acid	Ridge gourd, tomato, and potato samples	Buledi *et al* (2022)
Maneb	DPV	BDDE	24 nM	80–3000 nM	$Fe^{2+/3+}$, Mg^{2+}, Ca^{2+}, Na^+, K+, NO_3^-, Cl^-, SO_4^{2-}	Artifical water samples	Stankovic (2017)
Mercazole	SWV	CS-CNPs/GCE	17.7 nM	0.04–1.0 μM	—	River water samples	Ghalkhani *et al* (2018)
Oxycarboxin	SWV	BDDE	1.6 μM	8.0–100.0 μM	Chlornitrofen, dichlorophen, closantel, bromacil, cadmium, zinc, and copper	River water samples	Brycht *et al* (2017)
Pentachlorophenol	DPV	$NiFe_2O_4$/GCE.	0.0016 μM	0.01–90.0 μM	Cu^{+2}, Mg^{+2}, Ca^{+2}, K^+, SO_3^{2-}, Cl^-, ascorbic acid, trichlorophenol, endosulfan, hydroquinone, carbofuran	Canal, tap, and river water samples	Taqvi *et al* (2022)
Pentachlorophenol	DPV	AgNPs-rGO/GCE	0.001 μM	0.008–10.0 μM	Mn^{2+}, Cu^{2+}, Fe^{3+}, Zn^{2+}, NO_3^-, CO_3^{2-}	Vegetable samples	Wang *et al* (2020)

Picoxystrobin	SWV	BDDE	0.2 μM	0.7–20.0 μM	Pyraclostrobin, trifloxystrobin, kresoxim-methyl	Natural water samples	Dornellas *et al* (2015)
Posaconazole	SWV	BDDE	7.78×10^{-9} M	5.7×10^{-8}–8.44×10^{-7} M	K^+, Na^+, Cu^{2+}, Fe^{2+}, Ca^{2+}, Cd^{2+}, Mg^{2+}, Pb^{2+}, Zn^{2+}, Cl^-, NO_3^-, SO_4^{2-}, Triton X-100, tetrabutylamonium bromide, sodium dodecyl sulfate, methylparaben, triclosan, voriconazole, ketoconazole, clotrimazole	Biala and tap water samples	Mielech-Łukasiewicz and Starczewska (2019)
Propineb	SWV	Cu^{2+}-Mt-CPE	1.0 μM	5–30 μM	Cysteine, phenol, methomyl, ascorbic acid, glucose, endosulfan	Sea and river water samples	Abbaci *et al* (2014)
Prochloraz	DPSV	MSDDE	8.4×10^{-10} M	6.0×10^{-9}–8.0×10^{-5} M	Fe^{3+}, Na^+, Al^{3+}, K^+, Zn^{2+}, Cu^{2+}, Ca^{2+}, Pb^{2+}, Ag^+, Mg^{2+}	Orange rind samples	Li *et al* (2010)
Pyraclostrobin	SWV	BDDE	8.2×10^{-7} M	8.2×10^{-7}–2.0×10^{-5} M	Azoxystrobin, Trifluoxystrobin, Fluoxastrobin	Mineral and natural water samples	Dornellas *et al* (2014)
Pyrimethanil	DPV	TiC/CNF/GCE	33.0 nM	0.1–600 μM	NO_3^-, SO_4^{2-}, Mg^{2+}, Na^+, Zn^{2+}, K^+, Ca^{2+}, PO_4^{3-}, glucose, vitamin C, thiabendazole	Water, apple, cucember samples	Sui *et al* (2019)
Pyrimethanil	DPV	$NiCo_2O_4$/rGO/IL/GCE	11.0 nM	20.0–140.0 μM	Triadimenol, thiabendazole, benomyl, tricyclazole, glucose, and vitamin C and, Na^+, Mg^{2+}, Zn^{2+}, K^+, Cl^-, SO_4^{2-}, NO_3^-, and PO_4^{3-}	Seawater, water, cucumber, apple, and fragrant-flowered garlic samples	Yang *et al* (2016)
Pyrimethanil	DPV	$NiCo_2S_4$/GCNF	20.0 nM	0.06–800 μM	$SO4^{2-}$, Mg^{2+}, NO_3^-, Na^+, Zn^+, Ca^{2+}, K^+, PO_4^{3-}, vitamin C, glucose, thiabendazole	Water, apple, tomato and pea samples	He *et al* (2020)

(Continued)

Table 8.1. (*Continued*)

Analyte	Method	Electrode	Limit of detection	Linearity range	Interferences	Applications	References
Pyrimethanil	CV	β-CD/MWCNT/GCE	1.04 μM	10–80 μM	NO_3^-, Ca^{2+}, PO_4^{3-}, Mg^{2+}, 2-Amino-4,6-dimethylpyrimidine, Fluquinconazole	Apple samples	Garrido *et al* (2016)
Pyrimethanil	DPV	MWCNTs–IL/GCE	1.6×10^{-8} M	1.0×10^{-5}–1.0×10^{-4} M	Paranitrophenol, tricyclazole, thiabendazole, carbendazim, ascorbic acid, K^+, Mg^{2+}, Na^+, Ca^{2+}, Cu^{2+}, Zn^{2+}, SO_4^{2-}, PO_4^{3-}, Cl^-, HPO_4^{2-}, NO_3^-	Orange, apple peel samples and water samples	Yang *et al* (2015)
Sodium diethyldithiocarbamate	SWAdSV	AgNP-SAE	7.26×10^{-8} M	2.83×10^{-7}–6.89×10^{-6} M	Na^+, K^+, NH_4^+, Fe^{2+}, Ca^{2+}, Cu^{2+}, Mg^{2+}, OH^-, Cl^-, F^-, PO_4^{-2},SO_4^{-2} molinate, atrazine, terbutryn, endosulfan, linuron, trifluralin	River water samples	Lucca *et al* (2018)
Sulcotrione	DPV	GCE	50 nM	2–50 μM	Ca^{2+}, K^+, Mg^{2+}, Na^+, Cu^{2+}, Hg^{2+}, Bi^{3+}, As^{3+}, Cr^{6+}, Pb^{2+}, Ni^{2+}, Cl^-, mesotrione	River water samples	Stankovic *et al* (2015)
Tebuconazole	DPV	m-CuSAE	0.21 μM	0.1–1.5 μM	—	Commercial formulation samples	Novokova *et al* (2013)
Thiabendazole	DPV	$ZnFe_2O_4$/SWCNTs/GCE	0.05 μM	1.0–100 μM	K^+, Na^+, NH_4^+, NO_3^-, Cl^-, HCO_3^-, $H_2PO_4^-$, CO_3^{2-}, SO_4^{2-}, Mg^{2+} ascorbic acid, anthocyanin, catechin, Pb^{2+}, Zn^{2+}, Cu^{2+}, Cd^{2+}, pyrimethanil, triadimenol, paranitrophenol, tricyclazole	Tomatoes, apple, leeks, and water samples	Dong *et al* (2017)

Thiabendazole	SWV	BDDE	1.27×10^{-7} M	4.98×10^{-7}–1.12×10^{-5} M	Ca^{2+}, Na^{+}, Cu^{2+}, Mg^{2+}, Al^{3+}, Ni^{2+}, Zn^{2+}, NH_4^{+}, Cl^{-}, SO_4^{-2}, NO_3^{-}, ascorbic acid, glycine uric acid, glucose, lactose,urea, diclofenac, dopamine, glycerol, carbendazim	Sugar cane, mango, pharmaceutical formulations, raw natural water samples	Ribeiro *et al* (2020)
Thiram	SWV	Zeolit modified CPE	6.74×10^{-9} M	0.36×10^{-7}–4.99×10^{-7} M	Na^{+}, Ca^{2+}, NH_4^{+}, Mg^{2+}, Cu^{2+}, Al^{3+}, OH^{-}, Cl^{-}, SO_4^{2-}, atrazine, endosulfan, cyproconazole, linuron	Grape juice and agricultural formulation samples	Maximiamo *et al* (2018)
Trifloxystrobin	SWV	BDDE	1.4×10^{-7} M	1×10^{-7}–1×10^{-5} M	Cyproconazole, Tebuconazole, Kresoxim-methyl, Picoxystrobin, Fluoxastrobin	Natural, mineral and orange juice samples	Almeida *et al* (2016)
Trifluralin	SWV	p-AgSAE	4.56×10^{-3} μM	0.25–1.75 μM	—	Eggplant, tomato, carrot, orange juice and grape juice samples	Gonçalves Filho and Souza (2021)

Abbreviations: BDDE: Boron doped diamond electrode, CHTCN: nitrogen-doped hydrothermal carbon, CNF: Carbon nanofiber, CNP: Carbon nanoparticles, CNT: Carbon nano tube, CS: Chitosan, CV: Cyclic voltammetry, DPAdSV: Differential pulse adsorptive stripping voltammetry, DP-ASV: Adsorptive differential pulse anodic stripping voltammetry, DPSV: Differential pulse stripping voltammetric, DPV: Differential pulse voltammetry, GCE: Glassy carbon electrode, GCNF: graphitized carbon nanofiber film, GO: Graphene oxide, Hg(Ag)FE: Silver amalgam film electrode, HF: Hollow fiber, HMDE: Hanging mercury drop electrode, IL: Ionic liquid, m-AgSAE: mercury meniscus-modified silver solid amalgam electrode, m-CuSAE: mercury meniscus-modified copper solid amalgam electrode, MIP: Molecularly imprinted polymers, Mt: Montmorillonit, MSDDE: medical stone doped disposable electrode, MWCNT: Multi-walled carbon nano tube, NPs: Nanoparticles, p-AgSAE: polished silver solid amalgam electrode, PGE: Pencil graphite electrode, rGO: Reduced graphene oxide, SAE: Solid amalgam electrode, SWAdSV: square wave adsorptive stripping voltammetry, SWCNT: Single walled carbon nano tube, SWV: Square wave voltammetry, TCP-CPE: Carbon paste electrode based on tricresyl phosphate, TiC: titanium carbide, TRGOPE: thermally reduced graphene oxide QDs: Quantum dots, β-CD: β-cyclodextrin.

of exerted deposition potential, and pulse amplitude. The modified electrode demonstrated a wide linear range from 1×10^{-6} to 1×10^{-9} M with a detection limit of 4.8×10^{-10} M. The designed sensor was applied to detect the dicloran pesticide in river water, tap water, and human urine samples without any specific sample preparation (Khadem *et al* 2016).

The nanostructured materials are exceptionally engaging to modify electrodes using electrochemical determination of organic and inorganic pollutants owing to their peculiar chemical, electronic, mechanical, and thermal features in comparison with traditional electrode materials (Kalaivani and Narayanan 2021). The extensive role for the modified electrode is to enhance the surface area of the electrode substrate, increase the preconcentration capability, and develop the charge transfer kinetics throughout the electrochemical procedure (Xiao *et al* 2021). Electrodes can be modified using nanomaterials with different modification techniques such as electrochemical deposition, chemical vapor deposition, chemical attachment with/without crosslinking agents, and also the mixing of nanoparticles with electrode materials (Ozcelikay *et al* 2018).

Carbendazim, with very widespread pollution compared to the conventional fungicides, has caused critical danger to human health and the environment. In addition, selective and sensitive determination of fungicide carbendazim in the environmental matrix is a major problem for their complicated chemical constituents. Zhang *et al* developed a new electrode for the determination of fungicide carbendazim based on MoS_2 quantum dots over MWCNTs by a straightforward assembly technique (figure 8.2). Some morphological analysis involving Raman spectroscopy, high-resolution transmission electron microscopy, x-ray photoelectron spectroscopy, and thermogravimetric analysis showed that anchoring structure advances the conductivity and nanohybrid surface area of the material. Therefore, the electroanalytical sensor designed based on the MoS2 QDs@MWCNT nanohybrid has perfect catalytic activity to oxidation of carbendazim. Under optimized experimental conditions, the electrochemical sensor showed a linear voltammetric response to carbendazim concentration between 0.04 and 1.00 μM, with a low detection limit of 2.6×10^{-8} M, as well as good reproducibility, high selectivity, and long-term stability. Also, the prepared sensor was successfully applied to detect

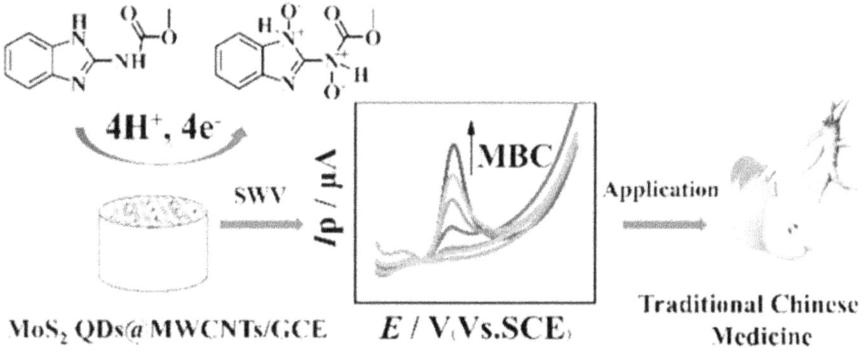

Figure 8.2. Electrochemical determination of carbendazim (Zhang *et al* 2021).

carbendazim in *Platycodon grandiflorum* and pear samples and the acquired results are in good accordance with the data reached by high-performance liquid chromatography (HPLC) (Zhang *et al* 2021).

Nowadays, boron doped diamond electrode (BDDE), as a notionally novel-posterity environmentally friendly electrode, demonstrates a potential electrode material which has useful electrochemical features, like a very stable and low background current, high current density, and a wide working potential scale. Because of excellent chemical and physical robustness, high thermal conductivity, hardness, and chemical inertness of the BDDE, it is well used as electrode for the electrochemical applications of diversified fungicides and herbicides (Brycht *et al* 2016). Šelešovská and co-workers studied the oxidation of fungicide azoxystrobin via CV with a BDDE in aqueous buffer solutions. They observed two irreversible anodic current peaks controlled by diffusion in wide pH window at potentials as +1600 and +2150 mV. Mechanism of the oxidation of azoxystrobin was proposed and promoted with HPLC/MS analysis of azoxystrobin solutions electrolyzed on carbon fiber (CF) electrode. The essential product of the first oxidation step with two-electron was defined as methyl 2-(2-{[6-(2-cyanophenoxy)pyrimidin-4-yl]oxy} phenyl)−2-hydroxy-3-oxopropanoate. The proposed method was applied for the detection of azoxystrobin in drinking and river water samples and commercial fungicide preparation by DPV with BDDE (table 8.2) (Šelešovská *et al.*, 2019).

Rajaram *et al,* developed an electrochemical method for the determination of herbicide paraquat using AuNPs-incorporated MWCNTs (figure 8.3). The electrode material was prepared and characterized using morphological characterization techniques like scanning electron microscopy (SEM), transmission electron microscopy, x-ray diffraction, and Raman spectroscopy. The characterization data disclosed the successful incorporation of AuNPs on the MWCNT and this material was used in the electrochemical detection of paraquat as electrode. The GCE was modified with AuNPs/MWCNT material reversibly using CV analysis. Electrochemical results showed that the designed electrode is an efficient catalyst for the effective electrochemical sensing of paraquat. Using CSSWV, Rajamar and co-workers calculated the limit of detection and sensitivity as 32 nM and 0.197 mA μM^{-1}, respectively with the linear concentration range between 1.0 and 2.0 μM (Rajaram *et al* 2022).

Carbon paste electrode (CPE) is a mixture of an electrically conductive a pasting liquid and graphite powder. It has been greatly utilized in electrochemical methods as a working electrode. This is because CPE possesses diversified benefits, for example: straightforward and fast sample preparation, a wide potential range, appropriate surface regeneration, modification and low residual current (Hernandez-Jimenez *et al* 2016). Modified CPEs have maintained significant attention during the last years and a great amount of electrochemical investigation has been dedicated to the improvement and implementation of various kinds of modified CPEs. Modification of CPEs with appropriate materials streamlines the electrochemical redox reactions of the compounds or ions to proceed without obstacles. The modification process ordinarily results in increased sensitivity and selectivity of the determinations. Modifying materials can catalyze the

Table 8.2. Some reported electrochemical applications for the detection of different herbicides.

Analyte	Method	Electrode	Limit of detection	Linearity range	Interferences	Applications	References
Aclonifen	DPV	$GdNbO_4$ modified GCE	1.15 nM	0.02–78 μM	Clothianidin, imidacloprid, thiamethoxam, metronidazole, chloramphenicol, metribuzin	Soil and river water samples	Gopi *et al* (2021)
Aclonifen	DPV	CPE	2×10^{-6} M	2×10^{-6}–1×10^{-4} M	—	—	Novotný and Barek (2015)
Alachlor	SWV	MIP/*o*-PD/GCE	0.8 nM	1 nM–1 μM	Propachlor, metalaxyl	Tap water samples	Elshafey and Radi (2018)
Atrazine	Chronoamperometry	PtNPs-BDD	0.0035 nM	0.0045–45 nM	—	—	Medina-Sanchez *et al* (2016)
Atrazine	SWV	PANI/Gr-α-ATZ	0.043 nM	2–20 000 nM	—	—	Van Chuc *et al* (2016)
Bentazone	DPV	BDDE	0.5 μM	2–100 μM	K^+, Mg^{2+}, Na^+, Ni^{2+}, Ca^{2+}, Zn^{2+}, Cu^{2+}, Pb^{2+}, Co^{2+}, SO_4^{2-}, NO_3^-, CH_3COO^-, CO_3^{2-}, sulcotrione, mesotrione	River water samples	Jevtic *et al* (2018b)
Bentazone	SWV	CB-CTS-ECH/GCE	1.4 μM	1.99–65.4 μM	Ba^{2+}, Mg^{2+}, vermicompost, humic acid	Natural water samples	Vaz *et al* (2021)
Bifenox	SWV	$MnFe_2O_4$@CTS/GCE	0.09 μM	0.3–4.4 μM	Isoproturon, diuron, pirimicarb, 4-nitrophenol, pendimethalin	Tap and river water analysis	Morawski *et al* (2021)
Difenzoquat	DPV	m-AgSAE	0.41 μM	0.2–4.0 μM	—	Water samples	Gajdár *et al* (2019)
Dinitramine	SWV	Au/AuNPs–S–$(CH_2)_3$–Si–SiO_2/Fe_3O_4	0.3×10^{-9} M	5×10^{-9}–1×10^{-7} M1 $\times 10^{-7}$–7×10^{-6} M	trifluralin, 2-methyl-4,6-dinitropheol, Al^{3+}, Fe^{3+}, Pb^{2+}, NO_3^-, Br^-, SO_4^{2-}, I^-, CO_3^{2-}	Wastewater samples	Irandoust and Haghighi (2016)
Diuron	DPAdSV	PtNPs/CS/GCE	17 μg l^{-1}	40–500 μg l^{-1}	Bifenox, 4-nitrophenol, Parathion, Pendimethalin, Roxarsone, Glyphosate, Fluometuron, Cu^{2+}; Zn^{2+}, Pb^{2+}, Cd^{2+}	River water samples	Morawski *et al* (2020)
Diuron	Amperometry	GO–MWCNT	1.49 μM	9 μM–0.38 mM	Na^+, Ca^{2+}, K^+, Mg^{2+}, Fe^{2+}, Ba^{2+}, Ni^{2+}, Co^{2+}, F^-, Cl^-, I^-, Br^-, $(COO)_2^{2-}$	Well, lake and irrigation ditch samples	Mani *et al* (2015)

Diuron	AdDPV	rGO–AuNPs/Nafion/GCE	0.3 nM	1×10^{-9}–1×10^{-7} M	Na^+, Ba^{2+}, Ca^{2+}, Al^{3+}, K^+, Cl^-, Co^{+2}, NO_3^-, Mg^{2+}, BrO_3^-, F^-, Cd^{2+}, Fe^{2+}, Ni^{2+}, $C_2O_4^{2-}$, Mn^{+2}, Cu^{2+}, biotin, 1-nitroso-2-naphtol, pyrocatechol violet, nitropyrene, amitraz, aminopyrene, bendiocarb, carbaryl malathion,	Orange, mineral and tap water samples	Zarei and Khodadadi (2017)
Fenuron	Amperometry	GO–MWCNT	—	0.9–47 μM	Na^+, Ca^{2+}, K^+, Mg^{2+}, Fe^{2+}, Ba^{2+}, Ni^{2+}, Co^{2+}, F^-, Cl^-, I^-, Br^-, $(COO)_2^{2-}$	Well, lake and irrigation ditch samples	Mani *et al* (2015)
Fluometuron	DPV	GCE	63 μg l^{-1}	207–3846 μg l^{-1}	Pirimicarb, pendimethalin, isoproturon, diuron, Pb^{2+}, Cd^{2+}, Cu^{2+}	River water and serum sample	Sousa *et al* (2022)
Isoproturon	DPAdSV	PtNPs/CS/GCE	7 μg l^{-1}	40–1000 μg l^{-1}	Bifenox, 4-nitrophenol, Parathion, Pendimethalin, Roxarsone, Glyphosate, Fluometuron, Cu^{2+}; Zn^{2+}, Pb^{2+}, Cd^{2+}	River water samples	Morawski *et al* (2020)
Isoproturon	DPV	AB/GCE	0.0956 μM	0.5–20 μM	Bentazone, Diquat, Imidacloprid, Paraquat, Mg^{2+}, NO_3^-, Zn^{2+}, Ca^{2+}	Tomato and river water samples	Liu *et al* (2022)
Linuron	DPV	Activated PGE	5.8×10^{-7} M	0.75–10 10–100 μM	—	Enriched tap water	Buleandra *et al* (2019)
Maleic hydrazide	DPV	SrS/Bi_2S_3/SPCE	1.8 nM	0.01–104 μM and 104–814 μM	paraquat, glyphosate, Carbendazim, 2,4,6-tri-chlorophenol, Iron, potassium, sodium.	Potato and river water samples	Akilarasan *et al* (2022)
Monolinuron	DPV	Activated PGE	3.7×10^{-7} M	0.75–7.5 7.5–100 μM	—	Enriched tap water	Buleandra *et al* (2019)
Metamitron	DPV	m-AgSAE	3.5×10^{-8} M	5.0×10^{-8}–2.0×10^{-5} M	Atrazine, Linuron, Glyphosate, Triasulfuron, Triclopyr, Picloram, Clopyralid, Cd^{2+}, Cu^{2+}, Pb^{2+}, Tl^+, Zn^{2+}	River water and Goltix Top samples	Šelešovská *et al* (2015)

(Continued)

Table 8.2. (*Continued*)

Analyte	Method	Electrode	Limit of detection	Linearity range	Interferences	Applications	References
Metoxuron	DPV	MIP-PGE	2.75 $\mu g\ ml^{-1}$		Diuron, 3,4-dikloroanilin, Propanil, urea		Sensoy *et al* (2020)
Metribuzin	DPV	HMDE	1.9×10^{-8} M	2.5×10^{-8}–8.0×10^{-7} M	Triclopyr, picloram, clopyralid, chlorpyrifos, glyphosate, linuron, triasulfuron, tribenuron, bentazon	Tap and river water, pesticide preparation samples	Janíková *et al* (2016)
Metribuzin	DPV	m-AgSAE	6.0×10^{-8} M	2.5×10^{-8}–8.0×10^{-7} M	Triclopyr, picloram, clopyralid, chlorpyrifos, glyphosate, linuron, triasulfuron, tribenuron, bentazon	Tap and river water, pesticide preparation samples	Janíková *et al* (2016)
Metribuzin	DPV	MWCNTs/MIP-based CPE	0.1 $pg\ ml^{-1}$	0.2 $ng\ ml^{-1}$–21.429 $\mu g\ ml^{-1}$	Chloropy rifos, glyphosate, dimethoate, thiamethoxam	Tomato, potato samples and commercial formulation samples	Fatah *et al* (2023)
Paraquat	DPV	GR/ZnO-modified SPE	21.0 nM	0.05–2.0 μM	—	Soil samples	Liu *et al* (2017)
Paraquat	DPV	micro-Cu_2O/PVP-GNs modified GC-RDE	2.65×10^{-7} M	1.0×10^{-6}–2.0×10^{-4} M	K^+, NH_4^+, Na^+, Ca^{2+}, Cu^{2+}, Zn^{2+}, Acetate, Cl^-, SO_4^{2-}, NO_2^-, NO_3^-, CO_3^{2-} thiocarbamide, p-aminobenzoic acid, oxalic acid, o-aminobenzoic acid	Cabbage samples	Ye *et al* (2012)
Paraquat	CSSWV	AuNP-MWCNT/GCE	32 nM	1.0–2.0 μM	—	—	Rajaram *et al* (2022)
Pethoxamid	SWV	BDDE	1.37 μM	3–100 μM	K^+, Ca^{2+}, Na^+, Mg^{2+}, Ni^{2+}, Cu^{2+}, NO_3^-, SO_4^{2-}, CO_3^{2-}, HCO_3^-, Cl^-, CH_3COO^-	Commercial herbicide formulation and river water samples	Jevtic *et al* (2018)

Propham	SWV	MWCNT/GCE	7.59×10^{-7} M	3.00×10^{-6}–3.86×10^{-5} M	—	River water samples	Leniart *et al* (2016)
Propham	SWAdSV	MWCNT/GCE	3.65×10^{-7}	2.00×10^{-6}–4.78×10^{-5} M	—	River water samples	Leniart *et al* (2016)
Simazine		MIP/ATP@AuNPs/ATP/Au	0.013 μM	0.03–140 μM	—	Tap water, river water and soil samples	Zhang *et al* (2017)
Sulfentrazone	SWV	MWCNT/SPE	0.8 μM	1.0–25 μM	Linuron, emamectin benzoate, clomazone, propanil, trifluralin, tebuthiuron, atrazine, glyphosate, carbaryl, humic acid, Na^+, Ba^{2+}, Ca^{2+}, Cu^{2+}, Ni^{2+}, Co^{2+}, Al^{3+}, NO_3^-, Cl^-,	Soy milk and Groundwater samples	Silva *et al* (2019b)
Terbutryn	DPV	HMDE	8.9×10^{-9} M	5×10^{-9}–1×10^{-6} M	Triclopyr, clopyralid, glyphosate, triasulfuron, imidacloprid, atrazine, chlorpyrifos, chloridazon, bentazon, metamitron, picloram, tribenuron, metribuzin	River and drinking water samples	Šelešovská *et al* (2016)
Terbutryn	DPV	m-AgSAE	2.9×10^{-8} M	1×10^{-7}–1×10^{-5} M	Triclopyr, clopyralid, glyphosate, triasulfuron, imidacloprid, atrazine, chlorpyrifos, chloridazon, bentazon, metamitron, picloram, tribenuron, metribuzin	River and drinking water samples	Šelešovská *et al* (2016)
Terbutryn	DPV	p-AgSAE	4.3×10^{-9} M	2×10^{-8}–5×10^{-5} M	Triclopyr, clopyralid, glyphosate, triasulfuron, imidacloprid, atrazine, chlorpyrifos, chloridazon, bentazon, metamitron, picloram, tribenuron, metribuzin	River and drinking water samples	Šelešovská *et al* (2016)

(*Continued*)

Table 8.2. (*Continued*)

Analyte	Method	Electrode	Limit of detection	Linearity range	Interferences	Applications	References
Terbutryn	DPV	BDDE	1.8×10^{-7} M	5×10^{-7}–5×10^{-5} M	triclopyr, clopyralid, glyphosate, triasulfuron, imidacloprid, atrazine, chlorpyrifos, chloridazon, bentazon, metamitron, picloram, tribenuron, metribuzin	River and drinking water samples	Šelešovská *et al* (2016)
Trifluralin	FFT-SWV	CuNW/CPE	0.15 nM	100–0.02 nM	—	Urine and plasma samples	Mirabi-semnakolaii *et al* (2011)
Ziram	SWV	p-AgSAE	2.4×10^{-7} M	5.00×10^{-7}–7.50×10^{-4} M	—	River water samples	Silva and De Souza (2017)
2,4-D	EIS	PPy/MIP/PGE	0.02 μg l^{-1}	0.06–1.25 μg l^{-1}	—	Water samples	Prusty and Bhand (2017)
4-chloro-2-methyl phenoxyacetic acid	CV	PANI-β-CD/MWCNT/GCE	0.99 μM	10.0–100 μM	Ca^{2+}, NO_3^-, Mg^{2+}, PO_4^{3-}, 2,4-D; 4-chloro-2-methylphenol,	River water samples	Rahemi *et al* (2012)

Abbreviations: AB: acetylene black, AdDPV: adsorptive differential pulse voltammetry, ATP: aminothiophenol, BDDE: Boron doped diamond electrode, CB: Carbon black, CS: Chitosan, CSSWV: cathodic stripping square wave voltammetry, CV: Cyclic voltammetry, DPAdSV: Differential pulse adsorptive stripping voltammetry, DP-ASV: Adsorptive differential pulse anodic stripping voltammetry, DPSV: Differential pulse stripping voltammetric, DPV: Differential pulse voltammetry, ECH: epichlorohydrin, EIS: Electrochemicai impedance spectroscopy, FFT: Fourier transformation, GCE: Glassy carbon electrode, GNs: graphene nanosheets, GO: Graphene oxide, GR: Graphene, HMDE: Hanging mercury drop electrode, IL: Ionic liquid, m-AgSAE: mercury meniscus-modified silver solid amalgam electrode, MIP: Molecularly imprinted polymers, MWCNT: Multi-walled carbon nano tube, NPs: Nanoparticles, NW: Nanowire, *o*-PD: *o*-Phenylenediamine, p-AgSAE: polished silver solid amalgam electrode, PANI: polyaniline, PGE: Pencil graphite electrode, PPy: polypyrrole, PVP: pyrrolidone, RDE: rotating disk electrode, rGO: Reduced graphene oxide, SWAdSV: square wave adsorptive stripping voltammetry, SPCE: Screen printed carbon electrode, SPE: Screen printed electrode, SWV: Square wave voltammetry, α-ATZ: anti-atrazine antibody, β-CD: β-cyclodextrin.

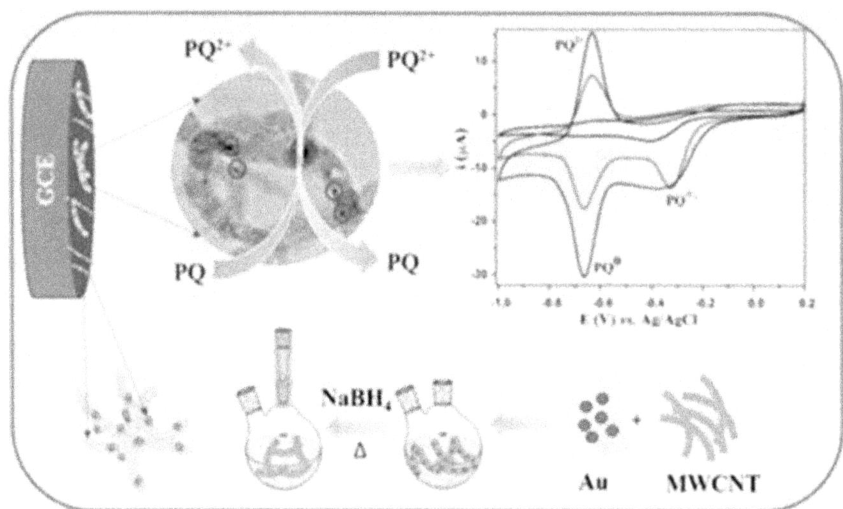

Figure 8.3. Electrochemical detection of herbicide paraquat via AuNPs-incorporated MWCNTs (Rajaram *et al* 2022).

electroreduction/electro-oxidation of some biologically and chemically significant ions and compounds (Joseph and Kumar 2009).

Novotný and Barek developed a novel electrochemical method for the detection of herbicide aclonifen at a CPE modified with tricresyl phosphate. The optimum electrochemical response was obtained using DPV in the negative potential range between −200 and −1600 mV. The limit of detection achieved for the proposed method was 2×10^{-6} M. The developed method was found appropriate for the determination of aclonifen in the concentration range between 2×10^{-6} and 1×10^{-4} M (Novotný and Barek 2015).

A capacitive electrochemical sensor for determination of 2,4-D (2,4-dichloro phenoxy acetic acid) in drinking water samples was developed using MIP-based polypyrrole on PGE (figure 8.4). MIP-coated PGE was arranged by electropolymerization of pyrrole monomer via chronopotentiometry technique in the availability of 2,4-D as the template molecule. The obtained electrode was characterized by SEM, CV, and electrochemical impedance spectroscopy (EIS). In the availability of 2,4-D, the capacitance change value of MIP electrodes were evaluated using EIS method. The designed capacitive sensor displayed a linear concentration range 0.06–1.25 $\mu g\ l^{-1}$ with detection limit of 0.02 $\mu g\ l^{-1}$ and good selectivity towards 2,4-D in water samples with recoveries between 92% and 110%.

8.4 Conclusions

Residuals of fungicide and herbicide compounds in environmental areas and agricultural products are inducing intense fatalities to human health and their hazardous effects are increasing day by day. Recently, most scientific works have centered on developing the design and improvement of traditional fungicide and

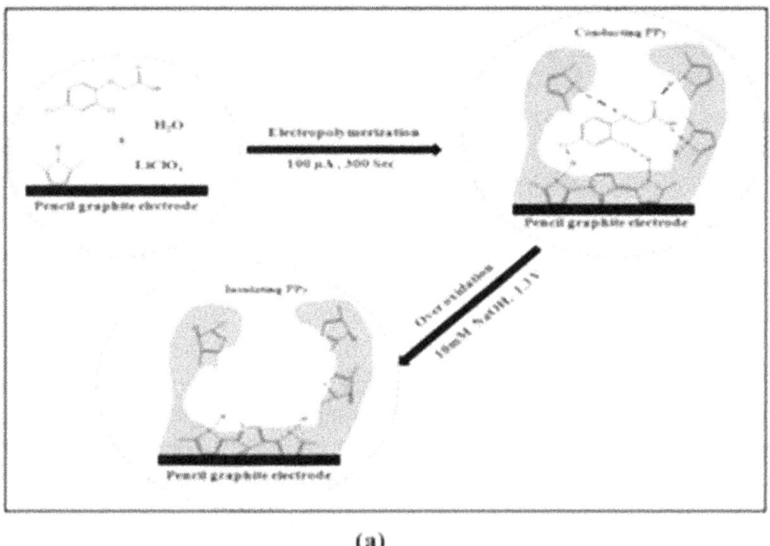

(a)

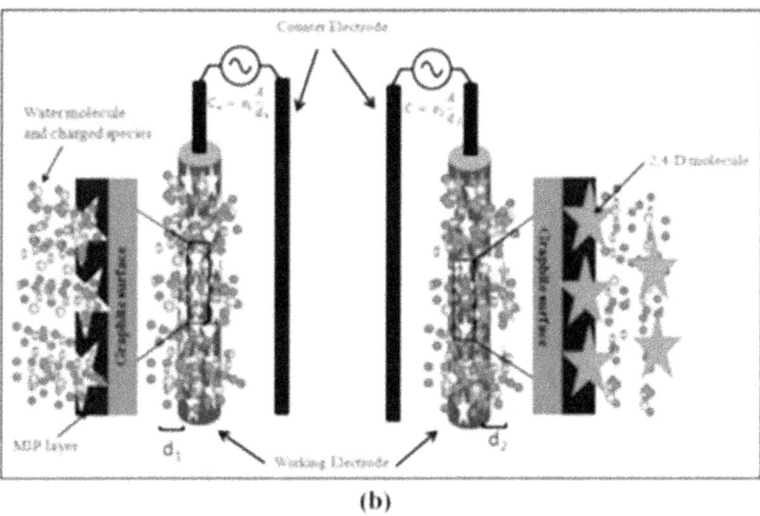

(b)

Figure 8.4. Schematic for preparation of MIP electrode using electropolymerization process. (b) Schematic of capacitive sensing (Prusty and Bhand 2017).

herbicide residue determination techniques. As a result of this situation, the complex strategies, long-time sample preparation processes, and time-consuming analysis associated with the traditional methods such as chromatography and spectroscopy are being changed by electrochemical on-site sensing techniques. In this chapter, developments and applications of electrochemical methods for fungicides and herbicides determination over the last 10 years were summarized. All these studies, as well as a large number of works not mentioned in this chapter, have shown that different fungicides and herbicides can be determined using electrochemical methods.

Though much progress has been made with the electrochemical determination of fungicides and herbicides, there is still a need to understand the challenges in their applications. Based on a literature survey of applications in electrochemistry area on fungicides and herbicides, we can say that future research is likely to focus on the following issues:

- Diversification of self-powered implements and varied disposable electrodes are extensive and these electrode systems will give remarkable status to low-cost, simplistic, and *in situ* analysis of fungicides/herbicides in decentralized laboratories.
- Samples taken from the environmental matrices, exclusively soil, river water and industrial wastewater, are very complicated as they implicate many types of contaminants along with pesticides. Simultaneous detection of different fungicides and herbicides in environmental samples, particularly contaminated or hazardous matrix, will be realized effectually outside the laboratory by electrochemical sensors modified with nanomaterials.
- Recent innovations in investigation in the nanostructured materials-based electrodes will also encourage more reforms in the building of a novel quantitative system, which would overcome the causes of interference and augment the analytical selectivity and sensitivity.
- The potent incorporation of several nano-scaled materials with each other will improve a new track for operating novel nanocomposite materials as the scaling up elements to design platforms of electrochemical electrodes with high efficiency.
- Sensors, as miniaturized appliances, are the convenient and interested apparatus to monitor trace and even ultra-trace effluents. They are highly eligible tools for on-site monitoring because of appropriate dimensions, portability, and low-cost. The improvement of sensor technology is significant to satisfy the ever-growing requisition for on-site high throughput detections. The implementations of these intelligent sensors to fungicide and herbicide analysis are tightly related to the advancement of portable devices. Portability properties will be achieved by connecting the developed electrochemical sensors with tablets and smartphones using suitable software.
- Together with the instructions of electrochemical oxidation and reduction mechanisms and a better comprehension of electrode designing with micro/nano-scale electrodes and tests, the success of electrode materials will become noteworthy in terms of both selectivity, sensitivity, quantification, and linearity limits. In particular, SPEs and paper electrodes will make prospective determination methods combined with transportable systems, a significant necessity for non-invasive and real-time analysis of an analyte, without any pretreatment and changing of the natural sample conditions of the environment.

References

Abbaci A, Azzouz N and Bouznit Y 2014 A new copper doped montmorillonite modified carbon paste electrode for propineb detection *Appl. Clay Sci.* **90** 130–4

Akilarasan M, Maheshwaran S, Chen S M, Tamilalagan E, Albaqami M D, Alotabi R G and Arumugam R 2022 In-situ synthesis of bimetallic chalcogenide SrS/Bi_2S_3 nanocomposites as an efficient electrocatalyst for the selective voltammetric sensing of maleic hydrazide herbicide *Process Saf. Environ. Prot.* **165** 151–60

Almeida J M S, Toloza C A T, Dornellas R M, Silva A R and Aucelio R Q 2016 Electrooxidation of trifloxystrobin at the boron-doped diamond electrode: electrochemical mechanism, quantitative determination and degradation studies *Int. J. Environ. Anal. Chem.* **96** 959–77

Ashrafi A, Milicevic J, Guzsvány V, Švancara I, Trtić-Petrović T, Purenović M and Vytřas K 2012 Trace determination of carbendazim fungicide using adsorptive stripping voltammetry with a carbon paste electrode containing tricresyl phosphate *Int. J. Electrochem. Sci.* **7** 9717

Bijekar S, Padariya H D, Yadav V K, Gacem A, Hasan M A, Awwad N S, Yadav K K, Islam S, Park S and Jeon B-H 2022 The state of the art and emerging trends in the wastewater treatment in developing nations *Water* **14** 2537

Bitew Z and Amare M 2020 Recent reports on electrochemical determination of selected antibiotics in pharmaceutical formulations: a mini review *Electrochem. Commun.* **121** 106863

Bozal-Palabiyik B and Uslu B 2016 A novel electroanalytical nanosensor based on MWCNT/ Fe_2O_3 nanoparticles for the determination of antiparkinson drug ropinirole *Ionics* **22** 115–23

Brycht M, Ozmen T, Burnat B, Kaczmarska K, Leniart A, Taskesen M, Kılıc E and Skrzypek S 2016 Voltammetric behavior, quantitative determination, and corrosion investigation of herbicide bromacil *J. Electroanal. Chem.* **770** 6–13

Brycht M, Skrzypek S K K, Burnat B, Leniart A and Gutowska N 2015a Square-wave voltammetric determination of fungicide fenfuram in real samples on bare boron-doped diamond electrode, and its corrosion properties on stainless steels used to produce agricultural tools *Electrochim. Acta* **169** 117–25

Brycht M, Burnat B, Skrzypek S, Guzsvány V, Gutowska N, Robak J and Nosal-Wiercinska A 2015b Voltammetric and corrosion studies of the fungicide fludioxonil *Electrochim. Acta* **158** 287–97

Brycht M, Skrzypek S, Robak J, Guzsvány V, Vajdle O, Zbiljic J, Nosal-Wiercinska A, Guziejewski D and Andrijewski G 2015c Ultra trace level determination of fenoxanil by highly sensitive square wave adsorptive stripping voltammetry in real samples with a renewable silver amalgam film electrode *J. Electroanal. Chem.* **738** 69–76

Brycht M, Kowalewska K, Skrzypek S K K and Mirceski V 2022 Electroanalytical study of fungicide bixafen on paste electrode based on the thermally reduced graphene oxide synthesized in air conditions and its determination in river water samples *Electroanalysis* **34** 1–11

Brycht M, Leniart A, Robak J, Burnat B, Kaczmarska K, Sipa K and Skrzypek S 2017 First electrochemical study of the fungicide oxycarboxin *Int. J. Environ. Anal. Chem.* **97** 1298–314

Brycht M, Skrzypek S, Nosal-Wiercinska A, Smarzewska S, Guziejewski D, Ciesielski W, Burnat B and Leniart A 2014 The new application of renewable silver amalgam film electrode for the electrochemical reduction of nitrile, cyazofamid, and its voltammetric determination in the real samples and in a commercial formulation *Electrochim. Acta* **134** 302–8

Brycht M, Burnat B, Nosal-Wiercinska A and Skrzypek S 2016 New sensitive square-wave adsorptive stripping voltammetric determination of pesticide chlornitrofen, and an evaluation of its corrosivity towards steel agricultural equipment *J. Electroanal. Chem.* **777** 8–18

Brycht M, Lukawska A, Frühbauerova M, Pravcova K, Metelka R, Skrzypek S and Sys M 2021 Rapid monitoring of fungicide fenhexamid residues in selected berries and wine grapes by square-wave voltammetry at carbon-based electrodes *Food Chem.* **338** 127975

Buleandra M, Popa D E, David I G, Bacalum E, David V and Ciucu A A 2019 Electrochemical behavior study of some selected phenylurea herbicides at activated pencil graphite electrode. Electrooxidation of linuron and monolinuron *Microchem. J.* **147** 1109–16

Buledi J A, Mahar N, Mallah A, Solagi A R, Palabıyık I M, Qambrani N, Karimi F, Vasseghian Y and Karimi-Maleh Y 2022 Electrochemical quantification of mancozeb through tungsten oxide/reduced graphene oxide nanocomposite: a potential method for environmental remediation *Food Chem. Toxicol.* **161** 112843

Carvalho R M, Pedão E R, Guerbas F M R, Tronchini M P, Ferreira V S, Petroni J M and Lucca B G 2023 Electrochemical study and forensic electroanalysis of fungicide benzovindiflupyr using disposable graphite pencil electrode *Talanta* **252** 123873

Chaabani A, Ben Jabrallah T and Belhadj Tahar N 2022 Electrochemical oxidation of ciprofloxacin on COOH-functionalized multi-walled carbon nanotube–coated vitreous carbon electrode *Electrocatalysis* **13** 402–13

Raluca C-S I, Frederick van Staden J K and Stefan-van Staden R-I 2022 Minireview-recent developments in electrochemical detection of atrazine *Anal. Lett.* **56** 847–69

Deep A, Sharma A L, Tuteja S K and Paul A K 2014 Phosphinic acid functionalized carbon nanotubes for sensitive and selective sensing of chromium (VI) *J. Hazard. Mater.* **278** 559–65

Dornellas R M, Nogueira D B and Aucelio R Q 2014 The boron-doped diamond electrode voltammetric method for ultra-trace determination of the fungicide pyraclostrobin and evaluation of its photodegradation and thermal degradation *Anal. Methods* **6** 944–50

Demir E and Silah H 2020 Development of a new analytical method for determination of veterinary drug oxyclozanide by electrochemical sensor and its application to pharmaceutical formulation *Chemosensors* **8** 25

Dindar C K, Erkmen C and Uslu B 2021 Electroanalytical methods based on bimetallic nanomaterials for determination of pesticides: past, present, and future *Trends Environ. Anal. Chem.* **32** e00145

Dong S, Bi Q, Qiao J, Shao S and Lu X 2020 Simultaneous determination of three nitroaniline isomers by β-cyclodextrins (β-CDs) and graphene quantum dots (GQDs) composite modified glassy carbon electrodes *Int. J. Electrochem. Sci.* **15** 8552–62

Dong Y, Yang L and Zhang L 2017 Simultaneous electrochemical detection of benzimidazole fungicides carbendazim and thiabendazole using a novel nano-hybrid material modified electrode *J. Agric. Food Chem.* **65** 727–36

Dornellas M, Tormin T F, Richter E M, Aucelio R Q and Muñoz R A A 2014 Electrochemical oxidation of the fungicide dimoxystrobin and its amperometric determination by batch-injection analysis *Anal. Lett.* **47** 492–503

Dornellas R M, Franchini R A, Da Silva A R, Matos R C and Aucelio R Q 2013 Determination of the fungicide kresoxim-methyl in grape juices using square-wave voltammetry and a boron-doped diamond electrode *J. Electroanal. Chem.* **708** 46–53

Dornellas M, Muñoz R A A and Aucelio R Q 2015 Electrochemical determination of picoxystrobin on boron-doped diamond electrode: square-wave voltammetry versus BIA-multiple pulse amperometry *Microchem. J.* **123** 1–8

Durodola S S, Adekunle A S, Olasunkanmi L O *et al* 2022 A review on graphene quantum dots for electrochemical detection of emerging pollutants *J. Fluoresc.* **32** 2223–36

Elshafey R and Radi A E 2018 Electrochemical impedance sensor for herbicide alachlor based on imprinted polymer receptor *J. Electroanal. Chem.* **813** 171–7

Ensafi A A, Razaloo F and Rezaei B 2017 Electrochemical determination of fenitrothion organophosphorus pesticide using polyzincon modified-glassy carbon electrode *Electroanalysis* **29** 2839–46

Erkmen C, Kurbanoglu S and Uslu B 2020 Fabrication of poly(3,4-ethylenedioxythiophene)-iridium oxide nanocomposite based Tyrosinase biosensor for the dual detection of catechol and azinphos methyl *Sens. Actuators* B **316** 128121

Fatah M A A, El-Moghny M G A, El-Deab M S and El Nashar R M 2023 Application of molecularly imprinted electrochemical sensor for trace analysis of Metribuzin herbicide in food samples *Food Chem.* **404** 134708

Gajdár J, Barek J and Fischer J 2019 Electrochemical microcell based on silver solid amalgam electrode for voltammetric determination of pesticide difenzoquat *Sens. Actuators* B **299** 126931

Garrido J M P J, Rahemi V, Borges F, Brett C M A and Garrido E M P J 2016 Carbon nanotube b-cyclodextrin modified electrode as enhanced sensing platform for the determination of fungicide pyrimethanil *Food Control* **60** 7–11

Ghalkhani M, Salehi M and Khaloo S S 2018 Fabrication of a biocompatible nanocomposite-based voltammetric sensor for sensitive determination of mercazole *J. Electron. Mater.* **47** 6251–9

Gholivand M, Akbari A and Norouzi L 2018 Development of a novel hollow fiber-pencil graphite modified electrochemical sensor for the ultra-trace analysis of glyphosate *Sens. Actuators* B **272** 415–24

Gonçalves Filho D and de Souza D 2021 Electrochemical determination of trifluralin herbicide using silver solid amalgam electrode: application in fresh food samples *J. Braz. Chem. Soc.* **32**

Gopi P K, Mutharani B, Chen S M, Chen T W, Eldesoky G E, Ali M A, Wabaidur S M, Shaik F and Tzu C Y 2021 Electrochemical sensing base for hazardous herbicide aclonifen using gadolinium niobate ($GdNbO_4$) nanoparticles-actual river water and soil sample analysis *Ecotoxicol. Environ. Saf.* **207** 111285

Guziejewski D, Brycht M, Nosal-Wiercinska A, Smarzewska S, Smarzewska S, Ciesielski W and Skrzypek S 2014 Electrochemical study of the fungicide acibenzolars-methyl and its voltammetric determination in environmental samples *J. Environ. Sci. Health.* B **49** 550–6

Hassan M M, Xu Y, He P, Zareef M, Li H and Chen Q 2022 Simultaneous determination of benzimidazole fungicides in food using signal optimized label-free HAu/Ag NS-SERS sensor *Food Chem.* **397** 133755

He Y, Wu T, Wang J, Ye J, Xu C, Li J and Guo Q 2020 A sensitive pyrimethanil sensor based on porous $NiCo_2S_4$/graphitized carbon nanofiber film *Talanta* **219** 121277

Hernandez-Jimenez A, Roa-Morales G, Reyes-Perez H, Balderas-Hernandez P, Barrera-Diaz C E and Bernabe-Pineda M 2016 Voltammetric determination of metronidazole using a sensor based on electropolymerization of α-cyclodextrin over a carbon paste electrode *Electroanalysis* **28** 704–10

Janíková L, Šelešovská R, Rogozinská M *et al* 2016 Sensitive voltammetric method for determination of herbicide metribuzin using silver solid amalgam electrode *Monatsh. Chem.* **147** 219–29

Jevtic S, Vukojeviic J, Djurdjic S, Pergal M V, Manojlovic D D, Petkovic B B and Stankovic D M 2018 First electrochemistry of herbicide pethoxamid and its quantification using electroanalytical approach from mixed commercial product *Electrochim. Acta* **277** 136–42

Jevtić S, Stefanović A, Stanković D M, Pergal M V, Ivanović A T, Jokić A and Petković B B 2018 Boron-doped diamond electrode—a prestigious unmodified carbon electrode for simple and fast determination of bentazone in river water samples *Diam. Relat. Mater.* **81** 133–7

Joseph R and Kumar K G 2009 Electrochemical reduction and voltammetric determination of metronidazole benzoate at modified carbon paste electrode *Anal. Lett.* **42** 2309–21

Kailas L W, Surinder S and Kumar S K 2020 Process intensification of treatment of inorganic water pollutants *Inorganic Pollutants in Water* (Amsterdam: Elsevier) ch 13 pp 245–71

Kalaivani A and Narayanan S S 2021 A novel electrochemical immobilization of Hg in CdSe QDs —modified graphite electrode and its improved performance in voltammetric determination of lead (II) ion *Mater. Today Proc.* **36** 863–6

Kalijadis A, Dordevic J, Papp Z, Jokic B, Spasojevic B, Babic B and Petrovic T T 2017 A novel carbon paste electrode based on nitrogen-doped hydrothermal carbon for electrochem ical determination of carbendazim *J. Serb. Chem. Soc.* **82** 1259–72

Khadem M, Faridbod F, Norouzi P, Rahimi Foroushani A, Ganjali M R, Shahtaheri S J and Yarahmadi R 2017 Modification of carbon paste electrode based on molecularly imprinted polymer for electrochemical determination of diazinon in biological and environmental samples *Electroanalysis* **29** 708–15

Khadem M, Faridbod F, Norouzi P, Foroushani A R, Ganjali M R and Shahtaheri S J 2016 Biomimetic electrochemical sensor based on molecularly imprinted polymer for dicloran pesticide determination in biological and environmental samples *J. Iran. Chem. Soc.* **13** 2077–84

Killedar L, Ilager D, Malode S J and Shetti N P 2022 Fast and facile electrochemical detection and determination of fungicide carbendazim at titanium dioxide designed carbon-based sensor *Mater. Chem. Phys.* **285** 126131

Koukouvinos G, Karachaliou C-E, Raptis I, Petrou P, Livaniou E and Kakabakos S 2021 Fast and sensitive determination of the fungicide carbendazim in fruit juices with an immunosensor based on white light reflectance spectroscopy *Biosensors* **11** 153

Krishnapandi A, Babulal S M, Chen S M, Palanisamy S, Kim S C and Chiesa M 2023 Surface etched carbon nanofiber companied ytterbium oxide for pinch level detection of fungicides carbendazim *J. Environ. Chem. Eng.* **11** 109059

Kumunda C, Adekunle A S, Mamba B B, Hlongwa N W and Nkambule T T I 2021 Electrochemical detection of environmental pollutants based on graphene derivatives: a review *Front. Mater.* **7** 616787

Ilager D, Malode S J and Shetti N P 2022 Development of 2D graphene oxide sheets-based voltammetric sensor for electrochemical sensing of fungicide, carbendazim *Chemosphere* **303** 134919

Irandoust M and Haghighi M 2016 Fabrication of a new modified gold electrode based on gold nanoparticles and nanomagnetic $Fe_3O_4/SiO_2–(CH_2)_3–SH$ core–shell for electrochemical evaluation and determination of dinitramine herbicide in water *RSC Adv.* **6** 49798

Lavanya A L, Kumari K G B, Prasad K R S and Brahman P K 2021 Development of pen-type portable electrochemical sensor based on Au–W bimetallic nanoparticles decorated graphene-chitosan nanocomposite film for the detection of nitrite in water, milk and fruit juices *Electroanalysis* **33** 1096–106

Leniart A, Brycht M, Burnat B and Skrzypek S 2016 Voltammetric determination of the herbicide propham on glassy carbon electrode modified with multi-walled carbon nanotubes *Sens. Actuators* B **231** 54–63

Li H, Li J, Hou C, Du S, Ren Y, Yang Z, Xu Q and Hu X 2010 A sub-nanomole level electrochemical method for determination of prochloraz and its metabolites based on medical stone doped disposable electrode *Talanta* **83** 591–5

Liu H, Chen M, Lin Y and Liu Y 2017 Electrochemical study of the herbicide paraquat based on a graphene-zinc oxide nanocomposite *Int. J. Electrochem. Sci.* **12** 8599–608

Liu R, Hu X, Cao Y, Pang H, Hou W, Shi Y, Li H, Yin X and Zhao H 2022 Highly sensitive electrochemical determination of isoproturon based on acetylene black nanoparticles modified glassy carbon electrode *Int. J. Electrochem. Sci.* **17** 220676

Lopez-Fernandez O, Barroso M F, Fernandes D M, Rial-Otero R, Simal-Gandara J, Morais S, Nouws H P A, Freire C and Matos C D 2015 Voltammetric analysis of mancozeb and its degradation product ethylenethiourea *J. Electroanal. Chem.* **758** 54–8

Lucca B G, Petroni J M and Ferreira V S 2018 Electrochemical study and voltammetric determination of sodium diethyldithiocarbamate using silver nanoparticles solid amalgam electrode *Int. J. Environ. Anal. Chem.* **98** 859–73

Mani V, Devasenathipathy R, Chen S M *et al* 2015 High-performance electrochemical amperometric sensors for the sensitive determination of phenyl urea herbicides diuron and fenuron *Ionics* **21** 2675–83

Maximiamo M E, Lima F, Cardoso C A L and Arruda G J 2018 Modification of carbon paste electrodes with recrystallized zeolite for simultaneous quantification of thiram and carbendazim in food samples and an agricultural formulation *Electrochim. Acta* **259** 66–76

Medina-Sanchez M, Mayorga-Martinez C, Watanabe T, Ivandini T, Honda Y, Pino F, Nakata K, Fujishima A, Einaga Y and Merkoci A 2016 Microfluidic platform for environmental contaminants and degradation based on boron-doped electrodes *Biosens. Bioelectron.* **75** 365–74

Mekeuo G A F, Despas C, Monseu-Njki C P, Walcarius A and Ngameni E 2022 Preparation of functionalized *Ayous* sawdust-carbon nanotubes composite for the electrochemical determination of carbendazim pesticide *Electroanalysis* **34** 667–76

Mercan H and Inam R 2010 Determination of cymoxanil fungicide in commercial formulation and natural water by square-wave stripping voltammetry *CLEAN—Soil Air Water* **38** 558–64

Mielech-Łukasiewicz K and Starczewska B 2019 The use of boron-doped diamond electrode for the determination of selected biocides in water samples *Water* **11** 1595

Mirabi-semnakolaii A, Daneshgar P, Moosavi-Movahedi A A *et al* 2011 Sensitive determination of herbicide trifluralin on the surface of copper nanowire electrochemical sensor *J. Solid State Electrochem.* **15** 1953–61

Morawski F M, Winiarski J P, Campos C E M, Parize A L and Jost C L 2020 Sensitive simultaneous voltammetric determination of the herbicides diuron and isoproturon at a platinum/chitosan bio-based sensing platform *Ecotoxicol. Environ. Saf.* **206** 111181

Morawski F M, Caon N B, Sousa K A P, Faita F L, Parize A L and Jost C L 2021 Hybrid chitosan-coated manganese ferrite nanoparticles for electrochemical sensing of bifenox herbicide *J. Environ. Chem. Eng.* **9** 106298

Noori J S, Mortensen J and Geto A 2021 Rapid and sensitive quantification of the pesticide lindane by polymer modified electrochemical sensor *Sensors* **21** 393

Novokova K, Navratil T, Dytrtová J J and Chýlková J 2013 Application of copper solid amalgam electrode for determination of fungicide tebuconazole *Int. J. Electrochem. Sci.* **8** 1–16

Novotný V and Barek J 2015 A voltammetric technique using a modified carbon paste electrode for the determination of aclonifen *Ecol. Chem. Eng.* S **22** 451–8

Ozcelikay G, Kurbanoglu S, Palabiyik B B, Uslu B and Ozkan S A 2018 MWCNT/CdSe quantum dot modified glassy carbon electrode for the determination of clopidogrel bisulfate in tablet dosage form and serum samples *J. Electroanal. Chem.* **827** 51–7

Prusty A K and Bhand S 2017 A capacitive sensor for 2,4-D determination in water based on 2,4-D imprinted polypyrrole coated pencil electrode *Mater. Res. Express* **4** 035306

Pardeshi S and Dhodapkar R 2022 Advances in fabrication of molecularly imprinted electrochemical sensors for detection of contaminants and toxicants *Environ. Res.* **212** 113359

Rahemi V, Vandamme J J, Garrido J M P J, Borges F, Brett C M A and Garrido E M P J 2012 Enhanced host–guest electrochemical recognition of herbicide MCPA using a β-cyclodextrin carbon nanotube sensor *Talanta* **99** 288–93

Rahmani T, Bagheri H, Behbahani M *et al* 2018 Modified 3D graphene-Au as a novel sensing layer for direct and sensitive electrochemical determination of carbaryl pesticide in fruit, vegetable, and water samples *Food Anal. Methods* **11** 3005–14

Rajaram R, Gurusamy T, Ramanujam K and Neelakantan L 2022 Electrochemical determination of paraquat using gold nanoparticle incorporated multiwalled carbon nanotubes *J. Electrochem. Soc.* **169** 047522

Ribeiro F W P, Oliveira R C, Oliveira A G, Nascimento R F, Becker H, Lima-Neto P and Correia A N 2020 Electrochemical sensing of thiabendazole in complex samples using boron-doped diamond electrode *J. Electroanal. Chem.* **866** 114179

Rouabhi R 2010 *Introduction and toxicology of fungicides, Fungicides* ed O Carisse (InTech) http://intechopen.com/books/fungicides/introduction-andtoxicology-of-fungicides

Šelešovská R, Herynková M, Skopalová J *et al* 2019b Reduction behavior of insecticide azoxystrobin and its voltammetric determination using silver solid amalgam electrode *Monatsh. Chem.* **150** 419–28

Šelešovská R, Schwarzova-Peckova K, Sokolova R, Krejcova K and Martinkova-Keliskova P 2021 The first study of triazole fungicide difenoconazole oxidation and its voltammetric and flow amperometric detection on boron doped diamond electrode *Electrochim. Acta* **381** 138260

Šelešovská R, Janíková L, Pithardtová K *et al* 2016 Sensitive voltammetric determination of herbicide terbutryn using solid electrodes based on silver amalgam and boron-doped diamond *Monatsh. Chem.* **147** 207–17

Šelešovská R, Herynkova M, Skopalova J, Kelíšková-Martinková P, Janikova L and Cankar P 2019 Oxidation behavior of insecticide azoxystrobin and its voltammetric determination using boron-doped diamond electrode *Electroanalysis* **31** 363–73

Šelešovská R, Janíková L and Chýlková J 2015 Green electrochemical sensors based on boron-doped diamond and silver amalgam for sensitive voltammetric determination of herbicide metamitron *Monatsh. Chem.* **146** 795–805

Sensoy K G, Muti M and Karagözler A E 2020 Highly selective molecularly imprinting polymer-based sensor for the electrochemical determination of metoxuron *Microchem. J.* **158** 105178

Silva R O, Silva E A, Fiorucci A R and Ferreira V S 2019b Electrochemically activated multi-walled carbon nanotubes modified screen-printed electrode for voltammetric determination of sulfentrazone *J. Electroanal. Chem.* **835** 220–6

Silva L M and De Souza D 2017 Ziram herbicide determination using a polished silver solid amalgam electrode *Electrochim. Acta* **224** 541–50

Stankovic D M 2017 Electroanalytical approach for quantification of pesticide maneb *Electroanalysis* **29** 352–7

Stankovic D M, Mehmeti E, Svorc L and Kalcher K 2015 Simple, rapid and sensitive electrochemical method for the determination of the triketone herbicide sulcotrione in river water using a glassy carbon electrode *Electroanalysis* **27** 1587–93

Sousa K A P, Morawski F M, de Campos C E M, Parreira R L T, Piotrowski M J, Nagurniak G R and Jost C L 2022 Electrochemical, theoretical, and analytical investigation of the phenylurea herbicide fluometuron at a glassy carbon electrode *Electrochim. Acta* **408** 139945

Sui L, Wu T, Liu L, Wang H, Wang Q, Hou H and Guo Q 2019 A sensitive pyrimethanil sensor based on electrospun TiC/C film *Sensors* **19** 1531

Taqvi S I H *et al* 2022 Plant extract-based green fabrication of nickel ferrite ($NiFe_2O_4$) nanoparticles: an operative platform for non-enzymatic determination of pentachlorophenol *Chemosphere* **294** 133760

Van Chuc N, Binh N H, Thanh C T, Van Tu N, Huy N L, Dzung N T, Minh P N, Thu V T and Lam T D 2016 Electrochemical immunosensor for detection of atrazine based onpolyaniline/graphene *J. Mater. Sci. Technol.* **32** 539–44

Vaz F C, Silva T A, Fatibello-Filho O, Assumpção M H M T and Vicentini F C 2021 A novel carbon nanosphere-based sensor used for herbicide detection *Environ. Technol. Innov.* **22** 101529

Wang L, Li X, Yang R *et al* 2020 A highly sensitive and selective electrochemical sensor for pentachlorophenol based on reduced graphite oxide-silver nanocomposites *Food Anal. Methods* **13** 2050–8

Wong A, Anderson M S, Rafael F A, Fernando C V, Orlando F F and Maria D P T S 2021 Simultaneous determination of direct yellow 50, tryptophan, carbendazim, and caffeine in environmental and biological fluid samples using graphite pencil electrode modified with palladium nanoparticles *Talanta* **222** 121539

Xiao J, Wang H, Li C, Deng K and Li X 2021 A simple dopamine sensor using graphdiyne nanotubes and shortened carbon nanotubes for enhanced preconcentration and electron transfer *Microchem. J.* **160** 105755

Xing Y, Wu G, Ma Y, Yu Y, Yuan X and Zhu X 2019 Electrochemical detection of bisphenol B based on poly(Prussian blue)/carboxylated multiwalled carbon nanotubes composite modified electrode *Meas. J. Int. Meas. Confed.* **148** 106940

Yang J, Wang Q, Zhang M, Zhang S and Zhang L 2015 An electrochemical fungicide pyrimethanil sensor based on carbon nanotubes/ionic-liquid construction modified electrode *Food Chem.* **187** 1–6

Yang L, Hu Y, Wang Q, Dong Y and Zhang L 2016 Ionic liquid-assisted electrochemical determination of pyrimethanil using reduced graphene oxide conjugated to flower-like $NiCo_2O_4$ *Anal. Chim. Acta* **935** 104e112

Yola M L 2022 Carbendazim imprinted electrochemical sensor based on $CdMoO_4$/g-C_3N_4 nanocomposite: application to fruit juice samples *Chemosphere* **301** 134766

Ye X, Gu Y and Wang C 2012 Fabrication of the Cu_2O/polyvinyl pyrrolidone-graphene modified glassy carbon-rotating disk electrode and its application for sensitive detection of herbicide paraquat *Sens. Actuators* B **173** 530–9

Zarei K and Khodadadi A 2017 Very sensitive electrochemical determination of diuron on glassy carbon electrode modified with reduced graphene oxide–gold nanoparticle–Nafion composite film *Ecotoxicol. Environ. Saf.* **144** 171–7

Zhang X, Du J, Wu D, Long X, Wang D, Xiong J, Xiong W and Liao X 2021 Anchoring metallic MoS_2 quantum dots over MWCNTs for highly sensitive detection of postharvest fungicide in traditional chinese medicines *ACS Omega* **6** 1488–96

Zhang J, Wang C, Niu Y, Li S and Luo R 2017 Electrochemical sensor based on molecularly imprinted composite membrane of poly(o-aminothiophenol) with gold nanoparticles for sensitive determination of herbicide simazine in environmental samples *Sens. Actuators* B **249** 747–55

IOP Publishing

Real-Time Applications of Advanced Electrochemical Sensing Devices

Jamballi G Manjunatha

Chapter 9

Modified voltammetric approaches for the analysis of soil

M Kavitha, B Vijaya, P Karpagavinayagam and C Vedhi

Numerous efforts have been undertaken to keep track of the examination of the soil's chemical and biological components. This chapter describes analysis of soil by using modified voltammetry techniques. Electrodes are modified on their surface by nanoparticles (NPs), nanostructures (carbon nanotubes (CNTs), graphene) or other innovative materials such as boron-doped diamond. Electrodes can be chemically modified electrodes, especially with conducting polymers. This chapter is also focused on bio-modified electrodes. Special attention will be paid to strategies using biomolecules (DNA, peptide or proteins), enzymes or whole cells. Heavy metals have been extensively detected using electrochemical techniques. The modification of electrochemical sensors with metal and metal oxide NPs for the voltammetric detection of heavy metals, however, appears to be more promising. Taking advantage of the unique properties of NPs along with the advantages of electrochemical detection over conventional detection techniques, the analytical performance of all the reported electrodes is enhanced. The result was a rapid response time, increased sensitivity, very low limits of detection (LODs), simplified operational procedures and enhanced reproducibility.

Different kinds of electrode materials have effectively provided electrochemical analyses for monitoring the environment. Voltammetry techniques' applicability and advantages over traditional methods of herbicide analysis have aslo been established. Modification of an electrode increases its properties allowing detection or quantification of low concentrations of a specific analyte, for metals or metal oxides-based electrode it becomes possible to detect herbicides that are not electroactive. Most of the discussed works used real samples like tap and lake water, spiked soil and food (fruit and vegetables) to determine toxic compounds. But the necessity remains to integrate those new devices to commercial analysis or industries.

doi:10.1088/978-0-7503-5377-9ch9

Potentiometric electrochemical sensors have several advantages, which are piqueing people's interest in using them to monitor soil nutrients. They have the capacity to quickly and automatically identify soil nutrients across many targets. This chapter gives a survey of the application of voltammetric trace metal analysis to soil, by focusing on the challenging task to perform reliable *in situ* measurements.

9.1 Introduction

In recent years, increased industrialization and urbanization have contaminated the air, land, and water. The determination of pollution by heavy metals such as copper, mercury, lead, zinc and cadmium is of special concern because of the formation of complexes with proteins and their high toxicity [1–4]. Heavy metal ions are hazardous to ecosystems and can cause serious danger to human population because of their accumulation in organs including liver, heart, brain etc [5]. For this reason, a number of techniques have been proposed up to this point for the determination of heavy metals, including atomic absorption spectrometry, UV-Vis spectroscopy, colorimetric analysis, ion chromatography, inductively coupled plasma mass spectrometry, and electroanalytical techniques [6, 7]. One of the most important elements of the Earth is the soil, along with the atmosphere and water [8]. The following is a definition of soil that largely satisfies the objectives of environmental study for nature conservation: the complex, three-phase, multifunctional open system known as soil is what develops through time on the surface of the Earth's crust as a result of interactions between the parent mineral ingredients and organisms, occasionally with anthropogenic effects [9]. Its geographical and temporal heterogeneity, diversity, and occasionally multidirectional operations of soil-forming elements are what make soil unique. The heterogeneity of parent geological rocks, vegetation, terrain, animal, and human activity all contribute to the regional variability of soils. 'Soil memory' refers to certain characteristics that soils retain over a long time. 'Soil moments' are defined as fast changes in soil properties over a period of hours or days [10]. Natural soils are the most fundamental component of the Earth's biosphere, but because anthropogenic activities are continually changing them, they can be entirely altered and keep their new qualities for a very long period [11]. These new features aren't always good for ecosystem stability.

Soils are currently regarded as non-renewable natural resources (within practical periods of time). Because these metals are hazardous to a wide range of creatures, including humans, soil contamination by heavy metals is a major environmental concern. In order to measure heavy metals in soils, spectroscopic analytical methods such as atomic absorption, induced coupled plasma (ICP), and fluorimetry are typically used. The instrumentation itself is expensive and requires significant maintenance and operator expertise. Therefore, it is particularly important to develop alternative analytical methods to enable quick, affordable assessment and treatment of polluted soils. The traditional wet chemistry techniques and a number of new instrumental techniques, including molecular emission spectroscopy, atomic absorption spectrometry (AAS), nuclear magnetic resonance (NMR) spectroscopy, high performance liquid chromatography (HPLC), and gas chromatography combined with mass spectrometry (GC-MS), are among the standard methods of soil analysis. Due to the extraction and pretreatment procedures, these approaches are regrettably frequently expensive and time-consuming and call for specialized tools and trained workers. New, efficient, inexpensive, quick, and nondestructive methods of soil analysis are thus required. These new analytical methods need to be quick, affordable, and suited for mass assaying. Sensors of the physical, biological (including bioassays), and chemical types can be used to track soil quality. In the last two decades, these devices have attracted a lot of attention from scientists due to their ease of use, relatively low costs, and sufficient selectivity. Additionally, sensors are increasingly in demand for precision agricultural applications because they can detect many analytes concurrently, deliver findings in real time, and monitor the chemical composition of soil *in situ* in real time [12]. The following section includes a list of the primary sensor types used in soil analysis. Using anodic stripping voltammetry (ASV) and under potential deposition (UPD) techniques to identify trace metals, particularly copper in soil matrices, is recommended [13].

On the other hand, several electroanalytical techniques have been used to examine miniature solid electrodes like ultramicroelectrode arrays (UMEAs) for the detection or screening of heavy metals in samples of various kinds [14]. Chemically modified electrodes in electrochemical methods have been extensively used in sensitive and focused analytical techniques for the detection of minute quantities of physiologically significant chemicals [15–17]. A subfield of chemical analysis known as electroanalytical chemistry uses electroanalytical methods to learn more about chemical species' amounts, characteristics, and environments. Izaak Maurits Kolthoff defined electroanalytical chemistry as the integration of analytical chemistry and electrochemistry. It is appropriate to think of electroanalytical chemistry as a field of chemical analysis where the electrode is used as a probe to directly or indirectly determine different chemical species. Therefore, the creation of straightforward, dependable, quick, sensitive, and accurate analytical techniques is needed. Such electrochemical approaches as voltammetric techniques are offered by electroanalytical chemistry. The simplicity, low instrument cost, great sensitivity, strong stability, environmental friendliness, and on-site monitoring of these electrochemical technologies define them. Heavy metals (HMs) cannot be degraded since they are persistent in the environment (waters and soils). HMs are primarily produced by human activity, such as mining, smelting, or the disposal of various pollutants. Note that although some HMs (such as iron, selenium, cobalt, copper, manganese, molybdenum, and zinc) is essential for life, the majority of them are hazardous. For instance, mercury (Hg), which is known to primarily harm the neurological system, can enter the environment from mining or industrial waste as well as coal combustion; lead (Pb), which can harm the neurological system, is found in things like old paint, mining waste, incinerator ash, and water from lead pipes; the mining and electroplating industries produce cadmium (Cd), which damages the kidneys. Arsenic (As), which damages the skin, eyes, and liver, is produced by herbicides or the mining sector. Chromium, nickel, tin, and thallium are further significant heavy metals. It has become vital to find and measure HMs in soils for hygiene reasons. Monitoring of environmental contaminants in soil and water is linked to a number of human endeavours.

In order to address pollution challenges, there is a growing demand for environmental analysis of developing pollutants, which necessitates the use of more sensitive and focused analytical techniques. For the quantitative evaluation of pollutants like the herbicide aclonifen, electroanalytical techniques such as cyclic voltammetry (CV) and square wave voltammetry (SWV) techniques are the methods that are most frequently used (ACF). One of the most significant difficulties in recent years has been the rehabilitation of environmental degradation. Global research efforts in this direction are concentrating on precise environmental monitoring and harmful chemical analysis methods. Human activities are directly linked to the environmental monitoring of developing pollutants and toxins, such as harmful pesticides [18–23]. Due to efforts to increase food production, industrialization and agricultural expansion have caused these operations to pick up speed. However, due to continual exposure to various pollutants, including pesticides, herbicides, and HMs, the amount of pollutants introduced by such activities into the environment will have a dangerous influence on human health and the immune system [24–27]. The creation of better electroanalytical devices has a lot of potential thanks to the chemical alteration of electrode surfaces. Improved sensitivity and selectivity of the electrode as well as the ability to identify species that exhibit weak electroactivity at unmodified electrode surfaces are potential benefits of utilizing chemically modified electrodes.

Chemically modified carbon paste electrodes (CMCPEs) have gained a lot of attention in recent years due to their use in analytical chemistry. These electrodes come in two varieties: those with a conducting binder and those without. A low background current, a broad range of useful potential, quick renewability, and ease of fabrication are only a few benefits of CMCPEs [28, 29]. Because the analyte can be easily recognised by its voltammetric peak potential, voltammetric techniques exhibit great selectivity and sensitivity. The interference of accompanying substances is the main issue with these approaches, though. Overlapping voltammetric responses are obtained as a result of their extremely comparable oxidation peak potentials. Electrodes changed with various materials and methods are used to solve this issue. [30, 31]. 'An electrode made up of conducting or semiconducting material and covered with a film of chemical modifier, reveals chemical, electrochemical or optical features of a film by means of faradaic reactions or interfacial potential differences,' is how the IUPAC defines a chemically modified electrode (CME). CMEs provide a number of advantages over unmodified electrodes, including the ability to reduce peak potentials by catalysing the oxidation or reduction of species that exhibit significant overvoltages at unmodified electrodes. [32–35].

The creation and use of various types of CMEs has been the focus of extensive electrochemical study throughout the years. For the modification of the electrode surface, a variety of materials were used, including carbon nanotubes (CNTs), gold nanoparticles (AuNPs), platinum nanoparticles (PtNPs), molecularly imprinted conducting polymers (MIPs), L-cysteine (Lcys), Prussian blue, graphene, metal complexes, zeolites, etc. There have been numerous reviews on the creation of modified electrodes and their use in various analytical domains. The electrochemical sensors based on non-covalent MIPs were evaluated by Blanco Lopez *et al* [36], who provided information on various recognition elements, electrochemical transduction, and integration methodologies.

9.2 Electrochemical analysis

9.2.1 Voltammetric techniques

Voltammetry is the most often used electrochemical technique for the detection of HM ions. All electrochemical systems that rely on measurements of potential-dependent current are collectively referred to as voltammetric systems. A working electrode, a counter electrode, and a reference electrode make up a three-electrode electrochemical setup. While measuring the current between the working and counter electrodes, the voltage is applied between the working and reference electrodes. Different strategies result from changing the potential change method. The simplest method for linearly sweeping the potential with time is called linear sweep voltammetry (LSV) [37]. In order to perform CV, a working electrode's potential is first linearly scanned in one direction and then reversed [38]. To put it another way, one or more triangular potential waveforms [39] are at play. The fundamental idea behind pulsed voltammetry is the utilization of a voltage pulse signal. Different types of pulsed voltammetry can be created by changing the pulse's shape and amplitude [38]. Fixed magnitude pulses superimposed on a linear potential ramp are used in differential pulse voltammetry (DPV) [39]. SWV is the process of applying a working electrode with a waveform of a symmetrical square wave overlaid on a base staircase potential [39]. ASV, which is a subset of stripping voltammetry, is based on a two-step procedure. The HM is first pre-concentrated or electrodeposited at the electrode surface using a reduction of metal ions. The metal is oxidized again to produce the ion in the second process, which is called stripping. After taking the two phases into account, a number of variables, including the electrode material, deposition potential, and deposition time [40], are known to affect the analysis. An alternative approach is obtained when the pre-concentration step is non-electrolytic and the analyte physically adsorbes at the electrode surface: AdSV, or adsorptive stripping voltammetry summarizes how the potential for the signal-producing processes of CV, LSV, DPV, SWV, and ASV varies with time [41]. Sensitivity and detection limits are increased when several of these techniques are combined. Square wave anodic stripping voltammetry (SWASV), differential pulse anodic stripping voltammetry (DPASV), and linear sweep anodic stripping voltammetry are some of the combinations (LSASV).

9.2.1.1 Cyclic voltammetry

By using electroanalysis, this method is frequently employed to detect metals [42]. It pinpoints the system's redox potentials and assesses how media elements affected the redox process. The working electrode is subjected to a linear sweep potential, Ei, until it reaches a switching potential, Ef. The sweep is inverted and reset to its initial values when the potential reaches its maximum value, creating the voltammogram [43]. The experiment's timeframe is controlled by an instrumental parameter called scan rate,

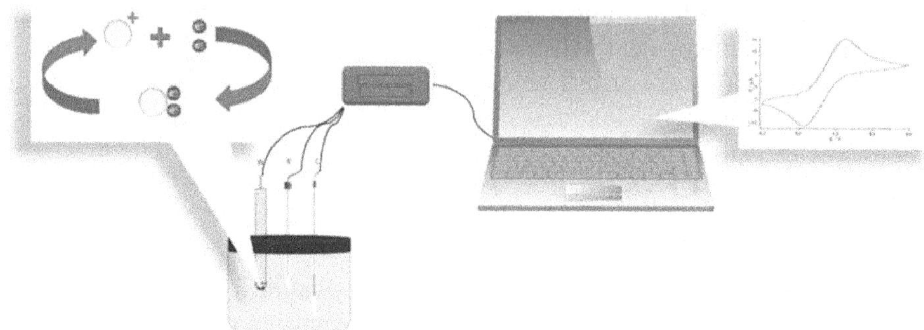

Figure 9.1. A schematic representation of CV technique.

and a current derivative from the applied potential is recorded. A voltammogram's features can be summed up as cathode and anode potential, as well as cathodic and anode currents [44] (figure 9.1).

Songa *et al* [45] employed a biosensor based on the enzyme horseadish peroxidase (HRP) to detect glyphosate in solutions. This biosensor was created by electrochemically depositing poly (2,5-dimethoxyaniline, or PDMA), doped with poly(4-styrenesulfonic acid, or PSS, onto the surface of a gold electrode. Additionally, the HPR enzyme binding to the PDMA-PPS composite film was encouraged. The experiment examined the HPR reaction to the substrate both before and after its interaction with glyphosate. It was based on the demonstration of HPR activity by glyphosate using H_2O_2 as substrate. The biosensor's response to H_2O_2 was reduced as a result of the herbicide's inhibition of HPR activity. Using CV, the constructed electrode exhibited a detection limit for glyphosate of 170 gl^{-1} (0.01 M).

9.2.1.2 Square wave voltammetry

The most efficient and practical pulse voltammetric method is SWV. The current potential curve's waveform is derivated by superimposing a potential ramp with steps, where the direct square wave pulse falls on the first step of the ramp. The square wave's reverse pulse also occurs at the halfway point of the ramp steps [46]. In order to measure the current in SWV, the identical square wave is sampled twice, one at the conclusion of the direct pulse and the other at the conclusion of the reverse pulse. According to the potential of the ramp step, the current difference between the two samplings is recorded [47–49]. Inam *et al* [50] investigated the SWM approach for the detection of the herbicide aclonifen. The herbicide displayed electroactivity on the working electrode made of glassy carbon, which made it possible to construct an analysis method employing SWV. They confirmed that using SWV in conjunction with square wave stripping voltammetry (SWSV) for the determination of cyclosulfamurons seems to be a good analytical methodology that allows for simple application, high levels of reproducibility, low detection limits, accuracy, and selectivity [51].

9.2.1.3 Differential pulse voltammetry

In DPV, the working electrode is subjected to pulses with fixed amplitudes superimposed on a ramp of increasing potential [52]. Two measurements of the current are taken: one before the pulse application (S1) and the other after the pulse (S2). Instrumentally subtracting the first current from the second, the difference in current is plotted versus the applied potential. The resulting voltammogram is made up of current peaks with a Gaussian shape, whose area is exactly proportional to the analyte concentration [53]. In many agricultural locations around the world, paraquat is a herbicide used to control weeds. Some nations have banned it and placed restrictions on it due to its high levels of toxicity [54]. Using a silver rotating electrode (SRE) in 0.1 mol l^{-1} Na_2SO_4 solution, Farahi *et al* [55] developed a quick and accurate approach to identify this substance. Two negative peaks of paraquat were visible at 0.7 and 1.0 V. They researched a number of factors, including deposition time, frequency, amplitude, and step amplitude, and selected the best ones for the DPV study. The author approach is effective in identifying low concentrations of Pq. Diuron, 2, 4-D, Tebuthiuron [56], Triasulfuron [57], Bromacil [58], Hexazinone [59], and Picloran [60] were all detected utilizing the DPV approach by unmodified and/or modified sensors.

9.2.1.4 Anodic stripping voltammetry

Metals that combine with mercury to create amalgams can be easily identified using anodic stripping voltammetry. When carbon (especially glassy carbon) is used as the working electrode and mercury(II) ions are added to the analyte solution in relatively low concentrations, a 'film' of mercury droplets with a diameter of a few micrometres is formed. At the same time, the analyte metals are deposited on the surface of the mercury droplets, where they typically dissolve a small amount of the analyte metal. Because these mercury droplets are so tiny, when they are oxidized, the metals have short diffusion lengths. As a result, relatively small peaks of strong currents are produced, enhancing the faradaic currents. In particular, ASV can be used to measure elements like Cu, Bi, Pb, Zn, Cd, and Sb that dissolve in mercury. However, many new techniques have been created recently that substitute carbon and bismuth electrodes for mercury [61].

9.2.1.5 Cathodic stripping voltammetry (CSV)

A material that was deposited on the electrode during the pre-concentration step is dissolved using the cathodic, or reductive, process in the CSV technique. When the potential of the mercury electrode is kept sufficiently positive to oxidize mercury to Hg_2^{2+} ions, which precipitate these salts, ions like chloride, bromide, and iodide can be pre-concentrated in the form of sparingly soluble salts, like Hg_2Cl_2, Hg_2Br_2, and Hg_2I_2 [62].

9.3 Types of modified electrodes

The creation of new techniques utilizing a variety of modified electrodes to be used with varied areas and samples is currently the subject of numerous research articles.

9.3.1 Chemically modified electrode materials

The choice of the necessary metallic oxide NPs was made since they are harmless at a wide range of doses and are essential for plant growth. NPs of Zn, Cu, Fe, and Mn metals were included in this group. Due to ZnO and CuO's high solubility, metal and metal oxide NPs have the ability to change or dissolve in soil. When determining whether a NP is effective or toxic, dissolution and transformation pathways must be taken into account. Although the long-term effects of NPs in plant systems are still unknown due to their extremely small size and potential to enter, translocate, and penetrate physiological barriers to travel within the plant tissues, some researchers have hypothesized that the NPs may cause morphological, physiological, genetic, and epigenetic changes that could affect plant growth and nutritional status. An electrode can generally be any electroconducting material that was initiated by a metal like platinum or gold, especially in the past. Later, glassy carbon was employed with a number of benefits, especially its simplicity and broad application. Following that, carbon paste was used because it is simple to prepare. To draw analytes, particularly metal ions, to the electrode surface and boost sensitivity, various chemicals have been blended. The surface areas for preconcentrating metal ions have significantly risen with the development of nanomaterials and conducting polymers, for instance, making the method ideal for trace metal analysis in accordance with the convenience of use and low cost of electrochemical methods. As a result, there are currently many research articles that involve the creation of new techniques employing a variety of modified electrodes that may be used with diverse areas and samples. This chapter intentionally avoids discussing the use of enzymes as biosensors in order to keep it clear and concise. Those who are interested can find those exact stories in-depth in the many sources that are readily available [63]. Furthermore, although a wide range of analytes can be used with modified electrodes, the focus of this study is on metal ions. However, the results of additional analytes will be succinctly presented for the benefit of the numerous applications that are already available and the promising qualities in adapting to metal ion analysis.

9.3.1.1 Micro and nanostructured materials in electrochemical sensors

In recent years, there has been a lot of interest in the incorporation of micro (MPs) or NPs into various matrices to create nanocomposite films, mostly because of their benefits such as low cost and special size-dependent features [64]. M/NPs are also exceptional because of their extraordinarily high surface area per mass, optical, mechanical, electrical, catalytic, and magnetic capabilities. These materials have been extensively used in a variety of electrochemical, electroanalytical, and bioelectrochemical applications. The development of strong, trustworthy electrical devices for effective process and pollution control is made possible by this new technology when combined with contemporary electrochemical techniques. According to their chemical makeup, the various M/NPs materials can be categorized into six groups: (1) metals; (2) metal oxides; (3) carbonaceous; (4) polymeric; (5) dendrimeric; and (6) composites [65]. Metals and metal oxides were the focus of the most pertinent research in developing an electrochemical sensor. For instance, zirconia (ZrO_2)-NPs were utilized to identify certain herbicides, and the material's application was justified by its high affinity for phosphorous groups. Additionally, the ZrO_2-NP offers a sizable surface area and improves the interaction of molecules containing phosphorous groups [66]. It was suggested to conduct similar research using AuNPs to detect hydrazine. The sensor was created using electrochemically reduced graphene oxide composite sheet made of CNTs and AuNPs (AuNPs-CNTs-ErGO) on a glassy carbon electrode (GCE). The results were obtained using CV and potential amperometry, and the improvement in sensor performance was attributed to the interaction between the AuNPs and CNTs-ErGO film as well as the exceptional catalytic impact of the AuNPs. The pre-concentration of metal ions as well as other chemicals can be accomplished using a variety of materials, which distinguishes electrochemistry as a special field in analytical chemistry and beyond. Metal NPs like silver or gold are the most often employed nanomaterials nowadays, especially in the beginning, to enhance surface areas and, as a result, the locations where metal ions can deposit. For a very long time, electrode surfaces have been altered to increase their capacity to sustain metal ions by using both conducting and nonconducting polymers. Numerous metal ions have been identified using mesoporous silica, which also offers the benefit of surface areas. Chitosan, a material from shrimp, is an example of a neutral compound having higher surface areas in collecting metal ions. It is certain that there is currently a chance and that many compounds are being investigated or are even awaiting discovery. Last but not least, it has also been demonstrated that combining different materials can increase the amount of metal ions that are received. Following the application of the aforementioned materials to electrodes, parameters are tuned before the techniques are applied to actual samples. Typically, the outcomes are assessed against accepted practices or verified using accepted materials. It is also possible to utilize a variety of spectroscopic and electrochemical techniques to provide further information about the analysis. Additionally, a variety of substances have been studied in the form of layers and sublayers as well as specialized holes as a specific substrate for particular analytes, hence the new term 'molecular imprinted,' which makes the process incredibly specific.

9.3.1.2 Conducting polymer-modified electrodes

CMEs, which improve the sensitivity and selectivity of electrochemical analytical techniques, have drawn more attention recently. Conducting polymers (CPs), among many modification techniques, have drawn a lot of attention because of their high electrical conductivities, good adhesive qualities, and simplicity of manufacture. The anti-fouling property of CPs was also established, which is a significant practical advantage over traditional electrode materials.

9.3.1.2.1 Surface modification of conducting polymers

Conducting polymers can be also modified to enhance the functionality of nanocomposites. The surface modifications of conducting polymers have some concerns including:

- enhancement of the charge transport of carriers between the implant and tissue;
- mediating the large difference in mechanical modulus;
- improvement of biodegradation;
- decreasing the impedance to enhance the sensitivity of the recording site;
- cell response enhancement;
- bioactivity enhancement.

9.3.1.2.2 Modified conducting polymers

As was previously demonstrated, some CPs have inherent characteristics that may confer sensitivity and selectivity toward HMs; however, in order to enhance these characteristics, ligands must be added. For instance, ligands can be added by creating copolymers, trapping known ionophores, or using specific counter ions [67].

9.3.1.3 CNT-modified conducting polymers

Sumio Iijima, a Japanese physicist, first identified CNTs as tubes with a nanometer-level diameter in 1991. They fall into two categories: single-walled (SWCNTs) and multi-walled (MWCNTs), each with distinct characteristics, particularly in terms of metallic and magnetic behaviour. They can be created via arc discharge, laser vaporization, or chemical vapour deposition. They can be used in many different contexts, particularly modified electrodes. CNTs can be put on any substrate electrode, but GCE is preferred. They can be mounted either alone or together with other materials. Due to its advantages of a highly ordered structure, light weight strength, as well as thermal and electrical conductivity, MWCNT is typically more satisfactory. In particular, the multi-walled type have been widely used in the determination of metal ions or organic compounds by spectroscopy or electrochemistry [68–71]. Their ability to absorb metal ions is primarily what gives them an edge in analysis [72]. This characteristic makes it appropriate for use in the field of energy [73]. Additionally, because of the vast surface areas of CNTs, a variety of single-layer or multi-layer materials can be put on them to improve the capacity to preconcentrate metal ions prior to their determination [74].

9.3.1.4 Electrodes modified by biomolecules

This is a brand-new electrochemical method for identifying bioavailable zinc (II) and cadmium (II) in actual soil samples. The successive deposition of bismuth and gallium thin films on altered screen-printed carbon electrodes (SPEs) allowed for this . A variety of graphitic modifications were tested, and the alteration that was most effective was reduced graphene oxide/graphitic carbon nitride (rGO/g-C_3N_4). The functional groups of GO were credited with the material's improved stability, while the intercalated g-C_3N_4 was responsible for its higher electron transfer rate. A bismuth thin film was used to measure the amount of cadmium in acetate buffer (pH 4.6) for voltammetric analysis (BiTF). After adjusting the pH to 5.1, Zn was next measured in the same cell using a gallium thin film (GaTF). The justification for two distinct thin films is given. In the matrix of actual samples, optimizations were carried out for the content of bismuth(III), gallium(III), potassium ferrocyanide, pH, and deposition potentials. The collected soil matrix's LODs and limits of quantification for bioavailable Cd and Zn were found to be 0.01 and 0.03 mg kg^{-1} and 0.01 and 0.04 mg kg^{-1}, respectively. When compared to the well-known method of ICP-OES, Cd and Zn levels in many soil samples showed good agreement. This method creates the opportunity for on-site, portable measurement of Cd and Zn concentrations in actual soil samples to assess the likelihood of Cd uptake by crops. This is the electrochemical method for voltammetric analysis of ACF herbicide in soil and water samples. A GCE (g-C_3N_4/GCE) covered with graphitic carbon nitride was used to create the sensing device for the detection of ACF. The SWV method was used to find ACF in traces of concentration. With an LOD of the analyte at 1.28 nM, linearity was seen over the range of 0.01–1.2 M. The approach devised was effectively applied to the monitoring of ACF herbicide (aclonifen) in wastewater as well as in actual soil samples because of its biological, agronomic, and environmental relevance. For these goals, CV and DPV were both used. Utilizing CV and elimination voltammetry linear scan (EVLS), the reaction process was studied. The detection threshold of 0.2 mol l^{-1} was attained by applying the prolonged accumulation time (60 s). Two samples of actual soil solutions were used to test the applicability of the devised method for the determination of Teb (Tebuconazole).

9.3.1.5 Organically modified clay minerals

Clay mineral surface changes have drawn a lot of attention because they enable the development of novel materials and uses [75]. Clay minerals that have undergone organic modification are now necessary for the creation of polymer nanocomposites. In addition, modified clays are utilized in a variety of other products, including paints, cosmetics, refractory varnish, thixotropic fluids, adsorbents for organic pollutants in soil, water, and the air, rheological control agents, and more. Clays and clay minerals can be altered by a number of different methods:

- adsorption;
- ion exchange with inorganic cations and organic cations;
- binding of inorganic and organic anions (mainly at the edges);
- grafting of organic compounds, reaction with acids;
- pillaring by different types of poly(hydroxo metal) cations;
- intraparticle and interparticle polymerization;
- dehydroxylation and calcinations, delamination and reaggregation of smectites;
- lyophilization;
- ultrasound; and
- plasma.

Organoclays are typically prepared using the well-known ion exchange process using alkylammonium ions. The creation of the organoclays is typically described in the articles in laboratory scale, under various experimental settings, using clays from various locations and suppliers, as well as various types of organic compounds.

9.3.1.6 Clay modified electrode

Clays are substances that fall into one of two categories: anionic clays, which have positively charged hydroxide layers, and cationic clays, which have negatively charged alumino silicate layers. Their crystalline configuration determines how they are categorized. The most prevalent mineral on the surface of the planet is cationic, and it is used to make ceramics, cosmetics, and has other significant uses including catalysts, adsorbents, and ion exchangers [76]. When electrode surfaces are synthesized for analytical applications, clays exhibit desirable qualities, such as stability and affordability. Compared to other modifications, clays are attractive materials due to their clearly defined layered structures, flexible adsorptive characteristics, and potential as catalysts or catalytic supports. Ion exchange and adsorption properties can be used to construct biosensors as well as for the electrochemistry analysis of medicines, heavy metals, pesticides, and other chemicals [77]. Clay-modified electrodes (CLMEs) have been intensively researched since 1990, particularly the phenomenon of electron transfer. Moroccan montmorillonite, kaolinite, and goethite are frequently used clays [78]. To evaluate paraquat in food samples, Kasmi *et al* [79] attempted to validate Moroccan clay as a modifier of graphite electrode (potato, lemon, orange). The interaction between clay and CPE, according to the authors, who also claimed that the modifier showed strong paraquat ion binding and retention. Natural Cameroonian smectite-type clay was transformed into an organoclay by exchanging its cationic surfactants for cetyltrimethylammonium (CTA) and didodecyldimethyl ammonium (DDA). This procedure was carried out in order to create a novel GCE based on organoclay (as a modifier) for the SWV measurement of mesotrione herbicide. This electrode was also utilized for the commercial formulation CALLISTO, which is sold in the European maize market, and has an LOD of 0.26 mM [80]. Isoproturon is a medically toxic herbicide that is selective and is classified as a toxic class III herbicide. A GCE modified with a hetero polyacid montmorillonite clay (HPMM) and a surfactant called cetyltrimethylammonium bromide (CTAB) was used to measure it. The process was successful in reaching limits of determination at ng ml^{-1} levels. The technology can be used to assess spiked soil and water samples because of advantages including high sensitivity, good reproducibility, and easy apparatus [81].

9.3.1.7 Screen-printed electrode

There are many options available for SPEs. The most often used substance is still carbon. Small size, low cost, simplicity, and less sample waste are all benefits of SPE. Electrode modification can be used to address the issue of decreased sensitivity, highlighting its applications in a wider range of fields [82].

9.3.1.8 Carbon paste electrode

For a variety of reasons, including the excellent quality of carbon as an electrode, its affordability, and its simplicity, carbon paste is still frequently utilized in the construction of modified electrodes [83]. Additional advantages like stability, reproducibility, and quick response time can be attained with intelligent design. It has been discovered that this substance is useful for determining both chemicals and metal ions.

9.4 Applications of voltammetry

9.4.1 Graphene oxide

The process of making graphene can be changed to produce novel forms with advantageous features, such as thermally reduced graphene, partly reduced graphene, or even electrochemically reduced graphene (ErGO). Graphene oxide, which is quite beneficial in collecting metal ions and offering superior selectivity, resolution, and precision, is often the arrangement of oxygen in the structure. Both electrocatalization and electroluminescence (ECL) can be promoted by the inclusion of another chemical that can create the bond via conjugation with graphene. This modified graphene derivatives can be used satisfactorily in both wastewater treatment via adsorption [84] as well as analysis in only one step [85] in addition to the development of new batteries [86, 87] and improvement of antibacterial properties [88].

9.4.2 Metals

Although the uses for metal ions are concentrated here, metal and metal alloys can also be employed in the examination of several species, including nitrite. Additionally, bismuth—the only metal that may successfully replace mercury—is highlighted as an example. As a replacement for mercury electrodes in ASV, the bismuth film electrode (BiFE), which consists of a thin bismuth film placed on a carbon substrate, was proposed in 2000. The main benefit of bismuth film electrodes' electrochemical properties over mercury film electrodes (MFEs) is that, in addition to being easier to prepare, having higher sensitivity, being clearly defined and having separated stripping signals, and being insensitive to dissolved oxygen, Bi is also less toxic to the environment. Bismuth-based electrodes perform better at stripping because they can combine other metals with mercury to generate 'fused' alloys [89]. There are three common ways to produce a bismuth film: (i) by *ex situ* preparation, which is the preparation of the bismuth from an acidic solution; (ii) by *in situ* setup, which is the codeposition of the analyte; and (iii) by electrode modification of a film, such as Bi_2O_3(s) or BiF_3, to produce the Bi(s) coating [85]. *Ex situ* plating was found to be simpler to control because there are no interferences during depositing and the conditions can differ from analytical or stripping conditions. However, it is more prone to changes in the electrode surface during electrode transfer, and the method requires more steps, which increases the time required. The ability to replenish the electrode at any time is another benefit of *ex situ* techniques. Additionally, the potential can be better regulated because, for *in situ* preparation, bismuth must be stripped at a potential higher than that required for bismuth oxidation before being replated [90].

9.4.3 Metal complexes

On the substrate, a variety of metal complexes are immobilized in order to attract or interact with other materials. This sort of modification substance is typically employed for the electrocatalytic determination of organic and inorganic compounds because it already contains metals [91]. For the study of ascorbic, diethyl stilbestol, and acetaminophen, cobalt phthalocyanin has been extensively and constantly researched and applied [92]. Numerous studies on manganese porphyrins have been conducted [93]. Considering how efficiently porphyrins themselves can house metal ions, they should be able to be utilized in the analysis of metal ions as surface areas rise [94].

9.4.4 Metal nanoparticles

A fantastic review on the use of metal NPs to measure arsenic, chromium, lead, cadmium, and antimony was published [95]. The analytical performances of the technology can be considerably enhanced in several ways, most notably sensitivity due to a greater volume of analytes collected, by combining metal NPs with a variety of chemicals.

9.4.5 Metal compound nanoparticles

Only modified magnetic iron oxide NPs (M-MIONPs) for mercury detection are presented here as an example due to the large diversity of metal compound NPs that have been used in metal ion analysis, especially recently [96]. Because of its toxicity and bioaccumulation in the body, mercury is known to be harmful to human health even at low quantities. It was discovered that modified magnetic iron oxide nanoparticles (M-MIONPs) with 2-mercaptobenzothiazole (MBT) were capable of quickly, affordably, simply, effectively, capably, with easy prepartation, and safely absorbing mercury (II) ions from polluted surface water [97]. When combined with 2-mercaptobanzothiazole, MIONPs can enhance their absorption percentage to 98.6%, compared to 43.47% when used alone. With a high loading capacity of 590 g/g, salt concentrations and pH were shown to have little impact on mercury ion buildup. This demonstrates that by mixing metal compound NPs with other chemicals, their propensity to attract analytes can be significantly enhanced. With an LOD of 0.02 ppb, they were able to simultaneously detect arsenic (III) and mercury (II) in the presence of copper (II) using SWASV [98]. The same device also allowed for the detection of Cr (VI) with an LOD of 0.1 ppb [99]. In order to better control the density of the nanoelectrodes, Mordegan *et al* constructed gold nano-electrode ensembles utilizing a polycarbonate membrane as a template. For the detection of arsenic, an LOD of 5 ppt was attained [100].

9.4.6 Metal oxide nanoparticles

Over the past few years, metal oxide NPs have been the subject of substantial research in electrochemical detection. To achieve varying sizes, stability, and morphologies, they have been synthesized utilizing various techniques. These variations enable them to display a range of electrical and photochemical properties, leading to a variety of applications [101]. Electrodes have been modified using a variety of metal oxides, primarily transition metals, for the detection of a variety of analytes, including heavy metals. Even though practically all transition metals were used to create these oxides, only a few were really employed for the detection of HMs (table 9.1).

Table 9.1. Comparison of different electrodes and methods for the detection of soil.

Sl. No.	Type of methods	Modified electrode	Metals found	Applicability to specific samples	References
1	Square wave anodic striping voltammetry (SWASV)	Bi/GCPE and Bi/ MWCNT-Naîon/GCE	Cd(II) Pb(II)	Soil	[102]
2	Voltammetric method	Au/UMEAs ultramicroelectrode arrays,	Cu(II)	soil	[103]
3	Differntial pulse voltammetry (DPASV)	Pencil graphite electrode (PGE)	Cu(II)	Soil	[104]
	DPCSV		Mn(II)		
	Square wave stripping voltammetry (SWASV)		Zn(II)		
4	Square wave anodic striping voltammetry (SWASV)	Mercury drop electrode (HMDE)	Cu(II), Pb(II) and Zn (II)	Soil	[105]
			Ni(II)		
	Square wave cathodic striping voltammetry (SWCSV)				
5	SWASV	Hg-electrode-Pt microelectrode	Cu(II), Mn(II) and Zn(II)	Soil	[106]
	SWV		Fe(II)		
6	DPASV	Screen-printed electrode (SCE)	Pb(II), Cd(II), Zn(II), Ni (II), Hg(II) and Cu(II)	Soil	[107]
7	DPASV	Hanging mercury drop electrode (HMDE)	Zn(II), Pb(II) and Cd(II)	Soil	[108]
	ADPCSV		Cu(II)		
8	ASV with Osteryoung Square Wave (OSWSV)	CPE CME-HgC2O4 75:25	Pb(II)	Soil	[109]
9	SWASV	modified screen-printed carbon electrodes (SPEs).	Cd(II), Zn(II)	Soil	[110]
10	DPV	bismuth film/Fe_3O_4/IL composite modified screen-printed electrode	Cd(II)	soil	[111]

Iron oxide in its various forms is the most prevalent metal oxide utilized for the detection of heavy metals ($MnFe_2O_4$, Fe_2O_3 and Fe_3O_4). In contrast to $MnFe_2O_4$ and Fe_2O_3, where iron is only found as Fe^{3+} in the first two species, Fe_3O_4 contains both Fe^{2+} and Fe^{3+}. This allows for an electron hopping process between the two ions, boosting electrical conductivity. It is also important to note that the two methods most frequently employed when working with Fe_3O_4 NPs are SWV and GCEs. Micronutrients copper (Cu), zinc (Zn), and manganese (Mn) are used in soil samples for effective fertilizer application. Plant-available soil micronutrients were collected using the DTPA extraction method, and then Cu(II), Zn(II), and Mn concentrations were determined using the differential pulse stripping voltammetry (DPSV) and SWSV techniques. For these ions, variables like the deposition potential, the deposition period, the pH, and the concentration of the supporting electrolyte were optimized. Mn(II), Cu(II), and Zn(II) in soil samples were successfully determined using the stated stripping voltammetry procedures [112–115].

The CV method was used to analyse the electrooxidation of copper ions in solution using a novel CPE enhanced with CuO NPs. Copper's overlapping voltammetric response was strongly resolved by the redesigned electrode into a single, distinct peak. Silver, zinc oxide, titanium dioxide, and iron oxide are a few examples of metal-based NPs that are frequently employed in the nanotechnology sector. It is expected that NPs will reach the soil compartment due to the release of particles from NP-containing goods, particularly through the land application of sewage sludge produced by wastewater treatment. This review provides a summary of the research on the impacts and destiny of metal-based NPs in soil [116–120].

Voltammetry is a very sensitive electrochemical technique with the advantages of sensitivity and speed that has been used extensively for HM detection. However, NPs have the benefits of great selectivity and high surface area. In order to emphasise this use of metallic and metallic oxide NPs for the voltammetric detection of HMs, this review will do just that. The NPs were either utilized as the only coating on the electrode or in combination with other materials as modifiers. In every instance, the synthesized devices demonstrated improved analytical performance, resulting in lower LODs and higher sensitivities when compared to voltammetric systems without NPs. Additionally, some of these systems' applicability was examined using actual samples.

The CV method was used to analyse the electrooxidation of zinc ions in solution using a novel CPE enhanced with ZnO NPs. Zinc's overlapping voltammetric response was strongly resolved by the redesigned electrode into a single, distinct peak. For the sensitive and precise determination of Pb(II) utilizing DPASV, both in batch and in a flow injection analysis system, a dithizone (DTZ)-modified CPE was created. A medium exchange to a clean solution and a proper anodic stripping were done after that. Regarding the amount of modifier in the paste, accumulation time, background electrolyte, Pb(II) content, and other factors, the analytical performance was assessed [121–124].

A single modified electrode surface can be used in various analytical determinations thanks to an easy and quick replacement of the electrode surface. Cd(II), Hg (II), Cu(II), and Zn(II) were among the coexisting metal ions that did not interfere with the detection of Pb (II). The proposed method was used to determine the presence of lead in soils near a metallurgical transformation plant.

9.5 Conclusions

The modified electrodes can accurately identify a variety of metal ions in the analysis of soil. Many approaches can be successfully applied to their analysis at the trace level thanks to stripping voltammetry's speed, ease of use, and sensitivity. The modified voltammetric technique is used to identify mixtures of different substances to increase the sensitivity for tracking the quantities of significant metal ions. Additionally, stripping voltammetry might have a promising future with the discovery of novel nanomaterials. Since they may be effectively manufactured using a variety of electrode materials, electrochemical sensors are excellent options for environmental monitoring. For metal or metal oxide based electrodes, it becomes possible to detect herbicides that are not electroactive since the electrode modification improves its characteristics and enables detection or quantification of low quantities of a specific analyte. However, there is still a need to incorporate those novel technologies into commercial analyses or companies. The majority of the reviewed works employed real samples of spiked soil to determine harmful chemicals.

9.6 Future trends

Long-standing applications and research in electrochemistry have provided solid foundations for the creation of more modern electrochemical techniques. By applying changed electrodes, valuable prior discoveries are waiting for their advancements. Improvements are being made to examine metal ions more effectively and to accommodate simultaneous readings. To enhance surface areas and serve species imprints, new compounds can be studied, combined, or immobilized. This necessitates further research into the drawbacks and interactions between the changed substrate and analytes. Spectroscopic, separation, and other techniques should all operate well with modified electrodes in a number of ways. Spectroelectrochemical research has already used modified electrodes. New theoretical explanations may be modified for improved comprehension and implementations, which would pave the way for future inventions of ever-greater scope in the analysis of soil.

References

[1] Petrlova J, Krizkova S, Zitka O, Hubalek J, Prusa R, Adam V, Wang J, Beklova M, Sures B and Kizek R 2007 Utilizing a chronopotentiometric sensor technique for metallothionein determination in fish tissues and their host parasites *Sens. Actuators* B **127** 112

[2] Zhao H, Jiang Y, Ma Y, Wu Z, Cao Q, He Y, Li X and Yuan Z 2010 Poly(2-amino-4-thiazoleacetic acid)/multiwalled carbon nanotubes modified glassy carbon electrodes for the electrochemical detection of copper(II) *Electrochim. Acta* **55** 2518

[3] Cauchi M, Bessant C and Setford S 2008 Simultaneous quantitative determination of cadmium, lead, and copper on carbon-ink screen-printed electrodes by differential pulse anodic stripping voltammetry and partial least squares regression *Electroanalysis* **20** 2571

[4] Malel E, Sinha J K, Zawisza I, Wittstock G and Mandler D 2008 Electrochemical detection of Cd^{2+} ions by a self-assembled monolayer of 1,9-nonanedithiol on gold *Electrochim. Acta* **53** 6753
[5] Bagal-Kestwal D, Karve M S, Kakade B and Pillai V K 2008 Invertase inhibition based electrochemical sensor for the detection of heavy metal ions in aqueous system: Application of ultra-microelectrode to enhance sucrose biosensor's sensitivity *Biosens. Bioelectron.* **24** 657
[6] Senthilkumar S and Saraswathi R 2009 Electrochemical sensing of cadmium and lead ions at zeolite-modified electrodes: Optimization and field measurements *Sens. Actuators B Chem.* **141** 65
[7] Burshtain D, Mandler D and Electroanal J 2005 *Chem.* **581** 310
[8] Doran J W and Zeiss M R 2000 Soil health and sustainability: managing the biotic component of soil quality *Appl. Soil Ecol.* **15** 3
[9] Lvova L and Nadporozhskaya M 2017 Chemical sensors for soil analysis: principles and applications *New Pesticides and Soil Sensors* ed A M Grumezescu *Series Nanotechnology in the Agri-Food Industry* (Amsterdam: Elsevier) vol 10 pp 637–78
[10] Targulian V O and Goryachkin S V (ed) 2008 *Soil Memory: Soil as a Memory of Biosphere-Geosphere-Anthroposphere Interaction; Institute of Geography* (Moscow: Russian Academy of Sciences) (in Russian)
[11] Schoenholtz S H, Miegroet H V and Burger J A 2000 A review of chemical and physical properties as indicators of forest soil quality: challenges and opportunities *For. Ecol. Manag.* **138** 335
[12] Marios S and Georgiou J 2017 Precision agriculture: challenges in sensors and electronics for real-time soil and plant monitoring *BioCAS 2017 Proc., Turin, Italy, 19–21 October 2017; Proc. 2017 IEEE Biomedical Circuits and Systems Conf.* p 135591
[13] Beni V, Newton H V, Arrigan D W M, Hill M, Lane W A and Mathewson A 2004 Voltammetric behaviour at gold electrodes immersed in the BCR sequential extraction scheme media: Application of underpotential deposition–stripping voltammetry to determination of copper in soil extracts *Anal. Chim. Acta* **502** 195
[14] Brand M, Eshkenazi I and Kirova Eisner E 1997 The silver electrode in square-wave anodic stripping voltammetry. Determination of Pb^{2+} without removal of oxygen *Anal. Chem.* **69** 4660
[15] Xie P, Chen X, Wang F *et al* 2006 Electrochemical behaviors of adrenaline at acetylene black electrode in the presence of sodium dodecyl sulfate *Colloids Surf.*, B **48** 17–23
[16] Chen X, Zhao J, Zhao G *et al* 2012 A review on the synthesis of TiO_2 nanoparticles by solution route *Cent. Eur. J. Chem.* **54** 279–94
[17] Pradhan P, Mascarenhas R J, Thomas T *et al* 2014 Electropolymerization of bromothymol blue on carbon paste electrode bulk modified with oxidized multiwall carbon nanotubes and its application in amperometric sensing of epinephrine in pharmaceutical and biological samples *J. Electroanal. Chem.* **732** 30–7
[18] Compton R, Ford J and Marken F 2003 Electroanalysis at diamond-like and doped diamond electrodes *Electroanalysis* **15** 1349–63
[19] Welch C and Compton R 2006 The use of nanoparticles in electroanalysis: a review *Anal. Bioanal. Chem.* **384** 601–19
[20] O'Connor K, Arrigan D and Gy S 1995 Calixarenes in electroanalysis *Electroanalysis* **7** 205–15

[21] Wang W K, Chen J J, Gao M, Huang Y X, Zhang X and Yu H Q 2016 Photocataytic degradation of atrazine by boron doped TiO_2 with a tunable rutile/anatase ratio *Appl. Catal.* B **195** 69–76
[22] Agraz R, Sevilla M and Hernandez L 1995 Voltammetric quantification and speciation of mercury compounds *J. Electroanal. Chem.* **390** 47–57
[23] Vidal A, Dinya Z, Mogyorodi F and Mogyorodi F 1999 Photocatalyic degradation of thiocarbamate herbicide active ingredients in water *Appl. Catal.* B **21** 259–67
[24] lkasir R, Ganesana M, Won Y, Stanciu L and Andreescu S 2010 Enzyme functionalize nanoparticles for electrochemical biosensors: a comparative study with applications for the detection of bisphenol A *Biosens. Bioelectron.* **26** 43–9
[25] Li M, Li Y, Li D and Long Y 2012 Recent developments and applications of screen-printed electrodes in environmental assays—a review *Anal. Chim. Acta* **734** 31–44
[26] Centi S, Marrazza G and Mascini M 2007 Procedure 25PCB analysis using immunosensors based on magnetic beads and carbon screen-printed electrodes in marine sediment and soil samples *Compr. Anal. Chem.* **49** 179–84
[27] Gajdar J, Barek J and Fischer J 2016 Antimony film electrodes for voltammetric determination of pesticide trifluralin *J. Electroanal. Chem.* **778** 1–6
[28] Kalcher K 1990 Chemically modified carbon paste electrodes in voltammetric analysis *Electroanalysis* **2** 419–33
[29] Kalcher K, Kauffmann J M, Wang J, Švancara I, Vytřas K, Neuhold C and Yang Z 1995 Sensors based on carbon paste in electrochemical analysis: a review with particular emphasis on the period 1990–1993 *Electroanalysis* **7** 5–22
[30] Sun Y, Fei J, Hou J, Zhang Q, Liu Y and Hu B 2009 Simultaneous determination of dopamine and serotonin using a carbon nanotubes-ionic liquid gel modified glassy carbon electrode *Microchim. Acta* **165** 373
[31] Ates M, Castillo J, Sezai Sarac A and Schuh_mann W 2008 Carbon fiber microelectrodes electrocoated with polycarbazole and poly(carbazole-co-p-tolylsulfonyl pyrrole) films for the detection of dopamine in presence of ascorbic acid *Microchim. Acta* **160** 247
[32] Gupta V K, Pal M K and Singh A K 2010 Drug selective poly(vinyl chloride)-based sensor of desipramine hydrochloride *Electrochim. Acta* **55** 1061
[33] Raoof J B, Ojani R and Chekin F 2007 Electrochemical analysis of D-Penicillamine using a carbon paste electrode modified with ferrocene carboxylic acid *Electroanalysis* **19** 1883
[34] Gupta V K, Goyal R N and Sharma R A 2009 Novel PVC membrane based alizarin sensor and its application; determination of vanadium, zirconium and molybdenum *Int. J. Electrochem. Sci.* **4** 156
[35] Codognoto L, Winter E, Paschoal J A R, Suffredi_ni H B, Cabral M F, Machado S A S and Rath S 2007 Electrochemical behavior of dopamine at a 3,3′-dithiodipropionic acid self-assembled monolayers *Talanta* **72** 427
[36] Blanco Lopez M C, Lobo Castanon M J, Miran da Ordieres A J and Tunon_Blanco P 2004 Electrochemical sensors based on molecularly imprinted polymers *Trends Anal. Chem.* **23** 36
[37] Kissinger P T and Heineman W R 1996 Laboratory techniques in electroanalytical chemistry *Division of Analytical Chemistry* 2nd edn (Washington, DC: American Chemical Society)
[38] Turdean G L 2011 Design and development of biosensors for the detection of heavy metal toxicity *Int. J. Electrochem.* **2011** 343125
[39] Verma N and Singh M 2005 Biosensors for heavy metals *Biometals* **18** 121e129

[40] WHO 2011 *Guidelines for Drinking-Water Quality* 4th edn (Geneva: WHO)
[41] Corr J J and Larsen E H 1996 Arsenic speciation by liquid chromatography coupled with ionspray tandem mass spectrometry *J. Anal. At. Spectrom.* **11** 1215
[42] de Gil S, Andrade C H, Barbosa N L, Braga R C and Serrano S H P 2012 Cyclic voltammetry and computational chemistry studies on the evaluation of the redox behavior of parabens and other analogues *Braz. Chem. Soc.* **23** 565–72
[43] Kulandainathan M A, Kulangiappar K, Raju T and Muthukumaran A 2005 Cyclic voltammetry and RRDE studies on the electrochemical behavior of azetidinone ester *Port. Electrochim. Acta* **23** 335
[44] Gulaboski R and Pereira C M 2008 Electroanalytical techniques and instrumentation in food analysis *Handbook of Food Analysis Instruments* (Boca Rotan, FL: Taylor and Francis)
[45] Songa E A, Arotiba O A, Owino J H O, Jahed N J, Baker P G L and Iwuoha E I 2009 Electrochemical detection of glyphosate herbicide using horseradish peroxidase immobilized on sulfonated polymer matrix *Bioelectrochemistry* **75** 117
[46] Souza D, Codognoto L, Malagutti A R, Toledo R A, Pedrosa V A, Oliveira R T S and Machado S A S 2004 Square wave voltammetry. Second part: applications *Quím. Nova* **27** 790
[47] Wang J 2006 Controlled-potential techniques *Analytical Electrochemistry* 3rd edn (Hoboken, NJ: Wiley)
[48] Souza D, Machado S A S and Avaca L A 2003 Square wave voltammetry. Part I: theoretical aspects *Quím. Nova* **26** 81
[49] Kilinc O, Grasset R and Reynaud S 2011 The herbicide aclonifen: The complex theoretical bases of sunflower tolerance *Pest. Biochem. Phys.* **100** 193
[50] Inam R and Çakmak Z 2013 A simple square wave voltammetric method for the determination of aclonifenherbicide *Anal. Methods* **5** 3314–20
[51] Sarıgül T and Inam R 2009 Study and determination of the herbicide cyclosulfamuron by square wave stripping voltammetry *Electrochim. Acta* **54** 5376–80
[52] Catrinck M N, Okumura L L, Silva A A, Saczk A A and Oliveira M F 2015 New and sensitive electroquantification of sulfentrazone in soil by differential-pulse voltammetry *J. Braz. Chem. Soc.* **26** 1751–9
[53] Tan S H and Kounaves S P 1998 Determination of selenium (IV) at a microfabricated gold ultramicroelectrode array using square wave anodic stripping voltammetry *Electroanalysis* **10** 364–8
[54] Poppenga R H and Oehme F W 2010 *Hayes' Handbook of Pesticide Toxicology* 3rd edn (Riverside, CA: Elsevier)
[55] Farahi A, Lahrich S, Achak M, El Gaini L, Bakasse M and El Mhammedi M A 2014 Parameters affecting the determination of paraquat at silver rotating electrodes using differential pulse voltammetry *Anal. Chem. Res.* **1** 16–21
[56] Duarte E H, Casarin J, Sartori E R and Tarley C R T 2018 Highly improved simultaneous herbicides determination in water samples by differential pulse voltammetry using boron-doped diamond electrode and solid phase extraction on cross-linked poly(vinylimidazole) *Sens. Actuators* B **255** 166
[57] Bandzuchová L, Selesovská R, Navrátil T and Chylková J 2013 Sensitive voltammetric method for determination of herbicide triasulfuron using silver solid amalgam electrode *Electrochim. Acta* **113** 1

[58] Brycht M, Özmen T, Burnat B, Kaczmarska K, Leniart A, Taştekin M and Skrzypek S 2016 Voltammetric behavior, quantitative determination, and corrosion investigation of herbicide bromacil *J. Electroanal. Chem.* **770** 6
[59] Toro M J U, Marestoni L D and Sotomayor M D P T 2015 A new biomimetic sensor based on molecularly imprinted polymers for highly sensitive and selective determination of hexazinone herbicide *Sens. Actuators* B **208** 299
[60] Bandzuchová L, Selesovská R, Navrátil T and Chylková J 2013 Voltammetric method for sensitive determination of herbicide picloram in environmental and biological samples using boron-doped diamond film electrode *Electrochim. Acta* **111** 242
[61] Wang J 2005 Stripping analysis at bismuth electrodes: a review *Electroanalysis* **17** 1341–6
[62] Scholz F and Kahlert H 2015 The calculation of the solubility of metal hydroxides, oxide-hydroxides, and oxides, and their visualisation in logarithmic diagrams *ChemTexts* **1** 7
[63] Turdean G L 2011 Design and development of biosensors for the detection of heavy metal toxicity *Int. J. Electrochem.* **2011** 1–15
Chooto P 2017 Modified Electrodes for Determining Trace Metal Ions (IntechOpen)
[64] Rassaei L, Amiri M, Cirtiu C M, Sillanpaa M, Marken F and Sillanpaa M 2011 Nanoparticles in electrochemical sensors for environmental monitoring *Trends Anal. Chem.* **30** 1704
[65] Mello R L S, Mattos-Costa F I, Villullas H M and Bulhões L O S 2003 Preparation and electrochemical characterization of Pt nanoparticles dispersed on niobium oxide *Eclé. Quím.* **28** 69
[66] Liu G and Lin Y 2005 Electrochemical sensor for organophosphate pesticides and nerve agents using zirconia nanoparticles as selective sorbents *Anal. Chem.* **77** 5894
[67] Ghasemi-Moarakeh L, Prabhakaran M P, Morshed M, Nasr Esfahani M H, Baharvand H, Kiani S, Al-Deyab S S and Ramakrishna S 2011 Application of conductive polymers, scaffolds and electrical stimulation for nerve tissue engineering *J. Tissue Eng. Regen. Med.* **5** 17–35
[68] Khalil M M and Abed El-aziz G M 2016 Multiwall carbon nanotubes chemically modified carbon paste electrodes for determination of gentamicin sulfate in pharmaceutical preparations and biological fluids *Mater. Sci. Eng.* C **59** 838–46
[69] Gooding J J 2005 Nanostructuring electrodes with carbon nanotubes: a review on electrochemistry and applications for sensing *Electrochim. Acta* **50** 3049–60
[70] Liang P, Liu Y, Guo L, Zenga J and Lua H 2004 Multiwalled carbon nanotubes as solid-phase extraction adsorbent for the preconcentration of trace metal ions and their determination by inductively coupled plasma atomic emission spectrometry *J. Anal. At. Spectrom.* **19** 1489–92
[71] Tuzena M, Saygia K O and Soylakb M 2008 Solid phase extraction of heavy metal ions in environmental samples on multiwalled carbon nanotubes *J. Hazard. Mater.* **152** 632–9
[72] Rao G P, Lu C and Su F 2007 Sorption of divalent metal ions from aqueous solution by carbon nanotubes: a review *Sep. Purif. Technol.* **58** 224–31
[73] Che G, Lakshmi B B, Fisher E R and Martin C R 1998 Carbon nanotubule membranes for electrochemical energy storage and production *Nature* **393** 346–9
[74] Pérez-Ràfols C, Serrano N, Díaz-Cruz J M, Ariño C and Esteban M 2016 Glutathione modified screen-printed carbon nanofiber electrode for the voltammetric determination of metal ions in natural samples *Talanta* **155** 8–13

[75] Betega de Paiva L, Morale A R and Diaz F R V 2008 Organoclays: properties, preparation and applications *Appl. Clay Sci.* **42** 8–24

[76] Mousty C 2004 Sensors and biosensors based on clay-modified electrodes—new trends *Appl. Clay Sci.* **27** 159

[77] Xiang Y and Villemure G 1995 Electrodes modified with synthetic clay minerals: Evidence of direct electron transfer from structural iron sites in the clay lattice *J. Electroanal. Chem.* **381** 21

[78] Guimarães A M F, Ciminelli V S T and Vasconcelos W L 2007 Surface modification of synthetic clay aimed at biomolecule adsorption: synthesis and characterization *Mater. Res.* **10** 37

[79] El Kasmi S, Lahrich S, Farahi A, Zriouil M, Ahmamou M, Bakasse M and El Mhammedi M A 2016 Electrochemical determination of paraquat in potato, lemon, orange and natural water samples using sensitive-rich clay carbon electrode *J. Taiwan Inst. Chem. Eng.* **58** 165–72

[80] Wagheu J K, Forano C, Besse-Hoggan P, Tonle I K, Ngameni E and Mousty C 2013 Electrochemical determination of mesotrione at organoclay modified glassy carbon electrodes *Talanta* **103** 337

[81] Manisankar P, Selvanathan G and Vedhi C 2006 Determination of pesticides using heteropolyacid montmorillonite clay-modified electrode with surfactant *Talanta* **68** 686

[82] Saengsookwaow C, Rangkupan R, Chailapakul O and Rodthongkum N 2016 Nitrogen-doped grapheme-polyvinylpyrrolidone/gold nanoparticles modified electrode as a novel hydrazine sensor *Sens. Actuators* B **227** 524–32

[83] Cazula B B and Lazarin A M 2017 Development of chemically modified carbon paste electrodes with transition metal complexes anchored on silica gel *Mater. Chem. Phys.* **186** 470–7

[84] Peng W, Li H, Liu Y and Song S 2017 A review on heavy metal ions adsorption from water by graphene oxide and its composites *J. Mol. Liq.* **230** 496–504

[85] Thiruppathi A R, Sidhureddy B, Keeler W and Chen A 2017 Facile one-pot synthesis of fluorinated graphene oxide for electrochemical sensing of heavy metal ions *Electrochem. Commun.* **76** 42–6

[86] Wu Y, Zhan L, Huang K, Wang H, Yu H, Wang S, Peng F and Lai C 2016 Iron based dual-metal oxides on graphene for lithium-ion batteries anode: effects of composition and morphology *J. Alloys Compd.* **684** 47–54

[87] Sahraei R and Ghaemy M 2017 Synthesis of modified gum tragacanth/graphene oxide composite hydrogel for heavy metal ions removal and preparation of silver nanocomposite for antibacterial activity *Carbohydr. Polym.* **157** 823–33

[88] Christos K, Anastasios E, Ioannis R and Constantinos E E 2008 Lithographically fabricated disposable bismuth-film electrodes for the trace determination of Pb(II) and Cd(II) by anodic stripping voltammetry *Electrochim. Acta* **53** 5294–9

[89] Yang D, Wang L, Chen Z, Megharaj M and Naidu R 2014 Voltammetric determination of lead (II) and cadmium (II) using a bismuth film electrode modified with mesoporous silica nanoparticles *Electrochim. Acta* **132** 223–9

[90] Karim A-Z and Fariba M 2014 Bismuth and bismuth-chitosan modified electrodes for determination of two synthetic food colorants by net analyte signal standard addition method *Cent. Eur. J. Chem.* **12** 711–8

[91] Leonardi S G, Bonyani M, Ghosh K, Dhara A K, Lombardo L, Donato N and Neri G 2016 Development of a novel Cu(ii) complex modified electrode and a

portable electrochemical analyzer for the determination of dissolved oxygen (DO) in water *Chemosensors* **4** 7–16

[92] Foster C W, Pillay J, Matters J P and Banks C E 2014 Cobalt phthalocyanine modified electrode utilized in electroanalysis: nono-structured modified electrodes vs. bulk screen-printed electrodes *Sensors* **14** 21905–22

[93] Sebarchievici I, Tăranu B O, Birdeanu M, Rus S F and Fagadar-Cosma E 2016 Electrocatalytic behaviour and application of manganese porphyrin/gold nanoparticle-surface modified glassy carbon electrodes *Appl. Surf. Sci.* **390** 131–40

[94] Tung H C, Chooto P and Sawyer D T 1991 Electron-transfer thermodynamics, valence-electron hybridization, and bonding of the meso-tetrakis(2,6-dichlorophenyl)porphinato complexes of manganese, iron, cobalt, nickel, copper, silver, and zinc and of the P+Mn(O) and .bul.P+Fe(O) oxene adducts *Langmuir* **7** 1635–41

[95] Metters J P and Banks C E 2014 Nanoparticle modified electrodes for trace metal ion analysis *Nanosensors for Chemical and Biological Applications* 1st edn ed K C Honeychurch (Amsterdam: Elsevier) pp 54–79

[96] Karim-Nezhad G, Khorablou Z, Zamani M, Dorraji P S and Alamgholiloo M 2016 Voltammetric sensor for tartrazine determination in soft drinks using poly (p-amino-benzenesulfonic acid)/zinc oxide nanoparticles in carbon paste electrode *J. Food Drug Anal.* **25** 293–301

[97] Parham H, Zargar B and Shiralipour R 2012 Fast and efficient removal of mercury from water samples using magnetic iron oxide nanoparticles modified with 2-mercaptobenzo-thiazole *J. Hazard. Mater.* **205–6** 94–100

[98] Jena B K and Raj C R 2008 Gold Nanoelectrode ensembles for the simultaneous electro-chemical detection of ultratrace arsenic, mercury, and copper *Anal. Chem.* **80** 4836–44

[99] Jena B K and Raj C R 2008 Highly sensitive and selective electrochemical detection of sub-ppb level chromium (VI) using nano-sized gold particle *Talanta* **76** 161–5

[100] Mardegan A, Scopece P, Lamberti F, Meneghetti M, Moretto L M and Ugo P 2012 Electroanalysis of trace inorganic arsenic with gold nanoelectrode ensembles *Electroanalysis* **24** 798–806

[101] George J M, Antony A and Mathew B 2018 Metal oxide nanoparticles in electrochemical sensing and biosensing: a review *Mikrochim. Acta* **185** 358

[102] Zhao G, Wang H and Liu G 2018 Sensitive determination of trace Cd(II) and Pb(II) in soil by an improved stripping voltammetry method using two different *in situ* plated bismuth-film electrodes based on a novel electrochemical measurement system *RSC Adv.* **8** 5079–89

[103] Orozco-Messana J, Mahon M and Iborra Lucas M 2017 Nucleation modelling on ZnO electrodeposition. Material-ES, 1(5), 83–86

[104] Dilara Kiliç H, Seda Deveci K B D, Çetinkaya E, Karadağ S and Doğ M 2018 Application of stripping voltammetry method for the analysis of available copper, zinc and manganese contents in soil samples *Int. J. Environ. Anal. Chem.* **98** 308–22

[105] Colombo C and Van Den Berg C M G 2006 Determination of trace metals (Cu, Pb, Zn and Ni) in soil extracts by flow analysis with voltammetric detection *Int. J. Environ. Anal. Chem.* **71** 1–17

[106] Toledo R A, Simões M L, Silva W T, Martin-Neto L and Pedrovaz C M 2009 Determination of metal ions extracted by DTPA in a soil treated with effluent using an Hg-electroplated-Pt microelectrode *Int. J. Environ. Anal. Chem.* **89** 1099

[107] Christidis K, Robertson P, Gow K and Pollard P 2007 Voltammetric in situ measurements of heavy metals in soil using a portable electrochemical instrument *Measurement* **40** 960

[108] Nedeltcheva T, Atanassova M, Dimitrov J and Stanislavova L 2005 Determination of mobile form contents of Zn, Cd, Pb and Cu in soil extracts by combined stripping voltammetry *Anal. Chim. Acta* **528** 143

[109] Beristain-Ortíz E G, Sumbarda-Ramos M, García García M T and Oropeza-Guzmán 2008 Pb detection in soil samples by electroanalysis: cyclic voltammetry and osteryoung square wave stripping voltammetry *GECS Trans.* **15** 527–33

[110] Mc Eleney a C, Alves b S and Mc Crudden D 2020 Novel determination of Cd and Zn in soil extract by sequential application of bismuth and gallium thin films at a modified screenprinted carbon electrode *Anal. Chim. Acta* **1137** 94e102

[111] Wang H, Zhao G, Yin Y, Wang Z and Liu G 2018 Screen-printed electrode modified by bismuth /Fe_3O_4 nanoparticle/ionic liquid composite using internal standard normalization for accurate determination of Cd(II) in Soil *Sensors* **18** 6

[112] Hareesha N, Manjunatha J G, Amrutha B M, Pushpanjali P A, Charithra M M and Prinith Subbaiah N 2021 Electrochemical analysis of indigo carmine in food and water samples using a poly (glutamic acid) layered multi-walled carbon nanotube paste electrode *J. Electron. Mater.* **50** 1230–8

[113] Tigari G and Manjunatha J G 2020 Optimized voltammetric experiment for the determination of phloroglucinol at surfactant modified carbon nanotube paste electrode *Instrum. Exp. Tech.* **63** 750–7

[114] Manjunatha J G 2020 Poly (Adenine) modified graphene-based voltammetric sensor for the electrochemical determination of catechol, hydroquinone and resorcinol *Open Chem. Eng. J.* **14** 52–62

[115] Manjunatha Charithra M and Manjunatha J G 2020 Electrochemical sensing of paracetamol using electropolymerised and sodium lauryl sulfate modified carbon nanotube paste electrode *ChemistrySelect* **5** 9323–9

[116] Pushpanjali P A and Manjunatha J G 2020 Development of polymer modified electrochemical sensor for the determination of alizarin carmine in the presence of tartrazine *Electroanalysis* **32** 2474–80

[117] Tigari G and Manjunatha J G 2020 Poly (glutamine) film-coated carbon nanotube paste electrode for the determination of curcumin with vanillin: an electroanalytical approach *Monatsh. Chem.* **151** 1681–8

[118] Manjunatha J G, Raril C, Hareesha N, Charithra M M, Pushpanjali P A, Tigari G, Ravishankar D K, Mallappaji S C and Gowda J 2020 Electrochemical fabrication of poly (niacin) modified graphite paste electrode and its application for the detection of riboflavin *Open Chem. Eng. J.* **14** 90-8

[119] 2022 *Carbon Nanomaterials-Based Sensors: Emerging Research Trends in Devices and Applications* ed J G Manjunatha and C M Hussain (Elsevier)

[120] Raril C, Manjunatha J G, Ravishankar D K, Fattepur S, Siddaraju G and Nanjundaswamy L 2020 Validated electrochemical method for simultaneous resolution of tyrosine, uric acid, and ascorbic acid at polymer modified nano-composite paste electrode *Surf. Eng. Appl. Electrochem.* **56** 415–26

[121] Prinith N S, Manjunatha J G and Hareesha N 2021 Electrochemical validation of L-tyrosine with dopamine using composite surfactant modified carbon nanotube electrode *J. Iran. Chem. Soc.* **18** 3493–503

[122] Pushpanjali P A, Manjunatha J G and Srinivas M T 2020 Highly sensitive platform utilizing poly (l-methionine) layered carbon nanotube paste sensor for the determination of voltaren *FlatChem.* **24** 100207
[123] Raril C and Manjunatha J G 2020 Fabrication of novel polymer-modified graphene-based electrochemical sensor for the determination of mercury and lead ions in water and biological samples *J. Anal. Sci. Technol.* **11** 1–10
[124] Manjunatha J G 2020 A surfactant enhanced graphene paste electrode as an effective electrochemical sensor for the sensitive and simultaneous determination of catechol and resorcinol *Chem. Data Collect.* **25** 100331

Chapter 10

Advances in electrochemical detection of pathogens

Geethanjali Parivara Appaji

Accurate detection of pathogenic microorganisms still remains a big challenge in the analytical and biomedicals field due to their abundance and variety. Therefore, development of strategies that are fast, reliable, inexpensive, sensitive and specific for detection by employing nanomaterials that are integrated with microfluidics devices, PCR amplification methods, or by combining these strategies, is attracting significant attention. Among all of them, electrochemical methods are bringing promising advantages as they exhibit versatile detection of organisms with various quantification measurements providing broader applications. Electrochemical biosensors used for pathogen detection are reviewed here in terms of transduction elements, biorecognition elements, electrochemical techniques and measurement methods. Transduction elements refer to different electrode materials used, whereas biorecognition elements include different biological components opted for in terms of availability, immobilization and production approaches. For measurement formats incorporated for pathogen detection that are classified as sample preparation, binding steps are discussed. Here we also discuss recent advances for detection of pathogens using electrochemical biosensors, also throwing light on emerging areas of electrochemical biosensor designs such as electrode modification and transducer integration. In addition to basic characters like analyte, biorecognition and transduction elements, different fabrication techniques, detection principles and applications are discussed. All this information highlights the applications of electrochemical biosensors for pathogen detection in food, water safety, medical diagnostics, environmental monitoring etc.

10.1 Advent of pathogens

Pathogens represent infectious agents causing diseases, which include microorganisms like bacteria, fungi, protozoans, minute agents such as viruses, prions etc.

doi:10.1088/978-0-7503-5377-9ch10

Today, pathogenic infections remain one of the principal causes of a high percentage of morbidity and mortality worldwide. The numbers of deaths occuring due to infections are declining each year by 1% with a forecast of 13 million deaths by 2050. Pathogens such as foodborne, airborne and waterborne enter the body through many modes of infection being responsible for spreading of diseases [1]. Amongst 1400 recognized human pathogens so far, the majority of infectious diseases are caused by mere few, about 20 pathogens only. This class includes common pathogens like norovirus, influenza virus, COVID-19 virus and bacterias like *Escherichia coli* and *Staphylococcus aureus*. All these pathogens have varied factors like virulence, transmission mode, contagiousness, infectious doses etc. To tackle this problem there are techniques available for rapid and sensitive detection of pathogens present in complex matrices like body fluids, aerosols, on surfaces etc, that are critical for treating infectious diseases and further to control the spread of diseases [2].

10.1.1 Ways to tackle pathogens

So far the development of vaccines and novel antibacterial drugs with improved treatments are continuing at pace. The research labs are financially supported by policy makers, and act in a manner to limit and control infectious outbreaks through targeting the development and implementation at budding stages of pathogens. Contamination through food and water is a potential threat to human health, significantly impacting socioeconomic status. Therefore, detection of pathogenic bacteria remains not only a scientific challenge but also a practical problem worldwide. There are various analytical methods followed that include conventional culturing and staining techniques. Also, improved molecular methods that are based on polymerase chain reaction amplification and immunological assays have emerged over the years to identify pathogenic agents. The advanced techniques that are employed to both identify and quantify pathogens are broadly distinguished as immunoassays or DNA-based assays. The accuracy of immunoassays over DNA-based assays depends on various factors such as stage of infection, availability of antibodies, DNA sequence data of virus, toxin-producing genes, species specific and strain selective genes etc. Immunoassays are found ubiquitously among medical diagnostics and food safety applications, where pathogens can be spotted through the presence of generated antibodies in an organism that are present both during and after the infection. These techniques are highly sensitive in most cases, and these approaches are highly labor-, time- and cost-consuming. They also demand trained personnel to carry out frequently complex assays.

10.1.2 Challenges in pathogen detection

In the case of immunoassays and DNA-based assays, the only way to identify pathogens is through detecting the presence of generated antibodies in an organism that may be present both during and after an infection. The capability of detecting pathogens directly or indirectly through generated antibodies and epitopes of pathogen shows immunoassays to be flexible techniques [3]. In cases where limited antibodies are available, there is a need for accurate and sensitive results as those infections do not generate a significant level of antibodies in the organism though pathogens are present, so DNA-based assays are generally employed [4]. These DNA-based assays demand the presence of adequate pathogens in the sample in the form of antibodies or toxin-producing genes and their expression. Thus, the targets that are associated with detection of pathogens are nucleic acids, toxins, viruses, oocysts, cells etc [5]. Much comprehensive data is available on pathogen detection through high-throughput bioanalytical techniques such as enzyme-linked immunosorbent assay and polymerase chain reaction, which are still considered the gold standard for pathogen detection [6–8].

Advancements in this have focused on emergence of label-free biosensors for detecting pathogens, which also provides useful characteristics in biomanufacturing processes, precision agriculture and environmental monitoring. The bioanalytical techniques make use of a selective biorecognition element called a molecular probe, which in combination with analytical systems like plate reader, PCR analyzer quantifies one or more components of a sample. Rather than being capable of highly sensitive and robust methods, they have become destructive testing methods requiring addition of reagents to the sample with extensive sample preparation steps, increasing the time to generate results. PCR-like bioanalytical techniques sometimes also encounter inhibition effects that are caused by background species present along with the sample, introducing measurement bias and increasing measurement uncertainty [9–13]. Therefore, by considering these limitations present in traditional plate-based bioanalytical techniques, there is a need to examine alternative bioanalytical techniques for real-time capabilities among various applications.

10.1.3 Electrochemical biosensors

Over the last two decades, biosensors have emerged to support PCR and ELISA for detecting pathogens. Biosensors work based on the direct combination of selective biorecognition element and sensitive transducer element, providing complementary platforms for pathogen identification and quantification in the case of PCR and ELISA. As per International Union of Pure and Applied Chemistry (IUPAC), a biosensor must have a biorecognition element present in direct spatial contact with a transduction element [14]. Also, a biosensor should generate quantitative or semi-quantitative analytical information and measurement beyond requiring additional processing steps and reagents. In addition, a biosensor should be a self-contained and integrated device, for which measuring varies from droplet formats to continuous flow formats which require associated fluid handling systems. Biosensors have so far achieved sensitive and selective real-time identification of pathogens in various environments beyond the need for sample preparation. For instance, biosensors have empowered the detection of abundant pathogens in various matrices and environments like foods, body fluids and surfaces of objects. They make sample preparation easy with common protocols and are compatible with label-free protocols [15–18]. Labels used here are often called reporters of molecular species, organic dyes or quantum dots that become attached to the target directly or through a biorecognition element, following a series of sample preparation steps or binding steps to facilitate detection via properties of the label [19]. Therefore, label-free biosensors avoid usage of reporter species to detect the target species [20, 21]. Also, label-free assays often have few sample preparation steps because of elimination procedures associated along with target labeling, lowering the cost over that of label-based assays considering applications where preparation facilities or trained personnel are unavailable [20, 21].

In this regard, various types of transducers are investigated for pathogen biosensing such as mechanical and optical transducers, cantilever biosensors, surface plasmon resonance (SPR)-based biosensors, electrochemical biosensors that are extensively applied to pathogen detection [22–26]. Electrochemical biosensors used for pathogen detection often utilize conducting and semiconducting materials to be the transducer, referred to as electrodes. This chemical energy involved in binding between target pathogens and electrode-immobilized biorecognition elements is thus converted into electrical energy by an electrochemical method involving the electrode and pathogen-containing electrolyte solution. To date, the electrochemical biosensors have validated sample preparation-free detection of active pathogens in various matrices [27–32]. Here, we have tried to give some insights into electrochemical biosensors for pathogen detection methods employed so far and as classified with respect to IUPAC-recommended definitions and classifications [14].

A biochemical sensor is a device which transforms chemical information like concentration of sample component or total composition into an analytically useful signal [14]. This electrochemical method used here is a distinguishing aspect of any electrochemical biosensor. Also in this method, sample handling and sensor signal readout formats provide determining aspects of a biosensor-based approach for detecting pathogens. Therefore, the electrochemical biosensors are built using a framework on transducer elements, biorecognition components and measurement formats. Here, the detection of bacterial pathogens remains the main area of focus followed by detection of viral pathogens, protozoa that have emerged with various matrices. Light is also thrown towards transduction elements, biorecognition components and measurement formats that are associated with electrochemical biosensors for detection of all kinds of pathogens.

10.1.3.1 Transduction elements

Transduction elements of any electrochemical biosensor are electrochemical cells forming the main component that works as the electrode. A three-electrode format viz. working, auxiliary and reference is commonly employed in a potentiostatic system, whereas a two-electrode format with working and auxiliary is employed for conductometry and electrochemical impedance spectroscopy (EIS). Such electrodes can be fabricated with multiple materials following various manufacturing processes. Basically, an electrode is an electronic conductor in which charge is transported through the movement of electrons and holes [33]. These electrodes can be fabricated using conducting and semiconducting materials of metals like gold (Au) and non-metals like carbon. Another process of manufacturing includes fabricate electrodes of varying sizes, bulk structures (>1 mm), micro and nano-structures. Therefore, electrodes can be classified based on type and form of material used for manufacturing process and designing. The designs of electrode can be classified based on planar, nanostructure, wired or array-based. The material used, approach used for fabrication and design used affect the electrode's structure and properties that ultimately regulate the performance of biosensors like sensitivity, selectivity, limit of detection (LOD) and dynamic range. These also influence the cost, disposability, manufacturability and measurement capacities of the biosensor.

Metal electrodes commonly used for pathogen detection are made of gold and platinum. Thick metals are fabricated from bulk structures through cutting process and thin-film metals are fabricated by deposition on insulating substrates through conventional microfabrication approaches like physical vapor deposition and screen printing [34]. The resultant conductive components are often embedded in insulating polymer or ceramic substrates, including Teflon, polyetherkeytone (PEK) and glass to finish fabrication of the transducer element. Conducting and semiconducting ceramics such as indium tin oxide (ITO), polysilicon and titanium dioxide (TiO_2) are also utilized for pathogen detection. Due to their high conductivity and transparent nature they present various measurement advantages, like correlating accurate biosensor response with pathogen surface coverage [35, 36]. This transparency of electrodes also enables *in situ* verification with target binding through microscopic techniques offering compatibility with optical methods [37]. Carbon electrodes that are based on various allotropes of carbon like graphite and glass-like carbon are classified as ceramic materials because of their mechanical properties and brittle nature.

Further polymer electrodes have been tested and proved to be compatible with biorecognition element immobilization techniques [38, 39]. Polymers also show mechanical properties enabling electrode-tissue mechanical matching which is an important criterion while designing implantable and wearable biosensors. Polymer electrodes are classified as conjugated and composite, where polyaniline and polypyrrole are the most commonly used conjugated polymers for pathogen detection because of their high conductivity in doped state [40]. In addition, polypyrrole has been proved to be biocompatible, exhibiting affinity for methylated nucleic acids [38]. The only drawback polyaniline films is that they lose electrochemical activity in solutions having pH greater than 4, presenting a measurement challenge with samples of varying pH [41]. These conjugated polymer electrodes generally show thin-film form factors and thus are deposited onto insulating substrates through layer-by-layer approaches, spin coating or electrochemical polymerization [42]. These electrodes are composed of non-conducting polymer matched with conducting or semiconducting dispersed phase. Micro- and nanomaterials like gold, graphite, graphene, carbon nanotubes are used as the dispersed phase in combination with polymers such as chitosan, polyethylenimine, polyallyamine etc [43–48].

There are advanced three-dimensional (3D) printing processes like inkjet printing, laser melting and microextrusion printing fabrication of electrochemical sensors and electrodes with various metals, that are yet to be applied for pathogen detection [49–54]. Gold electrodes of varying size and forms are used for pathogen detection. The incorporation of complex masks and programmable tool paths having lithographic and 3D printing processes will also enable the fabrication of complex electrode configurations [55, 56]. Along with the above features, lithographic processes, 3D printing processes and assembly operations add the fabrication of electrode arrays with electrode patterning [57]. Further, interdigitated microelectrodes and other patterned electrodes are developed to enhance the sensitivity and multiplexing abilities of biosensors. Interdigitated array microelectrodes (IDAMs) comprise alternating and parallel-electrode fingers arranged in an interdigitated pattern. These IDAMs have been proved to exhibit rapid response and increased signal-to-noise ratio [58]. These above-mentioned manufacturing processes are also integrated to construct electrode arrays that show geometries other than interdigitated designs in electrochemical sensing applications.

Transducers possessing physical dimensions that are comparable to target species are widely investigated to create sensitive biosensors [59–62]. When electrodes ranging from micrometers to nanometers are investigated, nanoscale planar electrodes are most commonly suggested for pathogen detection [63, 64]. The fabrication of nanoscale structures with conducting and semiconducting materials employing a wide range of bottom-up and top-down nanomanufacturing processes like nanowires has led to investigation of nanostructured electrodes for pathogen detection [65]. Nanostructuring can be performed simultaneously employing bottom-up electrode fabrication processes; also, it can be utilized for post-processing steps with a top-down electrode fabrication process. Nanowire-based electrodes are so far fabricated using many engineering materials adopting both bottom-up and top-down nanomanufacturing methods [66, 67]. Nanowires thus will exhibit hexagonal, circular and even triangular cross-sections, keeping aspect ratio, i.e. the ratio of the length to width, which often ranges from 1 to >10 [68, 69]. Already the metallic and ceramic microwires and nanowire-based electrodes are being examined for detecting pathogens. Wang *et al* used nanowire-bundled TiO_2 electrodes that are synthesized using bottom-up wet chemistry process for detecting *Listeria monocytogenes* [70]. Shen *et al* have fabricated silicon nanowire-based electrodes using chemical vapor deposition process to achieve rapid detection of human influenza A virus [71]. Although the polymer nanowires are more relatively applied for detecting non-pathogenic species, they also can be potent in pathogen detection [72]. Polymer nanowires can also be prepared using bottom-up and top-down nanomanufacturing methods using hard template, soft template or physical approaches. However, efficiency and large-scale synthesis still remain a challenge [73].

Summarizing research done using micro- and nanowire electrodes for pathogen detection, different metal microelectrode arrays that are modified with polypyrrole nanoribbons detecting virus are used [74]. On top of this topographical modification of electrode surfaces with micro- and nanostructured features far from wire-like structures are under investigation. The nanostructuring of an electrode increases its surface area without increasing the electrode volume, thus raising the ratio of electrode surface area to fluid volume analyzed [75]. Topographical modification of electrodes might affect their mechanical and electrical properties, like electrochemical deposition of compound on silicon electrodes reduced electrical impedance offering measurement advantages for neural monitoring and recording applications

[76]. Therefore, nanostructuring of electrodes for detecting pathogens above the fabrication of nanowire-based electrodes is accomplished primarily using bottom-up wet laboratory chemistry and electrochemical approaches. Among these methods available for electrode nanostructuring, the electrodes are often fabricated either by deposition or coupling of nanoparticles on the planar electrodes [77]. Also the gold nanoparticles that are commonly deposited on planar electrodes provide nano-structured surface for immobilization of the biorecognition element, wherein the particles are bound to the planar electrode through physical adsorption processes or chemical methods [78, 79]. Also, the fabricated nanostructured electrodes that are coupling with already processed nanomaterials to planar electrodes, incorporating electrochemical methods are also commonly used for bottom-up electrode nano-structuring processes to fabricate nanostructured electrodes. Hong *et al* fabricated a nanostructured gold electrode through electrochemical deposition of gold (III) chloride hydrates and detected norovirus in lettuce extracts [80]. Also, this physical or chemical deposition of metals on planar electrodes provides a useful nano-structuring approach that introduces porosity to the electrode, enabling electrode nanostructuring. For instance, Nguyen *et al* utilized nanoporous alumina-coated Pt microwires detecting West Nile virus. A few studies have reported improved biosensor performance incorporating electrode nanostructuring with improved sensitivity considering the positive effect on immobilization of the biorecognition element and target binding [81]. The nanostructured electrodes exhibited high-aspect-ratio structures showing other three-dimensional structures to enhance biomolecular steric hindrance effects, contributing implications in pathogen detection applications [82, 83]. Although a great deal of findings in this field are present, it is still unclear that how the structural features like topography, crystal structure, material properties, and electrical properties of nanostructured surfaces differ among mass-produced electrodes. How such variance happens in nanostructuring quality affecting the repeatability of biosensor performance is still to be elucidated.

Recently, it has been proved that biosensors containing integrated electrodes and complementary transducers give promising results for pathogen detection applications, wherein the electrodes are integrated with transducers enabling simultaneous mixing of fluid and monitoring molecular binding events [84]. Biosensors that are composed of multiple transducers can be thus referred to as hybrid biosensors offering unique opportunities for *in situ* confirmation of target binding with complementary analytical measurements. These hybrid electrochemical biosensors used for pathogen detection are developed by integrating the electrodes with optical and mechanical transducers. Along with providing complementary responses for verifying of binding events, the hybrid biosensors also generate fluid and particle mixing near the electrode–electrolyte interface, acoustic streaming or primary radiation effects of mechanical transducers [55, 85, 86]. Also being utilized are hybrid biosensor designs that are composed with combinations of electrodes and other transducers, and hybrid biosensor-based assays for detecting pathogens that are based on combining the effect of an electrochemical biosensor along with traditional bioanalytical technique. Electrochemical-colorimetric (EC-C) biosensing can be a combination of electrochemical method with colorimetric, luminescent or fluorescent detection method. These electrodes can detect the presence of target species, whereas colorimetric transduction pathway allows quantification of products that are associated within a reaction between target and active species [87]. The optical labels used here usually include biological fluorophores like green fluorescent protein, organic fluorophores that are non-protein fluorescein, rhodamine, nanoparticles like quantum dots—CdS, GaAs, CdSe etc [88, 89]. Therefore, use of such additional reagents in detecting target species becomes useful and economical in many such applications.

10.1.3.2 Biorecognition elements

A biosensor is a device that is composed of transducer elements integrated with biorecognition elements. We have to further see the biorecognition elements which are used for selective detection of pathogens and their immobilization techniques used for coupling with electrodes. Basically, the biorecognition elements used for electrochemical biosensors are biocatalytic or biocomplexing form. In the case of biocatalytic elements, the response of a biosensor is based on a reaction that is catalyzed by macromolecules. The most commonly used biocatalytic biorecognition elements are enzymes, whole cells and tissues. Enzymes generally provide biorecognition elements in many chemical sensing applications that are commonly introduced through secondary binding steps, whereas in the case of biocomplexing biorecognition elements, it is based on the interaction with analytes and macromolecules or organized molecular assemblies to which it shows a response. Also, the antibodies, peptides and phages are used mostly as biocomplexing biorecognition elements for detection of pathogens. In addition to biomacromolecules, imprinted polymers have also been examined to be a biocomplexing biorecognition element in pathogen detection using electrochemical biosensors.

Biosensors prepared using antibody-based biorecognition elements are called immunosensors. These antibodies show high selectivity and have good binding affinity against target species generating a wide range of infectious agents acting as a gold-standard element, as they have recognition sites that exclusively bind to antigens in the specific region of antigen called the epitope [90]. They can be labeled with fluorescent or enzymatic tags leading to designation of the above-mentioned approaches and measurement constraints are associated with usage of additional reagents and more processing steps; antibody labeling can also alter binding affinity affecting the biosensor's selectivity [3, 17, 20, 91–93]. Both the monoclonal and polyclonal antibodies lead to selective detection of pathogens having variation in production method, binding affinity and selectivity. Among them, monoclonal antibodies show a higher degree of selectivity, are more expensive and take longer time to procure than polyclonal antibodies. Some drawbacks using antibodies include high cost, storage conditions and stability challenges. The assays so far performed have involved secondary binding steps where monoclonal antibodies served as primary biorecognition element immobilized on electrode, while polyclonal antibodies act as secondary element facilitating labeling of target. There are antibody fragments like single-chain variable fragments (scFvs) which offer similar selectivity to antibodies achieving increased packing densities on the surface of the electrode due to its relatively smaller size [94]. Also, in addition to scFvs, re-engineered IgGs, Fabs and dimers could be potentially employed as biorecognition elements in a pathogen detection process [95].

Carbohydrate-binding proteins like lectins will also act as selective biorecognition elements based on their capability to bind ligands present on target species. These peptide-based biorecognition elements are obtained at low-cost, producing high yield automated synthesis processes that are modifiable [96]. Oligosaccharides and trisaccharides are the kind of carbohydrates that can selectively bind to carbohydrate-specific receptors present on pathogens. The only disadvantage is that they have weak affinity towards carbohydrate–protein interactions and hence low selectivity, mediated through secondary interactions [97]. Another biorecognition element that can be used is oligonucleotides single-stranded DNA (ssDNA) for DNA-based assays where ssDNA aptamers are utilized for pathogen detection as they are capable of binding to various molecules with high affinity and selectivity [98, 99]. These aptamers are isolated from the random sequence pool following a selection process utilizing systematic evolution of ligands through exponential enrichment called SELEX [100]. With this suitable binding sequence it can be thus isolated and amplified for use, which exhibits high selectivity among target species [100]. Regardless, the utilization of aptamers is not yet replaced by traditional biorecognition elements like antibodies due to doubts about its stability, cross-reactivity, degradation and reproducibility achieved through alternative processing approaches [98]. Phages are viruses that infect and replicate inside bacteria via selective binding through tail-spike proteins, and are also examined as biorecognition elements for detection of pathogens using electrochemical biosensors [101]. These bacteriophages have varying morphologies that are classified by selectivity and structure, making them potent as biorecognition elements in water safety and environmental monitoring applications, which requires chronic monitoring of liquids.

Cell- and molecularly-imprinted polymers focus on creating engineered molecular biorecognition elements like scFvs. They exploit the principle of target-specific morphology for selective capture and are based on cell- and molecularly-imprinted polymers, abbreviated as CIPs and MIPs [102, 103]. Imprints can be created using processes like bacteria-mediated lithography, colloid imprints, micro-contact stamping etc [102, 104]. Although they perform with the absence of highly selective molecular biorecognition element, CIPs and MIPs show selectivity when they are exposed to samples containing multiple analytes that are non-target species [105–107]. They mainly highlight the biosensor regeneration which might show adverse effects like structure mutations.

The biosensors thus are self-contained devices containing integrated transducer-biorecognition elements, where immobilization of biorecognition elements on electrodes is important taking into consideration design, performance and fabrication of electrochemical biosensors. The goal is to achieve a stable and irreversible bond during immobilization between the biorecognition element and electrode having suitable packing density and orientation, maintaining high accessibility and binding affinity towards target species, which can be achieved through the biorecognition layer. Here the rate of biosensor response is typically governed by a mass transfer-limited heterogeneous reaction that happens between immobilized biorecognition element and target species. The net change produced in the biosensor response involves thermodynamics, here binding affinity occurring between biorecognition element and target species like antibody antigen is reported in terms of dissociation constant [108]. The binding affinity of antibodies varies by order of magnitude depending on pathogen of interest along with clonality of antibody, which highlights the importance of thermodynamic analysis.

10.1.3.3 Measurement patterns

Along with a physical device made of an integrated transduction element and biorecognition element, the electrochemical biosensor-based assay pattern involves a few steps associated with sample preparation, physical systems for biosensor housing and handling of samples. The protocols associated for sample preparation and handling are referred to as the measurement formats. Several points followed during measurement format can be considered and they vary based on design of assay, biosensor performance like sensitivity, volume, material properties, composition sample, application etc. The use of DNA-based assays for detection of pathogen requires sample preparation steps that are associated with extraction of genetic material. Likewise, use of label-based biosensing method requires sample preparation steps that are associated with labeling of target, pre-concentration steps where sample concentration is low.

Sample preparation involves filtration and pre-concentration preparation of sample for amplifying target species, reducing the concentration of disturbing inhibitory species, and reducing the heterogeneity in the sample's composition and properties [5]. Sample filtration is usually done to overcome size discrepancy between target pathogen and background species, involving membranes, channels and fibers. Biorecognition elements that exhibit affinity to a broad group of pathogens like lectins are used in pre-concentration steps. Physical properties of cell-based pathogens are leveraged in biofiltration processes using electropositive filters, involving separation processes like manual handling, filtration processes integrated with microfluidic-based biosensing platforms [109]. Centrifugal separation using centrifugation method can be used as a density gradient-based separation for concentrating target pathogens in a sample, usually employed in applications of complex matrices like body fluids. This centrifugation-based separation technique is potentially applied in microfluidic-based biosensing platforms for detection of pathogens.

Broth enrichment is one more technique employed to increase the concentration of target species present in a sample through growth or replication of target species before measurement, to increase the number present for detection. Though enrichment can be a useful step, it is naturally restricted to viable and cultural organisms. Magnetic separation is one more technique used to separate target species from a sample by using magnetic beads. Target pre-concentration through magnetic bead-based separation method involves binding of antibody-functionalized magnetic beads to the target species; complexes thus formed are subsequently separated from the solution through externally-applied magnetic fields. These magnetic-assisted separation methods are helpful when the target species has similar properties to other analytes or background species present in the sample with similar size, chemical properties and density [110]. The complexes formed with bead and target are then introduced to the biosensor to quantify target pathogens present in the initial sample, mainly used for general substrates in traditional immunoassays detecting a variety of pathogens like bacteria and viruses [71].

Followed by sample preparation, sample handling format is also highly determined in biosensor application. Pathogens have the ability to be aerosolized, acting as one of the significant modes of disease transmission seen with viral pathogens such as influenza and COVID-19. Handling of samples can be generally categorized as droplet, flow or surface-based, where droplet formats indulge sampling from a larger volume and analyzed by depositing them on a functionalized transducer to a fluidic delivery system used with colorimetric biosensors, being simple handling formats that can even be performed by unskilled users. Also, this format will avoid technical and methodological barriers for measurement, like elimination of physical systems associated with biosensor housing and samples exhibiting measurement challenges with mass transport and target sampling. On the other hand, flow formats involve the detection of target species in flow fields like continuously-stirred systems-tank bioreactors, flow cells, microfluidics etc, having the advantage of exposing biosensor to samples in a controlled and repeatable way, benefiting exposure through convective mass transfer methods. They are typically compatible when sample volumes are large in liters, where flow cells are fabricated through milling and extrusion processes with Teflon or Plexiglas, referred to as rigid three-dimensional biosensors. Additionally, the flow formats are commonly used for microfluidic devices that exhibit thin two-dimensional form factors called planar electrodes offering many measurement advantages. They are consistently fabricated from machinable polymers and microfluidics can be typically fabricated using polydimethylsiloxane (PDMS) and polymethyl methacrylate (PMMA) contributing to low cost and similarity with microfabrication methods. The main advantage of microfluidic devices is their ability to handle integrated sample preparation steps eliminating the need for additional steps involved in a sample-to-result process [11]. Therefore, the presence of pathogens on the object surfaces can be analyzed using droplet and flow-based handling formats incorporating material transfer processes like swabbing; *in situ* detection directly on object surfaces acts as vital measurement in medical diagnostic, food safety and infection control applications. The sample handling format also takes into consideration a biosensor's reusability which classifies the above into single or multi-use biosensors that can be repeatedly recalibrated signifying its importance in industrial applications [14].

There are various electrochemical methods that can be performed where methods differ in electrode configuration, applied and measured signals, mass transport regimes, binding information, target size-selectivity etc [16]. They are basically classified as potentiometric, conductometric, amperometric, impedimetric, ion-charge or field-effect, signifying the measured signal [14]. Potentiometric methods are referred to as controlled-current methods where electrical potential is measured in response to applied current with low amplitude, the advantage being the ability to use low-cost instrumentation [33]. Voltammetric methods are also referred to as controlled-potential methods in which current is measured with response to applied electrical potential driving redox reactions [33]. Here, the measured current indicates electron transfer within the sample on the electrode surface and thereby concentration of the analyte. In chronoamperometry the electrical potential is applied in steps at the working electrode measuring the resulting current as a function of time. There are various types of biosensors found compatible with voltammetry-based measurements, and field-effect transistor (FET)-based biosensors utilize amperometric-based methods for detection of pathogens [111, 112]. They detect pathogens through measured changes in source–drain channel conductivity arising from the electric field of the environment present in the sample. Linear sweep and cyclic voltammetry methods are those in which the current is measured in response to applied electrical potential, which is kept constant across a range of electrical potentials [33]. Cyclic voltammetry is a commonly used linear-sweep method where electrical potential is swept into both forward and reverse directions in the case of partial, full or a series of cycles, being one of the most widely used voltammetric methods for detecting pathogens. Pulse voltammetry is a type where electrical potential is applied in pulses, where improved speed and sensitivity act relative to traditional voltammetric techniques [33]. In the case of staircase voltammetry, the electrical potential is pulsed in a step-wise manner and current is measured after each step, reducing the effect of capacitive charging on the current signal. Square wave voltammetry (SWV) works similar to staircase voltammetry applying a symmetric square-wave pulse that is superimposed on a staircase potential waveform. In the case of differential pulse voltammetry (DPV), the electrical potential is often scanned with a series of fixed amplitude pulses that can be superimposed on a changing base potential, measuring current before and after pulse application and end of pulse, respectively, allowing decay of nonfaradaic current [113]. Stripping voltammetry is a modified kind to include a step where targets can be preconcentrated on the electrode surface and stripped from the surface by applying electrical potential. Electrochemical impedance spectroscopy involves a response based on step changes or continuous sweeps in the applied current or voltage that drives the electrode to a condition that is far from equilibrium, providing advantages of measuring a wide range of times, frequencies and high precision in time-averaged responses.

10.1.4 Applications of electrochemical biosensors

Implementation of electrochemical biosensors for detecting pathogens is certainly reviewed with respect to the target pathogen, biosensor design, sample matrix, measurement format, fabrication method and biosensor performance. Future directions of electrochemical biosensors for improvising pathogen detection, including present technological and methodological challenges and possible application areas are so far discussed. The applications of electrochemical biosensors for detecting pathogens are mainly seen in food and water safety, infection control, environmental monitoring, medical diagnostics, bio-threat defenses etc.

Foodborne and waterborne pathogenic organisms originate from a variety of sources and infect human beings through their exposure. Waterborne pathogens cause 2.2 million deaths annually worldwide and food-related deaths amount to around 420 000 annually [114]. Therefore, the selective detection of pathogens present in food and water remains a global healthcare challenge, and if extensive use of immunoassays like ELISA is done, it helps to obtain commercially-available monoclonal and polyclonal antibodies for a large number of foodborne and waterborne pathogens. These biosensor applications that are associated with process monitoring applications require biosensor designs and measurement formats that expedite high-throughput analysis, biosensor reusability and continuous monitoring capability, and detection should be done at various stages of the processing operation [115]. The amplification approaches used so far with the secondary binding of enzyme-labeled secondary antibodies are electron transfer mediation, nanostructuring of surface for increased rate of charge transfer kinetics, conversion of electrochemically inactive substrate into a detectable electroactive product and catalysis of oxidation of glucose for production of hydrogen peroxide for electrochemical detection [116].

Environmental monitoring and infection control applications are also an important aspect of healthcare where there are diseases associated with environmental pathogens leading to death in low-income economies. They typically spend a substantial part of their lifecycle residing outside human hosts, and when they come in contact with humans spread disease with measurable frequency. Here also we require biosensors that are capable of analyzing pathogen-containing complex matrices like the water or food matrix, for environmental pathogens that are present in multiple types of matrices. As these pathogens enter the body through direct physical contact, through aerosols or through organisms as vectors for infectious agents, detection of environmental pathogens requires analysis of matrices like air, and object surfaces within healthcare facilities [117]. As it is difficult to obtain antibodies for various environmental pathogens like protozoa, nematodes, traditional bioanalytical techniques like PCR are employed. These, biosensor-based assays also utilize measurement formats analyzing liquids for detection of aerosolized pathogens. Without doubt, the medical diagnostics field heavily relies on identification and quantification of infective pathogens present in body fluids, whole blood, urine, stool, mucus, saliva, sputum etc. Here, biosensors offer a complementary platform to diagnose and enable rapid, cost-effective measurements, high sensitivity, ability to make measurements in complex matrices that are posing challenges to traditional bioanalytical techniques [118]. The concentration of pathogens should be diagnostically-relevant in each type of matrix and should be taken into consideration while designing biosensors.

To overcome biological defense and bio-threat applications the potential weapon drives demand for rapid and sensitive biosensors for applications on biological defense. They are not new and are related to the above-mentioned applications in food-water safety, environmental monitoring and medical diagnostics, considering dangerous pathogens. These pathogens may be native and endogenous agents, and in some cases might be exogenous agents that are weaponized and dispersed intentionally [119]. Thus, these biosensor-based assays designed for bio-threat applications should be low-cost, portable to allow integration with existing physical facilities, and move with drones or combatants on a battlefield etc.

With the above information on types of transduction factors, biorecognition elements, electrochemical methods, measurement formats and pathogen detection, it would be relevant to understand a few applications, present challenges and future directions in the field of electrochemical biosensor-based pathogen detection. These recent technologies only stand if they have the ability to create low-cost, robust biosensors for detection of pathogens. A primary method is to reduce cost and size by switching on to carbon-based electrodes—graphite, graphene, CNTs etc—as an alternative to expensive metallic or ceramic electrodes [120]. Polymer-based electrodes also act as low-cost alternatives to metal electrodes; 3D printing can be employed for fabrication of various components of electrochemical biosensors like electrodes, fluid handling components, substrates, device packaging etc.

Wearable biomedical devices are emerging as promising tools in diagnostics and health monitoring, and are used to detect small molecules like lactate, electrolytes, glucose. Here challenges such as biocompatibility, skin irritation, biosensor-tissue mechanical-geometric matching and device power consumption are an issue [121]. Mainly, the size of pathogen shows significant impact on an electrochemical biosensor's performance as the pathogens can be small, thereby having relatively smaller effect on transferring charge at the electrode–electrolyte interface. Also, the majority of infectious agents detected so far using electrochemical biosensors are human pathogens and emerging agricultural pathogens, which gives hope for this new technology. In addition, multiplexed detection of pathogens has emerged for phenotype identification of multiple pathogenic threats via various approaches involving use of multiple transducers exhibiting different biorecognition elements [122]. Saturation-free and continuous monitoring formats are necessary to monitor biosensor-based process and control applications, which should remove the earlier captured target *in situ* using a chemical-free approach preserving biorecognition layer for subsequent measurements [123]. For low-cost and single-use portable biosensors, disposable substrates are a present challenge, where paper-based substrates have emerged as attractive alternatives to costlier ceramic substrates [124]. Sample preparation-free protocols can help to improve measurement confidence, readability, repeatability, reducing time, which contributes to important aspects of healthcare decision-making [125]. Wireless transduction approaches also add up in creating portable and wearable biosensing platforms for detection of pathogens distributing sensing systems towards infection control and process monitoring [126]. They are also essential to create implantable and integrated biosensors, acting as an emerging area for pathogen detection.

10.2 Conclusion

Electrochemical biosensors offer great potential serving as resources for improving global healthcare to prevent the spread of highly contagious diseases with detection of pathogens. Facing a great challenge in this field is to develop sensitive, rapid, specific and miniaturized devices validating the presence of pathogens in a cost- and time-efficient manner. Therefore, electrochemical sensors are well accepted powerful tools used for detecting disease-related biomarkers and for answering environmental-organic hazards. Their widespread interest has been extended in recent years towards detection of foodborne and waterborne pathogens due to their marker-free character and excessive sensitivity. Here we have focused on current electrochemical-based approaches for microorganism recognition and bringing them into the context of other conventional methods and sensing devices for pathogens like microorganism culturing on agar plates and polymer chain reaction used to identify the DNA of the causative pathogen. Recent breakthroughs are also highlighted, including utilization of microfluidic devices and immunomagnetic separation among multiple pathogens analysis using a single device. Overall, there is currently no adequate solution available which recognizes selective and sensitive binding to a particular microorganism, that is fast in detection, screening, cost-effective, and can be scaled up for a wide range of biologically matching targets etc.

References

[1] Dye C 2014 After 2015: infectious diseases in a new era of health and development *Philos. Trans. R. Soc. Lond. B Biol. Sci.* **369** 20130426–206

[2] Cesewski E and Johnson B N 2020 Electrochemical biosensors for pathogen detection *Biosens. Bioelectron.* **159** 112214

[3] Alahi E E M and Mukhopadhyay C S 2017 Detection methodologies for pathogen and toxins: a review *Sensors* **17** 8

[4] Lazcka O, Del Campo F J and Munoz F X 2007 Pathogen detection: a perspective of traditional methods and biosensors *Biosens. Bioelectron.* **22** 1205–17

[5] Zourob M, Elwary S and Turner A P 2008 *Principles of Bacterial Detection: Biosensors, Recognition Receptors and Microsystems* (Berlin: Springer Science & Business Media)

[6] Law J W F, Ab Mutalib N S, Chan K G and Lee L H 2015 Rapid methods for the detection of foodborne bacterial pathogens: principles, applications, advantages and limitations *Front. Microbiol.* **5** 770

[7] Klein D 2002 Quantification using real-time PCR technology: applications and limitations *Trends Mol. Med.* **8** 257–60

[8] Malorny B, Tassios P T, Rådstr€om P, Cook N, Wagner M and Hoorfar J 2003 Standardization of diagnostic PCR for the detection of foodborne pathogens *Int. J. Food Microbiol.* **83** 39–48

[9] Justino C I L, Duarte A C and Rocha-Santos T A P 2017 Recent progress in biosensors for environmental monitoring: a review *Sensors* **17** 12

[10] Scognamiglio V, Rea G, Arduini F and Palleschi G 2016 *Biosensors for Sustainable Food-New Opportunities and Technical Challenges* (Amsterdam: Elsevier Science)

[11] Sin M L, Mach K E, Wong P K and Liao J C 2014 Advances and challenges in biosensor-based diagnosis of infectious diseases *Expert Rev. Mol. Diagn.* **14** 225–44

[12] Clark K D, Zhang C and Anderson J L 2016 Sample preparation for bioanalytical and pharmaceutical analysis *Anal. Chem.* **88** 11262–70

[13] Silverman J D, Bloom R J, Jiang S, Durand H K, Mukherjee S and David L A 2019 Measuring and mitigating PCR bias in microbiome data *bioRxiv* **604025**

[14] Thevenot D R, Toth K, Durst R A and Wilson G S 2001 Electrochemical biosensors: recommended definitions and classification *Biosens. Bioelectron* **16** 121–31

[15] Daniels J S and Pourmand N 2007 Label-free impedance biosensors: opportunities and challenges *Electroanalysis* **19** 1239–57

[16] Rapp B E, Gruhl F J and Länge K 2010 Biosensors with label-free detection designed for diagnostic applications *Anal. Bioanal. Chem.* **398** 2403–12

[17] Sang S, Wang Y, Feng Q, Wei Y, Ji J and Zhang W 2016 Progress of new label-free techniques for biosensors: a review *Crit. Rev. Biotechnol.* **36** 465–81

[18] Vestergaard M, Kerman K and Tamiya E 2007 An overview of label-free electrochemical protein sensors *Sensors* **7** 3442–58

[19] Resch-Genger U, Grabolle M, Cavaliere-Jaricot S, Nitschke R and Nann T 2008 Quantum dots versus organic dyes as fluorescent labels *Nat. Methods* **5** 763–75

[20] Cooper M A 2009 *Label-free Biosensors: Techniques and Applications* (Cambridge: Cambridge University Press)

[21] Syahir A, Usui K, Tomizaki K Y, Kajikawa K and Mihara H 2015 Label and label-free detection techniques for protein microarrays *Microarrays* **4** 228–44

[22] Singh R, Das Mukherjee M, Sumana G, Gupta R K, Sood S and Malhotra B D 2014 Biosensors for pathogen detection: a smart approach towards clinical diagnosis *Sens. Actuators* B **197** 385–404

[23] Yoo S M and Lee S Y 2016 Optical biosensors for the detection of pathogenic microorganisms *Trends Biotechnol.* **34** 7–25

[24] Felix F S and Angnes L 2018 Electrochemical immunosensors—a powerful tool for analytical applications *Biosens. Bioelectron.* **102** 470–8

[25] Pereira da Silva Neves M M, Gonzalez-García M B, Hernandez-Santos D and Fanjul-Bolado P 2018 Future trends in the market for electrochemical biosensing *Curr. Opin. Electrochem.* **10** 107–11

[26] Saucedo N M, Srinives S and Mulchandani A 2019 Electrochemical biosensor for rapid detection of viable bacteria and antibiotic screening *J. Anal. Testing* **3** 117–22

[27] Amiri M, Bezaatpour A, Jafari H, Boukherroub R and Szunerits S 2018 Electrochemical methodologies for the detection of pathogens *ACS Sens.* **3** 1069–86

[28] Duffy G and Moore E 2017 Electrochemical immunosensors for food analysis: a review of recent developments *Anal. Lett.* **50** 1–32

[29] Furst A L and Francis M B 2019 Impedance-based detection of bacteria *Chem. Rev.* **119** 700–26

[30] Mishra G, Sharma V and Mishra R 2018 Electrochemical aptasensors for food and environmental safeguarding: a review *Biosensors* **8** 28

[31] Monzo J, Insua I, Fernandez-Trillo F and Rodriguez P 2015 Fundamentals, achievements and challenges in the electrochemical sensing of pathogens *Analyst* **140** 7116–28

[32] Rastogi M and Singh S K 2019 Advances in molecular diagnostic approaches for biothreat agents *Defense Against Biological Attacks: Volume II* ed S K Singh and J H Kuhn (Cham: Springer International Publishing) pp 281–310

[33] Bard A J and Faulkner L R 2000 *Electrochemical Methods: Fundamentals and Applications* 2nd edn (New York: Wiley)

[34] Hierlemann A, Brand O, Hagleitner C and Baltes H 2003 Microfabrication techniques for chemical/biosensors *Proc. IEEE* **91** 839–63

[35] Aydın E B and Sezgintürk M K 2017 Indium tin oxide (ITO): a promising material in biosensing technology *TrAC, Trends Anal. Chem.* **97** 309–15
[36] Yang L and Li Y 2005 AFM and impedance spectroscopy characterization of the immobilization of antibodies on indium-tin oxide electrode through self-assembled monolayer of epoxysilane and their capture of *Escherichia coli* O157:H7 *Biosens. Bioelectron.* **20** 1407–16
[37] Wenzel T, H€artter D, Bombelli P, Howe C J and Steiner U 2018 Porous translucent electrodes enhance current generation from photosynthetic biofilms *Nat. Commun.* **9** 1299
[38] Arshak K, Velusamy V, Korostynska O, Oliwa-Stasiak K and Adley C 2009 Conducting polymers and their applications to biosensors: emphasizing on foodborne pathogen detection *IEEE Sensor. J.* **9** 1942–51
[39] Guimard N K, Gomez N and Schmidt C E 2007 Conducting polymers in biomedical engineering *Prog. Polym. Sci.* **32** 876–921
[40] Kaur G, Adhikari R, Cass P, Bown M and Gunatillake P 2015 Electrically conductive polymers and composites for biomedical applications *RSC Adv.* **5** 37553–67
[41] Wan M 2008 *Conducting Polymers with Micro or Nanometer Structure* (Berlin: Springer)
[42] Xia L, Wei Z and Wa M 2010 Conducting polymer nanostructures and their application in biosensors *J. Colloid Interface Sci.* **341** 1–11
[43] Dong J, Zhao H, Xu M, Ma Q and Ai S 2013 A label-free electrochemical impedance immunosensor based on AuNPs/PAMAM-MWCNT-Chi nanocomposite modified glassy carbon electrode for detection of Salmonella typhimurium in milk *Food Chem.* **141** 1980–6
[44] Lee D, Chander Y, Goyal S M and Cui T 2011 Carbon nanotube electric immunoassay for the detection of swine influenza virus H1N1 *Biosens. Bioelectron.* **26** 3482–7
[45] Lee I and Jun S 2016 Simultaneous detection of *E. coli* K12 and *S. aureus* using a continuous flow multijunction biosensor *J. Food Sci.* **81** N1530–6
[46] Li Y, Cheng P, Gong J, Fang L, Deng J, Liang W and Zheng J 2012 Amperometric immunosensor for the detection of *Escherichia coli* O157:H7 in food specimens *Anal. Biochem.* **421** 227–33
[47] Viswanathan S, Rani C and Ho J A 2012 Electrochemical immunosensor for multiplexed detection of food-borne pathogens using nanocrystal bioconjugates and MWCNT screen-printed electrode *Talanta* **94** 315–9
[48] Güner A, Cevik E, Senel M and Alpsoy L 2017 An electrochemical immunosensor for sensitive detection of *Escherichia coli* O157:H7 by using chitosan, MWCNT, polypyrrole with gold nanoparticles hybrid sensing platform *Food Chem.* **229** 358–65
[49] Bhat K S, Ahmad R, Yoo J Y and Hahn Y B 2018 Fully nozzle-jet printed non- enzymatic electrode for biosensing application *J. Colloid Interface Sci.* **512** 480–8
[50] Medina-Sanchez M, Martínez-Domingo C, Ramon E and Merkoçi A 2014 An inkjet-printed field-effect transistor for label-free biosensing *Adv. Funct. Mater.* **24** 6291–302
[51] Pavinatto F J, Paschoal C W A and Arias A C 2015 Printed and flexible biosensor for antioxidants using interdigitated ink-jetted electrodes and gravure-deposited active layer *Biosens. Bioelectron.* **67** 553–9
[52] Ambrosi A, Moo J G S and Pumera M 2016 Helical 3D-printed metal electrodes as custom-shaped 3D platform for electrochemical devices *Adv. Funct. Mater.* **26** 698–703
[53] Loo A H, Chua C K and Pumera M 2017 DNA biosensing with 3D printing technology *Analyst* **142** 279–83

[54] Foo C Y, Lim H N, Mahdi M A, Wahid M H and Huang N M 2018 Three-dimensional printed electrode and its novel applications in electronic devices *Sci. Rep.* **8** 7399

[55] Cesewski E, Haring A P, Tong Y, Singh M, Thakur R, Laheri S, Read K A, Powell M D, Oestreich K J and Johnson B N 2018 Additive manufacturing of three-dimensional (3D) microfluidic-based microelectromechanical systems (MEMS) for acoustofluidic applications *Lab Chip* **18** 2087–98

[56] Xu Y, Wu X, Guo X, Kong B, Zhang M, Qian X, Mi S and Sun W 2017 The boom in 3D-printed sensor technology *Sensors* **17** 1166

[57] Hintsche R, Paeschke M, Wollenberger U, Schnakenberg U, Wagner B and Lisec T 1994 Microelectrode arrays and application to biosensing devices *Biosens. Bioelectron* **9** 697–705

[58] Varshney M and Li Y 2009 Interdigitated array microelectrodes based impedance biosensors for detection of bacterial cells *Biosens. Bioelectron* **24** 2951–60

[59] Gupta A, Akin D and Bashir R 2004 Single virus particle mass detection using microresonators with nanoscale thickness *Appl. Phys. Lett.* **84** 1976–8

[60] Pumera M, Sanchez S, Ichinose I and Tang J 2007 Electrochemical nanobiosensors *Sens. Actuators B* **123** 1195–205

[61] Singh K V, Whited A M, Ragineni Y, Barrett T W, King J and Solanki R 2010 3D nanogap interdigitated electrode array biosensors *Anal. Bioanal. Chem.* **397** 1493–502

[62] Wei D, Bailey M J A, Andrew P and Ryhänen T 2009 Electrochemical biosensors at the nanoscale *Lab Chip* **9** 2123–31

[63] Hong S A, Kwon J, Kim D and Yang S 2015 A rapid, sensitive and selective electrochemical biosensor with concanavalin A for the preemptive detection of norovirus *Biosens. Bioelectron.* **64** 338–44

[64] Peh A E and Li S F 2013 Dengue virus detection using impedance measured across nanoporous alumina membrane *Biosens. Bioelectron.* **42** 391–6

[65] Patolsky F and Lieber C M 2005 Nanowire nanosensors *Mater. Today* **8** 20–8

[66] Hu J, Odom T W and Lieber C M 1999 Chemistry and physics in one dimension: synthesis and properties of nanowires and nanotubes *Acc. Chem. Res.* **32** 435–45

[67] Yogeswaran U and Chen S M 2008 A review on the electrochemical sensors and biosensors composed of nanowires as sensing material *Sensors (Basel)* **8** 290–313

[68] Vaseashta A and Dimova-Malinovska D 2005 Nanostructured and nanoscale devices, sensors and detectors *Sci. Technol. Adv. Mater.* **6** 312–8

[69] Wanekaya A K, Chen W, Myung N V and Mulchandani A 2006 Nanowire-based electrochemical biosensors *Electroanalysis* **18** 533–50

[70] Wang R, Dong W, Ruan C, Kanayeva D, Tian R, Lassiter K and Li Y 2008 TiO_2 nanowire bundle microelectrode based impedance immunosensor for rapid and sensitive detection of *Listeria monocytogenes* *Nano Lett.* **8** 2625–31

[71] Shen F *et al* 2012 Rapid flu diagnosis using silicon nanowire sensor *Nano Lett.* **12** 3722–30

[72] Travas-Sejdic J, Aydemir N, Kannan B, Williams D E and Malmstrom J 2014 Intrinsically conducting polymer nanowires for biosensing *J. Mater. Chem.* **B2** 4593–609

[73] Xia L, Wei Z and Wan M 2010 Conducting polymer nanostructures and their application in biosensors *J. Colloid Interface Sci.* **341** 1–11

[74] Chartuprayoon N, Rheem Y, Ng J C K, Nam J, Chen W and Myung N V 2013 Polypyrrole nanoribbon based chemiresistive immunosensors for viral plant pathogen detection *Anal. Methods* **5** 3497–502

[75] Soleymani L, Fang Z, Sargent E H and Kelley S O 2009 Programming the detection limits of biosensors through controlled nanostructuring *Nat. Nanotechnol.* **4** 844–8

[76] Ludwig K A, Uram J D, Yang J, Martin D C and Kipke D R 2006 Chronic neural recordings using silicon microelectrode arrays electrochemically deposited with a poly (3, 4-ethylenedioxythiophene)(PEDOT) film *J. Neural Eng.* **3** 59

[77] Eftekhari A, Alkire R C, Gogotsi Y and Simon P 2008 *Nanostructured Materials in Electrochemistry* (New York: Wiley)

[78] Wang Y, Ping J, Ye Z, Wu J and Ying Y 2013 Impedimetric immunosensor based on gold nanoparticles modified graphene paper for label-free detection of *Escherichia coli* O157:H7 *Biosens. Bioelectron* **49** 492–8

[79] Attar A, Mandli J, Ennaji M M and Amine A 2016 Label-free electrochemical impedance detection of rotavirus based on immobilized antibodies on gold sononanoparticles *Electroanalysis* **28** 1839–46

[80] Hong S A, Kwon J, Kim D and Yang S 2015 A rapid, sensitive and selective electrochemical biosensor with concanavalin A for the preemptive detection of norovirus *Biosens. Bioelectron* **64** 338–44

[81] Nguyen B T, Koh G, Lim H S, Chua A J, Ng M M and Toh C S 2009 Membrane-based electrochemical nanobiosensor for the detection of virus *Anal. Chem.* **81** 7226–34

[82] Lam B, Fang Z, Sargent E H and Kelley S O 2012 Polymerase chain reaction-free, sample-to-answer bacterial detection in 30 min with integrated cell lysis *Anal. Chem.* **84** 21–5

[83] Mahshid S S, Vallee-Belisle A and Kelley S O 2017 Biomolecular steric hindrance effects are enhanced on nanostructured microelectrodes *Anal. Chem.* **89** 9751–7

[84] Choi S, Goryll M, Sin L Y M, Wong P K and Chae J 2011 Microfluidic-based biosensors toward point-of-care detection of nucleic acids and proteins *Microfluid. Nanofluidics* **10** 231–47

[85] Johnson B N and Mutharasan R 2012 Sample preparation-free, real-time detection of microRNA in human serum using piezoelectric cantilever biosensors at attomole level *Anal. Chem.* **84** 10426–36

[86] Johnson B N and Mutharasan R 2013a Electrochemical piezoelectric-excited millimeter-sized cantilever (ePEMC) for simultaneous dual transduction biosensing *Analyst* **138** 6365–71

[87] Hou Y H, Wang J J, Jiang Y Z, Lv C, Xia L, Hong S L, Lin M, Lin Y, Zhang Z L and Pang D W 2018 A colorimetric and electrochemical immunosensor for point-of- care detection of enterovirus *Biosens. Bioelectron.* **99** 186–92

[88] Mungroo N and Neethirajan S 2016 Optical biosensors for the detection of food borne pathogens *Nanobiosensors for Personalized and Onsite Biomedical Diagnosis* ed P Chandra (IET) pp 179–206

[89] Pires N M, Dong T, Hanke U and Hoivik N 2014 Recent developments in optical detection technologies in lab-on-a-chip devices for biosensing applications *Sensors* **14** 15458–79

[90] Patris S, Vandeput M and Kauffmann J M 2016 Antibodies as target for affinity biosensors *TrAC, Trends Anal. Chem.* **79** 239–46

[91] Ahmed A, Rushworth J V, Hirst N A and Millner P A 2014 Biosensors for whole-cell bacterial detection *Clin. Microbiol. Rev.* **27** 631–46

[92] Bozal-Palabiyik B, Gumustas A, Ozkan S A and Uslu B 2018 Biosensor-based methods for the determination of foodborne pathogens *Foodborne Diseases* ed A M Holban and A M Grumezescu (New York: Academic) pp 379–420

[93] Leonard P, Hearty S, Brennan J, Dunne L, Quinn J, Chakraborty T and O'Kennedy R 2003 Advances in biosensors for detection of pathogens in food and water *Enzym. Microb. Technol.* **32** 3–13
[94] Sharma H and Mutharasan R 2013 Half antibody fragments improve biosensor sensitivity without loss of selectivity *Anal. Chem.* **85** 2472–7
[95] Byrne B, Stack E, Gilmartin N and O'Kennedy R 2009 Antibody-based sensors: principles, problems and potential for detection of pathogens and associated toxins *Sensors* **9** 4407–45
[96] Pavan S and Berti F 2012 Short peptides as biosensor transducers *Anal. Bioanal. Chem.* **402** 3055–70
[97] Zeng X, Andrade C A S, Oliveira M D L and Sun X L 2012 Carbohydrate–protein interactions and their biosensing applications *Anal. Bioanal. Chem.* **402** 3161–76
[98] Lakhin A V, Tarantul V Z and Gening L V 2013 Aptamers: problems, solutions and prospects *Acta Naturae* **5** 34–43
[99] Reverdatto S, Burz D S and Shekhtman A 2015 Peptide aptamers: development and applications *Curr. Top. Med. Chem.* **15** 1082–101
[100] Stoltenburg R, Reinemann C and Strehlitz B 2007 SELEX—a (r)evolutionary method to generate high-affinity nucleic acid ligands *Biomol. Eng.* **24** 381–403
[101] Haq I U, Chaudhry W N, Akhtar M N, Andleeb S and Qadri I 2012 Bacteriophages and their implications on future biotechnology: a review *Virol. J.* **9** 9
[102] Pan J, Chen W, Ma Y and Pan G 2018 Molecularly imprinted polymers as receptor mimics for selective cell recognition *Chem. Soc. Rev.* **47** 5574–87
[103] Zhou T, Ding L, Che G, Jiang W and Sang L 2019 Recent advances and trends of molecularly imprinted polymers for specific recognition in aqueous matrix: preparation and application in sample pretreatment *TrAC, Trends Anal. Chem.* **114** 11–28
[104] Chen L, Wang X, Lu W, Wu X and Li J 2016a Molecular imprinting: perspectives and applications *Chem. Soc. Rev.* **45** 2137–211
[105] Golabi M, Kuralay F, Jager E W H, Beni V and Turner A P F 2017 Electrochemical bacterial detection using poly(3-aminophenylboronic acid)-based imprinted polymer *Biosens. Bioelectron* **93** 87–93
[106] Jafari H, Amiri M, Abdi E, Navid S L, Bouckaert J, Jijie R, Boukherroub R and Szunerits S 2019 Entrapment of uropathogenic *E. coli* cells into ultra-thin sol-gel matrices on gold thin films: a low cost alternative for impedimetric bacteria sensing *Biosens. Bioelectron.* **124–125** 161–6
[107] Qi P, Wan Y and Zhang D 2013 Impedimetric biosensor based on cell-mediated bioimprinted films for bacterial detection *Biosens. Bioelectron* **39** 282–8
[108] Squires T M, Messinger R J and Manalis S R 2008 Making it stick: convection, reaction and diffusion in surface-based biosensors *Nat. Biotechnol.* **26** 417
[109] Song Y A, Jianping F, Wang Y C and Han J 2013 Biosample preparation by lab-on-a-chip devices *Encyclopedia of Microfluidics and Nanofluidics* ed D Li (Springer) 1–19
[110] Chen Y T, Kolhatkar A G, Zenasni O, Xu S and Lee T R 2017 Biosensing using magnetic particle detection techniques *Sensors* **17** 2300
[111] Huang Y X, Dong X C, Liu Y X, Li L J and Chen P 2011 Graphene-based biosensors for detection of bacteria and their metabolic activities *J. Mater. Chem.* **21** 12358–62
[112] Liu F, Kim Y H, Cheon D S and Seo T S 2013 Micropatterned reduced graphene oxide based field-effect transistor for real-time virus detection *Sens. Actuators* B **186** 252–7

[113] Scott K 2016 Electrochemical principles and characterization of bioelectrochemical systems *Microbial Electrochemical and Fuel Cells* ed K Scott and E H Yu (Boston, MA: Woodhead Publishing) pp 29–66

[114] Pandey P K, Kass P H, Soupir M L, Biswas S and Singh V P 2014 Contamination of water resources by pathogenic bacteria *Amb. Express* **4** 51

[115] Ye Y, Guo H and Sun X 2019 Recent progress on cell-based biosensors for analysis of food safety and quality control *Biosens. Bioelectron.* **126** 389–404

[116] Kokkinos C, Economou A and Prodromidis M I 2016 Electrochemical immunosensors: critical survey of different architectures and transduction strategies *TrAC, Trends Anal. Chem.* **79** 88–105

[117] Lai K, Emberlin J and Colbeck I 2009 Outdoor environments and human pathogens in air *Environ. Health* **8** S15

[118] Rappo U, Schuetz A N, Jenkins S G, Calfee D P, Walsh T J, Wells M T, Hollenberg J P and Glesby M J 2016 Impact of early detection of respiratory viruses by multiplex PCR assay on clinical outcomes in adult patients *J. Clin. Microbiol.* **54** 2096

[119] Mirski T, Bartoszcze M, Bielawska-Drozd A, Cieslik P, Michalski A J, Niemcewicz M, Kocik J and Chomiczewski K 2014 Review of methods used for identification of biothreat agents in environmental protection and human health aspects *Ann. Agric. Environ. Med.* **21** 2

[120] Afonso A S, Uliana C V, Martucci D H and Faria R C 2016 Simple and rapid fabrication of disposable carbon-based electrochemical cells using an electronic craft cutter for sensor and biosensor applications *Talanta* **146** 381–7

[121] Singh M *et al* 2017 3D printed conformal microfluidics for isolation and profiling of biomarkers from whole organs *Lab Chip* **17** 2561–71

[122] Li Y, Xiong Y, Fang L, Jiang L, Huang H, Deng J, Lian W and Zheng J 2017 An electrochemical strategy using multifunctional nanoconjugates for efficient simultaneous detection of *Escherichia coli* O157: H7 and *Vibrio cholerae* O1 *Theranostics* **7** 935–44

[123] Zelada-Guillen G A, Bhosale S V, Riu J and Rius F X 2010 Real-time potentiometric detection of bacteria in complex samples *Anal. Chem.* **82** 9254–60

[124] Martinez A W, Phillips S T, Whitesides G M and Carrilho E 2009 *Diagnostics for the Developing World: Microfluidic Paper-Based Analytical Devices* (Washington, DC: ACS Publications)

[125] Davenport M, Mach K E, Shortliffe L M D, Banaei N, Wang T H and Liao J C 2017 New and developing diagnostic technologies for urinary tract infections *Nat. Rev. Urol.* **14** 296–310

[126] Ghafar-Zadeh E 2015 Wireless integrated biosensors for point-of-care diagnostic applications *Sensors* **15** 3236–61

Chapter 11

Electrochemical assessment of toxic dyes and pesticides

M G Gopika, Bhama Sajeevan, Surya Gopidas and Beena Saraswathyamma

11.1 Introduction

Chemical technologies for the manufacture and processing of organic compounds as well as their use in the agricultural field for the production of high-yielding, high-quality and reliable food are a major cause of the gradual build-up of organic compounds in water bodies. For several industrial purposes, large quantities of dyes with varying chemical compositions as well as many heavy metal ions like lead, mercury and cadmium are employed [1]. Wastewater makes up a sizable fraction of the total and is released into the environment. Wastewater treatment is very necessary for all sectors in order to follow environmental requirements. The three anthropogenic activities that have had the most impact on freshwater environments are human settlements, industry, and agriculture. For instance, there are already more than 100 000 chemicals listed, the majority of which are connected to our daily lives and can certainly infiltrate freshwater [2–5]. As a result, more than a third of the world's population lacks access to clean drinking water [5]. For the conservation of marine ecospheres and the provision for safe drinking water, monitoring water quality is crucial.

One of the most prominent signs of water pollution is color, and discharging brightly colored pollutants that include dyes might harm the receiving bodies of water. Complicated and diverse forms of waste are produced by the textile industry primarily relying on treated fabric and the type of procedure used, such as washing, bleaching, dyeing etc [6]. In the textile and printing industries, where synthetic dyes are widely utilized, they are frequently released into the environment untreated [7, 8]. Although dyes are frequently not the main contributor to textile wastewater, dye molecules frequently garner the most focus because of their color and the toxicity of certain of the basic ingredients used for their synthesis. Though dyes typically have a

doi:10.1088/978-0-7503-5377-9ch11

lower concentration in effluents than any other chemical present in these wastewaters, their vibrant color makes them apparent even in extremely low quantities, which creates major aesthetic issues with wastewater treatment and disposal [9, 10]. Dyes and recalcitrant pollutants, which can inhibit photosynthesis and cause varying levels of toxic effects, radiation, and cancer in life forms, are present in effluents from the textile sector. Sizable classes of chemicals used in the textile industry are reactive dyes. An important environmental concern is the prevalence of synthetic dyes in the environment. The washing cycle, one of the final phases in the textile coloring process, helps to both establish the color in the fabric and remove any excess color. After the fabric dyeing cycle, the contaminated water and pigments are dumped into nearby water bodies. A considerable amount of the water that is contaminated with colored dye is dumped into our environment by inefficient and non-environment-friendly contaminated water treatment plants. The contaminated water released from the plants is lethal and even highly cancer causing because of the incorporation of several chromophoric groups in the dye structure, such as N = N–, = C = O, C–NH, –CH = N–, C–S [11–15]. The most significant class of industrial dyes are acid blue 29 (AB 29) which is an azo dye and acid blue 40 (AB 40), which is an anthraquinone dye, and they are the most commonly used commercial dyes in the textile industry. These are the most frequently utilized colors in modern sectors such as textile manufacture, leather production, and paper making [16, 17].

Pesticide surplus and metabolites in food, water, and soil are now the most pressing challenges in environmental chemistry. Because of their expanding usage in agriculture, pesticides are among the most significant environmental contaminants [18–20]. Pesticides are the combinations of chemicals or compounds intended to prevent, control, or diminish hazardous organisms that may affect the plant growth. These Pesticides are required for the production of high-yielding, high-quality, and dependable food. Pesticides, on the other hand, constitute a high risk to all living forms owing to their chemical qualities. This sparks intense debate in the scientific community, particularly because of the carcinogenic consequences of pesticides with hazardous qualities. It is critical to regulate the usage of these pesticides. According to the World Health Organization, pesticide poisoning is responsible for around 3 million illnesses and 220 000 fatalities per year in poor nations. A surge in the number of pesticides linked to homicides and accidents or suicide deaths is also concerning [21–23]. Organophosphorus (OP) and carbamate insecticides are among the most hazardous pesticides; their toxicity is dependent on the inhibition of acetylcholinesterase (AChE). The toxicity of OP and carbamate pesticides varies greatly based on the pesticide's chemical composition. Pesticides enter the soil, air, and water bodies after being sprayed on agricultural crops, where their build-up relies on their durability. Regulatory agencies are attempting to address this environmental concern by establishing standard levels of pesticide content in drinking water [24].

Since the presence of both the toxic dyes and the pesticides in the food and drinking water are severely affecting all living forms, their removal is presently a major task for environmental engineers working on soil remediation and water treatment. The authorities have assigned the safe amount of these chemicals that can be present in the food samples and drinking water. The development of electrochemical processes which can detect the presence of these toxic dyes as well as toxic pesticides may aid in addressing this problem. In this chapter, we are highlighting the different electrochemical methods for the identification and quantification of toxic dyes as well as the pesticides that are harmful to every lifeform. The traditional detection methods include chromatographic techniques such as gas chromatography (GC) and high performance liquid chromatography (HPLC) combined with mass spectrometry (MS). These procedures are exceedingly sensitive and efficient, but they have significant drawbacks, such as complicated and time-consuming sample processing, such as pesticide extraction, extract cleansing, solvent replacement, and so on [25–31] (figure 11.1).

Electrochemical biosensors, compared to other analytical methods, provide more flexibility for the construction of portable devices, detect extremely low concentrations of analytes with high sensitivity, and are less costly [32, 33]. Electroanalytical techniques use an electrochemical cell to analyze an analyte via potential and/or current measurements. The research provides every critical, even minute, detail on the analyte, such as its chemical nature and characteristics, amount, and so on. The core premise of electroanalytical technique is that specific chemical changes must occur to the analyte molecule, i.e., the analyte molecule must respond directly or indirectly through several connected processes, or it must be adsorbed on the electrode surface for electroanalytical study [34]. Such electroanalytical methods have a wide variety of uses and applications in clinical, pharmaceutical, biomedical, environmental, and industrial domains, among others, due to their precision, accuracy, enhanced sensitivity, cheap cost, greater selectivity, specificity, accuracy, miniaturization, and low detection limit (LOD) [35–38]. Electroanalytical processes are categorized based on which feature of the electrochemical cell is regulated and which aspect is assessed. For simple, sensitive, and selective onsite detection, biosensors centered on the transduction principle can also

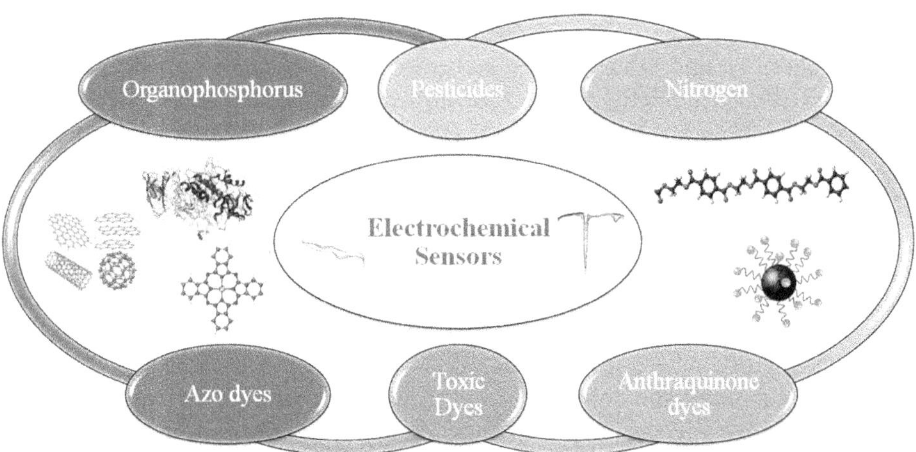

Figure 11.1. Graphical abstract showing the various applications of electrochemical sensors.

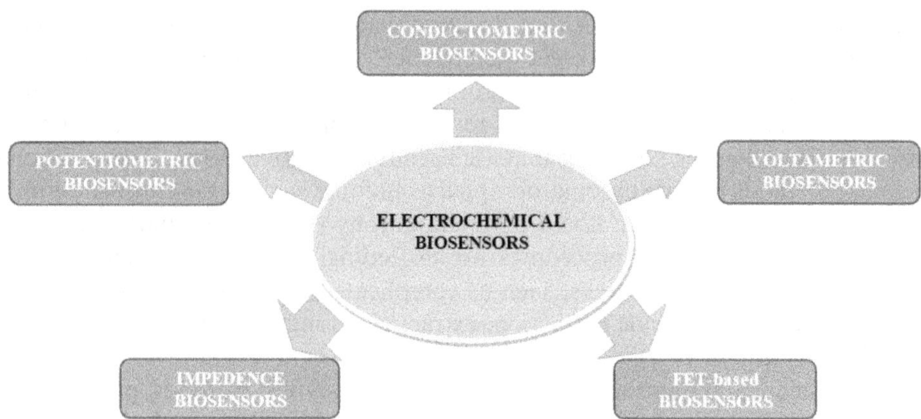

Figure 11.2. Most common types of electrochemical biosensors.

be safely employed. Because of their ultra-sensitivity and simplicity, biosensors have gained prominence in recent years. Furthermore, the incorporation of biocompatible nanomaterials into biosensors increases their efficacy by increasing the surface area, improving electrical conductivity and sensitivity, and shortening reaction time. In light of this, numerous electrochemical sensors for toxic dyes and biosensors for OP pesticide detection have been developed, utilizing a varied variety of simple or composite nanomaterials. Amperometric biosensors, potentiometric biosensors, conductometric biosensors, impedance biosensors, and FET-based biosensors are the most common types of electrochemical biosensors, as shown in figure 11.2.

11.2 Electrochemical assessment of toxic dyes

Synthetic dyes are chemical compounds that are used in a range of sectors such as textiles, cosmetics, food, leather, pharmaceuticals, and fuel labeling. They appear colored due to the sophisticated organic structures that may contain both chromophoric groups and auxochromic groups (like –OH, $-SO_3H$, $-NH_2$,) that enhance both the solubility, and affinity to the substrate, and they can also be electrochemically oxidizable or reducible [39, 40]. The most popular commercially used dyes that find application in the textile industry are acid blue 40 (AB 40) and acid blue 29 (AB 29), which are anthraquinone [41] and azo dyes. These colors are commonly used in modern industries such as textile manufacturing, paper manufacturing, and leather processing. Azo dyes are unquestionably the most common dye type used in industrial applications. More than 3000 types of azo dyes are used globally, accounting for nearly 65% of the industrial dyes. The main reason for this is that they can be easily prepared using a wide range of donor and acceptor groups. Azo dyes are distinguished because of the fact that they contain azo groups (–N=N–) and may produce a broad spectrum according to their structure. They are also highly chromophoric owing to the presence of electronic transitions that take place between the non-bonding electrons present in the nitrogen atom and the energy levels of the corresponding conjugated systems attached to the pi molecular orbitals [42].

Figure 11.3. General azo dye electrochemical oxidation and reduction mechanism.

However, azo dyes may easily be oxidized, in anaerobic conditions, to generate carcinogenic aromatic amines (AAs) [43]. It was observed that the azo groups near an auxochrome group underwent oxidation resulting in the conversion of the azo group to the diazonium group [44], whereas it was noted that in azo dyes such as tartrazine the azo group underwent reduction to give the corresponding secondary amine group [45]. The azo group that is usually found in the dyes used in textile manufacturing, leather and food coloring, among other things belongs to the class of chromophores (figure 11.3).

It is believed that at least 20% of dyes generated are discarded into effluents due to manufacturing or dyeing process losses. Because dyes include a wide range of chemicals, precise procedures for identifying, quantifying, and degrading these compounds are necessary. UV/visible spectrophotometry is the most widely used detection method for dyes; this technique is used to evaluate dye absorption. Nevertheless, dyes are usually diluted in some samples, which necessitates the use of sensitive detection techniques [46]. Due to their low limit of detection in many matrices, chromatographic techniques have been frequently employed for dye detection. But these methods encourage the use of various organic solvents and time-consuming sample preparation; in addition, there can be damage to columns and detectors that may necessitate expensive repair [47, 48]. As a result, there has been a lot of interest in using electrochemical approaches for the detection of dyes that are extremely selective, sensitive, and low-cost by monitoring both the oxidation and or reduction of the respective functional groups electrochemically [49]. They employ lesser-quantity samples with very little or no sample treatment and create a reduced quantity of waste, contributing to a lowering in environmental deterioration. This chapter explores the recent evolution of the various types of electroanalytical techniques for the detection of dyes, emphasizing breakthroughs in the working electrode modifications and critically assessing the developments made in this field. The most prevalent method for electroanalytical dye determination is the reduction of anthraquinone, azo, and nitro chromophores found in a majority of the dyes and the oxidation of auxochrome groups such as phenol and amine compounds [50–53].

11.2.1 Electroanalysis of textile dyes

Textile dyes are categorized based on their fixing capabilities, regardless of the dye chromophore group. Acidic, basic, sulfur, reactive, dispersion, metallic complex, direct, and VAT dyes are among them. Because certain dyes are hazardous and mutagenic, they are currently monitored and controlled by a variety of regulatory authorities. Touzi H *et al* highlighted the utilization of an impedimetric sensor which was based on a gold electrode modified using Cu^{2+}-methylnaphthyl cyclen complex for the measurement of Acid Yellow 25 (AY25). This approach resulted in a LOD of

1.0×10^{-12} mol l^{-1} [54]. Bouzayani *et al* [55] in 2018 used linear sweep voltammetry for evaluating Reactive Black 5 diazo along with various byproducts of electro-Fenton treatment with a screen-printed carbon electrode (SPCE). In the same year, Materón *et al* used a graphite epoxy electrode modified with magnetic nanoparticles to explore the processes of interaction between glutathione-s-transferase and azo textile dye Disperse Orange 3 (DO37). In a procedure performed using glutathione-s-transferase, reduced glutathione was employed to bleach DO37. An oxidation peak corresponding to a potential of 0.91 V was found, which was linked to glutathione-DO37 conjugation that showed a linear increase with DO37 amount [56]. Pemmatte *et al* developed a poly(Glycine) modified carbon nanotube paste electrode (PGMCNTPE) for electrochemically examining the textile dye Alizarin Carmine (AC) using the cyclic voltammetry (CV) method. At 0.782 V, AC underwent electrochemical oxidation with a peak height of 26.96 A, and Tartrazine was also identified at PGMCNTPE at the same time. PGMCNTPE and bare carbon nanotube paste electrode surface properties were investigated utilizing electrochemical impedance spectroscopy and pictures from field emission scanning electron microscopy. The application of this suggested sensor to measure AC in sewage water proved successful [57].

11.2.2 Electroanalysis of hair dyes

Synthetic dyes are used in hair coloring to repair and/or modify the natural color of the hair. Temporary, keratin treatment, and permanent colors are available. When in an alkaline oxidative media, amines and phenolic compounds react, permanent dyes are produced, whereas temporary and keratin treatment dyes are made from either direct, reactive acidic, or basic, dyes [58]. Corrêa *et al* in 2016 described an approach that used carboxyl groups functionalized magnetic nanoparticles as a tool for both pre-concentration and the sensing of Basic Brown 16 which is an azo dye, and it yields a detection limit of 10.1 nmol l^{-1} [59]. Bessegato *et al* [60] employed nanotube electrodes modified by self-doped TiO_2 to determine the p-phenylenediamineda dye which was the precursor used for hair colorings This article describes the usage of TiO_2 nanotube (TiO_2NTs) electrodes as anodes for electroanalytical applications without UV radiation. Although n-type semiconductors, like TiO_2NTs, do not give any potential electrochemical performance at the anodic region, they can be 'activated' once they have already undergone a straightforward cathodic polarization (P-TiO_2NTs) and used to monitor the oxidation reaction without any additional modifications.

11.2.3 Electroanalysis of food dyes

Food dyes are compounds which impart, improve, or retain the color of edibles and are employed as synthetic colorants that contribute no nutritional value to food. Given the growing concern about the dangers presented by these dyes, it is critical to introduce accurate, selective, sensitive, and reliable electroanalytical techniques for analyzing and determining these dyes in order to assist in limiting their harmful effects.

Shweta *et al* in 2018 [61] used a montmorillonite nanoclay-modified carbon paste electrode (CPE) to determine anodic Sunset yellow (SY) and received the minimum LOD. This study's great sensitivity was ascribed to nanoclay surface features, which included strong adsorption capacity and better stability. The sensor was used in pharmaceutical samples as well as body fluids before preprocessing and achieved a total recovery rate of even more than 95%. Another research used CPE modified with copper-based metal–organic frameworks to determine Ponceau 4R and got a LOD of 1.08×10^{-9} mol l^{-1} [62]. Cu-1,3,5-benzenetricarboxylic acid frameworks' huge surface area and many micropores contributed to increased aggregation performance and much more active sites for Ponceau 4R oxidation. On the Cu-BTC/CPE surface, the dye revealed oxidation and reduction peaks. After 1 min accumulation, the procedure was applied to many using differential pulse voltammetry (DPV). Many investigations have described the usage of sensors for tartrazine (TZ) measurement, with LODs ranging from 3.5×10^{-9} to 4×10^{-8} mol l^{-1} [63, 64]. The key contribution worth highlighting is that the modification of the glassy carbon electrode (GCE) using molecularly imprinted copolymers that were electropolymerized that selectively bond to a tartrazine molecule, attaining LOD of 3.5×10^{-9} mol l^{-1}. A molecularly imprinted polymer (MIP)-PmDB/PoPDe modified sensor was used for tartrazine analysis by using 5×10^{-3} mol l^{-1} $K_3Fe(CN)_6$. The redox system of $K_3Fe(CN)_6$ was disrupted, and corresponding current was reduced in proportion to the amount of tartrazine. There is the presence of an interactivity between the aromatic groups of tartrazine and sulfonated groups present in 3,4-ethylenedioxythiophene that increases the anodic current for tartrazine detection [65]. Because of their high conductivity, multi-walled carbon nanotubes (MWCNTs) were used in another study to greatly improve the current response for tartrazine determination. Further, the ionic liquid facilitated tartrazine pre-concentration, whereas the Pt nanoparticles (PtNPs) demonstrated catalytic activity, speeding electron transfer and giving the required conduction channel. In 2018, He Q *et al* introduced a sensor using GCE modified using MnO_2 nanorods-ErGO nanocomposites to determine Amaranth (AM) food dye. The approach produced an LOD of 1.0×10^{-9} mol l^{-1} and the current intensity obtained was found to be 38 fold greater than that obtained with a bare electrode [66]. After grinding, dispersion, centrifugation, and dilution, food items such as candy, watermelon, and juice were evaluated. Allura Red (AR) and Adsorption of Amaranth (AM) on the electrode surface was observed using differential pulse adsorptive stripping voltammetry having two kinds of silver solid amalgam electrodes [67]. Likewise, for erythrosine measurement employing DPV Shetti *et al* proposed and developed a Au electrode enhanced with TiO_2 nanoparticles [68]. In the study, TiO_2 nanoparticles were used to change the gold electrode surface and highlight a practical method for erythrosine measurement. The improvement enhanced the active sensing area. Thus, the dye's oxidation signal amplification was seen. The sensor demonstrated excellent selectivity, respectable accuracy, and great sensitivity.

There are also methods corresponding to the simultaneous detection using electroanalysis methods on food dyes. In 2017 [69] Sierra-Rosales P *et al* determined carmoisine(CAR), tartrazine, and SY, concurrently on a carbon ceramic electrode modified using MWCNTs, yielding 0.22, 0.12, and 0.11 × 10^{-6} mol l^{-1}, respectively as LODs. Amaranth and tartrazine were also measured concurrently utilizing square-wave adsorptive stripping voltammetry on double-stranded copper (I) helicate (H) on an SPCE modified using single-walled carbon nanotubes. The detection limits for Amaranth and tartrazine were 30.0 and 60.0 nmol l^{-1}, respectively. In 2016 Ji L *et al* developed a sensor with CPE modified with copper-based metaleorganic skeletons via DPV that had the best LODs, with oxidation maxima at 0.62 and 0.85 V for SY and tartrazine, respectively [70]. The benefits of this study include significantly increased oxidation activity toward Tartrazine and Sunset Yellow, as well as high accuracy and significant potential for application in actual sample analysis. Deroco *et al* also identified indigo carmine and AR in candies employed a cathodic pretreatment of diamond electrode dopped using boron in conjunction with flow injection detection by multiple pulse amperometry. Indigo carmine and AR both produced distinct oxidation peaks at 0.67 and 0.87 V, respectively, as LODs. Hareesha *et al* using DPV and CV techniques, developed a simple electrochemically polymerized glutamic acid layered multiwalled carbon nanotube paste electrode (P(GA)LMWCNTPE) for the detection of indigo carmine (IC). The P(GA)LMWCNTPE exhibits an appreciably high rate of electrocatalytic activity toward the redox behavior of IC under the optimized experimental conditions. With methyl orange, the predicted P(GA)LMWCNTPE exhibits a respectable selectivity for IC. With accurate detection limits of 4.2 and 0.36 M, respectively, the improved sensor exhibits an acceptable linear increase between oxidative peak current and concentration in both CV and DPV techniques. The created sensor was also successfully used to find IC in water and food samples [71]. Tigari and Manjunatha created a carbon nanotube paste electrode with a poly (glutamine) film modification (PGAMCNTPE). Utilizing CV) and DPV, it was employed to measure the electrochemical activity of curcumin in phosphate buffer at pH 7.5. While curcumin could not be detected by the bare carbon nanotube paste electrode, PGAMCNTPE can detect curcumin oxidation at 0.168 V and reduction at 0.098 V with high current sensitivity. The PGAMCNTPE exhibits a linear voltammetric response for curcumin from 0.4 to 6 and 6 to 10 M under calibrated circumstances, with a lower LOD of 2.79 108 M. With great recovery from 90% to 97%, the intended curcumin sensing approach was applied to dietary supplements [72]. Some reported works for the detection of important dyes are included in tables 11.1–11.3.

Most techniques, including those that used carbon-based CPE, GCE, and SPCE electrodes, were successful in raising efficiency by surface modification [92]. CNTs captured significant attraction in the past years because of their excellent electrical properties, mechanical, and chemical resistance, as well as their current famous position in electrochemical investigations. They promote charge-transfer processes and significantly raises the electrocatalytic behavior of the surfaces, acting as a sensor [93]. By modifying electrochemical parameters, which may regulate the film thickness (a polymer film exhibits high stability and repeatability), and charge transport properties, electropolymerization is frequently employed to make polymer-modified electrodes [94]. Studies that commonly found a reduction in LOD used

Table 11.1. Recent works on electroanalytical determination of Sunset yellow (SY).

Electrode	Electrochemical technique	LOD/mol l^{-1}	Linear range/ mol l^{-1}	Peak potential V^{-1}	References
Molecularly imprinted copolymer (MIP-PmDB/PoPD-GCE)	DPV	3.5×10^{-9}	5.0×10^{-9}–1.1×10^{-6}	+0.25	[63]
Imprinted polymer—multi-walled carbon nanotubes—ionic liquid supported Pt nanoparticles composite film oated (MIP–MWNTs-IL@PtNPs/GCE)	DPV	8.0×10^{-9}	0.03×10^{-6}–20.0×10^{-6}	+0.92	[73]
Co-polymerization of Tartrazine and acrylamide on the carbon nanotubes (NIP/MWCNTs/GCE)	DPV	2.7×10^{-8}	8.0×10^{-8}–1.0×10^{-6}	+0.93	[74]
Poly(3,4-ethylenedioxythiophene) (PEDOT)@Terbium hexacyanoferrate (TbHCF) composite on GCE (PEDOT@TbHCF/GCE)	DPV	3.2×10^{-8}	0.1×10^{-6}–206×10^{-8}	+0.9	[65]
Poly p-aminobenzenesulfonic acid ((Pp-ABSA)/ZnO NPs (Pp-ABSA/ZnO NPs-CPE)	DPV	8.0×10^{-8}	0.35×10^{-6}–5.44×10^{-6}	+1.0	[75]
Laccase conjugated microspheres and gold nanoparticles (AuNPs) coated on SPCE	DPV	0.04×10^{-6}	0.2×10^{-6}–14×10^{-6}	+1.1	[64]
Chitosan/N-doped graphene natively grown on hierarchical porous carbon nanocomposite (N-PC-G/CS)	DPV	3.6×10^{-6}	0.05×10^{-3}–15.0×10^{-3}	+0.41	[76]

Table 11.2. Recent works on electroanalytical determination of tartrazine (TZ).

Electrode	Electrochemical technique	LOD/mol l^{-1}	Linear range/mol l^{-1}	Peak potential V^{-1}	References
Zinc oxide/reduced graphene on Zn foil	DPV	3×10^{-9}	0.01×10^{-6}–5×10^{-6}	+0.64	[77]
Cuprous oxide-electrochemically reduced graphene nanocomposite (Cu_2O-ErGO/GCE)	SDLSV	6.0×10^{-9}	2.0×10^{-8}– 1.0×10^{-4}	+0.79	[78]
Chitosan/graphene nanocomposite (EXF/GCE)	CV	6.66×10^{-8}	2×10^{-7}–1×10^{-4}	+0.72	[79]
Graphene oxide decorated with silver nanoparticles–molecular imprinted polymers (GO/AgNPs–MIPs/GCE)	CV	0.02×10^{-6}	0.1×10^{-6}–12×10^{-6}	+0.73	[80]
Graphene, multi-walled carbon nanotubes, gold nanoparticles, and nanocomposite membrane of chitosan (CHIT/GO/MWCNTs/AuNPs/GCE)	DPV	7×10^{-5}	0.02–0.2	+0.74	[81]
Pd–Pt bimetallic nanocages (PDDA-Gr-(Pd–Pt), Pt–Cu bimetallic nanoframes (PDDA-Gr-(Pt–Cu)) and Co–Ni bimetallic nanoflowers (PDDA-Gr-(Co–Ni)) and poly (diallyldimethylammonium chloride) (PDDA)-dispersed graphene (Gr)	DPV	PDDA-Gr-(Pd–Pt) $6.0 \times 10–9$ PDDA-Gr-(Pt–Cu) PDDA-Gr-(Co–Ni)	PDDA-Gr-(Pd–Pt) 0.02×10^{-9}–10.0×10^{-9} PDDA-Gr-(Pt–Cu) 0.02×10^{-9}–10.0×10^{-9} PDDA-Gr-(Co–Ni) 0.008×10^{-9}–10.0×10^{-6}	+0.85	[82]
Nanoclay particles on CPE	SWV	0.2×10^{-9}	0.001×10^{-6}–0.1×10^{-6}	+0.80	[61]
Zinc oxide nanoflower (ZnONF/CPE)	SWV	2.0×10^{-1}	1.1×10^{-9}–1.5×10^{-7}	+0.70	[83]
Palladium-ruthenium nanoparticles incorporated carbonaerogel (Pd–Ru/CA) on SPCE	DPV	7.1×10^{-9}	0.02×10^{-6}–110.2×10^{-6}	+0.72	[84]

Table 11.3. Recent works on electroanalytical determination of other toxic dyes.

Dye	Electrode	Electrochemical technique	LOD mol l^{-1}	Linear range mol l^{-1}	Peak potential V^{-1}	References
Amaranth	Graphene nanomeshes (GNM)	SWV	7.0×10^{-9}	5.0×10^{-9}–1.0×10^{-6}	+0.74	[85]
Amaranth	Manganese dioxide nanorods-electrochemically reduced graphene oxide nanocomposites (MnO_2NRs-ErGO/GCE)	SDLSV	1.0×10^{-9}	0.02×10^{-6}–400×10^{-6}	+0.78	[86]
Amaranth	RuO2 nano-rod (RuO2/NR) and 1,3-dipropylimidazolium bromide (DPIBr) (RuO_2/NR/DPIBr/CPE)	SWV	3.0×10^{-9}	0.008×10^{-9}–550×10^{-9}	+0.82	[87]
Amaranth	Poly(sodium p-styrenesulfonate) (PSS)-functionalized graphene supported palladium nanoparticles (PSS-GR-Pd/GCE)	DPV	7×10^{-9}	1×10^{-9}–9×10^{-6}		[88]
Ponceau 4R	Cu-BTC framework (Cu-BTC/CPE)	DPV	1×10^{-9}	$0,2 \times 10^{6}$–1×10^{-6}	+0.59	[89]
Ponceau 4R	Reduced graphene oxide (rGO/GCE)	SWV	2.84×10^{-8}	0.200×10^{-6}–20.0×10^{-6}	+0.61	[90]
Erythrosine	TiO_2 nanoparticles on gold electrode	DPV	2.6×10^{-9}	0.1×10^{-6}–10.0×10^{-6}	+0.72	[68]
Carmoisine	GCE	DPV	4×10^{-8}	1×10^{-7}–1×10^{-6}	−0.15	[91]

conducting polymer materials, which are known to assist charge transfer, and large surface areas. The creation of disposable, economical, straightforward, and long-lasting electrodes; the use of reasonably priced and easily repairable materials; and the need to make electrochemistry more accessible are the main obstacles to the widespread adoption of electroanalytical techniques.

11.3 Electrochemical assessment of pesticides

The majority of fertilizers and pesticides are extremely nonbiodegradable and do not dissolve with passage of time; nonetheless, they may get absorbed into the soil particles and slowly seep into groundwater, producing ongoing pollution. When consumed by humans, fertilizers and pesticides can have major negative impacts. Nitrate-containing ones have been linked to both cancer and the blue baby syndrome, while several insecticides are known to be carcinogenic [95–99]. There are chemical sensors that convert chemical data into an analytically legible and usable signal, such as the concentration of a particular pesticide. The development of pesticide biosensors for tracking pesticide residue in food and drinking water has received a lot of attention in recent years. Pesticide residue detection is viewed as a difficulty for managing food and water safety as well as environmental protection. Chemical and biological pesticides are the two categories into which pesticides fall. The biopesticides are obtained organically from natural sources, whereas chemical pesticides are manufactured compounds that selectively target insects [100]. Insecticides, herbicides, fungicides, rodenticides, and nematicides are the five main types of chemical pesticides that are categorized according to their application, shown in figure 11.4. However, chemical pesticides, particularly insecticides, are the most often utilized in the food and agriculture industries. Pesticides for insects are grouped into four main groups: carbamates, organochlorines, pyrethroids, and organophosphorus [101]. OCPs, organochlorine pesticides are a class of insecticides used mostly in agriculture to safeguard cultivated plants which can seriously impact the nerve system. Esters made from phosphoric

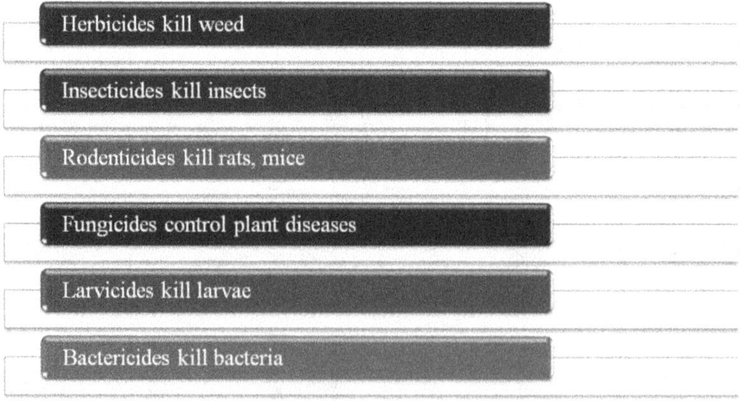

Figure 11.4. Main types of pesticides.

acid are known as organophosphorus pesticides (OPs), and they are often used as insecticides to control pests and miticides on plants. OPs enter people by touch and consumption. Esters formed from acids and carbamic acid are carbamates. These pesticides' low toxicity to mammals, quick disappearance, and broad-range action has made them increasingly significant in recent years. This chapter examines and illustrates the most current developments in important chemical and pesticide detection technologies.

11.3.1 Electrochemical sensors for the detection of pesticides

There are different types of sensors for pesticide identification and determination, including electrochemical, optical, and mechanical detection techniques, shown in figure 11.4. Field-effect transistor (FET), screen-printed electrodes, and capacitive-based approaches are among the electrochemical detecting methods. There are several features of pesticide sensors, like biorecognition materials such as enzymes and antibodies, and an affinity biosensor is the pattern of the chemical interaction among the residual pesticide and the respective sensing materials. Electrochemical biosensors are marked to be effective instruments for detecting pesticide residues. Because of their amazing qualities such as low cost, ease of operation, portability, and rapid reaction, electrochemical methods are typically chosen over other analytical detection methods. The electrochemical biosensors are classed based on the signal measured, which are current, impedance, or potential; hence, these sensors are amperometric, impedimetric, or potentiometric. Some important reported works of different pesticides detection are listed in table 11.4 (figure 11.5).

Table 11.4. Electrochemical sensors for detection of pesticides.

Target pesticide	Material used	Detection principle	Limit of detection	References
Atrazine	N/A	Change in resistance	8.3 $\mu g\ l^{-1}$	[102]
Atrazine	N/A	Change in impedance	0.19 $\mu g\ l^{-1}$	[103]
Atrazine	AuNPs	Change in impedance	50 $\mu g\ Kg^{-1}$	[104, 105]
Atrazine	N/A	Change in impedance	10 ng ml^{-1}	[106]
Acetamiprid	AuNPs	Change in impedance	1 nM	[107]
Carbendazim	N/A	Change in impedance	0.9 ng ml^{-1}	[108, 109]
Carbofuran	N/A	Change in impedance	N/A	[110]

(*Continued*)

Table 11.4. (*Continued*)

Target pesticide	Material used	Detection principle	Limit of detection	References
Acetamiprid	Au/MWCNT-rGONR	Change in impedance	1.7×10^{-14} M	[111]
Chlorpyrifos	N/A	Change in impedance	0.014 ng ml^{-1}	[112]
Carbaryl	Chitosan	Change in impedance	1 ng ml^{-1}	[113]
Tetracycline	MWCNT	Change in resistance	10^{-9} M	[114]
Pathogen	N/A	Change in capacitance	1.5 aM	[115]
Bisphenol A	N/A	Change in capacitance	152.93 aM	[116]
Malathion, Cadusafos	rGO,AuNPs	Change in resistance	0.1 nM l^{-1}	[117]
Acetamiprid Atrazine	Platinum nanoparticles	Change in impedance	10 pM	[118, 119]
Paraoxon	Chitosan, AuNPs, MWCNT	Change in current	0.03 μg l^{-1}	[120]
Atrazine	CNT,ZnO	Change in resistance	21.61 K Ω μg^{-1} ml^{-1}	[120]
Imidacloprid	AuNPs	Change in currant	22 pM	[121]

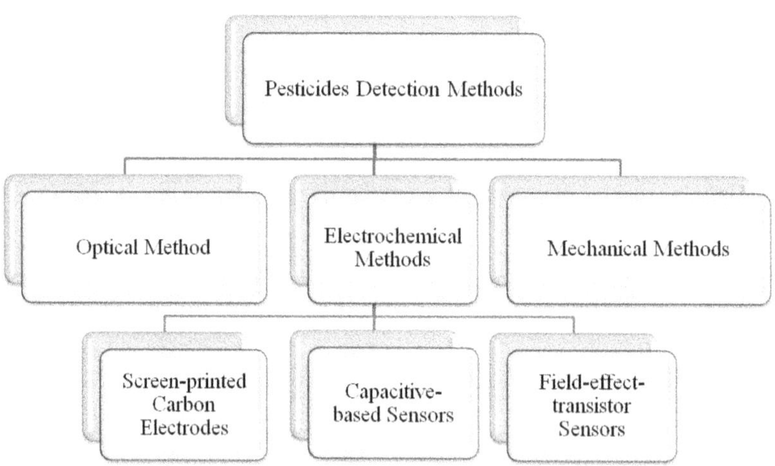

Figure 11.5. Different pesticides detection methods.

11.3.2 Capacitive biosensors for the detection of pesticides

Capacitive biosensor generally refers to biosensors that are based on electrochemical impedance spectroscopy (EIS), which measures capacitance fluctuations. The biosensors based on EIS primarily operate on Faradaic and non-Faradaic modes where the sensitivity of the non-Faradaic sensor is negligibly low in comparison to a Faradaic sensor [122]. A label-free biosensor was fabricated by Madianos *et al* for accurate sensing of pesticides like acetamiprid and atrazine by using the Faradic EIS approach, two-dimensional platinum nanoparticle films, and pesticide-specific aptamers. They compared the results of the impedimetric biosensor by using bare interdigitated electrodes without and with nanoparticles demonstrating a significant improvement in the performance for detection of atrazine with a limit of detection of 40 pM [118, 123]. Marrakchil *et al* [124] used interdigitated gold microelectrodes to create a sensitive, and label-free immunosensor for detection of atrazine herbicide which was based on immobilization approach. This immobilization approach was found to be a low-cost gold-functionalization method that can be used instead of SAM. The interaction between antibody and atrazine was assessed throughout a dynamic range of 10–150 ng ml^{-1}, with atrazine detection confined to 10 ng ml^{-1} in the PBS buffer. The immunosensor used antibody physisorption as a method of immobilization. Electrochemical impedance spectroscopy was used to describe each stage of the immunosensor development process. This technique exhibited little to no cross-reactivity to other related chemicals like terbutylazine, but was extremely selective for atrazine and propazine. Cao *et al* [125] developed an electrochemical immunosensor centered on interdigitated array microelectrodes (IDAMs) for the sensitive, definite, and quick sensing of pesticides containing chlorpyrifos. The EIS method was employed to identify chlorpyrifos pesticide by detecting the changing impedance. The impedance change was proportionate to chlorpyrifos values in the 10^0–10^5 ng ml^{-1} range, with a detection limit of 0.014 ng ml^{-1}. Furthermore, using electrochemical impedance spectroscopy, Fan *et al* [126] developed a highly sensitive and selective biosensor based on aptamer for the determination of acetamiprid pesticide. To improve the sensitivity, they electrodeposited the bare electrode surface with AuNPs. The developed gold electrode was employed as a substrate for acetamiprid pesticide-specific aptamer immobilization. A large linear range of 5–600 nM with a modest 1 nM detection limit was observed. The control studies show that the aptasensor only recognizes acetamiprid specifically, leading to great selectivity of the aptasensor, when using pesticides that may coexist with it or have a similar structure. Additionally, Facure *et al* [127] used impedance spectroscopy to construct an electronic tongue by using graphene hybrid nanocomposites for measuring low quantities of organophosphate pesticides. The generated sensor was made using interdigitated electrodes and comprised four sensing units which were layered by graphene hybrid nanomaterials utilizing drop casting procedure to improve the device's sensitivity.

11.3.3 Field-effect transistor-based biosensors for the detection of pesticides

Charge carriers of the substrate materials determine the working principle of FET biosensors. As a result, there are typically two types of FETs: n-type FET biosensors

Table 11.5. FET-based sensors for detection of pesticides.

Target pesticide	Material used	Detection principle	Limit of detection	References
Acetylcholine	Graphene	Change in current	2.3 μM	[139]
Chlorpyrifos	Graphene	Change in resistance	1.8 fM	[139].
Carbaryl	Graphene	Change in current	10^{-8} μg ml^{-1}	[138]
Atrazine	CNT	Change in current	0.001 ng ml^{-1}	[140]

with electrons as the primary charge carriers and p-type FET biosensors with holes as the primary charge carriers [128–130]. The introduction of nanomaterials like carbon nanotubes, graphene, and metal oxides has increased the functionality of FET biosensors. Because of their appealing features, mainly excellent performance in aqueous solution, quick response, increased sensitivity, and ability to operate at minute voltages, nanomaterial FET-based biosensors have received a lot of attention recently [131–135]. Graphene FETs immobilized using acetylcholinesterase(AchE) have been demonstrated by Fenoy *et al* [136] for acetylcholine detection. The technique relies on the electrosynthesis of an amino polymer layer on a graphene channel substrate. In a flow mode, our constructed biosensor devices demonstrated a lower detection limit of 2.3 M with a monitoring Ach range of 5–1000 M. The RSD of the described biosensor devices was 2.6%, indicating good device repeatability. Islam *et al* successfully built a microfluidic biosensor for chlorpyrifos pesticide detection in actual samples using a graphene FET. The biosensor was built on a Si/SiO_2 substrate and included a single-layer graphene nanomaterial that demonstrated exceptional sensitivity in the determination of pesticides. The fabricated graphene FET biosensor displayed great sensitivity, stability, and specificity for the detection of chlorpyrifos pesticide [137]. Thanh *et al* [138] used low-pressure chemical vapor deposition (LPCVD) to produce graphene layers on polycrystalline copper foil. This biosensor was designed to recognize enzymes by carbaryl enzymatic inhibition of urease. This biosensor's response was detected based on enzymatic reaction activities, with a modest current response produced by reduced urease enzymatic activity with carbaryl. Furthermore, with a detection limit of 10^{-8} g ml^{-1}, the disclosed ISFET biosensor demonstrated outstanding sensitivity to carbaryl insecticide. Some important works are listed in table 11.5.

11.3.4 Electrochemical detection of some important pesticides

Glyphosate and bentazone, which are the two most common pesticides, are frequently found in water sources in excess of the administrative limits established by the Environmental Protection Agency (EPA) while, due to its high toxicity, lindane is prohibited from use in agriculture. The legal limit for pesticides that can be present in drinkable water is 0.10 g l^{-1}, or 0.59 nM for glyphosate, 0.34 nM for lindane, and 0.42 nM for bentazone, respectively [141, 142].

Using bare and modified electrodes constructed of carbon, gold, copper, mercury and platinum, glyphosate has been detected. Prasad *et al* in 2014 [143] developed an electrode by modifying the pencil graphite electrode with AuNPs and a doubly imprinted nanofilm for the detection of glyphosate and glufosinate simultaneously, utilizing N-nitroso glyphosate and glufosinate as the template molecules. Here, the overlapping decrease peaks corresponding to glyphosate and glufosinate may be separated by 265 mV, allowing for a more accurate pesticide identification. For glyphosate, the sensor's detection limit was 2.0 nM, and for glufosinate, it was 1.0 nM, with a linear range of 0.024–1.04 M. In order to detect glyphosate in 2015, Do *et al* treated the gold electrodes with an MIP. In the presence of glyphosate template molecules, p-aminothiophenol-functionalized AuNPs were electropolymerized [144]. After removing the template, voids with glyphosate-like shapes developed, enabling the precise identification of glyphosate that would bond to aniline groups. By using linear sweep voltammetry (LSV), it was possible to identify the bound glyphosate molecules, with a linear range of 5.91 nM–5.91 M and a quantitative limit of 4.73 nM. In addition, Zhang *et al* [145] used CV for electropolymerized glyphosate and pyrrole to create a glyphosate molecularly imprinted polypyrrole-modified gold electrode (MIPPy). Following polymerization, an overoxidation technique was used to remove the contained glyphosate molecule from the polypyrrole membrane. DPV in a 0.10 KCl solution was then effectively used to identify glyphosate presence in cucumber as well as water samples using the imprinted modified electrode. The computed LOD was 1.60 nM, while the sensor showed a linear response between 0.03 and 4.73 M. For glyphosate sensing, a pencil graphite-based electrode (PGE) modified with multi-walled carbon nanotubes-ionic liquid (MWCNTs-IL) and copper oxide (CuO) NP composite was also used. The superior sensitivity and efficacy of the electrode were due to the exceptional electrical properties of CuO and its ability to form a complex with glyphosate when combined with the electrical conductivity and large surface area of MWCNTs-IL [146]. A simple CPE was employed by Oliveira *et al* to oxidize the glyphosate in the form of an isopropylamine salt. When glyphosate was present in a buffer solution having pH of 5, CV at the electrode potrayed a distinct oxidation peak at 0.95 V (versus Ag/AgCl). With a detection limit of 2.0 nM and under ideal conditions, it was feasible to identify the pesticide in milk, orange juice, and agricultural formulations [147]. Moraes *et al* determined glyphosate utilizing electrochemical oxidation and DPV, a GCE coated with copper phthalocyanine/MWCNT (GCE/MWCNT/CuPc) film has been utilized. According to the authors, the indirect detection of glyphosate at 50.0 mV versus SCE based on Cu(I)/Cu(II) couple was made possible by the high interaction between glyphosate and copper ions to produce a stable complex. With a limit of detection of 12.20 nM and concentration range of 0.83–9.90 M, glyphosate was identified using this approach [148]. For the purpose of glyphosate determination, electrochemical techniques based on biosensors were also reported. One such sensor was made by Songa *et al* by electrostatically attaching the enzyme horseradish peroxidase (HRP) to a revolving gold disk electrode that had been treated with NPs of poly(2,5-dimethoxyaniline)-poly(4-styrenesulfonic acid) (PDMA-PSS). This biosensor was employed to analyze glyphosate in spiked maize samples with detection limit of 0.59 nM and concentration range of 0.012–0.46 M [149]. Another study described the use of an electrophoresis-modified graphite-epoxy electrode for the measurement of glyphosate. This electrode had MWCNTs and HRP deposited on it. This sensor's claimed detection limit is 1.32 pM [150]. It was determined through a polarographic investigation of lindane at an platinum electrode coated using mercury in dimethyl sulfoxide that the reduction occurs via a six-electron transfer pathway since there is only one peak at 1.50 V versus SCE corresponding to reduction. Despite the fact that neither kind of reduction produced any further intermediate chlorinated compounds. However, it is known that several metals, including cobalt, may accelerate a stepwise dechlorination of lindane as well as other organohalides [151]. In order to directly reduce lindane in an aqueous-alcoholic media, Kumaravel *et al* [152] designed an electrochemical sensor utilizing cellulose acetate modified GCEs (CA/GCE). On this modified electrode, lindane's reduction potential was 1.50 V. With detection limit of 9.18 M, the peak current at the corresponding electrode was found to have a linear range from 50 to 180 M. For the electrochemical reduction of lindane, Fayemi *et al* [153] also assessed sensors based on PANI/Zn, Fe(III), and Nylon 6,6/MWCNT/Zn, and Fe(III) oxide nanofibers. Lowest limit of detection of 32.0 nM for lindane was reached at Nylon 6,6/ MWCNT/Fe_3O_4, and the range for determination of lindane was exhibited between 9.90 pM and 5.0 M. Birkin *et al* used 9,10 diphenylanthracene (DPA) as an electrochemical mediator in which there was an indirect reduction of lindane at a vitreous carbon disk electrode. In the presence of lindane, DPA revealed an irreversible decrease peak at 1.79 V (versus Ag/AgCl) which displaced by 30.0 mV and a considerable rise in peak current [154]. It has been claimed that a bare GCE may be used to detect bentazone in commercial herbicides. A plot of linear calibration of bentazone between 15.0 and 22.60 M was produced using SWV, with detection limit of 10.0 M [155]. Cerejeira *et al* [156] suggested an amperometric detection connected to a flow injection analysis (FIA) device at a GCE. Calibration curves were used to analyze bentazone in estuary waters with concentrations ranging from 2.50 to 50.0 M and an oxidation potential of 1.10 V in acetate buffer solution (ABS) with a pH of 4.5. Another study employed a GCE modified with a layer of polyaniline-carbon nanotubes-cyclodextrin (PANI-CD-MWCNT) to detect bentazone [157]. By CV, the electrode was used to determine bentazone in water samples in the range of 10.0–80.0 M, with LOD of 1.60 M. To create electrodes that may be used frequently without considerable activity loss, a simple low-cost manufacturing approach was adopted. Without the necessity for electrode modification, the SWV approach was employed to draw the calibration curve and measure bentazone based on its oxidation. The calibration plot has a linear range of 0.19–50.0 M and an LOD of 34.0 nM [158]. Further research has been carried out for the detection of these important pesticides of which relevant ones have been noted.

This chapter highlights several electrochemical approaches for pesticide detection that have been developed employing enzymes, antibodies, aptamers, and MIPs to assure food safety. Electrochemical biosensors have advanced significantly over the past years, and they now give a potent analytical tool for pesticide identification that is simple, fast, selective, sensitive, and affordable. While considerable progress has been achieved in boosting detection sensitivity through the use of nanomaterials, there is also potential to improve device reusability and mobility.

11.4 Conclusion

The use of electrochemical measuring techniques seems to be a workable answer that may satisfy consumer demand. Artificial electrolytes and sample preparation might be replaced by environmental samples. Utilizing electrochemistry for pollution detection can cause interference from other compounds which are present at the location. The ideal alternative in this situation would be to use biologically functionalized electrodes; however, this might increase environmental contamination already present due to the inherently unstable nature of biomolecules. By creating molecular imprinting techniques, which have a strong chance of functioning without biological alteration and can achieve the required selectivity, this might be avoided. Furthermore, as material science and nanotechnology advance, a wide range of innovative materials will be available for the production or modification of electrodes, allowing for a solution to the existing problems in the practical regular inspection of toxic dye as well as pesticide pollutants.

References

[1] Raril C and Manjunatha J G 2020 Fabrication of novel polymer-modified graphene-based electrochemical sensor for the determination of mercury and lead ions in water and biological samples *J. Anal. Sci. Technol.* **11** 1–10

[2] Mateo-Sagasta J, Zadeh S M, Turral H and Burke J 2017 Water pollution from agriculture: a global review *The Food and Agriculture Organization of the United Nations; the Water Land and Ecosystems Research Program* (Rome: The Food and Agriculture Organization of the United Nations)

[3] Li K, Zhang Z, Yang H *et al* 2018 Effects of instream restoration measures on the physical habitats and benthic macroinvertebrates in an agricultural headwater stream *Ecol. Eng.* **122** 252–62

[4] Mani T, Primpke S, Lorenz C, Gerdts G and BurkhardtHolm P 2019 Microplastic pollution in benthic midstream sediments of the Rhine River *Environ. Sci. Technol.* **53** 6053–62

[5] Schwarzenbach R P, Egli T, Hofstetter T B, von Gunten U and Wehrli B 2010 Global water pollution and human health *Annu. Rev. Environ. Resour.* **35** 109–36

[6] Torres N H *et al* 2019 Real textile effluents treatment using coagulation/flocculation followed by electrochemical oxidation process and ecotoxicological assessment *Chemosphere* **236** 124309

[7] Solís M, Solís A, Pére H I, Manjarrez N and Flores M 2012 Microbial decolouration of azo dyes: a review *Process Biochem.* **47** 1723–48

[8] Saratale R G, Gandhi S S, Purankar M V, Kurade M B, Govindwar S P, Oh S E and Saratale G D 2013 Decolorization and detoxification of sulfonated azo dye C.I. Remazol Red and textile effluent by isolated *Lysinibacillus* sp. RGS *J. Biosci. Bioeng.* **115** 658–67

[9] Reife A and Freeman H S 1996 *Environmental Chemistry of Dyes and Pigments* (Toronto, ON: Wiley)

[10] Zollinger H 2003 *Color Chemistry* (Zurich: Wiley-VCH)

[11] Kubra K T, Salman M S and Hasan M N 2021 Enhanced toxic dye removal from wastewater using biodegradable polymeric natural adsorbent *J. Mol. Liq.* **328** 115468
[12] Kubra K T *et al* 2021 Utilizing an alternative composite material for effective copper (II) ion capturing from wastewater *J. Mol. Liq.* **336** 116325
[13] Berradi M *et al* 2019 Textile finishing dyes and their impact on aquatic environs *Heliyon* **5** e02711
[14] Islam A *et al* 2013 Development of a procedure for spherical alginate–boehmite particle preparation *Adv. Powder Technol.* **24** 1119–25
[15] Islam K N *et al* 2013 A novel catalytic method for the synthesis of spherical aragonite nanoparticles from cockle shells *Powder Technol.* **246** 434–40
[16] Lin H, Zhang H and Hou L 2014 Degradation of C. I. acid orange 7 in aqueous solution by a novel electro/Fe3O4/PDS process *J. Hazard. Mater.* **276** 182–91
[17] Grčić I, Papić S, Koprivanac N and Kovačić I 2012 Kinetic modeling and synergy quan-tification in sono and photooxidative treatment of simulated dye house effluent *Water Res.* **46** 5683–95
[18] FAO 1993 Agriculture towards *Proc. C 93/94 Document of 27th Session of the FAO Conf.* (Rome: The Food and Agriculture Organization of the United Nations)
[19] Aspelin L 1994 *Pesticides Industry Sales and Usage, 1992 and 1993 Market Estimates* (Washington, DC: US Environmental Protection Agency)
[20] U.S. FDA Pesticide monitoring databases http://cfsan.fda.fda.gov/~lrd/pestadd.html
[21] Srinivas Rao C H, Venkateswarlu V, Surender T, Eddleston M and Buckley N A 2005 Pesticide poisoning in south India: opportunities for prevention and improved medical management *Tropical Med. Int. Health* **10** 581–8
[22] García-Repetto R 2018 Sample preparation for pesticide analysis in a forensic toxicology laboratory: a review *J. Forsenic. Sci. Dig. Invest.* **1** 27–45
[23] Eddleston M 2020 Poisoning by pesticides *Medicine* **48** 214–7
[24] Rodrigo M A, Oturan N and Oturan M A 2014 Electrochemically assisted remediation of pesticides in soils and water: a review *Chem. Rev.* **114** 8720–45
[25] American Public Health Association (ed) 1998 *Standard Methods for Examination of Water and Wastewater* 20th edn (Washington, DC: American Public Health Association) pp 6/85–90
[26] EPA Method 8141 A 2000 (US Environmental Protection Agency)
[27] Vicente A and Yolanda P 2004 Determination of pesticides and their degradation products in soil: critical review and comparison of methods *Trends Anal. Chem.* **23** 772–89
[28] Ferrer I, Garcia-Reyes J F and Fernandez-Alba A 2005 Identification and quantitation of pesticides in vegetables by liquid chromatography time-of-flight mass spectrometry *Trends. Anal. Chem.* **24** 671–82
[29] Hernandez F, Sancho J V and Pozo O J 2005 Critical review of the application of liquid chromatography/mass spectrometry to the determination of pesticide residues in biological samples *Anal. Bioanal. Chem.* **382** 934–46
[30] Tigari G and Manjunatha J G 2020 Optimized voltammetric experiment for the determination of phloroglucinol at surfactant modified carbon nanotube paste electrode *Instrum. Exp. Tech.* **63** 750–7
[31] Manjunatha J G 2020 Poly (Adenine) modified graphene-based voltammetric sensor for the electrochemical determination of catechol, hydroquinone and resorcinol *Open Chem. Eng. J.* **14** 52–62
[32] Yoon J, Cho H Y, Shin M, Choi H K, Lee T and Choi J W 2020 Flexible electrochemical biosensors for healthcare monitoring *J. Mater. Chem.* B **8** 7303–18

[33] Maheshwaran S, Akilarasan M, Chen S-M, Chen T-W, Tamilalagan E, Tzu C Y *et al* 2020 An ultra-sensitive electrochemical sensor for the detection of oxidative stress biomarker 3-nitro-l-tyrosine in human blood serum and saliva samples based on reduced graphene oxide entrapped zirconium (IV) oxide *J. Electrochem. Soc.* **167** 066517

[34] Manjunatha Charithra M and Manjunatha J G 2020 Electrochemical sensing of paracetamol using electropolymerised and sodium lauryl sulfate modified carbon nanotube paste electrode *ChemistrySelect* **5** 9323–9

[35] Manjunatha J G, Raril C, Hareesha N, Charithra M M, Pushpanjali P A, Tigari G and Gowda J 2020 Electrochemical fabrication of poly (niacin) modified graphite paste electrode and its application for the detection of riboflavin *Open Chem. Eng. J.* **14** 90–8

[36] Antherjanam S, Saraswathyamma B, Krishnan R G and Gopakumar G M 2021 Electrochemical sensors as a versatile tool for the quantitative analysis of vitamin B12 *Chem. Pap.* **75** 2981–95

[37] Pushpanjali P A, Manjunatha J G and Srinivas M T 2020 Highly sensitive platform utilizing poly (l-methionine) layered carbon nanotube paste sensor for the determination of voltaren *FlatChem.* **24** 100207

[38] Charithra M M, Manjunatha J G G and Raril C 2020 Surfactant modified graphite paste electrode as an electrochemical sensor for the enhanced voltammetric detection of estriol with dopamine and uric acid *Adv. Pharm. Bull.* **10** 247

[39] Hudari F F, Brugnera M F and Zanoni M V B 2017 Advances and trends in voltammetric analysis of dyes *Applications of the Voltammetry* ed M Stoytcheva and R Zlatev (London: IntechOpen) pp 75–108

[40] Gupta V K 2009 Suhas: application of low-cost adsorbents for dye removal—a review *J. Environ. Manag.* **90** 2313–42

[41] Prinith N S and Manjunatha J G 2019 Surfactant modified electrochemical sensor for determination of Anthrone—a cyclic voltammetry *Mater. Sci. Energy Technol.* **2** 408–16

[42] Vaghela S S, Jethva A D, Mehta B B, Dave S P, Adimurthy S and Ramachandraiah G 2005 Laboratory studies of electrochemical treatment of industrial azo dye effluent *Environ. Sci. Technol.* **39** 2848–55

[43] Ahlström L-H, Sparr Eskilsson C and Björklund E 2005 Determination of banned azo dyes in consumer goods *TrAC, Trends Anal. Chem.* **24** 49–56

[44] Desai N F and Giles C H 1949 The oxidation of azo dyes and its relation to light fading *J. Soc. Dyers Colour.* **65** 639–49

[45] Jain R, Bhargava M and Sharma N 2003 Electrochemical studies on a pharmaceutical azo dye: Tartrazine *Ind. Eng. Chem. Res.* **42** 243–7

[46] Heidarizadi E and Tabaraki R 2016 Simultaneous spectrophotometric determination of synthetic dyes in food samples after cloud point extraction using multiple response optimizations *Talanta* **148** 237–46

[47] Yamjala K, Nainar M S and Ramisetti N R 2016 Methods for the analysis of azo dyes employed in food industry—a review *Food Chem.* **192** 813–24

[48] Lewis S W 2009 Analysis of dyes using chromatography *Identification of Textile Fibres* ed M Houck (Woodhead) pp 203–23

[49] Manjunatha J G 2020 A surfactant enhanced graphene paste electrode as an effective electrochemical sensor for the sensitive and simultaneous determination of catechol and resorcinol *Chem. Data Collect.* **25** 100331

[50] de Oliveira R, Hudari F, Franco J and Zanoni M 2015 Carbon nanotubebased electrochemical sensor for the determination of anthraquinone hair dyes in wastewaters *Chemosensors* **3** 22–35
[51] Hudari F F, Ferreira S L C and Zanoni M V B 2016 Multi-responses methodology applied in the electroanalytical determination of hair dye by using printed carbon electrode modified with graphene *Electroanalysis* **28** 1085–92
[52] Gooding J J, Compton R G, Brennan C M and Atherton J H 1996 The mechanism of the electro-reduction of some azo dyes *Electroanalysis* **8** 519–23
[53] Kang N, Ji L, Zhao J, Zhou X, Weng X, Li H, Zhang X and Yang F 2019 Uniform growth of Fe_3O_4 nanocubes on the single-walled carbon nanotubes as an electrosensor of organic dyes and the study on its catalytic mechanism *J. Electroanal. Chem.* **833** 70–8
[54] Touzi H, Chevalier Y, Bessueille F, Ben Ouada H and JaffrezicRenault N 2018 Detection of dyestuffs with an impedimetric sensor based on Cu^{2+}-methyl-naphthyl cyclen complex functionalized gold electrodes *Sens. Actuators* B **273** 1211–21
[55] Bouzayani B, Bocos E, Elaoud S C, Pazos M, Sanromán MÁ and González-Romero E 2018 An effective electroanalytical approach for the monitoring of electroactive dyes and intermediate products formed in electro-Fenton treatment *J. Electroanal. Chem.* **808** 403–11
[56] Materón E M, Marchetto R, Araujo A R, Vega-Chacon J, Pividori M I, Jafelicci M, Shimizu F M, Oliveira O N and Zanoni M V B 2018 A simple electrochemical method to monitor an azo dye reaction with a liver protein *Anal. Biochem.* **553** 46–53
[57] Pushpanjali P A and Manjunatha J G 2020 Development of polymer modified electrochemical sensor for the determination of alizarin carmine in the presence of tartrazine *Electroanalysis* **32** 2474–80
[58] de Oliveira R A G, Zanoni T B, Bessegato G G, Oliveira D P, Umbuzeiro G A and Zanoni M V B 2014 The chemistry and toxicity of hair dyes *Quim. Nova* **37** 1037–46
[59] Corrêa G T, Tanaka A A, Pividori M I and Zanoni M V B 2016 Use of a composite electrode modified with magnetic particles for electroanalysis of azo dye removed from dyed hair strands *J. Electroanal. Chem.* **782** 26–31
[60] Bessegato G G, Hudari F F and Zanoni M V B 2017 Self-doped TiO_2 nanotube electrodes: a powerful tool as a sensor platform for electroanalytical applications *Electrochim. Acta* **235** 527–33
[61] Shetti N P, Nayak D S and Malode S J 2018 Electrochemical behavior of azo food dye at nanoclay modified carbon electrode-a nanomolar determination *Vacuum* **155** 524–30
[62] Yang X, Sun D, Zeng R, Guo L and Wu K 2017 Trace analysis of ponceau 4R based on the signal amplification of copperbased metal-organic framework modified electrode *J. Electroanal. Chem.* **794** 229–34
[63] Zhao X, Liu Y, Zuo J, Zhang J, Zhu L and Zhang J 2017 Rapid and sensitive determination of tartrazine using a molecularly imprinted copolymer modified carbon electrode (MIP-PmDB/ PoPD-GCE) *J. Electroanal. Chem.* **785** 90–5
[64] Mazlan S Z, Lee Y H and Hanifah S A 2017 A new Laccase based biosensor for tartrazine *Sensors* **17** 1–12
[65] Sakthivel M, Sivakumar M, Chen S M and Pandi K 2018 Electrochemical synthesis of poly (3,4-ethylenedioxythiophene) on terbium hexacyanoferrate for sensitive determination of tartrazine *Sens. Actuators* B **256** 195–203
[66] He Q, Liu J, Liu X, Li G, Deng P and Liang J 2018 Manganese dioxide Nanorods/ electrochemically reduced graphene oxide nanocomposites modified electrodes for cost-effective and ultrasensitive detection of Amaranth *Colloids Surf. B: Biointerfaces* **172** 565–72

[67] Tvorynska S, Josypcuk B, Barek J and Dubenska L 2018 Electrochemical behavior and sensitive methods of the voltammetric determination of food azo dyes amaranth and allura red AC on amalgam electrodes *Food Anal. Methods* **12** 409–21

[68] Shetti N P, Nayak D S and Kuchinad G T 2017 Electrochemical oxidation of erythrosine at TiO_2 nanoparticles modified gold electrode - an environmental application *J. Environ. Chem. Eng.* **5** 2083–9

[69] Sierra-Rosales P, Toledo-Neira C and Squella J A 2017 Electrochemical determination of food colorants in soft drinks using MWCNTmodified GCEs *Sens. Actuators* B **240** 1257–64

[70] Ji L, Cheng Q, Wu K and Yang X 2016 Cu-BTC frameworks-based electrochemical sensing platform for rapid and simple determination of Sunset yellow and Tartrazine *Sens. Actuators* B **231** 12–7

[71] Hareesha N, Manjunatha J G, Amrutha B M, Pushpanjali P A, Charithra M M and Prinith Subbaiah N 2021 Electrochemical analysis of indigo carmine in food and water samples using a poly (glutamic acid) layered multi-walled carbon nanotube paste electrode *J. Electron. Mater.* **50** 1230–8

[72] Tigari G and Manjunatha J G 2020 Poly (glutamine) film-coated carbon nanotube paste electrode for the determination of curcumin with vanillin: an electroanalytical approach *Monatsh. Chem.* **151** 1681–8

[73] Zhao L, Zeng B and Zhao F 2014 Electrochemical determination of tartrazine using a molecularly imprinted polymer—multi-walled carbon nanotubes - ionic liquid supported Pt nanoparticles composite film coated electrode *Electrochim. Acta* **146** 611–7

[74] Wang Z, Shan Y, Xu L, Wu G and Lu X 2017 Development and appli- cation of the tartrazine voltametric sensors based on molecularly imprinting polymer *Int. J. Polym. Anal. Charact.* **22** 83–91

[75] Karim-Nezhad G, Khorablou Z, Zamani M, Seyed Dorraji P and Alamgholiloo M 2017 Voltammetric sensor for tartrazine determi-nation in soft drinks using poly (p-aminobenzene-sulfonic acid)/zinc oxide nanoparticles in carbon paste electrode *J. Food Drug Anal.* **25** 293–301

[76] An Z Z, Li Z, Guo Y Y, Chen X L, Zhang K N, Zhang D X, Xue Z H, Bin Zhou X and Lu X Q 2017 Preparation of chitosan/N-doped graphenenatively grown on hierarchical porous carbon nano-composite as a sensor platform for determination of tartra-zine *Chin. Chem. Lett.* **28** 1492–8

[77] Wu X, Zhang X, Zhao C and Qian X 2018 One-pot hydrothermal synthesis of ZnO/RGO/ ZnO@Zn sensor for sunset yellow in soft drinks *Talanta* **179** 836–44

[78] He Q, Liu J, Liu X, Xia Y, Li G, Deng P and Chen D 2018 Novel electrochemical sensors based on cuprous oxide- electrochemically reduced graphene oxide nanocomposites modified electrode toward sensitive detection of sunset yellow *Molecules* **23** 2130

[79] Magerusan L, Pogacean F, Coros M, Socaci C, Pruneanu S, Leostean C and Pana I O 2018 Green methodology for the preparation of chitosan/graphene nanomaterial through electrochemical exfoliation and its applicability in Sunset Yellow detection *Electrochim. Acta* **283** 578–89

[80] Qin C, Guo W, Liu Y, Liu Z, Qiu J and Peng J 2017 A novel electrochemical sensor based on graphene oxide decorated with silver nanoparticles–molecular imprinted polymers for determination of sunset yellow in soft drinks *Food Anal. Methods* **10** 2293–301

[81] Rovina K, Siddiquee S and Shaarani S M 2017 Highly sensitive electrochemical determination of sunset yellow in commercial food products based on CHIT/GO/MWCNTs/ AuNPs/GCE *Food Control* **82** 66–73

[82] Li L, Zheng H, Guo L, Qu L and Yu L 2018 Construction of novel electrochemical sensors based on bimetallic nanoparticle functionalized graphene for determination of sunset yellow in soft drink *J. Electroanal. Chem.* **833** 393–400

[83] Ya Y, Jiang C, Li T, Liao J, Fan Y, Wei Y, Yan F and Xie L 2017 A zinc oxide nanoflower-based electrochemical sensor for trace detection of sunset yellow *Sensors* **17** 1–9

[84] Thirumalraj B, Rajkumar C, Chen S M, Veerakumar P, Perumal P and Bin Liu S 2018 Carbon aerogel supported palladium-ruthenium nanoparticles for electrochemical sensing and catalytic reduction of food dye *Sens. Actuators* B **257** 48–59

[85] Wang M, Cui M, Zhao M and Cao H 2018 Sensitive determination of Amaranth in foods using graphene nanomeshes *J. Electroanal. Chem.* **809** 117–24

[86] He Q, Liu J, Liu X, Li G, Deng P and Liang J 2018 Manganese dioxide Nanorods/electrochemically reduced graphene oxide nano-composites modified electrodes for cost-effective and ultrasensitive detection of Amaranth *Colloids Surf. B: Biointerfaces* **172** 565–72

[87] Sheikhshoaie M, Karimi-Maleh H, Sheikhshoaie I and Ranjbar M 2017 Voltammetric amplified sensor employing RuO2 nano-road and room temperature ionic liquid for amaranth analysis in food samples *J. Mol. Liq.* **229** 489–94

[88] Gao Y, Wang L, Zhang Y, Zou L, Li G and Ye B 2017 Electrochemical behavior of amaranth and its sensitive determination based on Pd-doped polyelectrolyte functionalized graphene modified electrode *Talanta* **168** 146–51

[89] Yang X, Sun D, Zeng R, Guo L and Wu K 2017 Trace analysis of ponceau 4R based on the signal amplification of copper-based metal-organic framework modified electrode *J. Electroanal. Chem.* **794** 229–34

[90] De Moraes P B, Hudari F F, Silva J P and Zanoni M V B 2018 Enhanced detection of ponceau 4R food dye by glassy carbon electrode modified with reduced graphene oxide *J. Braz. Chem. Soc.* **29** 1237–44

[91] Nuñez-Dallos N, Macías M A, García-Beltrán O, Calderón J A, Nagles E and Hurtado J 2018 Voltammetric determination of amaranth and tartrazine with a new double-stranded copper(I) helicate-single-walled carbon nanotube modified screen printed electrode *J. Electroanal. Chem.* **822** 95–104

[92] Manjunatha J G and Hussain C M 2022 *Carbon Nanomaterials-Based Sensors: Emerging Research Trends in Devices and Applications* (Elsevier)

[93] Raril C, Manjunatha J G, Ravishankar D K, Fattepur S, Siddaraju G and Nanjundaswamy L 2020 Validated electrochemical method for simultaneous resolution of tyrosine, uric acid, and ascorbic acid at polymer modified nano-composite paste electrode *Surf. Eng. Appl. Electrochem.* **56** 415–26

[94] Prinith N S, Manjunatha J G and Hareesha N 2021 Electrochemical validation of L-tyrosine with dopamine using composite surfactant modified carbon nanotube electrode *J. Iran. Chem. Soc.* **18** 3493–503

[95] Kumar N, Pathera A K and Kumar M 2012 The effects of pesticides on human health *Ann. Agri-Bio Res.* **17** 125–7

[96] Hiltbold A E 1974 *Persistence of Pesticides in Soil* 2nd edn (Madison, WI: Soil Science Society of America, Inc)

[97] Edwards C A 1975 Factors that affect the persistence of pesticides in plants and soils *Pure Appl. Chem.* **42** 39–56

[98] Cogger C G, Stark J D, Bristow P R, Getzin L W and Montgomery M 1998 Transport and persistence of pesticides in alluvial soils: II. Carbofuran *J. Environ. Qual.* **27** 551

[99] Hara J and Kawabe Y 2007 Long-term persistence of cyclodiene pesticide in soil *AIP Conf. Proc.* **898** 32–5

[100] Verma N and Bhardwaj A 2015 Biosensor technology for pesticides—a review *Appl. Biohem. Biotechnol.* **175** 3093–119

[101] Samsidar A, Siddiquee S and Shaarani S M 2018 A review of extraction, analytical and advanced methods for determination of pesticides in environment and foodstuffs *Trends Food Sci. Technol.* **71** 188–201

[102] Valera E, Ramón-Azcón J, Rodríguez Á, Castañer L M, Sánchez F-J and Marco M-P 2007 Impedimetric Immunosensor for atrazine detection using interdigitated μ-electrodes (IDμE's) *Sens. Actuators* B **125** 526–37

[103] Ramón-Azcón J, Valera E, Rodríguez Á, Barranco A, Alfaro B, Sanchez-Baeza F and Marco M-P 2008 An impedimetric immunosensor based on interdigitated microelectrodes (IDμE) for the determination of atrazine residues in food samples *Biosens. Bioelectron.* **23** 1367–73

[104] Valera E, Muñiz D and Rodríguez Á 2010 Fabrication of flexible inter-digitated μ-electrodes (FIDμEs) for the development of a conductimetric immunosensor for atrazine detection based on antibodies labelled with gold nanoparticles,' *Microelectron. Eng.* **87** 167–73

[105] Valera E, Ramón-Azcón J, Barranco A, Alfaro B, Sánchez-Baeza F, Marco M-P and Rodríguez Á 2010 Determination of atrazine residues in red wine samples.conductimetric solution *Food Chem.* **122** 888–94

[106] Marrakchi M, Sánchez I C, Helali S, Mejri N, Camino J S, Gonzalez-Martinez M A, Hamdi M and Abdelghani A 2011 A label-free interdigitated microelectrodes immunosensor for pesticide detec-tion,' *Sens. Lett.* **9** 2203–6

[107] Fan L, Zhao G, Shi H, Liu M and Li Z 2013 A highly selective electro-chemical impedance spectroscopy-based aptasensor for sensitive detec-tion of acetamiprid *Biosens. Bioelectron.* **43** 12–8

[108] Contreras Jiménez G, Eissa S, Ng A, Alhadrami H, Zourob M and Siaj M 2015 Aptamer-based label-free impedimetric biosensor for detection of progesterone *Anal. Chem.* **87** 1075–82

[109] Eissa S and Zourob M 2017 Selection and characterization of DNA aptamers for electrochemical biosensing of carbendazim *Anal. Chem.* **89** 3138–45

[110] Zhao W P, Zhao G, Chen D F, Mao Z H and Wang X Y 2014 Rapid detection technology for pesticides residues based on microelectrodes impedance Immunosensor *Sens. Transducers* **178** 56–62 https://sensorsportal.com/HTML/DIGEST/P_2352.htm

[111] Fei A, Liu Q, Huan J, Qian J, Dong X, Qiu B, Mao H and Wang K 2015 Label-free impdimetric aptasensor for detection of fem-tomole level acetamiprid using gold nanoparticles decorated multi-walled carbon nanotube-reduced graphene oxide nanoribbon compos-ites *Biosens. Bioelectron.* **70** 122–9

[112] Cao Y, Sun X, Guo Y, Zhao W and Wang X 2015 An electrochemi-cal immunosensor based on interdigitated array microelectrode for the detection of chlorpyrifos *Bioprocess Biosyst. Eng.* **38** 307–13

[113] Gong Z, Guo Y, Sun X, Cao Y and Wang X 2014 Acetylcholinesterase biosensor for carbaryl detection based on interdigitated array micro- electrodes *Bioprocess Biosyst. Eng.* **37** 1929–34

[114] Hou W, Shi Z, Guo Y, Sun X and Wang X 2017 An interdigital array microelectrode aptasensor based on multi-walled carbon nanotubes for detection of tetracycline *Bioprocess Biosyst. Eng.* **40** 1419–25

[115] Wang L, Veselinovic M, Yang L, Geiss B J, Dandy D S and Chen T 2017 A sensitive DNA capacitive biosensor using interdigitated electrodes *Biosens. Bioelectron.* **87** 646–53

[116] Mirzajani H, Cheng C, Wu J, Chen J, Eda S, Aghdam E N and Ghavifekr H B 2017 A highly sensitive and specific capacitive aptasensor for rapid and label-free trace analysis of bisphenol a (BPA) in canned foods *Biosens. Bioelectron.* **89** 1059–67

[117] Facure M H M, Mercante L A, Mattoso L H C and Correa D S 2017 Detection of trace levels of organophosphate pesticides using an elec-tronic tongue based on graphene hybrid nanocomposites *Talanta* **167** 59–66

[118] Madianos L, Tsekenis G, Skotadis E, Patsiouras L and Tsoukalas D 2018 A highly sensitive impedimetric aptasensor for the selective detection of acetamiprid and atrazine based on microwires formed by platinum nanoparticles *Biosens. Bioelectron.* **101** 268–74

[119] Madianos L, Skotadis E, Tsekenis G, Patsiouras L, Tsigkourakos M and Tsoukalas D 2018 Impedimetric nanoparticle aptasensor for selec-tive and label free pesticide detection *Microelectron. Eng.* **189** 39–45

[120] Hua Q T, Ruecha N, Hiruta Y and Citterio D 2019 Disposable electro-chemical biosensor based on surface-modified screen-printed electrodes for organophosphorus pesticide analysis *Anal. Methods* **11** 3439–45

[121] Pérez-Fernández B, Mercader J V, Abad-Fuentes A, Checa-Orrego B I, Costa-García A and Escosura-Muñiz A D L 2020 Direct competitive immunosensor for imidacloprid pesticide detection on gold nanoparticle-modified electrodes *Talanta* **209** 120465

[122] Tsouti V, Boutopoulos C, Zergioti I and Chatzandroulis S 2011 Capacitive microsystems for biological sensing *Biosens. Bioelectron.* **27** 1–11

[123] Madianos L, Skotadis E, Tsekenis G, Patsiouras L, Tsigkourakos M and Tsoukalas D 2018 Impedimetric nanoparticle aptasensor for selective and label free pesticide detection *Microelectron. Eng.* **189** 39–45

[124] Marrakchi M, Sánchez I C, Helali S, Mejri N, Camino J S, Gonzalez-Martinez M A, Hamdi M and Abdelghani A 2011 A labelfree interdigitated microelectrodes immunosensor for pesticide detection *Sens. Lett.* **9** 2203–6

[125] Cao Y, Sun X, Guo Y, Zhao W and Wang X 2015 An electrochemical immunosensor based on interdigitated array microelectrode for the detection of chlorpyrifos *Bioprocess. Biosyst. Eng.* **38** 307–13

[126] Fan L, Zhao G, Shi H, Liu M and Li Z 2013 A highly selective electrochemical impedance spectroscopy-based aptasensor for sensitive detection of acetamiprid *Biosens. Bioelectron.* **43** 12–8

[127] Facure M H M, Mercante L A, Mattoso L H C and Correa D S 2017 Detection of trace levels of organophosphate pesticides using an electronic tongue based on graphene hybrid nanocomposites *Talanta* **167** 59–66

[128] Kaisti M 2017 Detection principles of biological and chemical FET sensors *Biosens. Bioelectron.* **98** 437–48

[129] Ishige Y, Takeda S and Kamahori M 2010 Direct detection of enzymecatalyzed products by FET sensor with ferrocene-modified electrode *Biosens. Bioelectron.* **26** 1366–72

[130] Graef M, Hosenfeld F, Horst F, Farokhnejad A, Hain F, Iñíguez B and Kloes A 2018 Advanced analytical modeling of double-gate tunnel-FETs—a performance evaluation *Solid-State Electron* **141** 31–9

[131] Andronescu C and Schuhmann W 2017 Graphene-based field effect transistors as biosensors *Curr. Opin. Electrochem.* **3** 11–7

[132] Berninger T, Bliem C, Piccinini E, Azzaroni O and Knoll W 2018 Cascading reaction of arginase and urease on a graphene-based FET for ultrasensitive, real-time detection of arginine *Biosens. Bioelectron.* **115** 104–10

[133] Kucherenko I S, Soldatkin O O, Kucherenko D Y, Soldatkina O V and Dzyadevych S V 2019 Advances in nanomaterial application in enzymebased electrochemical biosensors: a review *Nanoscale Adv.* **1** 4560–77

[134] Hu P, Zhang J, Li L, Wang Z, O'Neill W and Estrela P 2010 Carbon nanostructure-based field-effect transistors for label-free Chemical/Biological sensors *Sensors* **10** 5133–59

[135] Ba Hashwan S S, Ruslinda A R, Fatin M F, Arshad M K M and Hashim U 2017 Reduced graphene oxide–multiwalled carbon nanotubes composites as sensing membrane electrodes for DNA detection *Microsyst. Technol.* **23** 3421–8

[136] Fenoy G E, Marmisollé W A, Azzaroni O and Knoll W 2020 Acetylcholine biosensor based on the electrochemical functionalization of graphene field-effect transistors *Biosens. Bioelectron.* **148** 111796

[137] Islam S, Shukla S, Bajpai V K, Han Y-K, Huh Y S, Ghosh A and Gandhi S 2019 Microfluidic-based graphene field effect transistor for femtomolar detection of chlorpyrifos *Sci. Rep.* **9** 1–7

[138] Thanh C T *et al* 2018 An interdigitated ISFET-type sensor based on LPCVD grown graphene for ultrasensitive detection of carbaryl *Sens. Actuators* B **260** 78–85

[139] Fenoy G E, Marmisollé W A, Azzaroni O and Knoll W 2020 Acetyl-choline biosensor based on the electrochemical functionalization of graphene field-effect transistors *Biosens. Bioelectron.* **148** 111796

[140] Belkhamssa N, Justino C I L, Santos P S M, Cardoso S, Lopes I, Duarte A C, Rocha-Santos T and Ksibi M 2016 Label-free disposable immunosensor for detection of atrazine *Talanta* **146** 430–4

[141] Standard and Certification Department 2015 UTZ Certified List of Banned Pesticides and Pesticides Watchlist

[142] Registration Eligebility Decision(RED) Lemonene 1994 United States Environmental Protection Agency, Office of prevention, pestices and toxic substances. EPA 738-R-94-034

[143] Prasad B B, Jauhari D and Tiwari M P 2014 Doubly imprinted polymer nanofilm-modified electrochemical sensor for ultra-trace simultaneous analysis of glyphosate and glufosinate *Biosens. Bioelectron.* **59** 81–8

[144] Do M H, Florea A, Farre C, Bonhomme A, Bessueille F, Vocanson F, Tran-Thi N T and Jarezic-Renault N 2015 Molecularly imprinted polymer-based electrochemical sensor for the sensitive detection of glyphosate herbicide *Int. J. Environ. Anal. Chem.* **95** 1489–501

[145] Zhang C, She Y, Li T, Zhao F, Jin M, Guo Y, Zheng L, Wang S, Jin F, Shao H *et al* 2017 A highly selective electrochemical sensor based on molecularly imprinted polypyrrole-modified gold electrode for the determination of glyphosate in cucumber and tap water *Anal. Bioanal. Chem.* **409** 7133–44

[146] Gholivand M B, Akbari A and Norouzi L 2018 Development of a novel hollow fiber-pencil graphite modified electrochemical sensor for the ultra-trace analysis of glyphosate *Sens. Actuators* B **272** 415–24

[147] Oliveira P C, Maximiano E M, Oliveira P A, Camargo J S, Fiorucci A R and Arruda G J 2018 Direct electrochemical detection of glyphosate at carbon paste electrode and its determination in samples of milk, orange juice, and agricultural formulation *J. Environ. Sci. Health* B **53** 817–23

[148] Moraes F C, Mascaro L H, Machado S A S and Brett C M A 2010 Direct electrochemical determination of glyphosate at copper phthalocyanine/multiwalled carbon nanotube film electrodes *Electroanalysis* **22** 1586–91

[149] Songa E A, Somerset V S, Waryo T, Baker P G L and Iwuoha E I 2009 Amperometric nanobiosensor for quantitative determination of glyphosate and glufosinate residues in corn samples *Pure Appl. Chem.* **81** 123–39

[150] Cahuantzi-Muñoz S L, González-Fuentes M A, Ortiz-Frade L A, Torres E, Talu S, Trejo G and Méndez-Albores A 2019 Electrochemical biosensor for sensitive quantification of glyphosate in maize kernels *Electroanalysis* **31** 927–35

[151] Paramo-Garcia U, Gutierrez-Grandos S, Garcia-Jimenez M G and Ibanez J G 2010 Catalytic behavior of cobalt(I) salen during the electrochemical reduction of lindane and hexachlorobenzene *J. New Mater. Electrochem. Syst.* **13** 356–60

[152] Kumaravel A, Vincent S and Chandrasekaran M 2013 Development of an electroanalytical sensor for-hexachlorocyclohexane based on a cellulose acetate modified glassy carbon electrode *Anal. Methods* **5** 931–8

[153] Fayemi O E, Adekunle A S and Ebenso E E 2016 A sensor for the determination of lindane using PANI/Zn, Fe(III) oxides and nylon 6,6/MWCNT/Zn, Fe(III) oxides nanofibers modified glassy carbon electrode *J. Nanomater.* **2016** 4049730

[154] Birkin P R, Evans A, Milhano C, Montenegro M I and Pletcher D 2004 The mediated reduction of lindane in DMF *Electroanalysis* **16** 583–7

[155] Manuela Garrido E, Costa Lima J L, Delerue-Matos C M and Maria Oliveira Brett A 1998 Electrochemical oxidation of bentazon at a glassy carbon electrode application to the determination of a commercial herbicide *Talanta* **46** 1131–5

[156] Cerejeira R P A G, Delerue-Matos C and Vaz M C V F 2002 Development of an FIA system with amperometric detection for determination of bentazone in estuarine waters *Anal. Bioanal. Chem.* **373** 295–8

[157] Rahemi V, Garrido J M P J, Borges F, Brett C M A and Garrido E M P J 2013 Electrochemical determination of the herbicide bentazone using a carbon nanotube-cyclodextrin modified electrode *Electroanalysis* **25** 2360–6

[158] Geto A, Noori J S, Mortensen J, Svendsen W E and Dimaki M 2019 Electrochemical determination of bentazone using simple screen-printed carbon electrodes *Environ. Int.* **129** 400–7

IOP Publishing

Chapter 12

Electrochemical (bio)sensors for detection of drug abuse in biological samples

Bruna Coldibeli, Gustavo Fix and Elen Romão Sartori

This chapter summarizes the current year's literature on the development of electrochemical (bio)sensors for the detection/determination of drugs of abuse, such as amphetamine-type stimulants, mephedrone, cocaine, fentanyl, methadone, pethidine, and morphine in biological samples, with an emphasis on real-time applications. The discussion is grouped around the different designs of each analytical device and the comparison of analytical features between each other.

12.1 Introduction

Drug abuse continues to be a major global problem that affects different spheres of society, such as health, criminality, economics, and the environment. The United Nations Office on Drugs and Crime (UNODC) report estimates that in 2020, 284 million people (5.6% of the world's population) used a drug in 12 months [1]. The misuse of controlled substances is responsible for several premature deaths annually worldwide, either from overdose or other drug-related disorders. Only in 2019, an estimate counts 494 000 lives lost due to drug use [1].

A drug of abuse is a controlled substance (both legal and illegal) that acts on the central nervous system altering mood, the level of perception, and brain functioning. They are abused for their various side effects, as they can relieve pain, anxiety, or depression, induce sleep, provide energy, and even cause euphoria or hallucinations. Drugs of abuse can be classified according to their properties, similar effects, and withdrawal symptoms. Thus, they can be categorized into stimulants, depressants, opioids, hallucinogens, and anabolic steroids [2, 3]. While cannabis remains the most consumed drug in the world in 2020 (209 million people), opioids

doi:10.1088/978-0-7503-5377-9ch12

are the most harmful due to deleterious effects and health complications, and accounted for 69% of drug-related deaths in 2019. Also, the number of opioid users doubled in a period of 10 years, totaling 61 million people in 2020, according to estimates [1].

It is known that drug abuse is directly associated with several negative consequences for the user and those around them. Some examples of physical and mental health disorders caused by drugs are addiction, violence, behavioral problems, crime, disease, overdose, and premature deaths [4, 5]. Different harmful consequences are related to each type of drug. Regarding the use of injecting drugs, users are still exposed to other risks in addition to the effects of drugs, such as infections and transmission of viruses. Collected data estimated that between 11.2 million people who injected drugs in 2020, 5.5 million are living with hepatitis C and 1.4 million living with HIV [1].

Substance abuse monitoring is critical for legal and clinical toxicological evaluation. Currently, detection of drug abuse is based on analysis of biological samples such as urine, blood, saliva, hair, breath, and sweat. In most cases, initial screening tests are performed using immunoassay platforms (analyzers and/or point-of-care) specific for each group of drugs with similar structures. Despite the advantages of their use, including the wide availability of immunoassay platforms, rapid response time, and simplicity of analytical procedures, immunoassays are qualitative or semiquantitative tests with limited specificity and sensitivity that can lead to false positive or negative results [6, 7]. Therefore, a confirmatory test is required using more accurate analytical techniques, such as mass spectrometry in conjunction with gas or liquid chromatography. These analyses are carried out in large central laboratories, due to the use of more sophisticated instruments, tedious sample treatment, the requirement for trained professionals, and costly and time-consuming assays [6, 8].

In recent years, more attention has been placed on developing instruments and procedures for field analysis. Obtaining real-time data with sensitivity and accuracy could be very useful for monitoring drug abuse in different situations. In this sense, electrochemical techniques (voltammetry and amperometry) have great potential for this purpose due mainly to their simple instrumentation with the possibility of miniaturization and portability, minimal use of organic solvents, simple treatment of samples required, and comparable sensitivity and accuracy. They have been proven to be a valuable tool for the development of portable electrochemical systems for on-site analysis, point-of-care tests, wearable devices, and disposable sensors for the determination of drug abuse in biological samples in a simple, rapid, reliable, low-cost, and eco-friendly way [8–10].

In view of this, this brief review presents a description of the electrochemical bio (sensors) developed this year for the detection and identification of drug abuse. Emphasis was given to some drugs of abuse classified as stimulants and opioids, as well as to devices developed coupled with voltammetric and amperometric techniques.

12.2 Detection and/or determination of drugs of abuse in biological samples employing electrochemical (bio)sensors

Chemical sensors are used as an analytical tool to measure and detect a target analyte in solution after an electrochemical reaction [11–13]. It usually leads to a measurable change in the electrical current that is proportional to the target analyte concentration in the sample. They can operate coupled with the voltammetric techniques, in which the measurement of current is made while a variable potential is applied, or coupled with amperometry, in which current is measured as a function of time at a fixed applied potential. In the stripping voltammetry method, the target analyte is first deposited on the electrode surface by physical adsorption and then oxidized from the electrode during the stripping step, which involves the dissolution of the deposited analyte [14]. It is highly sensitive, selective, and easy to run.

The most common electrodes used in electroanalysis are carbon-based or metallics, and their surface modification leads to improvements in sensitivity and selectivity for the analysis, ensuring a valuable device to identify or quantify low concentrations of a target analyte in solution [14]. Furthermore, they offer different benefits such as portability, can be used for on-site monitoring of complex samples with a short turnaround time, and are less expensive compared to other analytical instruments. Alternatively, the efficient modification of the electrode surface through the immobilization of (bio)recognition elements (e.g. antibodies or aptamers) becomes an effective strategy to achieve these factors. Briefly, aptamers are short chemically synthesized short DNA or RNA sequences that can selectively bind to a target analyte with high specificity, such as antibody–antigen interactions. Antibodies are proteins and their production requires an antigen, involving several steps and time [15]. Aptamers have been developed to mimic antibodies, offering the advantages of eliminating possible interference from endogenous antibodies and the risk of biological contamination during their obtention, among others. Therefore, analytical devices developed with both materials provide high selectivity and affinity to detect target analytes in complex samples [15].

Several electrochemical (bio)sensors were developed in 2022 for application in the measurements of drugs of abuse in body fluids such as blood, urine, sweat, and saliva. Many of these analytical devices are easy to manufacture, disposable, and portable, with potential for point-of-care testing. Thus, our review brings the most relevant studies on the analysis of drugs of abuse in biological samples using voltammetric/amperometric sensors, modified or not with nanostructured materials, and electrochemical biosensors developed with antibodies or aptamers. The drugs of abuse selected refer to stimulants, such as amphetamine-type stimulants, mephedrone, and cocaine, and opioids, such as fentanyl, methadone, pethidine, and morphine. The most relevant studies in this year's literature on the development of (bio)sensors for the determination of these drugs of abuse are summarized in table 12.1.

Table 12.1. Analytical features of (bio)sensors for the determination of drugs of abuse in biological samples.

Drug of abuse	Analytical device	Technique	LOD (μ mol l^{-1})	Possible interferents	Biological sample	Features of the method	References
Stimulants							
MAMP	Sensor: AgNDs/CNOs/GCE	DPV	0.030	Codeine Diclofenac Acetaminophen Methimazole Uric acid Epinephrine Glucose Glycine Cysteine Ascorbic acid	Human blood serum and urine	Surface of GCE modified; Electrode preparation is reasonably simple; Simple sample preparation: acetonitrile to precipitate proteins/centrifugation and dilution; Stability: more than 89% of the initial sensor response remained after one month; Analysis by the standard addition method.	[16]
	Biosensor: G-PEG-dial/GA/IL/mAb	DPV	Buffer solution: 0.037 Human saliva: 0.048	Cocaine Amphetamine Synthetic cannabinoid (JWH-073) Uric acid Benzoylecgonine	Human saliva	Screen-printed gold electrode used to make the biosensor; On-site detection; Time-consuming steps to prepare the biosensor (incubation overnight); Production on large scale. Same linear range in both supporting electrolyte and human saliva samples. High selectivity; Non-invasive approach.	[17]

Biosensor: Apta-4/GE	DPV	0.0031	Amphetamine Codeine Morphine 3-hydroxybutyl metabolite	Synthetic urine	Bare gold electrode used to fabricate the biosensor; High specificity; Time-consuming steps to prepare the biosensor (incubation/dried steps); Easy-to-use, detect low concentrations; Analysis by the standard addition method.	[18]
Biosensor: EAB	SWV	0.020	Cocaine Ketamine Morphine	Serum, urine, and saliva	Gold disk electrodes were used to fabricate the biosensor; Simple, cost-effective, and reagentless platform; Time-consuming steps to prepare the biosensor (incubation/dried steps); High specificity; No dilution of serum and urine samples; Analysis by the standard addition method.	[19]
Biosensor: MAMP-aptamer	SWV	1.7×10^{-9}	Cocaine Nicotine Methadone Uric acid Ascorbic acid	Human serum and urine	Gold electrodes were used to fabricate the biosensor; Time-consuming steps to prepare the biosensor (took on average 6 h); High specificity; Kept 4 °C;	[20]

(*Continued*)

Table 12.1. (*Continued*)

Drug of abuse	Analytical device	Technique	LOD (μ mol l^{-1})	Possible interferents	Biological sample	Features of the method	References
						Stability of the sensor: 1 month, decrease of 2.88% of the initial response; Analysis by the standard addition method.	
Amphetamine	Biosensor: Apt/AuNFs@Au	DPV	0.00051	Aspirin Benzaldehyde Aniline Anisole Benzoic acid	Urine	Gold electrodes were used to fabricate the biosensor; AuNFs were electrodeposited onto gold electrode; Label-free electrochemical sensor; Time-consuming steps to prepare the biosensor (took on average 6 h); Specificity (interference of benzoic acid); Analysis by the standard addition method.	[21]
MDMA and MDA	Sensor: 3D-printed CB/PLA	CSWV	MDMA: 0.6 MDA: 0.1	Amphetamine MAMP 25B-NBOMe Lysergic acid diethylamide Ephylone Acetaminophen Caffeine	Human saliva	Disposable, and easy large-scale production; Promise of on-site screening analyses; Low-cost sensor (< $0.20) Noninvasive and less susceptible to adulteration procedure; Simple sample preparation: dilution; Analysis by the standard addition method.	[22]

Mephedrone	Sensor: AgNPs@Sa/CPE	SWAdSV	3.43×10^{-6}	—	Human urine	Carbon paste electrode was used for the modification; Electrode preparation is reasonably simple; Promising tool for mephedrone's reliable and sensitive detection; Noninvasive method; Simple sample preparation: dilution; Similar values slopes of analytical curves in both supporting electrolyte and urine samples. Analysis by the standard addition method.	[23]
Cocaine	Sensor: SPE	SWAdSV	0.1	Albumin Ascorbate Urea Uric acid	Saliva	Promising potential in point-of-care testing; Simple sample preparation: dilution; Noninvasive method; Rapid and inexpensive method;	[24]
	Biosensor: DNA/MER/ITO	DPV	2.6×10^{-4}	Morphine ATP Diazepam Caffeine	Human serum and urine	ITO electrode was used for the modification; Layer of Au deposited on a flexible ITO substrate; Time-consuming steps to prepare the biosensor; Highly sensitive and selectivity detection; Stored in phosphate buffer solution;	[25]

(*Continued*)

Table 12.1. (*Continued*)

Drug of abuse	Analytical device	Technique	LOD (μ mol l^{-1})	Possible interferents	Biological sample	Features of the method	References
						Stability: 8 days, retained about 80% of its signal of the first day; Simple sample preparation: centrifugation and/or dilution;	
	Biosensor: β-CD-Fc-DNA/ITO	SWV	9.89×10^{-6}	Methadone Nicotine	Human serum	ITO electrode was used for the fabrication of biosensor; Time-consuming steps to prepare the biosensor; Highly sensitive and selectivity detection; Simple sample preparation: dilution; Analysis by the standard addition method.	[26]
	Biosensor: DNA/SPGE	DPV	1.5×10^{-5}	Diazepam Cocaine Propranolol Morphine Amoxicillin Lorazepam Diclofenac	Human serum	Commercial screen-printed gold electrode was used for the modification Disposable; High sensitivity and selectivity; Time-consuming steps to prepare the biosensor; Simple sample preparation: dilution;	[27]

						Stability of the sensor: decrease of 5.1% of the initial response for 10 days; Analysis by the standard addition method.	
Opioids							
Fentanyl	Sensor: (TGA-CdSe/ZnS@FCNT)	DPV	0.006	Tramadol Codeine Methadone Nicotine Caffeine Naproxen	Blood serum and human urine	Pencil graphite rods used to make working electrodes; Suspension of the nanomaterials prepared in N,N-dimethylformamide; Tedious preparation of the sensor; Simple sample preparation: centrifugation and filtration steps; Stability of the sensor: kept at room temperature for 30 days—RSD < 7%; Analysis by the standard addition method.	[28]
	Sensor: Cu-H3BTC/MWCNT-HA	DPV	0.003	Thiopental Medetomidine Ketanserin Ketamine Citric acid Caffeine L-cysteine H_2O_2	Human blood serum	GCE for modification; Hydroxyapatite was used to improve the dispersion of MWCNTs; Suspension of the nanomaterials prepared in N,N-dimethylformamide;	[29]

(*Continued*)

Table 12.1. (*Continued*)

Drug of abuse	Analytical device	Technique	LOD (μ mol l^{-1})	Possible interferents	Biological sample	Features of the method	References
				Inorganic ions		Modification of GCE surface by drop-casting method; Simple sample preparation: centrifugation; Stability of the sensor: 2 months—a decrease of 14% of the initial response; Comparison of the results by HPLC.	
	Sensor: LCE	SWV	1.0	Theophylline Acetaminophen Ascorbic acid Uric acid Caffeine	Human serum	Disposable sensor strip; Simple, rapid, cost-effective, and scalable manufacturing; No surface modification; Portable and accessible; Rapid electrochemical detection; Simple sample preparation: dilution in phosphate buffer solution; Stability of the sensor: 30 days—a decrease of 6.35% of the initial response; Similar analytical curves in both supporting electrolyte and serum samples.	[30]
	Sensor: microneedles	SWV	27.8	Uric acid Caffeine Ascorbic acid Acetaminophen Theophylline	Commercial Serum	Point-of-need; Potential for transdermal polymeric microneedle sensing platform; Minimally invasive;	[31]

						Similar sensing pattern in serum to that observed in phosphate buffer solution Necessity of dilution of the sample.	
	Immunosensor: MPSi/FEN-Ab	SWV	Phosphate buffer: 0.0178 Sweat: 0.0339	—	Human sweat	Rapid and on-site detection (potentiality); Noninvasive and point-of-care monitoring; Time-consuming steps to prepare the biosensor (incubation overnight); Different LOD values in phosphate buffer solution and human sweat. Application as wearable sweat sensor.	[32]
Methadone	Sensor: TGA@CdSe/GO	DPV	0.03	Tramadol Oxazepam Dopamine Diazepam	Human blood serum	Pencil graphite electrode used for modification; Time-consuming steps to prepare de sensor; Acceptable selectivity and high sensitivity; Simple sample preparation: acetonitrile to precipitate proteins/ centrifugation; Sensor response efficiency's decreased to 97.9% after 20 days.	[33]

(*Continued*)

Table 12.1. (*Continued*)

Drug of abuse	Analytical device	Technique	LOD (μ mol l^{-1})	Possible interferents	Biological sample	Features of the method	References
						Analysis by the standard addition method.	
	Sensor: (Gr/AgNPs)2/GCE	DPV	0.18	Ascorbic acid Uric acid Dopamine	Human blood serum	Time-consuming steps to prepare de sensor; Simple sample preparation: sulfuric acid to precipitate proteins/centrifugation; Analysis by the standard addition method.	[34]
	Sensor: L-arginine/GCE	DPV	0.032	Ascorbic acid Glucose Acetaminophen Sucrose Epinephrine Glysine Cysteine	Human serum and urine	Stable sensing platform; Simple, short time and cost-effective fabrication of the sensor; Simple sample preparation: organic solvent to precipitate proteins/centrifugation; Decrease of less than 4.1% in the peak currents for one week later of preparation. Analysis by the standard addition method.	[35]
	Sensor: CMK-5/GCE	SWV	0.029	Ibuprofen Amphetamine Ascorbic acid Codeine Glucose	Human urine	Suspension of the CMK-5 prepared in N, N-dimethylformamide; Modification of GCE surface by drop-casting method;	[36]

				Sucrose Fructose Inorganic ions		Simple sample preparation: HCl to precipitate proteins/centrifugation and filtration; Reusability of the sensor: RSD = 5.7% after 10 tests. Analysis by the standard addition method.	
	Sensor: AuNPs/PPyox/SPE	SWV	0.45	Ascorbic acid Uric acid Glucose Calcium Potassium	Urine and blood serum	Modification of SPE surface by electropolymerizing and electrochemical synthesis of nanoparticles; Simple sample preparation: acid to precipitate proteins/centrifugation, adjusted pH, filtration; Analysis by the standard addition method.	[37]
Pethidine	Sensor: CSe2NF/CCE	DPV	0.0193	Tryptophan Glycine Cysteine Methadone Codeine Diclofenac Methimazole Uric acid Epinephrine Glucose Ascorbic acid	Human blood, urine, and saliva	Conductive carbon cloth was used to prepare the sensor; Preparation time of sensor of more than 12 h; Several sensors are prepared at the same time. Rapid, noninvasive, and on-site measurement; High sensitivity; Flexible sensor; Simple sample preparation: acetonitrile to precipitate proteins/centrifugation, filtration, dilution;	[38]

(Continued)

Table 12.1. (*Continued*)

Drug of abuse	Analytical device	Technique	LOD (μ mol l^{-1})	Possible interferents	Biological sample	Features of the method	References
						Stability: kept 93.2% of the initial signal over 30 days; Sensor was validated by HPLC analysis.	
	Sensor: CoMn2O_4-rGO/IL/CPE	DPV	0.024	Ascorbic acid Glycine Valine Urea Uric acid Glucose Sucrose Starch Inorganic ions	Human plasma and urine	Facile preparation of the paste: a mixture of graphite, CoMn2O4, and rGO; Simple surface renewal: with soft paper; Simple sample preparation: dilution; Decrease of less than 7.0% in the peak currents after 3 weeks (in comparison with the initial value). Analysis by the standard addition method.	[39]
Morphine	Sensor: graphene-Co_3O_4/SPGE	DPV	0.007	Fructose Glucose Lactose Histidine Phenyl alanine Alanine Methionine Glycine Methanol/Ethanol Tryptophan Thiourea	Human urine	Screen-printed graphite electrode used for modification; Modification of electrode surface by drop-casting method; Dispersion of graphene-Co_3O_4 nanocomposite prepared in water; Portability; Stability: 20-day storage of the sensor at ambient temperature;	[40]

			Tyrosine Cysteine Acetaminophen Ascorbic acid Dopamine Uric acid Methyldopa Inorganic ions		decrease in the peak current to 97.1% of its original response. Simple sample preparation: filtration and dilution; Analysis by the standard addition method.	
Sensor: MoS2NPs/SPGE	DPV	0.03	—	Human urine	Screen-printed graphite electrode used for modification; Modification of electrode surface by drop-casting method; Portability; Stability: RSD of 1.6% for storage for two weeks; Simple sample preparation: centrifugation and dilution; Analysis by the standard addition method.	[41]
Sensor: CHM@MWCNTs/GCE	DPV	0.0092	Tartaric acid, L-cysteine Glucose Sucrose Ascorbic acid Oxycodone Dopamine Uric Acid Noscapine Methadone Thebaine	Human urine	GCE used to modification; Good selectivity; Suspension of CHM@MWCNTs was dropped onto the GCE surface; Dried at room temperature; Stability: stored in a humid environment for 7 days, 95.4% of initial oxidation peaks. Simple sample preparation: dilution.	[42]

(Continued)

Table 12.1. (*Continued*)

Drug of abuse	Analytical device	Technique	LOD (μ mol l^{-1})	Possible interferents	Biological sample	Features of the method	References
						Analysis by the standard addition method.	
	Sensor: $TbFeO_3/CuO/$SPGE	DPV	0.010	Dopamine Glucose Uric acid Caffeine Folic acid Ascorbic acid Codeine Cysteine	Human serum and urine	Screen-printed graphite electrode used for modification; Dispersion of $TbFeO_3/CuO$ nanocomposite prepared in ultrapure water; Modification of electrode surface by drop-casting method; Preparation time of sensor of 2 h; Simple sample preparation: centrifugation and dilution; Analysis by the standard addition method.	[43]

Abbreviations: AgNDs/CNOs/GCE = glassy carbon electrode modified with nanodiamond-derived carbon nano-onions decorated with the silver nanodendrites; G-PEG-dial/GA/IL/mAb = screen-printed gold electrode modified with gelatin-polyethylene glycol-dialdehyde/ glutaraldehyde/ionic liquid/anti-(+)-methamphetamine monoclonal antibody; Apta-4/GE = aptamer-4 immobilized on gold electrode; EAB = electrochemical aptamer-based biosensor; Apt/AuNFs@Au = Gold electrode modified with gold nanoflowers and aptamers; 3D-printed CB/PLA = 3D-printed carbon black and polylactic acid electrode; AgNPs@Sa/CPE = silver nanoparticles capped with saffron modified carbon paste electrode; DNA/SPGE = screen-printed gold electrode modified with DNA; DNA/MER/ITO = Indium tin oxide electrode modified with DNA-mercaptohexanoic acid; β-CD-Fc-DNA/ITO = indium tin oxide electrode modified with β–cyclodextrin-ferrocene-DNA;TGA-CdSe/ZnS@FCNT = thioglycolic acid-bonded cadmium selenide quantum dot-zinc sulfide quantum dot-functionalized carbon nanotubes modified pencil graphite electrode; GCE/MWCNT-HA/Cu-H3BTC = copper(II) benzene-1,3,5-tricarboxylate organometallic framework/hydroxyapatite/multi-walled carbon nanotube modified glassy carbon electrode; LCE = laser carbonized electrode; PANI = polyaniline; MPSi/FEN-Ab = nanoporous Si matrix functionalized with anti-fentanyl polyclonal antibodies; TGA@CdSe/GO = thioglycolic acid-decorated cadmium selenide doped graphene oxide-modified graphite electrode; GCE/Gr/(AgNPs)2 = glassy carbon electrode modified with two layers of graphene/Ag nanoparticles nanocomposite; GCE/L-arginine, glassy carbon electrode modified with L-arginine; GCE/CMK-5 = glassy carbon electrode modified with CMK-5; AuNPs/PPyox/SPE= screen-printed electrode modified with composite gold nanoparticles/overoxidized polypyrrole; CSe_2NF/CCE = carbon cloth modified with carbon selenide nanofilms; $CoMn_2O_4$-rGO/IL/CPE = carbon paste electrode modified with the $CoMn_2O_4$-rGO/IL = nanocomposite; graphene-Co_3O_4/SPGE = screen-printed graphite electrode modified with graphene-cobalt oxide nanocomposite; MoS_2NPs/SPGE = screen-printed graphite electrode was modified with MoS_2 nanoparticles; CHM@MWCNTs/GCE = glassy carbon electrode modified with Cu-Hemin metal–organic frameworks and functionalized multi-walled carbon nanotubes; $TbFeO_3$/CuO/SPE = screen-printed electrode modified with perovskite $TbFeO_3$ and copper oxide nanoparticles.

12.2.1 Stimulants

Stimulants are a group of drugs that increase energy and alertness. They include several drugs, such as methamphetamine (MAMP), amphetamine, 3,4-methylenedioxymethamphetamine (MDMA or 'ecstasy'), 3,4- methylenedioxyamphetamine (MDA), mephedrone, and cocaine. Figure 12.1 presents the chemical structures of these drugs. For their determination, carbon-based or metallic electrodes were developed, as well as several immunosensors and aptamer-based biosensors, which are presented below.

12.2.1.1 Amphetamine-type stimulants

Amphetamine-type stimulants are drugs of abuse with a common phenylethylamine core structure, such as MAMP (figure 12.1(a)), amphetamine (figure 12.1(b)), MDMA (figure 12.1(c)), and MDA (figure 12.1(d)) [44]. This group is known for its stimulating effect on the central nervous system. However, some of these drugs (e.g., MDA and MDMA) may also cause hallucinogenic effects because of their structural similarity to the hallucinogenic compound mescaline (methylenedioxy group attached to the aromatic ring) [45].

The first synthesis of amphetamine was carried out in 1887, whereas its N-methylated derivative MAMP was in 1893 [44]. Today, although approved for specific medical uses, both drugs present a high potential for abuse and are widely trafficked and illegally sold for recreational purposes [46–48]. MDA and MDMA were first synthesized in the early 1900s. They are illegal recreational drugs and do

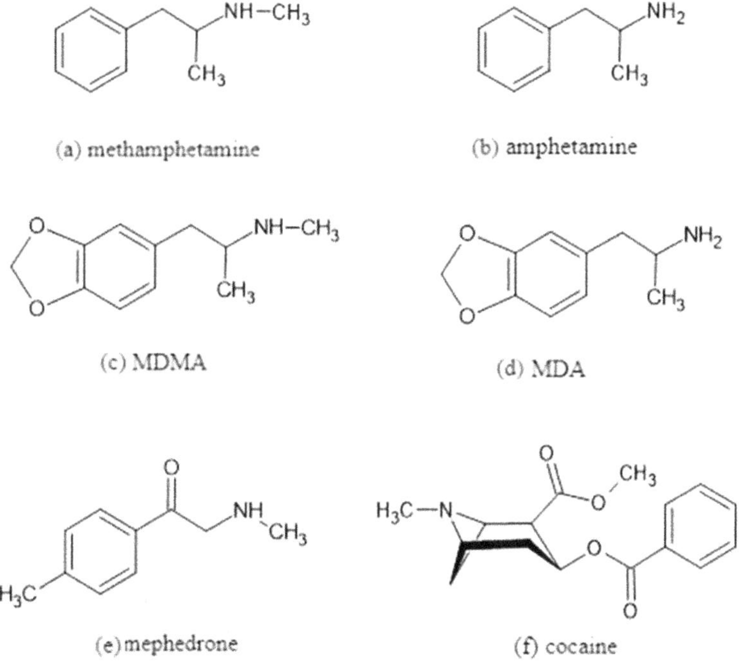

Figure 12.1. Chemical structures of some stimulant drugs.

not have approved medical uses. Generally, low and moderate doses of amphetamine-type stimulants produce symptoms including heightened alertness, euphoria, and arousal. On the other hand, high dosages and frequent use may cause psychotic episodes, tachycardia, hypertension, and peripheral hyperthermia [47, 49].

The most common ways to use amphetamine are by intravenous injection, orally, or intranasally. Its primary metabolites benzoic acid and 4-hydroxyamphetamine are excreted in the urine as well as unchanged amphetamine [50, 51]. MAMP can be injected, smoked, snorted, or ingested orally. Approximately 70% of the administered dose is excreted in urine within 24 h, mostly unchanged (30%–50%), metabolized to 4-hydroxymethamphetamine (15%) and amphetamine (10%) [52]. MDMA and MDA are mostly found as tablets of different colors, sizes, and shapes with characteristic logos, and in powder form. MDMA is rapidly metabolized via different pathways to form a variety of metabolites. Regardless of dose, 50% of MDMA is excreted in urine within 24 h along with its main metabolites MDA, 4-hydroxy-3-methoxymethamphetamine, and 4-hydroxy-3-methoxyamphetamine [53].

A glassy carbon electrode (GCE) was modified with the nanodiamond-derived carbon nano-onions (CNOs) decorated with silver nanodendrites (AgNDs) for the determination of MAMP [16]. This modification provided high conductivity, large surface area, and excellent electrocatalysis capability for the sensor. Its preparation is reasonably simple, the CNOs dispersion was drop-coated onto the cleaned GCE surface, which was heated in an oven (50 °C) for 1 h. Finally, AgNDs were electrodeposited on the CNOs/GCE surface. The sensor showed satisfactory selectivity in detecting MAMP in the presence of some compounds with a concentration 10-fold higher than the stimulant drug. Furthermore, the analysis of MAMP in human serum and urine samples demonstrates the efficiency of the sensor in routine analysis with sufficient accuracy.

Several electrochemical biosensors (immunosensor and aptamer) were developed for the determination of MAMP. For this purpose, monoclonal antibodies with a high affinity for MAMP were used due to their specificity for the drug. Ghorbanizamani *et al* [17] used a combination of hydrogel, ionic liquid 1-butyl-3-methylimidazolium bis(trifluoromethylsulfonyl)imide, and anti-(+)-methamphetamine monoclonal antibody to modify the surface of a screen-printed gold electrode. This immunosensor showed high selectivity to MAMP compared with other evaluated interferents (cocaine, amphetamine, JWH-073, uric acid, and benzoylecgonine). For validation, spiked human saliva was used to simulate real conditions, and a limit of detection (LOD) value of 0.048 μmol l^{-1} was determined.

Synthetic aptamer (apta-4) was fabricated and immobilized on the surface of the gold electrode via gold-thiol affinity by Bor *et al* [18]. The apta-4 sequence had the lowest LOD value of 0.0031 μmol l^{-1}, as well as a lower percentage of interference from other drugs, such as amphetamine, codeine, morphine, and the 3-hydroxybutyl metabolite. Xie *et al* also used a gold electrode to obtain an electrochemical aptamer-based biosensor for the determination of MAMP in several biological samples (serum, urine, and saliva) [19]. For the fabrication of the biosensor, one end of the gold electrode surface was modified with the synthetic 38-base aptamer sequence and the other methylene blue, where the target-induced conformational change. The LOD obtained at the nanomolar level (0.020 μmol l^{-1}) was far below the clinical detection threshold, which is essential for practical application in a preliminary screening of drugged driving. The biosensor did not respond to other illicit drugs, such as cocaine, ketamine, and morphine, because the binding affinity of MAMP for aptamer was much stronger.

In another work, click chemistry and atom transfer radical polymerization were used to develop a novel aptamer electrochemical biosensor for the highly sensitive detection of MAMP (LOD of 17 fmol l^{-1}) [20]. In addition, this biosensor showed specific recognition of this drug in the presence of cocaine, nicotine, and methadone, due to the binding of nucleic acid aptamers and target molecules. Uric acid and ascorbic acid had almost no effect on the detection of MAMP using this analytical device. The application of the biosensor in human serum and urine samples indicated practical prospects in toxicology detection.

An aptasensor was developed by the immobilization of thiolated amphetamine-specific aptamers on gold nanoflower (AuNFs) as a sensing interface for amphetamine detection [21]. Modification of the gold electrode with AuNFs provided more anchoring sites for the aptamers, resulting in high sensitivity and low LOD value (0.51 nmol l^{-1}). Among the substances tested as interferents (aspirin, benzaldehyde, aniline, and anisole), only benzoic acid exhibited a comparatively equal current due to a similar chemical structure to amphetamine. Analysis in real urine samples raised the prospect of its use as a rapid analytical tool to trace amphetamine in diverse biological applications.

The most consumed amphetamine-type stimulants worldwide, MDMA and MDA, have also been targets for the development of sensors for their determination in biological matrices. Faria *et al* used 3D-printed carbon electrodes and cyclic square-wave voltammetric (CSWV) for this purpose [22]. A commercially available filament made of polylactic acid and carbon black was used to manufacture this type of sensor. These are disposable devices that can be easily and inexpensively produced on a large scale. Also, LOD values lower than 1.8 μmol l^{-1} were obtained for the determination of the drugs. Analysis by the standard addition method demonstrated the potential of the sensor as a tool for the detection of the recent use of these substances.

12.2.1.2 Mephedrone

Mephedrone (figure 12.1(e)), also known as 4-methylmethcathinone or 4-methylephedrone, is a synthetic stimulant derived from cathinone, which has similar effects to MDMA and cocaine. It was first synthesized in 1928 and is now marketed as capsules, tablets of various designs, or white powder that can be taken orally, injected intravenously or insufflated. It is considered a recreational drug that has no medical use. Short-term effects of mephedrone include increased euphoria, alertness, restlessness, and a faster heartbeat. It also causes hallucinations, induces paranoia, nosebleeds, and even seizures [54]. After ingestion, the peak concentration of mephedrone is reached in approximately 1.25 h, with a concentration range of 51.7–218.3 ng ml^{-1} and a half-life of 2.15 h [55]. Mephedrone occurs unchanged along with its metabolites in urine samples. Its metabolites include nor-mephedrone, dihydro-mephedrone, and 4-carboxy-mephedrone, which are detected in urine and blood samples, whereas hydroxytolyl- and 4-carboxy-dihydro-mephedrone are primarily found in urine [56].

Papaioannou *et al* described a new electrochemical sensor using a carbon paste electrode (CPE) modified with silver nanoparticles capped with saffron (AgNPs@Sa) for monitoring mephedrone in urine samples by square-wave adsorptive stripping voltammetric (SWAdSV) [23]. The AgNPs@Sa were electrodeposited on the CPE surface. The use of this nanomaterial provided a greater electrochemical activity for the sensor, i.e., the peak current increased, while the value of the peak potential shifted to lower potential values. The authors performed the determination of mephedrone in human urine samples by using the standard addition method. They verified that the different slopes of the calibration curve in buffer solution and spiked urine samples were equal to 0.0027, indicating that the matrix effect was insignificant. This study demonstrated the potential applicability of the sensor in clinical analysis; in addition, the manufacture of this sensor is relatively simple and consists of a noninvasive method.

12.2.1.3 Cocaine

Cocaine (figure 12.1(f)), also known as benzoylmethylecgonine, is a natural alkaloid found in the leaves of *Erythroxylum coca*, which were used in ancient times for religious and medicinal reasons, whereas today it is mainly used as an illicit recreational drug [57]. It is a stimulant of the central nervous system and has short-term effects such as mood amplification (euphoria and dysphoria), and increasing feelings of well-being, energy, and alertness [58, 59]. Cocaine is available in salt form (cocaine hydrochloride) and freebase (popularly known as 'crack' or 'rock cocaine'). It can be consumed orally, intranasally, by intravenous injection, or by smoking. Plasma levels of cocaine are directly related to consumption, with faster absorption when smoked and injected intravenously. Therefore, simple doses taken by these routes present higher concentrations of circulating cocaine (500–1000 ng ml^{-1}) than when consumed orally or by snorting (100–500 ng ml^{-1}). Cocaine is rapidly metabolized to benzoylecgonine and ecgonine methyl ester (75%–90% of cocaine metabolism) and other minor metabolites (ecgonine and norcocaine), both of which are excreted renally [59].

A disposable screen-printed electrode (SPE) has been employed for the detection of cocaine by Joosten *et al* [24]. The strategy to mitigate the biofouling property of the protein present in the saliva matrix was to adjust the pH of the buffer solution from 9 to 10, together with the use of the surfactant sodium dodecyl sulfate. The LOD value obtained was 0.1 μmol l^{-1}. Saliva tests showed the applicability of the SPE for cocaine detection in the field.

In order to obtain better selectivity, sensitivity, and consequently lower LOD, some electrochemical aptasensors were developed. For this purpose, some authors used indium tin oxide (ITO) electrode-modified or screen-printed gold electrode [25–27] to immobilize commercial DNA aptamer to obtain the biosensors. These devices allowed the detection of cocaine in complex matrices at nanomolar/picomolar levels with high selectivity.

12.2.2 Opioids

Opioids are a class of drugs of abuse that not only relieve pain but also produce euphoria. This class includes several drugs such as fentanyl, methadone, pethidine, and morphine. Figure 12.2 shows the chemical structures of these opioids. This year, carbon-based electrodes modified with nanostructured materials and unmodified electrodes were developed for the analysis of these drugs, in addition to biosensors based on aptamers and a transdermal platform developed by polymeric microneedle array. These reported devices are presented below, separately by drug.

12.2.2.1 Fentanyl

Fentanyl (N-(1-phenethyl-4-piperidinyl) propionanilide), figure 12.2(a), is a very potent synthetic analgesic with rapid action that was first synthesized in 1960 and marketed as a pain-treating medicine [60, 61]. Since then, fentanyl has become one of the most widely used opioids in pain management, particularly because of its flexibility, potency, familiarity, and physical properties [62]. Fentanyl is available in soluble powder, transdermal patch, liquid, or tablet form, and can be taken by injection, ingestion, smoking, or inhalation [63]. Its short-term effects are analgesia, anesthesia, drowsiness, feelings of relaxation, and euphoria [61, 63]. After intravenous administration, approximately 80% of fentanyl is excreted within 72 h [60]. The major metabolite of fentanyl in urine is norfentanyl, and

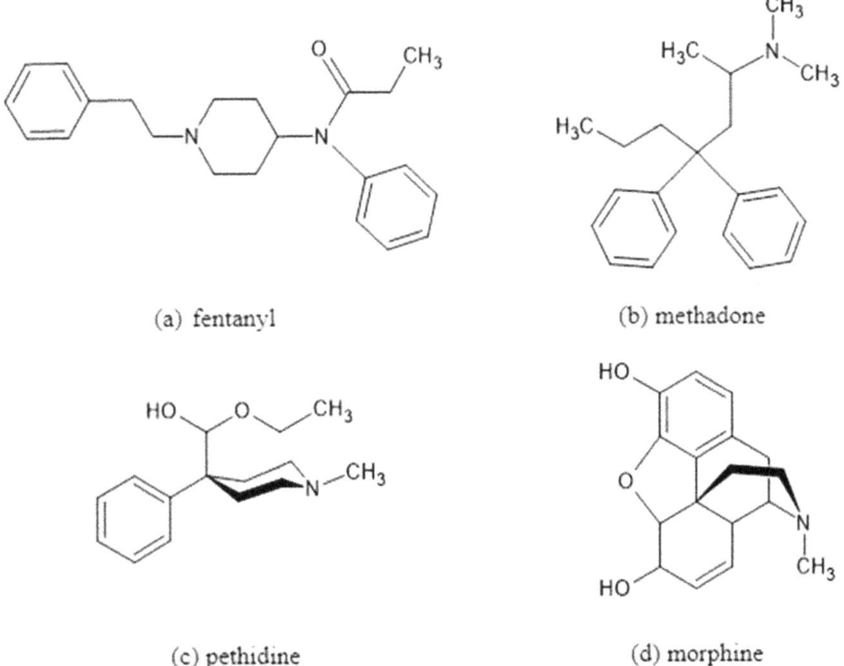

Figure 12.2. Chemical structures of some opioids.

other known metabolites are hydroxylfentanyl, hydroxynorfentanyl, and despropionylfentanyl [64]. Only a small fraction of fentanyl (8%–10%) is excreted unchanged [60, 61].

Carbon-based electrodes have been extensively developed recently for the determination of fentanyl in biological samples. A pencil graphite electrode (PGE) modified with functionalized carbon nanotubes (MWCNTs), thioglycolic acid-bonded cadmium selenide quantum dot, and zinc sulfide quantum dot was developed by Nazari and Eshaghi for the simultaneous determination of fentanyl and morphine using differential pulse voltammetry (DPV) [28]. The modification of the PGE surface with nanomaterials facilitated the charge transfer capacity, increased the peak current, and shifted the oxidation potential of fentanyl and morphine to lower values, with good peak-to-peak separation of these two drugs of abuse, allowing their simultaneous determination. Despite the excellent results, the preparation of this sensor is tedious, since the pencil graphite rod must be immersed in the suspension containing the nanomaterials three times, each time for 4 h, and each time, after immersion, it was dried at room temperature for 24 h. Successful analyzes were performed on both blood serum and human urine samples with good results.

MWCNTs were also used in the development of a novel electrochemical sensor for the determination of fentanyl. They were combined with a copper(II) benzene-1,3,5-tricarboxylate organometallic framework to modify the surface of the GCE [29]. Hydroxyapatite improved the dispersion of MWCNTs by acting as an ionic crosslinker and a biosurfactant. The combination of MWCNT-hydroxyapatite and Cu-H_3BTC increased the surface area of the modified electrode, which increased the extent of charge transfer. With these modifications, the LOD value decreased to 3 nmol l^{-1}. Tests were performed in blood serum samples and the results were compared with those obtained by high-performance liquid chromatography (HPLC), demonstrating the potential of the sensor for the determination of fentanyl in complex samples.

A rapid and low-cost surveillance system was developed by Mishra *et al* [30]. To achieve these characteristics, the authors used laser-induced nanoporous carbon structures directly onto commercially available polyimide sheets. This procedure resulted in a high level of porosity and surface roughness on the sensor surface and an improvement in sensitivity for the detection of fentanyl by square-wave voltammetry (SWV). Furthermore, it is a direct, and one-step laser process assuring scalable manufacturing of this device. The disposable sensor strip presented a relevant working range of 20–200 μmol l^{-1} with similar performance in both supporting electrolyte (phosphate buffer solution) and serum samples. The authors described that the oxidation of the fentanyl molecule occurs at the pyridine ring, leading to the formation of norfentanyl, in which two electrons are involved. The laser carbonization process produced a sensor with high selectivity to fentanyl determination against the investigated interfering compounds (theophylline, acetaminophen, ascorbic acid, uric acid, and caffeine). This novel technology stands out as a new route toward scalable manufacturing of cost-effective sensors for the rapid detection of other opioid drugs.

A transdermal platform was developed for fentanyl *in situ* monitoring. Joshi *et al* developed a 3D-printed polymeric microneedle array for rapid fentanyl-screening applications in human serum samples [31]. The working sensor was modified by graphene ink and subsequently with 4-(3-butyl-1-imidazole)–1-butanesulfonate ionic liquid. This sensor was able to differentiate the fentanyl from interfering agents tested, such as uric acid, caffeine, ascorbic acid, acetaminophen, and theophylline. This platform opens additional opportunities for opioids detection in emergencies.

An immunosensor for fentanyl determination in human sweat was developed by Tokranova *et al* [32]. The device consisted of a meso- and macroporous silicon substrate containing the gold working electrode and counter electrode deposited on it. Anti-fentanyl polyclonal antibodies were immobilized between these electrodes. Coupling with the SWV technique resulted in a very high analytical sensitivity for fentanyl in phosphate buffer solution (pH 4.4; 17.8 nmol l^{-1}) and human sweat (33.9 nmol l^{-1}). Tests were performed by adapting this sensor as the sensory chip integrated into a wearable bracelet for rapid and accurate detection of this opioid drug.

12.2.2.2 Methadone

Methadone, also known as (RS)–6-(dimethylamino)–4,4-diphenylheptan-3-one, is a synthetic opioid first synthesized in the 1940s and currently marketed as soluble tablets. Its short-term effects include sedation and euphoria. In addition, it causes lethargic apathy and a reduction in sexual interest and activity. Methadone can be administered by different routes, including oral, intramuscular, or intravenous. After oral administration, the peak plasma concentration of methadone is reached between 2.5 and 4 h. Oxidative biotransformation is the major route of methadone excretion. It is metabolized in 2-ethylidene-1,5-dimethyl-3,3-diphenylpyrrolidine, an inactive metabolite that is excreted renally. Other metabolites include 2-ethyl-5-methyl-3,3-diphenyl-1-pyrroline, and normethadone [65].

Nazari and Es'haghi reported an electrochemical analytical device for the simultaneous determination of methadone and morphine [33]. It was based on thioglycolic acid-decorated cadmium selenide (CdSe) doped graphene oxide-modified GPE. This sensor had a slightly longer preparation time. Graphite rods were placed in the suspension of graphene oxide for 6 h. Next, graphene oxide coating was placed inside the thioglycolic acid-capped cadmium selenide solution for 6 h, along with continuous stirring of the solution. CdSe provided an increased sensitivity to the sensor when compared with the graphene oxide-coated graphite pencil electrode. Then it dried for one day at room temperature. This platform presented selectivity for some interferent compounds investigated, such as tramadol, oxazepam, dopamine, and diazepam. The simultaneous determination of methadone and morphine in blood serum samples indicates the wide potential for applications of this sensor.

GCE was modified with different materials to determine methadone in biological samples. Baghayeri *et al* presented a GCE modified with two layers of graphene/Ag nanoparticles nanocomposite as a novel sensor for electrochemical methadone determination [34]. The preparation of this sensor was a little tedious because it involved two stages of modification of the GCE surface: first, graphene was placed by drop-casting method on the GCE surface that was dried for 10 min in an oven at 35 °C; second, the AgNPs were electrochemically deposited. This procedure was repeated twice. Graphene combined with Ag nanoparticles showed good electrocatalytic activity toward methadone oxidation and no impact on the methadone signal was observed for some interfering compounds tested. However, other modifications of the GCE surface provided better figures of merit for the voltammetric determination of methadone. Karim-Nezhad *et al* promoted an electropolymerization of L-arginine monomer on the surface of GCE [35]. This modification improved the conductivity and electron transfer of GCE for detecting methadone, with a LOD of 0.032 μmol l^{-1}. In another case, Habibi *et al* modified a GCE with CMK-5 mesoporous carbon [36]. The suspension of this material was prepared in N,N-dimethylformamide, next dropped onto the surface of the GCE, and dried under the IR lamp. CMK-5 presented a high ability to absorb methadone in its structure and then provide excellent sensitivity to the analytical determination, with an LOD of 0.029 μmol l^{-1}.

An SPE modified with the composite gold nanoparticles/overoxidized polypyrrole was also described for methadone determination [37]. Polypyrrole was electropolymerized on the SPE surface and AuNPs were synthesized electrochemically on the surface of overoxidized polypyrrole. The overoxidized polypyrrole acted as an adsorption layer and the AuNPs provided a large surface area with electrocatalytic activity toward methadone. The sensor showed a greatly enhanced response to methadone and no interference was observed with the presence of other species, including ascorbic acid, uric acid, glucose, calcium, and potassium.

12.2.2.3 Pethidine

Pethidine (ethyl 1-methyl-4-phenylpiperidine-4-carboxylate), figure 12.2(c), also known as meperidine, was first described in 1939. It is a potent analgesic extensively used that has morphine-like effects. Its short-term effects include analgesia, sedation, euphoria, and respiratory depression. Pethidine is usually administered by intravenous and intramuscular injection and reaches a peak concentration at approximately 4.2–11.4 min, with a blood concentration of 0.76 μg ml^{-1}. The half-life of pethidine varies from 3.1 to 4.1 h. Major metabolites include norpethidine, pethidinic acid, and norpethidinic acid, which can be found in urine samples, as well as unchanged pethidine and other metabolites at lower concentrations [66, 67].

A sensitive portable sensor for the screening of pethidine was developed by Khorablou *et al* [38]. For this, the authors modified a conductive carbon cloth with carbon selenide nanofilms, which increased the active surface area relative to the unmodified carbon cloth. This developed flexible sensor is promising for designing some sizable wearable sensors at a low cost. Further, its high sensitivity and selectivity are a bonus for the rapid and on-site measurement of pethidine, which may open up a route for noninvasive routine analysis in clinical samples.

A novel platform based on $CoMn_2O_4$-rGO modified CPE was described by Shahinfard *et al* using DPV [39]. This sensing platform was fabricated by using hydrophilic ionic liquid 1-ethyl-3-methylimidazolium chloride as a binder. After the modification, the sensor exhibited an increase in electrical conductivity and an increase in the surface area, and a good peak-to-peak separation of pethidine, morphine, and olanzapine. Tests were performed in human plasma and urine samples proving the potential of the sensor for clinical applications.

12.2.2.4 Morphine

Morphine (4R,4aR,7S,7aR,12bS)−3-methyl-2,4,4a,7,7a,13-hexahydro-1H-4,12-methanobenzofuro[3,2-e]isoquinoline-7,9-diol), figure 12.2(d), was first totally synthesized only in 1952. It is an opioid commonly found as a white powder [68]. The main side effects of morphine include respiratory depression and chemical dependence. Its major metabolites include morphine-3-glucuronide and morphine-6-glucuronide, both of which are present in plasma. Peak plasma levels of morphine depend on the route of administration and vary from 0.2 to 2.5 h, reaching concentrations of approximately 200–400 ng ml^{-1} (intravenous injection). The half-life of morphine can vary between 0.8 and 4.1 h [69].

Screen-printed graphite electrode (SPGE) was modified with graphene-Co_3O_4 nanocomposite for the voltammetric assay of morphine in the presence of diclofenac in pharmaceutical and biological samples [40]. This nanostructured sensor presented a high current response at a low overpotential, with a narrow LOD (0.007 μmol l^{-1}) and a broad linear range (0.02–575.0 μmol l^{-1}). Further, it presented the ability to detect morphine in the presence of diclofenac, with different oxidation potentials. Another sensor that yielded interesting figures of merit for the determination of morphine (LOD of 0.03 μmol l^{-1} and linear range of 0.05–600.0 μmol l^{-1}) in the presence of diclofenac was the SPGE modified with MoS_2 nanosheets [41]. This sensor showed high current intensity and more negative oxidation potentials compared to the bare SPE. Recovery from 97.5% to 103.4% reflects the potential application of this sensor for the determination of morphine in human fluids.

Mousaabadi *et al* used Cu-Hemin metal–organic frameworks@MWCNTs to modify a GCE and obtain an electrochemical sensor for the simultaneous determination of morphine and codeine in human urine [42]. The sensor showed the highest current (3–4 times higher) and the lowest overpotential compared to the unmodified GCE. This sensor showed a large surface area, due to the high electrical conductivity of functionalized MWCNTs, and the superior synergic effect of Cu in Cu-Hemin metal–organic frameworks. Further, the sensor allowed the determination of morphine at concentrations level from 0.09 to 30.0 μmol l^{-1} with an LOD of 0.0092 μmol l^{-1}, and satisfactory detection in real samples, even in the presence of other compounds commonly found in the analyzed samples, such as dopamine, uric acid, and codeine. The authors reported that the analytical signal is related to the oxidization of the phenol ring followed by the dimerization of the free radical to pseudo morphine.

The determination of morphine was performed using an SPGE modified with perovskite-type nanocomposite ($TbFeO_3/CuO$) [43]. The combination of CuO nanoparticles with the perovskite-type oxide, $TbFeO_3$, exhibited higher electrochemical activities and greater selectivity and sensitivity for the determination of morphine, with a LOD of 0.01 μmol l^{-1}. The authors also reported that the morphine molecule is oxidized in the phenol ring followed by the dimerization of the free radical to pseudo morphine. These features qualify this sensor for the determination of morphine in complex real samples such as urine or also other biological samples.

12.3 Conclusion

From this brief overview, it is clear that much attention has been paid to the development of electrochemical sensors and biosensors for real-time detection of drugs of abuse, such as stimulants and opioids, in biological samples. This research has been intensified because these devices are portable, disposable, and allow noninvasive analysis with speed, selectivity, and precision. In addition, these devices had low LODs and wide linear concentration ranges sufficient for the detection of stimulants and opioids in human urine, human serum, and sweat samples with simple sample pretreatment. Many of the devices presented here have the potential to be adapted to wearable systems to further facilitate the detection of these drugs of abuse in point-of-care testing.

Acknowledgments

The authors gratefully acknowledge financial support and scholarships from the Brazilian funding agencies CNPq [grant numbers 305764/2022–5, 305320/2019–0], Fundação Araucária do Paraná, and CAPES [grant number 88887.674846/2022–00].

References

[1] UNODC. World Drug Report 2022 *United Nations publication* (Vienna: United Nations publication) Sales No. 22.XI.8; 2022 82 p https://unodc.org/res/wdr2022/MS/WDR22_Booklet_2.pdf

[2] Milhorn H T 1994 *Drug and Alcohol Abuse* (Boston, MA: Springer)

[3] Schuckit M A 2006 *Drug and Alcohol Abuse: A Clinical Guide to Diagnosis and Treatment* 6th edn (New York: Springer)

[4] Yusu A M and Salehumar 2022 The impact of drug abuse on society: a review on drug abuse in the context of society *Int. J. Innov. Sci. Res. Technol.* **7** 199–201

[5] Wasilow-Mueller S and Erickson C K 2001 Reviews: drug abuse and dependency: understanding gender differences in etiology and management *J. Am. Pharm. Assoc.* **41** 78–90

[6] Liu L, Wheeler S E, Venkataramanan R, Rymer J A, Pizon A F, Lynch M J *et al* 2018 Newly emerging drugs of abuse and their detection methods *Am. J. Clin. Pathol* **149** 105–16

[7] Li Z and Wang P 2020 Point-of-care drug of abuse testing in the opioid epidemic *Arch. Pathol. Lab Med.* **144** 1325–34

[8] Teymourian H, Parrilla M, Sempionatto J R, Montiel N F, Barfidokht A, Van Echelpoel R *et al* 2020 Wearable electrochemical sensors for the monitoring and screening of drugs *ACS Sens.* **5** 2679–700

[9] Zanfrognini B, Pigani L and Zanardi C 2020 Recent advances in the direct electrochemical detection of drugs of abuse *J. Solid State Electrochem.* **24** 2603–16

[10] García-Miranda Ferrari A, Rowley-Neale S J and Banks C E 2021 Screen-printed electrodes: transitioning the laboratory in-to-the field *Talanta Open* **3** 100032

[11] Manjunatha J G 2020 Poly (adenine) modified graphene-based voltammetric sensor for the electrochemical determination of catechol, hydroquinone and resorcinol *Open Chem. Eng. J.* **14** 52–62

[12] Pushpanjali P A and Manjunatha J G 2020 Development of polymer modified electrochemical sensor for the determination of alizarin carmine in the presence of tartrazine *Electroanalysis* **32** 2474–80

[13] Manjunatha J G 2020 A promising enhanced polymer modified voltammetric sensor for the quantification of catechol and phloroglucinol *Anal. Bioanal. Electrochem.* **12** 893–903

[14] Bard A J and Faulkner L R 2001 *Electrochemical Methods: Fundamentals and Applications* 2nd edn (New York: Wiley) 833 p

[15] Thiviyanathan V and Gorenstein D G 2012 Aptamers and the next generation of diagnostic reagents *Proteomics Clin. Appl* **6** 563–73

[16] Khorablou Z, Shahdost-fard F and Razmi H 2022 Nanodiamond-derived carbon nano-onions decorated with silver nanodendrites as an effective sensing platform for methamphetamine detection *Surf. Interfaces* **31** 102061

[17] Ghorbanizamani F, Moulahoum H, Guler Celik E and Timur S 2022 Ionic liquid-hydrogel hybrid material for enhanced electron transfer and sensitivity towards electrochemical detection of methamphetamine *J. Mol. Liq.* **361** 119627

[18] Bor G, Bulut U, Man E, Balaban Hanoglu S, Evran S and Timur S 2022 Synthetic antibodies for methamphetamine analysis: design of high affinity aptamers and their use in electrochemical biosensors *J. Electroanal. Chem.* **921** 116686

[19] Xie Y, Wu S, Chen Z, Jiang J and Sun J 2022 Rapid nanomolar detection of methamphetamine in biofluids via a reagentless electrochemical aptamer-based biosensor *Anal. Chim. Acta* **1207** 339742

[20] Sun H, Liu J, Qiu Y, Kong J and Zhang X 2022 High sensitive electrochemical methamphetamine detection in serum and urine via atom transfer radical polymerization signal amplification *Talanta* **238** 123026

[21] Soni S, Jain U, Burke D H and Chauhan N 2022 A label free, signal off electrochemical aptasensor for amphetamine detection *Surf. Interfaces* **31** 102023

[22] de Faria L V, Rocha R G, Arantes L C, Ramos D L O, Lima C D, Richter E M *et al* 2022 Cyclic square-wave voltammetric discrimination of the amphetamine-type stimulants MDA and MDMA in real-world forensic samples by 3D-printed carbon electrodes *Electrochim. Acta* **429** 141002

[23] Papaioannou G C, Karastogianni S and Girousi S 2022 Development of an electrochemical sensor using a modified carbon paste electrode with silver nanoparticles capped with saffron for monitoring mephedrone *Sensors* **22** 1625

[24] Joosten F, Parrilla M and De Wael K 2022 Electrochemical detection of cocaine in authentic oral fluid *Eng. Proc* **16** 13

[25] Azizi S, Gholivand M B, Amiri M, Manouchehri I and Moradian R 2022 Carbon dots-thionine modified aptamer-based biosensor for highly sensitive cocaine detection *J. Electroanal. Chem.* **907** 116062

[26] Wang J, Qiu Y, Li L, Qi X, An B, Ma K *et al* 2022 A multi-site initiation reversible addition −fragmentation chain-transfer electrochemical cocaine sensing *Microchem. J.* **181** 107714

[27] Abnous K, Abdolabadi A khakshour, Ramezani M, Alibolandi M, Nameghi M A, Zavvar T *et al* 2022 A highly sensitive electrochemical aptasensor for cocaine detection based on CRISPR-Cas12a and terminal deoxynucleotidyl transferase as signal amplifiers *Talanta* **241** 123276

[28] Nazari Z and Eshaghi Z 2022 Carbon nanotube reinforced heterostructure electrochemical sensor for the simultaneous determination of morphine and fentanyl in biological samples *Iran. J. Anal. Chem.* **9** 63–77

[29] Akbari M, Mohammadnia M S, Ghalkhani M, Aghaei M, Sohouli E, Rahimi-Nasrabadi M *et al* 2022 Development of an electrochemical fentanyl nanosensor based on MWCNT-HA/Cu-H_3BTC nanocomposite *J. Ind. Eng. Chem.* **114** 418–26

[30] Mishra R K, Krishnakumar A, Zareei A, Heredia-Rivera U and Rahimi R 2022 Electrochemical sensor for rapid detection of fentanyl using laser-induced porous carbon-electrodes *Microchim. Acta* **189** 198

[31] Joshi P, Riley P R, Mishra R, Azizi Machekposhti S and Narayan R 2022 Transdermal polymeric microneedle sensing platform for fentanyl detection in biofluid *Biosensors* **12** 198

[32] Tokranova N, Cady N, Lamphere A and Levitsky I A 2022 Highly sensitive fentanyl detection based on nanoporous electrochemical immunosensors *IEEE Sens. J.* **22** 20165–70

[33] Nazari Z and Es'haghi Z 2022 A new electrochemical sensor for the simultaneous detection of morphine and methadone based on thioglycolic acid decorated CdSe doped graphene oxide multilayers *Anal. Bioanal. Electrochem.* **14** 228–45

[34] Baghayeri M, Nabavi S, Hasheminejad E and Ebrahimi V 2022 Introducing an electrochemical sensor based on two layers of Ag nanoparticles decorated graphene for rapid determination of methadone in human blood serum *Top. Catal.* **65** 623–32

[35] Karim-Nezhad G, Khorablou Z, Sadegh B and Mahmoudi T 2022 Electro-polymerized poly (L-arginine) film as an efficient electrode modifier for highly sensitive determination of methadone in real samples *Anal. Bioanal. Electrochem.* **14** 730–41

[36] Habibi M M, Ghasemi J B, Badiei A and Norouzi P 2022 Simultaneous electrochemical determination of morphine and methadone by using CMK-5 mesoporous carbon and multivariate calibration *Sci. Rep.* **12** 8270

[37] Shafaat A and Faridbod F 2022 Overoxidized polypyrrole/gold nanoparticles composite modified screen-printed voltammetric sensor for quantitative analysis of methadone in biological fluids *Anal. Bioanal. Electrochem.* **14** 319–30

[38] Khorablou Z, Shahdost-Fard F and Razmi H 2022 Voltammetric determination of pethidine in biofluids at a carbon cloth electrode modified by carbon selenide nanofilm *Talanta* **239** 123131

[39] Shahinfard H, Shabani-Nooshabadi M, Reisi-Vanani A and Ansarinejad H 2022 A novel platform based on CoMn2O4-rGO/1-ethyl-3-methylimidazolium chloride modified carbon paste electrode for voltammetric detection of pethidine in the presence morphine and olanzapine *Chemosphere* **301** 134710

[40] Beitollahi H, Garkani Nejad F, Tajik S and Di Bartolomeo A 2022 Screen-printed graphite electrode modified with graphene-Co3O4 nanocomposite: voltammetric assay of morphine in the presence of diclofenac in pharmaceutical and biological samples *Nanomaterials* **12** 3454

[41] Baezzat M R, Tavakkoli N and Zamani H 2022 Construction of a new electrochemical sensor based on MoS 2 nanosheets modified graphite screen printed electrode for simultaneous determination of diclofenac and morphine *Iran. Chem. Soc. Anal. Bioanal. Chem. Res.* **9** 153–62

[42] Mousaabadi K Z, Ensafi A A and Rezaei B 2022 Simultaneous determination of some opioid drugs using Cu-hemin MOF@MWCNTs as an electrochemical sensor *Chemosphere* **303** 135149

[43] Mahmoudi-Moghaddam H, Amiri M, Javar H A, Yousif Q A and Salavati-Niasari M 2022 A facile green synthesis of a perovskite-type nanocomposite using Crataegus and walnut leaf for electrochemical determination of morphine *Anal. Chim. Acta* **1203** 339691
[44] Tambaro S and Bortolato M 2015 Interactions of cannabis and amphetamine-type stimulants In: P Campolongo and L Fattore *Cannabinoid Modulation of Emotion, Memory, and Motivation* (New York: Springer) pp 409–42
[45] Kalant H 2001 The pharmacology and toxicology of 'ecstasy' (MDMA) and related drugs *Can. Med. Assoc. J.* **165** 928
[46] Uddin M S, Sufian M A, Kabir M T, Hossain M F, Nasrullah M, Islam I *et al* 2017 Amphetamines: potent recreational drug of abuse *J. Addict. Res. Ther.* **8** 330
[47] Chung H and Choe S 2019 Amphetamine-type stimulants in drug testing *Mass. Spectrom. Lett.* **10** 1–10
[48] Jayanthi S, Daiwile A P and Cadet J L 2021 Neurotoxicity of methamphetamine: main effects and mechanisms *Exp. Neurol.* **344** 113795
[49] Cao D-N, Shi J-J, Hao W, Wu N and Li J 2016 Advances and challenges in pharmacotherapeutics for amphetamine-type stimulants addiction *Eur. J. Pharmacol.* **780** 129–35
[50] Dring L G, Smith R L and Williams R T 1970 The metabolic fate of amphetamine in man and other species *Biochem.* J **116** 425–35
[51] Kramer J C 1967 Amphetamine abuse *JAMA* **201** 305
[52] Cruickshank C C and Dyer K R 2009 A review of the clinical pharmacology of methamphetamine *Addiction* **104** 1085–99
[53] Green A R, Mechan A O, Elliott J M, O'Shea E and Colado M I 2003 The pharmacology and clinical pharmacology of 3,4-methylenedioxymethamphetamine (MDMA, 'Ecstasy') *Pharmacol. Rev.* **55** 463–508
[54] Schifano F, Albanese A, Fergus S, Stair J L, Deluca P, Corazza O *et al* 2011 Mephedrone (4-methylmethcathinone; 'meow meow'): chemical, pharmacological and clinical issues *Psychopharmacology* **214** 593–602
[55] Papaseit E, Pérez-Mañá C, Mateus J-A, Pujadas M, Fonseca F, Torrens M *et al* 2016 Human pharmacology of mephedrone in comparison with MDMA *Neuropsychopharmacology* **41** 2704–13
[56] Pedersen A J, Reitzel L A, Johansen S S and Linnet K 2013 *In vitro* metabolism studies on mephedrone and analysis of forensic cases *Drug Test. Anal.* **5** 430–8
[57] Weiss R D, Mirin S M and Bartel R L 1994 *Cocaine.* 2nd edn (Washington, DC: American Psychiatric Press)
[58] Spronk D B, van Wel J H P, Ramaekers J G and Verkes R J 2013 Characterizing the cognitive effects of cocaine: a comprehensive review *Neurosci. Biobehav. Rev.* **37** 1838–59
[59] Carrera M R A, Meijler M M and Janda K D 2004 Cocaine pharmacology and current pharmacotherapies for its abuse *Bioorg. Med. Chem.* **12** 5019–30
[60] Poklis A 1995 Fentanyl: a review for clinical and analytical toxicologists *J. Toxicol. Clin. Toxicol.* **33** 439–47
[61] Armenian P, Vo K T, Barr-Walker J and Lynch K L 2018 Fentanyl, fentanyl analogs and novel synthetic opioids: a comprehensive review *Neuropharmacology* **134** 121–32
[62] Stanley T H 2014 The fentanyl story *J. Pain* **15** 1215–26
[63] Kuczyńska K, Grzonkowski P, Kacprzak Ł and Zawilska J B 2018 Abuse of fentanyl: an emerging problem to face *Forensic Sci. Int.* **289** 207–14

[64] Wu F, Slawson M H and Johnson-Davis K L 2017 Metabolic patterns of fentanyl, meperidine, methylphenidate, tapentadol and tramadol observed in urine, serum or plasma *J. Anal. Toxicol.* **41** 289–99

[65] Lugo R A, Satterfield K L and Kern S E 2005 Pharmacokinetics of methadone *J. Pain Palliat. Care Pharmacother* **19** 13–24

[66] Mather L E and Meffin P J 1978 Clinical pharmacokinetics pethidine *Clin. Pharmacokinet* **3** 352–68

[67] Edwards D J, Svensson C K, Visco J P and Lalka D 1982 Clinical pharmacokinetics of pethidine *Clin. Pharmacokinet* **7** 421–33

[68] Brook K, Bennett J and Desai S P 2017 The chemical history of morphine: an 8000-year journey, from resin to de-novo synthesis *J. Anesth. Hist* **3** 50–5

[69] Glare P A and Walsh T D 1991 Clinical pharmacokinetics of morphine *Ther. Drug Monit* **13** 1–23

Chapter 13

Sustainability of electrochemical sensors with reference forensic analysis

M B Shivaswamy, B S Madhukar, M J Deviprasad, B S Hemanth and H S Nagendra Prasad

Forensic chemistry is the application of analytical chemistry to forensic analysis, and it is currently one of the most debated topics in the scientific community because it can provide valuable and rapid information about crimes that have been committed and aid in their resolution. The literature in the field of forensic chemistry is rapidly expanding and covers a wide range of topics. This chapter focuses on potentially portable devices such as electrochemical sensors, field tests, and intelligent devices that can be used in various areas of forensic analysis and covers the following topics such as chemical detection, illicit drugs, alcohol sensing, explosives, toxicology and drugs.

13.1 Introduction

Several scientific techniques are used in forensic examinations to present scientific evidence about a case or as investigative tools to aid detectives [1]. The evaluation of a few variables is a prerequisite for selecting analytical techniques, methodologies, and procedures. Some of the important analytical performance qualities include sensitivity, selectivity, robustness, and the method's capacity to analyze a specific sample matrix. These needs are typically not satisfied by a single methodology, but rather frequently by a combination of techniques and methodologies. Particularly in the case of crime scene analysis, confirmatory analysis is frequently carried out in a forensic lab [2]. In recent years, electrochemical sensors have been a popular choice for forensic applications due to their adaptability, mobility, selectivity, sensitivity, and capacity to assess samples with little pretreatment and without [3, 4]. Electrochemical sensors are used in a variety of forensic science applications, such as the detection of poisons [5], narcotics [6], explosives [7], gunshot residue (GSR),

doi:10.1088/978-0-7503-5377-9ch13

and more surprisingly fingerprints [8, 9]. In this section, we provide a brief overview of recent electrochemical sensor applications in forensic drug investigation, highlighting the unique advantages that electrochemistry provides in this field as well as the most significant developments in the electrochemistry-based drug detection and analysis sector. The conventional analytical techniques currently used in forensic analysis won't likely be completely replaced by electrochemical approaches. This is partially attributable to the complicated and varied range of sample matrices, which frequently need the use of a variety of sample analysis methodologies as well as the legal requirements for presentation in courts [10]. However, they present a workable portable approach, whose application at crime scenes would ultimately enhance forensic investigations. While these can be viewed as the more obvious uses for forensic electrochemistry, additional uses, like fingerprint recognition and alcohol sensing, have also been made use of. Here, we will illustrate the unique opportunities that electrochemistry presents to the discipline of forensic chemical analysis while also highlighting the most important developments to date.

13.2 Materials method

13.2.1 Chemical detection

Using the Milli-Q system, all solutions were made using deionized water with a resistivity of at least 18.2 M cm (at 25 °C) (Millipore, USA). UNODC provided the MDPV analytical standard in powder form, which was solubilized in LC-MS grade methanol to create a 1.0×10^{-2} mol l^{-1} stock solution, which was diluted in a supporting electrolyte for electrochemical measurements. The electrochemical behavior of MDPV was investigated in Britton–Robinson (BR) buffer solution, which was generated using a mixture of boric, phosphoric, and acetic acids at various pH values (from 2 to 12), all with 10% (v/v) methanol. To alter the pH, sodium hydroxide was utilized. As a supporting electrolyte for MDPV detection at a glassy carbon electrode (GCE), acetate buffer solution (0.1 mol l^{-1}) in pH 5.0 with 10% methanol was also tested. All reagents were of analytical quality and obtained from Sigma-Aldrich (Lancashire, UK). The UNODC reference material was white crystals comprising 43.1% (w/w) MDPV and 56.9% excipients. A weighted aliquot of 111.5 mg of the substance was diluted in 2 ml of methanol and sonicated for 10 min. After homogenization, it was diluted in a 0.1 mol l^{-1} BR buffer solution (pH 6.0) and analyzed at SPE-Gr using the AdS DPV method [11].

13.2.2 Development of a simple and rapid screening method for the detection of 1-(3-chlorophenyl) piperazine in forensic samples

All solutions were made with deionized water (Millipore DirectQ3; Bedford, MA, USA) with a resistivity of at least 18 M cm at 25 °C. Synth (Diadema, Brazil) provided phosphoric acid (85% w/v) and acetic acid (99.7% w/v), and AppliChem Panreac (Barcelona, Spain) supplied sodium hydroxide (98% w/w) and boric acid (99% w/w). BR buffer solutions included 0.04 mol l^{-1} acetic, phosphate, and boric acids. Normal stock solutions of mCPP were made by dissolving it in methanol and storing it at 4 °C. Interference studies were carried out using standard solutions of other synthetic illegal drugs, which were also provided by the UNODC, such as amphetamine, methamphetamine, MDMA, 1-benzylpiperazine, methylone, mephedrone, ethylone, and 3,4-methylenedioxypyrovalerone, as well as caffeine (CAF), all purchased from Sigma-Aldrich (Lancashire, United Kingdom). The voltammetry tests were carried out using a PGSTAT 128 N potentiostat (MetrohmAutolab BV, Netherlands) operated by Nova 1.11 software and fitted with a cable connection for SPEs (CAC4MMH; MetrohmDropsens). After baseline correction, square-wave voltametric (SWV) scans were shown using the 'moving average' (peak width = 2) technique included in the Nova 1.11 program. All voltametric measurements (drop-casting method; 100 l) were made with commercially available screen-printed carbon electrodes (SPCEs). The LC-Q-TOF-MS technique was employed as a conclusive method for detecting (identifying) banned medicines in confiscated samples. The LC-Q-TOF-MS investigations were carried out with the help of a 1290 Infinity Ultra High-Performance Liquid Chromatography System linked to a 6540 Quadrupole Time of Flight Mass Spectrometer (Agilent Technologies, Santa Clara, CA, USA). The analytes were transferred from the LC equipment (ionization technique) to the MS using the Dual Agilent Jet Stream electrospray (Dual AJS ESI). Agilent Mass Hunter Qualitative Analysis software version B 06.00 and Personal Compound Database and Library version B 02.00 were used to detect and report the compounds from accurate mass scanning data (PCDL) [12].

13.2.3 Electrochemical detection of benzodiazepines—forensic investigations

Several benzodiazepines are thermally unstable and produce common breakdown products. As a result, procedures such as liquid chromatography (LC), which can be performed at room temperature or near room temperature, are common approaches for the separation and quantification of these medicines. Trojanowicza recently examined the use of electrochemical detection with flow injection analysis and liquid chromatography for a variety of chemicals. The fundamental theory and applications were discussed, including the determination of various benzodiazepines [13]. Emphasis was initially concentrated on liquid chromatography-electrochemical detection (LC-ED) in relation to neurochemical concerns, resulting in the production of the first commercially viable detectors in 1974. These systems provided a number of advantages, including improved selectivity, low detection limits, and inexpensive cost. Normal phase chromatography cannot be used in most cases because nonpolar solvents are incompatible with typical buffers. Nevertheless, subsequent research has demonstrated the feasibility of combining similar techniques such as hydrophilic contact LC with electrochemical detection, resulting in the production of the first commercially viable detectors in 1974. These systems provided a number of advantages, including improved selectivity, low detection limits, and inexpensive cost [14].

13.2.4 Toxicology and drugs

The chemicals used were of analytical quality and were utilized exactly as supplied, with no additional purification, from Sigma-Aldrich. All solutions were made using deionized water having a resistivity of at least 18.2 MU cm. Voltametric measurements were performed with an Auto lab III potentiostat (ECO-Chemie, The Netherlands). The measurements were carried out with the use of a screen-printed electrode configuration. SPCEs were created in-house utilizing a microDEK 1760RS screen-printing machine and suitable stencil patterns (DEK, Weymouth, UK). A previously used carbon-graphite ink formulation16 was first screen printed onto a polyester flexible (AutoStart, 250 mm thickness). This layer was cured for 30 min in a fan oven set to 60° [15]. The plastic substrate was then screen printed with Ag/AgCl paste (Gwent Electronic Materials Ltd, UK) to contain a silver/silver chloride reference electrode. Finally, a dielectric paste ink (Gwent Electronic Materials Ltd, UK) was printed to cover the connections and highlight the 3 mm diameter graphite working electrode. The screen-printed electrode is ready to use after 30 min of curing at 60°. Similar electrodes have previously been described and employed for detecting chromium (VI), 17 methionine, 18 cytochrome C19, and, more recently, selenium; 20 such applications emphasize the wide range of analytes that may be easily detected employing screen-printed graphite electrodes [16].

13.2.5 Alcohol sensing

The following ingredients were purchased from Sigma-Aldrich: alcohol oxidase (AOx, from Pichia pastoris, 10–40 units mg^{-1} protein), chitosan, bovine serum albumin (BSA), potassium phosphate monobasic (K_2PO_4), potassium phosphate dibasic (K_2HPO_4), ethanol, acetic acid, pilocarpine nitrate, sodium nitrate, L(+)-ascorbic acid (St. Louis, MO). All reagents were used without being purified further. For the electrochemical characterization in buffer media and on-body assessment, an Auto lab type III PGSTAT302N (Metrohm) operated by Auto lab NOVA software version 1.11.2 was used. Patterns for printing the sensor were created in AutoCAD (Autodesk, San Rafael, CA) and translated to stainless steel plates (1212 inch2) that were etched to create stencils (Metal Etch Services, San Marcos, CA). HPS Papilio provided temporary transfer tattoo paper kits (Rhome, TX). An MPM-SPM semi-automatic screen printer was used to print the silver/silver chloride (Ag/AgCl) ink. The tattoo sensor design, consists of iontophoresis and pseudo reference electrodes patterned from Ag/AgCl ink, and working and counter electrodes patterned from PB conductive carbon ink. The diameter of the printed working electrode was 3 mm. To limit the electrodes and contact regions, the print was across the surface of the electrode design. In a convection oven, the Ag/AgCl ink was cured at 90 °C for 10 min, while the PB conductive carbon ink was cured at 80 °C for 10 min [17].

13.2.6 Explosives

Accuse Standard Inc., New Haven, CT, provided explosive standards (reference materials) of hexamethylenetriperoxidediamine (HMTD) and triacetonetriperoxide (TATP) (www.accustandard.com). Both compounds were available as 0.1 mg ml^{-1} acetonitrile solutions (catalogue numbers M-8330-ADD-24 and M-8330-ADD-25). Fourier transform-infrared spectroscopy (FT-IR) was used to corroborate the identity of the confiscated (powder) components. The purity of all other materials utilized was greater than 98% (w/w). All the solvents utilized were either high performance LC (HPLC) or glass distilled quality. A Milli-Q/Organ ex Q system was used to purify the water (Millipore). The remaining reagents were of analytical grade. A sample volume of 10 μl acetone or methanol solution was injected into the HPLC system. Chromatographic separation was achieved with a Waters Nova-Pack 4 μm C18 3.9 × 150 mm HPLC cartridge column using an isocratic mobile phase of a methanol–water mixture (75:25) with 2.5 mM ammonium acetate at a flowrate of 0.4 ml/min. The API was operated in the atmospheric pressure chemical ionization positive ion mode (APCI+) with the vaporizing temperature being 360 °C, heated capillary 150 °C, vaporizing spray voltage 3.5 kV, and sheath gas 80 psi. The positive ions generated by the APCI ion source were detected by the MS or MS/MS system, where the current intensities of the mass to charge ratio (m/z) of the ionized components and fragments (ions) were recorded [16]. Instruments capable of detecting and identifying explosives from long distances are in high demand among security services in order to anticipate threats from a safe distance and ensure personal safety. Many analytical methodologies have been developed for this aim thus far, most of them are laser-based technology. Ifa *et al.* provided an excellent summary of stand-off detection systems for trace detection. The authors reviewed and discussed several laser-based methods, such as laser-induced \breakdown spectroscopy (LIBS), Raman spectroscopy, laser-induced fluorescence spectroscopy, and IR spectroscopy; but they also focused on bulk-detection technologies (i.e., millimeter-wave imaging and terahertz spectroscopy), since they complement the laser-based techniques [18].

13.3 Results and discussion

13.3.1 Chemical detection

Globally, the widespread use of illegal substances continues to be a serious social and health issue [11, 19]. In the domains of drug analysis and toxicology, there is a need for quick, sensitive, straightforward and affordable approaches for illicit substance detection [20]. Electrochemical sensors have shown a lot of potential as an alternative to the traditional chromatographic and spectroscopic techniques. Numerous illicit substances, including cocaine, cannabis [21, 22], amphetaminetype substances (ATS), opiates, diverted medicines, and novel psychoactive substances (NPS), can be detected by electrochemical methods using an approach specific to the illicit substance [23–25]. The vast range of concentration ranges, as well as the quantity and relative concentration of cutting agents, both diluents and adulterants, present a unique analytical difficulty for forensic analysis. For instance, a recent study indicated that cocaine samples from the UK ranged in purity from 1.365% to 78.8%, with a mean of 43.1% [26]. Expressed as cocaine base, this is a little higher than what was discovered in street samples of cocaine from Belgium, where cocaine purity was frequently reported to be above 30% [27]. These rates are much higher than those for ATS street samples, which were found to have ranges ranging from 1.4% to 6% [28]. These samples also contain a variety of contaminants and street samples that contain between 3 and 73.10 cutting agents, such as caffeine, paracetamol, and some other illegal drugs. The difficulty of conducting forensic analysis is highlighted by the continually shifting composition of street samples. The presence of metabolites in toxicological samples might also be viewed as an interference [29].

13.3.2 Development of a simple and rapid screening method for the detection of 1-(3-chlorophenyl) piperazine in forensic samples

Synthetic substance 1-(3-chlorophenyl) piperazine (mCPP), which has hallucinogenicproperties, is frequently discovered in samples that have been confiscated. In this situation, point-of-care tests that are simple to administer can be quite helpful in conducting preliminary forensic investigation. One author and their co-workers suggested an electrochemical approach for the quick, easy, and portable detection of mCPP in materials that have been seized. The technique is based on the use of disposable SPCEs and quick square-wave voltammetry screening methods with small sample volumes (100 l). At +0.65 V on SPCE (versus Ag), mCPP demonstrated an irreversible electrochemical oxidation process employing 0.04 mol l^{-1} BR buffer solution as the supporting electrolyte. Peak current and mCPP concentration were linearly correlated by the suggested approach (r = 0.998; limit of detection (LOD) = 0.1 mol l^{-1}) in the range of 1–30 mol l^{-1}. Interference tests were conducted for adulterants and other drug classes that were also present in the mCPP-containing samples that were seized, including caffeine, amphetamine, methamphetamine, 1-benzylpiperazine, 3,4-methylenedioxymethamphetamine, methylone, mephedrone, ethylone, and 3,4-methylenedioxypyrovalerone (figure 13.1). The method that has been developed has a lot of potential as a quick and easy screening tool to find mCPP in forensic materials. The author and their co-workers demonstrated a quick, low-cost, and easy approach for finding mCPP in samples that had been confiscated. The procedure relies on the drop-casting technique, a disposable device (SPCE), and SWV (portable features). Even in the presence of additional substances

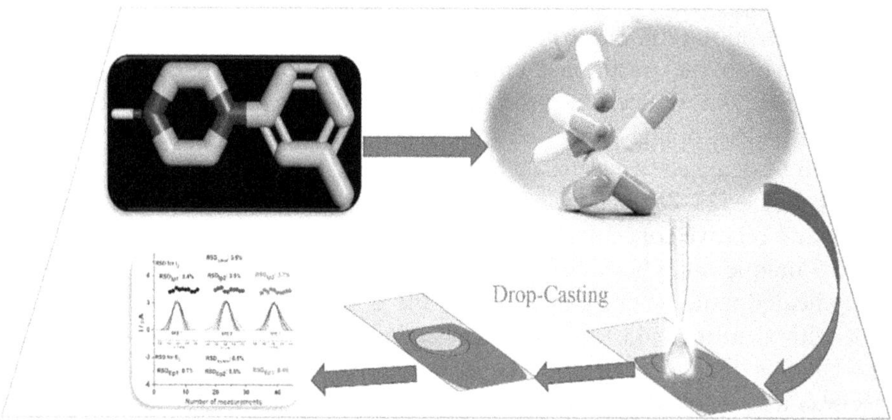

Figure 13.1. Graphical representation of development of a simple and rapid screening method for the detection of 1-(3-chlorophenyl) piperazine in forensic samples.

(possible interferents) frequently discovered in conjunction with mCPP, such as calcium fluoride (CAF), 3,4-methylenedioxy-methamphetamine (MDMA), and amphetamine-like compounds, quick screening tests can be conducted on samples that have been seized [30].

13.3.3 Electrochemical detection of benzodiazepines—forensic investigations

A significant class of medications with a high potential for abuse and environmental harm are the benzodiazepines. The LC electrochemical detection of the benzodiazepine class of pharmaceuticals features in this investigation [31]. These are distinguished by a rapidly reducible azomethine group by electrochemistry, but a variety of other electrochemically active groups may also be substituted. In biological, pharmacological, biomedical, and forensic investigations, LC using single and dual electrode detection has been described for a range of benzodiazepines and their metabolites. Electrochemistry has recently been used to simulate biological processes [32].

The benzodiazepine medication class has a long history in medicine and is currently frequently used for a variety of ailments. However, there have been instances of their use in criminal activity, and environmental pollution is increasingly a more often documented issue. Since it is now known that benzodiazepines have an easily reducible azomethine group and some also contain other electroactive groups including nitro, hydroxyl, and N-oxide, electrochemical detection methods using a variety of electrode materials have been developed to detect them [31].

Applications for the electrochemical measurement of 1,4-benzodiazepines by LC have used both dual electrode modes of detection and more recently, reductive techniques of detection. Both methods have been demonstrated to be effective for the analysis of challenging materials, including serum and forensic samples, and allow for low detection limits [32]. However, it has been demonstrated that typical interferences like oxygen negatively affect the reductive mode. As has already been demonstrated for the simulation of drug metabolism, new analytical advancements will continue to be produced for domains other than medical and pharmaceutical analysis, such as forensic and environmental analysis. Figure 13.2 indicates the ongoing development of electrochemical tests for this significant class of pharmaceuticals will also be driven by the introduction of new benzodiazepines and related drugs like zaleplon and zolpidem [33].

13.3.4 Toxicology and drugs

Consumption and trafficking of illegal drugs are widespread globally and provide local authorities with a growing issue. As the newest class of 'emerging contaminants,' forensic drugs and their metabolites are discharged into wastewaters as a result of human excretion following unlawful drug use, as well as sporadically as a result of the disposal of covert laboratory waste into sewage systems. Therefore, to lessen and avoid trafficking, consumption, and harmful impacts on aquatic systems, it is crucial to develop effective and precise procedures to detect these types of substances in confiscated street samples, biological fluids, and wastewater. For the

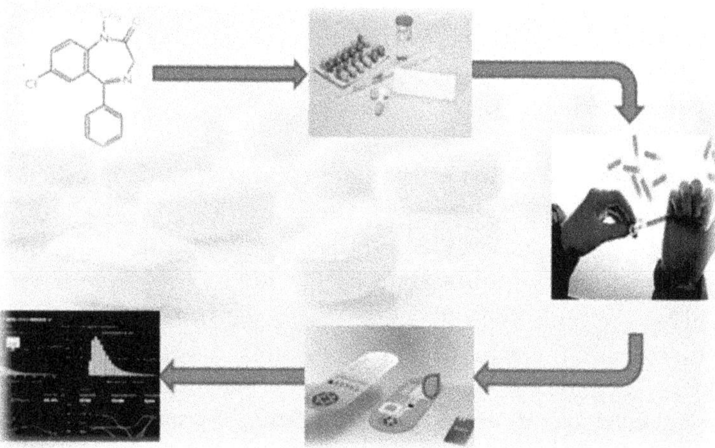

Figure 13.2. Graphical representation of development of a simple and rapid screening method for the detection of benzodiazepines in forensic samples.

study of forensic drugs and metabolites in various matrices, electrochemical procedures provide a quick, portable, affordable, and accurate alternative to chromatographic and spectrometric methods [34].

The improvements developed recently for the electrochemical detection of forensic substances are compiled in this reaction. Illegal drugs and their metabolites are only partially eliminated by treatment plants, exposing people and animals to their harmful effects. These pollutants have recently been classed as emerging contaminants. The paucity of literature on the detection of illegal substances in environmental samples highlights the necessity of creating novel electrochemical techniques for identifying, monitoring, and potentially decontaminating these pollutants from wastewater, surface water, and drinking water. To achieve the requisite detection limits, the few procedures described in the literature call for filtering and/or preconcentration by solid phase extraction of the materials, as well as modifications with aptamers or polymer composites. By monitoring drugs in wastewater on a weekly basis, one article even evaluated the value of electrochemical sensors for estimating drug consumption at the community level, demonstrating that these applications are certainly feasible for electrochemical technologies. There is more research accessible on methods to stop drug trafficking and usage, including the detection of illegal narcotics in body fluids and street samples. While some procedures employ unmodified electrodes to test unknown street samples for the presence of adulterants against a specific drug, the majority of approaches use electrode modifications, such as aptamers, antibodies, or nano-Molecularly imprinted polymers (nanoMIPs), to achieve low detection limits and great selectivity. The scientific community was interested in wearable sensors, such as gloves, as a practical means to conduct on-site analysis. Electrochemical approaches for illicit drug detection still have a long way to go before they are truly practical, despite their advantages. Preconcentration techniques or flow injection analysis, in particular, should be taken into consideration for wastewater analysis, along with improved electrode materials to enhance sensitivity and selectivity. There is a more potential for electrochemistry in the decontamination of illicit drugs [35].

13.3.5 Alcohol sensing

When it comes to substance addiction, alcohol is by far the most often abused substance. Exact blood alcohol content must be established during the analyses of a biological or breath sample at a suspected incident of drink-driving, which highlights the importance of being able to sense and compute alcohol (ethanol) from biological samples, which has significant applications ranging from workplace testing to policing. Blood, urine, and breath samples, in addition to transdermal testing, may all be used to calculate blood alcohol content (BAC). At the scene or on the roadside, breath analysis is usually the most frequent sample technique used. Henry's Law assumes that the concentration of ethanol in the alveolar air that is exhaled after drinking is exactly proportionate to the quantity of ethanol in the blood [36]. Breathalyzer testing is simple to do in public since it is a noninvasive method that can quickly, cheaply, and accurately determine BAC [17].

The most common and reliable use of electrochemical devices in a forensic setting is the detection of alcohol in exhaled air by electrochemistry. In 1935, Tom Parry devised the 'Alcolmeter,' a breath alcohol sensor based on a fuel cell, and this technology is still widely used in modern commercial breathalyzers. Protons, electrons, and byproducts like carbon dioxide (CO_2), acetaldehyde (CH_3CHO), and acetic acid (CH_3COOH) are produced during the catalytic electro oxidation of ethanol, the sensor's fuel. At the same time as protons are being shuttled through an ion exchange membrane from the anode to the cathode, the produced electrons travel via the external circuit to the opposite electrode at the cathode. As the cathode reacts with ambient oxygen, it produces water. The concentration of ethanol may be determined by measuring the current flowing between the anodic and cathodic compartments, which is proportional to the number of molecules present. Although commercial fuel cells have been in use for over 30 years, they still have a few design flaws that prevent them from being ideal. These flaws include a limited lifespan owing to the catalyst degrading and poisoning, a lack of electroactive area, and problems with specificity [37]. Despite these built-in flaws, the fuel cell technology has seen very little progress in the last 30 years. The advancement of transdermal sensing technologies is one area of heightened attention for the electrochemical detection of alcohol. Here, the user's perspiration may be tested for the presence of alcohol for continuous tracking. Figure 13.3 shows the design that the author and their co-workers presented in 2016 for a wearable epidermal electrochemical sensor based on 35 tattoos. Their sensor had a cathode and anode made of Ag/AgCl iontophoretic electrodes, with working, counter, and reference electrodes placed in the anode. We used Prussian blue conductive carbon ink to make both the working and counter electrodes, with the former being further modified using the alcohol-oxidase enzyme. The whole sensor's circuitry was printed on a flexible substrate and combined with a flexible circuit board to create a wearable sensor with exceptional flexibility and comfort. As a genuinely portable transdermal sensor system, it was

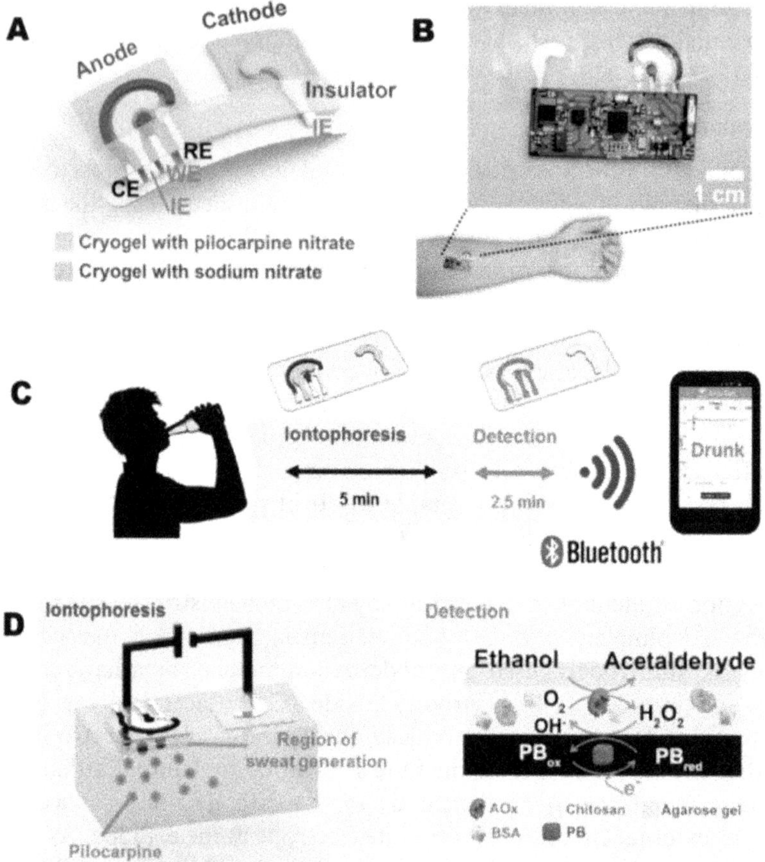

Figure 13.3. Tattoo-based transdermal alcohol sensors developed by Kim *et al* 2016. (A) 30 sensor schematic with iontophoretic electrodes (IE, anode and cathode) and amperometric electrodes (working WE, counter CE and reference RE). (B) Photograph of the sensor on the skin. (C) Diagram of the sensor's wireless functioning, with highlighted zones representing those active during iontophoresis and amperometric detection. (D) Diagram of the iontophoretic system's functioning (left) and amperometric alcohol sensing at the working electrode (right). Reprinted (adapted) with permission from [37]. Copyright (2016) American Chemical Society.

able to transmit data from an applied sensor to a laptop over Bluetooth. The presence of the alcohol-oxidase enzyme and the generation of perspiration by iontophoresis (with application of pilocarpine over the skin's surface) are the primary processes by which the sensor functions. By testing their prototype *in vitro* and, more importantly, on humans [37] they provided proof of concept for the creation of a portable transdermal sensor. Data revealed excellent agreement of BAC between the breathalyzer and transdermal findings, demonstrating that their sensor could determine whether the patient had taken alcohol due to the reduction in the amperometric current in the presence of ethanol. In order to calibrate their model, they took readings from each participant both before and after they imbibed. A constant baseline could be established thanks to the high degree of correlation between subsequent measures taken before intake, but no such connection was detected for subsequent data taken after consumption. Even though all 10 patients drank the same quantity of alcohol, there was noticeable variation in the measures. This variation, however, is not exceptional or surprising. The transdermal BAC assays showed a similar range of values, which may be explained by individual differences in metabolic rate and body weight [38].

Author and his colleges showed that the sensor they built could be used to measure blood alcohol content (BAC) by analyzing the amount of ethanol in the perspiration of an individual using the device. Further research into the possibility of device customization to account for changes in sweat composition and skin permeability is necessary to verify their approach on a larger subject pool. The technology developed by authors and his lab meats has the potential to be a low-cost, one-time-use, real-time monitoring sensor for non-invasively determining BAC. As opposed to the high production cost of the breathalyzer fuel cell system, which uses costly platinum catalysts, using screen-printed technology provides a low-cost option. The sensor's network capabilities open up several use cases and the possibility for 24/7 monitoring in those cases that call for it. Although these tests aren't likely to be used by police for roadside checks, they do demonstrate the potential of electrochemical sensors for mobile sensing in a forensic setting.

A paper-based electrochemical biosensor for the detection of ethanol was developed and reported by author and his coworker [38]. Rather than concentrating on human sample detection, as Authors and his co-workers estimate the ethanol content in a drink using a more direct method. In the food sector, where correct labeling is essential, this is a crucial topic of research. Forensic uses for simple ethanol concentration determination include spotting fake liquor and investigating cases of drink tampering. Author and his colleagues developed paper-based sensor that can be produced cheaply using regular office supplies like 80 g m^{-2} paper, wax, and screen-printing equipment. The alcohol oxidase enzyme, which, like Kim *et al*, generates H_2O_2, is used to add a bio-recognition component. Chronoamperometric data may be used as a proxy for ethanol concentrations by tracking H_2O_2 generation. Using the use of graphite screen printed electrode (SPE) modified with nanocomposites of carbon black and Prussian Blue on an 80 g m^{-2} paper substratewere able to obtain sensitive detection of ethanol down to 0.6 mM. By casting the alcohol oxidase enzyme onto the working electrode, they were able to reduce the detection threshold to 0.52 mM, and the enzyme remained stable for three weeks when stored at 4 °C [37, 38].

Like the use of aptasensors for drug detection, which included the use of a biosensor, the detection accuracy may be improved by using a biosensor [37, 38]. Increased specificity provided by bio-sensors was established by author and his lab meats using interference research with acetic acid, ascorbic acid, methanol, and glucose selected because of their abundance in commercial beer. Only methanol elicited a reaction, however it was much weaker than the one induced by ethanol, therefore no inference effects were shown for any of the other species. The alcohol oxidase enzyme is not ethanol-specific and hence catalyzes the breakdown of all primary alcohols, including ethanol. The computed concentrations of ethanol agreed with those provided by the sensor's manufacturer, demonstrating its precision and accuracy. With an analysis time of 40 s and the ability to quantify ethanol in real-world samples, the created sensor proved that a low-cost sensor could provide quick identification [38]. A paper-based support and single-use sensor provides a low-cost and disposable sensor, making it well suited for usage in the field. The sensor could be manufactured in a lab, stored, and then moved to a scene with ease thanks to its three-week stability postmanufacturer, which would further improve its infield deployment. While electrochemical methods for detecting alcohol have been around since 1935, the fuel cell system is now one of the most widely used and trusted electrochemical sensor applications in commercial and forensic settings. Fuel cell breathalyzer systems are widely used, but forensic electrochemistry is seldom considered as an alternative for detecting alcohol. Moreover, little R&D has been done to enhance the system and address its short comings, which include a high production cost. Recent advances in portable, low-cost sensors for alcohol detection have shown that existing detection methods may be improved. Wearable sensors would allow for continuous, non-intrusive tracking. These paper-based gadgets are inexpensive to produce and include durable, simple sensors. Recent advances certainly emphasize the potential of these other electrochemical systems, but further research into their resilience is needed before they can be evaluated as an acceptable alternative to the fuel cell system.

13.3.6 Sensing at your fingertips: glove based wearable chemical sensors

Wearable electrochemical sensors are a quickly developing digital health technology that allow for noninvasive monitoring of chemical indicators. There is a great deal of interest in using recent developments in wearable continuous glucose monitoring (CGM) systems in other crucial domains. This article discusses wearable electrochemical sensors for therapeutic and illicit drug monitoring for the first time [39]. To improve therapeutic outcomes while reducing drug side effects and associated medical costs, this rapidly developing class of wearable drugsensing devices answers the growing demand for individualized medicine. By realizing patientspecific dose regulation and tracking dynamic changes in pharmacokinetics behavior while assuring the medication is working, continuous, noninvasive monitoring of therapeutic drugs within bodily fluids enables clinicians and patients to correlate the pharmacokinetic properties with optimal results. Additionally, law enforcement officials can use wearable electrochemical drug monitoring devices as effective screening tools to stop drug trafficking and help on-site forensic investigations [40]. The paper scenario includes numerous wearable form factors created for on-site drug misuse screening as well as noninvasive monitoring of therapeutic medicines in various body fluids. With the goal of bringing exact real-time drug monitoring protocols and autonomous closed-loop platforms towards precise dose management and ideal therapeutic results, the prospects of such wearable drug monitoring devices are given. Finally, current issues and gaps are highlighted as potential catalysts for future technical advancements in individualized medicine. Due to the rapid rate of development and the enormous commercial potential for these wearable drug monitoring platforms to drive intense future research and commercialization efforts [41].

13.3.7 Explosive

Trace explosives must be found and identified after explosion in every suspected explosive device occurrence. Assessments and material identification must be done rapidly in explosives-suspected situations. Yet, time-consuming methods may make identification difficult, especially when only tiny amounts of material remain. Usually, scene evidence is taken to a forensic laboratory for investigation, delaying drug identification and identity confirmation. Rapid discovery at the site would give investigators time to act, which may save lives. Portable instrumentation for explosives scenarios has improved. Portable Raman, IR, and gas chromatographs are typical. While portable, they are expensive to run and understand and need skill. Electrochemical devices have a smaller footprint, are easy to use, and provide low-cost detection systems with single-use disposable electrodes, eliminating cross-contamination [42]. Forensics also seeks explosives. IEDs include urea nitrate. Using Ehrlich's reagent, p-dimethylaminocinnamaldehyde (p-DMAC) or p-dimethyl amino benzaldehyde (p-DMAB) is used to quickly detect urea nitrate [43]. Urea nitrate interacts with p-DMAC to form a red dye, whereas p-DMAB produces a yellow dye. Schiff bases may detect urea nitrate using these reagents. The red and yellow dyes are Schiff bases, and the intense pigments in the reaction medium come from protonated species from urea nitrate and p-DMAC or p-DMAB [44, 45]. Fluorescence emission is not a colorimetric approach, although Schiff Bases are interestingly linked to forensic applications. The tetra phenylethylene (TPE) macrocyclic Schiff base can detect nitrophenolic explosives through aggregation-induced emission (AIE) from nanospheres of nitrophenolic substances. Aqueous medium emits a yellow fluorescence [46]. Hence, the TPE macrocyclic Schiff base detects 2,4,6-trinitrophenol (TNP) and 2,4-dinitrophenol (DNP) at nanomolar quantities among other nitroaromatic substances. This Schiff base distinguishes TNP and DNP from other nitroaromatics. DNP and this Schiff base superamplify quenching. AIE moiety integration into a macrocyclic ring may also increase its explosive selectivity, providing a generic method for explosive sensor design. The procedure of identifying and transporting trace explosive leftovers after the explosion of an explosive device or during a criminal investigation involving explosives is an important one but may be time-consuming. It is possible to save a significant amount of time during the investigation by having the capacity to identify explosives at or near the site of an explosion. For this reason, one important area of study is the creation of field-deployable portable analytical devices. Within the scope of this study, the feasibility of using electrochemical sensors for on-site detection of trace explosives. This chapter places a significant emphasis on the use of room-temperature ionic liquids (RTIL), for the electrochemical detection of explosives. A comparison of the reaction pathways for the electrochemical reduction of TNT in aqueous solutions and in RTILs is shown in the review. A discussion on current work addressing the detection of explosives in aqueous, non-aqueous, and RTIL-based samples is presented. In conclusion, some commentary is offered on the predicted future direction of this area as well as the obstacles that it will face.

Electrochemical sensors have strong promise for examination 'at-the-scene' due to their cheap cost, mobility, quick analysis, and simplicity of use. If disposable electrodes were used to allay any contamination worries, this strategy would be very advantageous. Many of the methods covered in this review have built on well-established electrochemical principles and combined them with cutting-edge developments in chemical and material science, such as MIPs, graphene, and metal nanoparticles, as well as chemometrics as a method of analyzing the resulting data. These chemometric methods hold the most promise for the identification of specific explosives in an explosive's combination. RTILs have also been shown to have a great potential for application in electrochemical explosives detection since explosives like TNT are preferentially soluble and pre-concentrated in RTILs compared to aqueous samples (scheme 13.1). From a security perspective, this method may be exploited in the future for the detection of concealed pre-blast bulk explosives due to the low limits of detection for vaporsphase explosives detection in RTILs that have been achieved so far. Selectivity in mixed analyte mixtures may be improved through innovations in ionic liquid functionalization design. There is a lot of room for more research because the use of ionic liquids for the electrochemical detection of explosives is still a relatively new field of study. (Scheme 13.2). This could improve current capabilities and lead to the development of sensitive, selective, affordable, portable, and quick explosive electrochemical sensors. While the majority of this review's attention is given to explosives sensing for forensic applications, it should be emphasized that many of the methods and materials discussed here might also be used for environmental purposes, such as soil and water cleanup [49].

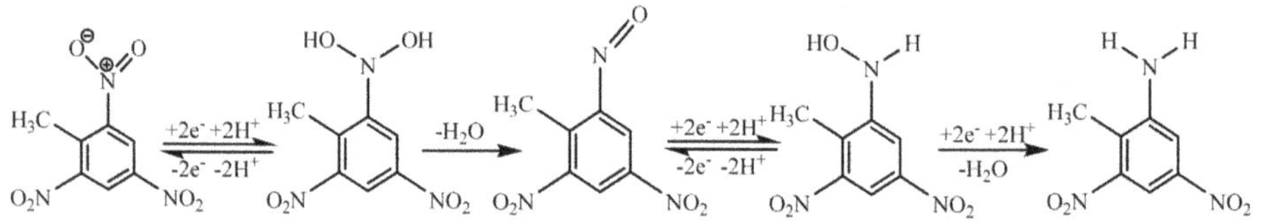

Scheme 1: TNT reduction mechanism (first peak) in protic solvents proposed by Chua et al. [47].

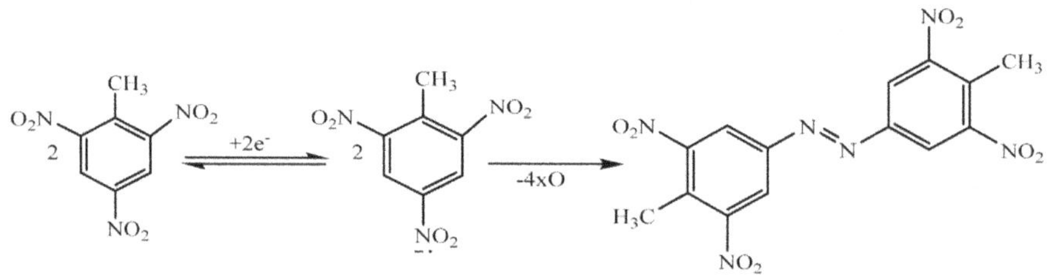

Scheme 2: TNT (first peak) reduction technique proposed in room temperature ionic liquids, taken from Kang et al., [48].

The following limitations with electrochemical explosives detection provide potential for further study:

- The various media vapor, solid, or dissolved samples in which explosives are present. Similar to other analytical methods for explosives detection, thorough sample type identification is necessary prior to the deployment of a specific electrochemical sensing device.
- In connection with the preceding point, numerous studies have shown that heating may produce enough explosive vapor for vapor phase explosive detection, however this method may be challenging to apply in 'at-the-scene' situations. This problem applies to all detection methods, not only electrochemical ones.
- Selecting the most effective electrochemical method to use. Cyclic voltmeter, SWV, differential pulse voltammetry, and chronoamperometry are just few of the many electrochemical methods available. This benefit is offset, however, by using a method that is appropriate for the sample type and concentration to be detected. It should also be noted that redox pathways may alter depending on the solvent environment.
- The reference electrode's composition; this factor is especially important for miniaturized sensor devices or those using RTIL solvents, where a stable reference system has not yet been created. In this scenario, the potential of the reduction peak would need to be determined using a voltammogram.
- When the RTIL anion or cation is changed, different analyte species behave in different ways. Although this may have a major impact on selectivity, the sheer variety of ionic liquids available may need several tests to identify the best solvent/electrolyte mixture for a given analyte.
- The potential for electrode fouling if non-disposable electrodes are utilized, or false positives from interfering species in a true mixed analyte sample. The parts-per-billion detection limits observed in all three solvents (aqueous, organic solvents, and ionic liquids) indicate great potential for employing electrochemical methods to detect very low trace quantities, even though this study has not focused on approaches with the lowest LODs. These kinds of sensors might be used as portable tools in the field with ease. It is evident that using electrochemical methods to detect explosives is a growing field of study that will undoubtedly continue in the years to come [50].

13.4 Conclusion

In this chapter the authors have made an attempt to describe the applications of electrochemical sensors in today's world. Electrochemical sensors offer great potential in the detection of alcohol sensing, narcotic drugs, toxic chemicals and explosives. Furthermore, in this chapter, different techniques used to detect the above-mentioned substances are also discussed, such as spectroscopic techniques, disposable SPCEs, SWV screening methods, alcolmeters, aptasensors and wearable electrochemical sensors. The content of narcotics drugs and explosives in the sample provides different information for forensic science. Illicit chemical composition in drug samples can indicate the toxicity of the drugs; their presence in medicine, food and explosives samples can indicate a danger to society, which can cause serious problems and causes life threats to society. However, narcotic drug samples provide evidence of a crime scene and can even help in identifying a suspect. Furthermore, this review shows that there are still many ways to improve it, using electronic sensor techniques or different sensing materials or substrates.

References

[1] Shaw L and Dennany L 2017 Applications of electrochemical sensors: forensic drug analysis *Curr. Opin. Electrochem.* **3** 23–8

[2] Raril C and Manjunatha J G 2020 Fabrication of novel polymer-modified graphene-based electrochemical sensor for the determination of mercury and lead ions in water and biological samples *J. Anal. Sci. Technol.* **11** 3

[3] Smith J P, Randviir E P and Banks C E 2016 An introduction to forensic electrochemistry *Forensic Science: A Multidisciplinary Approach* ed E Katz and J Halámek (Wiley)

[4] Hareesha N, Manjunatha J G, Amrutha B M, Pushpanjali P A, Charithra M M and Prinith Subbaiah N 2021 Electrochemical analysis of indigo carmine in food and water samples using a poly (glutamic acid) layered multi-walled carbon nanotube paste electrode *J. Electron. Mater.* **50** 1230–8

[5] Licata S C and Rowlett J K 2008 Abuse and dependence liability of benzodiazepine-type drugs: GABAA receptor modulation and beyond *Pharmacol. Biochem. Behav.* **90** 74–89

[6] Shahdost-Fard F and Roushani M 2016 Conformation switching of an aptamer based on cocaine enhancement on a surface of modified GCE *Talanta* **154** 7–14

[7] Prinith N S, Manjunatha J G and Hareesha N 2021 Electrochemical validation of L-tyrosine with dopamine using composite surfactant modified carbon nanotube electrode *J. Iran. Chem. Soc.* **18** 3493–503

[8] Bratin K, Kissinger P T, Briner R C and Bruntlett C S 1981 Determination of nitro aromatic, nitramine, and nitrate ester explosive compounds in explosive mixtures and gunshot residue by liquid chromatography and reductive electrochemical detection *Anal. Chim. Acta* **130** 295–311

[9] Pushpanjali P A, Manjunatha J G and Srinivas M T 2020 Highly sensitive platform utilizing poly (l-methionine) layered carbon nanotube paste sensor for the determination of voltaren *FlatChem.* **24** 100207

[10] van der Gouwe D, Brunt T M, van Laar M and van der Pol P 2017 Purity, adulteration and price of drugs bought on-line versus off-line in the Netherlands *Addiction* **112** 640–8

[11] Lima C D, Couto R A, Arantes L C, Marinho P A, Pimentel D M, Quinaz M B and dos Santos W T 2020 Electrochemical detection of the synthetic cathinone 3,4-methylenedioxypyrovalerone using carbon screen-printed electrodes: A fast, simple and sensitive screening method for forensic samples *Electrochim. Acta* **354** 136728

[12] Silva W P, Rocha R G, Arantes L C, Lima C D, Melo L M, Munoz R A and Richter E M 2021 Development of a simple and rapid screening method for the detection of 1-(3-chlorophenyl) piperazine in forensic samples *Talanta* **233** 122597

[13] Trojanowicz M 2011 Recent developments in electrochemical flow detections—a review: part II. Liquid chromatography *Anal. Chim. Acta* **688** 8–35

[14] Manjunatha J G, Kumara Swamy B E, Gilbert O, Mamatha G P and Sherigara B S 2010 Sensitive voltammetric determination of dopamine at salicylic acid and TX-100, SDS, CTAB modified carbon paste electrode *Int. J. Electrochem. Sci.* **5** 682–95

[15] Xu X, Van De Craats A M, Kok E M and De Bruyn P C A M 2004 Trace analysis of peroxide explosives by high performance liquid chromatography-atmospheric pressure chemical ionization-tandem mass spectrometry (HPLC-APCI-MS/MS) for forensic applications *J. Forensic Sci.* **49** 1230–6

[16] Pushpanjali P A and Manjunatha J G 2020 Development of polymer modified electrochemical sensor for the determination of alizarin carmine in the presence of tartrazine *Electroanalysis* **32** 2474–80

[17] Ramdani O, Metters J P, Figueiredo-Filho L C S, Fatibello-Filho O and Banks C E 2013 Forensic electrochemistry: sensing the molecule of murder atropine *Analyst* **138** 1053–9

[18] Ifa D R, Jackson A U, Paglia G and Cooks R G 2009 Forensic applications of ambient ionization mass spectrometry *Anal. Bioanal. Chem.* **394** 1995–2008

[19] UNODC World Drug Report 2016 unodc.org/doc/wdr2016/WORLD_DRUG_REPORT_2016_web.pdf (assessed 28 March 2017)

[20] European Drug Report 2016 http://emcdda.europa.eu/system/files/publications/2637/TDAT16001ENN.pdf (assessed 28 March 2017)

[21] Charithra M M, Manjunatha J G G and Raril C 2020 Surfactant modified graphite paste electrode as an electrochemical sensor for the enhanced voltammetric detection of estriol with dopamine and uric acid *Adv. Pharm. Bull.* **10** 247–53

[22] McGeehan J and Dennany L 2016 Electrochemiluminescent detection of methamphetamine and amphetamine *Forensic Sci. Int.* **264** 1–6

[23] Sengel T Y, Guler E, Gumus Z P, Aldemir E, Coskunol H, Akbulut H and Yagci Y 2017 An immunoelectrochemical platform for the biosensing of 'Cocaine use' *Sens. Actuators* B **246** 310–8

[24] Dennany L, Mohsan Z, Kanibolotsky A L and Skabara P J 2014 Novel electro chemiluminescent materials for sensor applications *Faraday Discuss.* **174** 357–67

[25] Tigari G and Manjunatha J G 2020 Optimized voltammetric experiment for the determination of phloroglucinol at surfactant modified carbon nanotube paste electrode *Instrum. Exp. Tech.* **63** 750–7

[26] Brockbals L, Karlsen M, Ramsey J D and Miserez B 2016 Single injection quantification of cocaine using multiple isotopically labeled internal standards *Forensic Toxicol.* **35** 153–61

[27] Eliaerts J, Dardenne P, Meert N, Van Durme F, Samyn N, Janssens K and De Wael K 2017 Rapid classification and quantification of cocaine in seized powders with ATR-FTIR and chemometrics *Drug Test. Anal.* **9** 1480–9

[28] Manjunatha Charithra M and Manjunatha J G 2020 Electrochemical sensing of paracetamol using electropolymerised and sodium lauryl sulfate modified carbon nanotube paste electrode *ChemistrySelect* **5** 9323–9

[29] Al Attas A S 2009 Construction and analytical application of ion selective bromazepam sensor *Int. J. Electrochem. Sci.* **4** 20–9

[30] Salem A E A, Barsoum B N, Saad G R and Izake E L 2002 Potentiometric determination of some 1, 4-benzodiazepines in pharmaceutical preparations and biological samples *J. Electroanal. Chem.* **536** 1–9

[31] Honeychurch K C and Hart J P 2014 Electrochemical detection of benzodiazepines, following liquid chromatography, for applications in pharmaceutical, biomedical and forensic investigations *Insciences J.* **4** 1–18
[32] Manjunatha J G 2020 A promising enhanced polymer modified voltammetric sensor for the quantification of catechol and phloroglucinol *Anal. Bioanal. Electrochem.* **12** 893–903
[33] Baumann A, Lohmann W, Schubert B, Oberacher H and Karst U 2009 Metabolic studies of tetrazepam based on electrochemical simulation in comparison to *in vivo* and *in vitro* methods *J. Chromatogr.* A **1216** 3192–8
[34] Manjunatha J G, Raril C, Hareesha N, Charithra M M, Pushpanjali P A, Tigari G and Gowda J 2020 Electrochemical fabrication of poly (niacin) modified graphite paste electrode and its application for the detection of riboflavin *Open Chem. Eng. J.* **14** 90–8
[35] Brown K and Dennany L 2019 Electrochemical devices for forensic chemical sensing *Forensic Anal. Methods* **13** 115
[36] Ozoemena K I, Musa S, Modise R, Ipadeola A K, Gaolatlhe L, Peteni S and Kabongo G 2018 Fuel cell-based breath-alcohol sensors: innovation-hungry old electrochemistry *Curr. Opin. Electrochem.* **10** 82–7
[37] Kim J, Jeerapan I, Imani S, Cho T N, Bandodkar A, Cinti S and Wang J 2016 Noninvasive alcohol monitoring using a wearable tattoo-based iontophoretic-biosensing system *ACS Sens.* **1** 1011–9
[38] Cinti S, Basso M, Moscone D and Arduini F 2017 A paper-based nanomodified electrochemical biosensor for ethanol detection in beers *Anal. Chim. Acta* **960** 123–30
[39] Saito M, Uchida N, Furutani S, Murahashi M, Espulgar W, Nagatani N and Tamiya E 2018 Field-deployable rapid multiple biosensing system for detection of chemical and biological warfare agents *Microsyst. Nanoeng.* **4** 1–11
[40] Bandodkar A J, Jeerapan I, You J M, Nuñez-Flores R and Wang J 2016 Highly stretchable fully-printed CNT-based electrochemical sensors and biofuel cells: combining intrinsic and design-induced stretchability *Nano Lett.* **16** 721–7
[41] Hubble L J and Wang J 2019 Sensing at your fingertips: glove-based wearable chemical sensors *Electroanalysis* **31** 428–36
[42] Manjunatha J G 2018 Highly sensitive polymer based sensor for determination of the drug mitoxantrone *J. Surf. Sci. Technol.* **34** 74–80
[43] Rozin R and Almog J 2011 Colorimetric detection of urea nitrate: the missing link *Forensic Sci. Int.* **208** 25–8
[44] Lemberger N and Almog J 2007 Structure elucidation of dyes that are formed in the colorimetric detection of the improvised explosive urea nitrate *J. Forensic Sci.* **52** 1107–10
[45] Feng H-T and Zheng Y-S 2014 Highly sensitive and selective detection of nitrophenolic explosives by using nanospheres of a tetraphenylethylene macrocycle displaying aggregation-induced emission *Chemistry* **20** 195–201
[46] Manjunatha J G 2020 A surfactant enhanced graphene paste electrode as an effective electrochemical sensor for the sensitive and simultaneous determination of catechol and resorcinol *Chem. Data Collect.* **25** 100331
[47] Chua C K, Pumera M and Rulíšek L 2012 Reduction pathways of 2,4,6-trinitrotoluene: an electrochemical and theoretical study *J. Phys. Chem.* C **116** 4243–51
[48] Kang C, Lee J and Silvester D S 2016 Electroreduction of 2,4,6-trinitrotoluene in room temperature ionic liquids: evidence of an EC2 mechanism *J. Phys. Chem.* C **120** 10997–1005

[49] Chuang M C, Windmiller J R, Santhosh P, Ramírez G V, Galik M, Chou T Y and Wang J 2010 Textile-based electrochemical sensing: effect of fabric substrate and detection of nitroaromatic explosives *Electroanalysis* **22** 2511–8
[50] Holly A Y, DeTata D A, Lewis S W and Silvester D S 2017 Recent developments in the electrochemical detection of explosives: towards field-deployable devices for forensic science *TrAC, Trends Anal. Chem.* **97** 374–84

IOP Publishing

Real-Time Applications of Advanced Electrochemical Sensing Devices

Jamballi G Manjunatha

Chapter 14

New approaches for glucose biosensors

Vinayak Adimule, Rangappa Keri, Santosh Nandi and Gangadhar Bagihalli

In recent years, inorganic metal oxides, organic–inorganic hybrid nanostructures (NS), carbon-based nanostructures attracted much attention in the development of nanoelectronic based devices. These NS possess exceptional features such as high surface-to-volume ratio, enhanced physical, chemical and electron transfer, thermal properties. Therefore, these NS can be effectively used in electrochemical sensors. The integration of the NS with different functional groups provides effective solid support and immobilization of the enzymes. In medical diagnostic and in food industries glucose level is measured by using biosensors. Glucose biosensors detect the glucose molecule by catalyzing it into gluconic acid and H_2O_2 (hydrogen peroxide) in the presence of O_2 (oxygen). Biosensors provide high accuracy and determination of the glucose molecule. In this chapter, use of sensors, gate voltage, *I–V*, characteristics were employed for the determination of the glucose response for the biosensors were presented. In the proposed model, existing experimental data were compared and analyzed. Further technical improvements required for the effective determination of the glucose molecule using biosensors, standardization of the analytical goals for their improvisation, basic principles, electrochemical sensor behaviour and sensing mechanisms are discussed in detail.

14.1 Introduction

There has been a dramatic increase in the development of nanosensors and nanoprobes for the effective detection and quantitative measurement of nano dimensioned particles and they are used for bio-molecular measurement techniques [1]. The different tools used for the detection, monitoring of the biological species in bio-molecular determination can create enormous development in the field of medical and diagnostic therapy in cell biology and medical science [2]. A biosensor typically consists of probe and transducer and is used for the varieties of applications in industries, environmental and biochemical diagnostics [3]. Blood glucose level detection finds extensive applications.

doi:10.1088/978-0-7503-5377-9ch14

Glucose oxide (GO_x) has the ability to identify the target molecule and quickly, and to accurately monitor its level [4, 5]. Larger constraints are involved in the detection of glucose in human blood due to their large detection range, high cost and knowledge complexity. Modern biosensors employ carbon-based nanomaterials due to their superior electrical properties [6, 7]. Recent biosensor performance and research based on the optimization of the electron transfer between enzyme and electrode surface are characteristics of improved biosensors [8, 9]. Significant efforts are made to improve biosensor performance and surface modifiers to improve the immobilization matrices which can improve prevention of leakage, and preserve enzyme activities [10]. In the present research, various nanocomposites, hybrid nanomaterials and nanoparticles have attracted much attention for modifying the surface properties of electrodes [11, 12]. Using special methods of organic synthesis, the properties of conducting polymers can be improved dramatically [13]. Furthermore quantification of glucose level is important for clinical analysis, food quality measurement and also in quality control [14].

14.2 Experimental section

14.2.1 Materials and methods

All the chemicals, reagents, were procured form Sigma–Aldrich, S-d fine chemicals Ltd, Spectro-chem Ltd etc and used without performing any further purification. Most of the nanomaterials were procured from Spectro-Chem and Sigma–Aldrich Ltd and all are in the form of a corresponding nitrate. GO_x procured from Sigma–Aldrich India Ltd. The inorganic/organic hybrids nanomaterials were fabricated using electrochemical metal deposition over a large conducting area of the substrate. Figure 14.1 displays the surface modification of the deposition of the nanoparticles over the substrate. This method of fabrication over the conducting substrate has been reported in previous literature [15]. Enzyme immobilization was carried out by the combination of biomolecules with nanostructures with their synergistic functions and properties [16]. There are different methods used for the immobilization of the enzymes over the substrate surface, one is enzymes on an inorganic substrate molecule, as reported in the literature [17]. The most established method is crosslinking and another method is physical adsorption of enzymes/biomolecules [18]. Sol–gel is the effective method for the immobilization

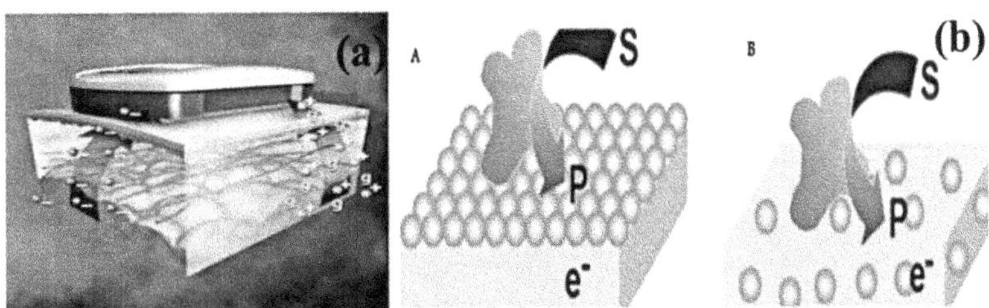

Figure 14.1. (a) Glucose watch operating principle. (b) Modification of electrode surface by AgNP (A) directly on the electrode surface and (B) embedded inside the matrix on the electrode surface. Part (a) reprinted from [19], Copyright (2005), with permission from Elsevier. Part (b) reprinted from [20], Copyright (2011), with permission from Elsevier.

of enzymes because it provides active proteins to get deposited over the inorganic/organic hybrid nanocomposites [21].

14.2.2 Basic principles of glucose biosensors

A typical biosensor consists of three main elements: (i) biorecognition elements, (ii) a transducer, (iii) a signal processing system that converts signals into readable form [22]. Enzymatic ampherometric glucose biosensors are the most commonly available and commercially used. The mechanism of glucose biosensors depends on enzymes like hexokinase and glucose oxidase [23–25]. GO_x is the standard used for the biosensors because of its cheap availability, greater selectivity, higher ionic strength etc. A glucose biosensor mechanism deals with the fact that immobilized GO_x catalyses the oxidation of glucose to gluconic acid and hydrogen peroxide [26]. Flavin adenine dinucleotide (FAD) acts as cofactor and will be reduced to $FADH_2$. The formed H_2O_2 oxidizes to O_2.

$$\text{Glucose} + GO_x - FAD^+ \rightarrow \text{Glucolactone} + GO_x - FADH_2 \tag{14.1}$$

$$GO_x - FADH_2 + O_2 \rightarrow GO_x - FAD + H_2O_2 \tag{14.2}$$

$$H_2O_2 \rightarrow 2H^+ + O_2 + 2e^- \tag{14.3}$$

Recent study has produced biosensors based on glucose oxidase and either GO-PQQ (pyro quinine quinone) or GO-NAD (nicotinamide adenine dinucleotide) [27, 28]. PQQ is used as recognition element and the enzymatic action of GO is represented as below

$$\text{Glucose} + \text{PQQ} \rightarrow \text{Glucolactone} + \text{PQQ} \tag{14.4}$$

$$\text{Glucose} + NAD^+ \rightarrow \text{Glucolactone} + \text{NADH} \tag{14.5}$$

$$\text{NADH} \rightarrow NAD^+ + H^+ + 2e^- \tag{14.6}$$

whereas GO with NAD produces NADPH which in turn accepts electrons and facilitates the oxidation of glucose. The reduced form of the carrier is called NADPH.

14.2.3 Types of glucose biosensors and their brief history

The history of glucose biosensors started with Clark and Lyons [23], who developed electrodes entrapped with glucose oxidase and oxygen

$$\text{Glucose} + O_2 \xrightarrow{\text{GO}} \text{Gluconic acid} + H_2O \tag{14.7}$$

The negative potential at the cathode is where O_2 is reduced to molecules of water

$$O_2 + 4H^+ + 4e^- \rightarrow 2H_2O \tag{14.8}$$

14.2.4 First-generation sensors

First-generation glucose biosensors involve the use of yellow glucose enzyme with FAD, subsequently reduced to $FADH_2$ with the following reactions.

$$GO(FAD) + Glucose \rightarrow GO(FADH_2) + Gluconolactone \tag{14.9}$$

The reoxidation of flavin regenerates the $FADH_2$

$$GO(FADH_2) + O_2 \rightarrow GO(FAD) + H_2O_2 \tag{14.10}$$

One of the greatest disadvantages of glucose biosensors (first-generation) is the measurement of H_2O_2 which requires greater potential and leads to various reducing species like ascorbic acid and uric acid [29]. The interference of the H_2O_2 can be minimized by use of a selective membrane and reducing the potentials applied across the electrodes. Another problem to overcome is the mass transport of the polyurethane or polycarbonate which was tailored to transport the glucose and oxygen [30].

14.2.5 Second-generation sensors

Second-generation sensors work on the principle of non-physiological electron transport acceptor to transfer the electrons and to reduce the oxygen deficiencies in the electrodes. In this case gold nanoparticles and carbon nanoparticles were used as electrode connectors. Patolsky *et al* studied the biosensor properties using single-walled carbon nanotubes (SWCNTs) with enzyme on the electrodes [31]. In the study, FAD was first attached to SWCNTs and glucose oxidase was reconstituted over the FAD, as displayed in figure 14.2.

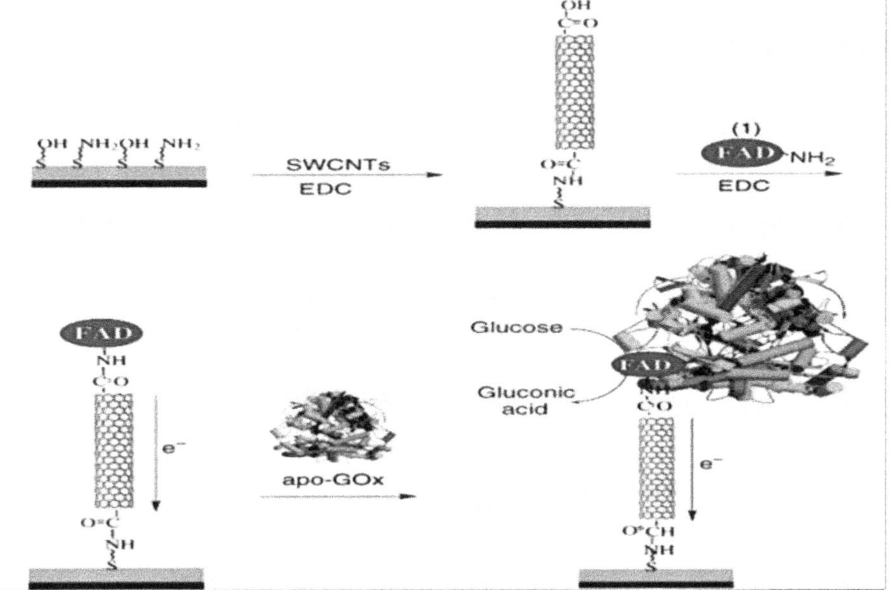

Figure 14.2. Assembly of the SWCNT electrically contacted glucose oxidase electrode. A 2-thioethanol/cystamine mixed monolayer. Adapted from [32]. CC BY 4.0.

The reactions occurring in the non-physiological determination of glucose oxidase are shown below

$$\text{Glucose} + GO_{(ox)} \rightarrow \text{Gluconic acid} + GO_{(red)} \tag{14.11}$$

$$GO_{(red)} + 2M(ox) \rightarrow GO_{(ox)} + 2M_{(red)} + 2H^{+} \tag{14.12}$$

$$2M_{(red)} \rightarrow 2M_{(ox)} + 2e^{-} \tag{14.13}$$

where, $M_{(ox)}$ and $M_{(red)}$ are the oxidized and reduced form of the mediators. There are several artificial mediators used such as ferrocine, transition metal complexes, ferricynide which are of particular interest [33]. A good electron carrier between FAD and glucose is required to fulfill the following criteria: (i) it must react rapidly with reduced enzyme; (ii) must possess good electrochemical properties; (iii) must possess low solubility, nontoxic and chemical stability [34]. Manjunatha *et al* discussed widely the applications of electroanalytical techniques in the effective detection of various substances [35–42].

14.2.6 Third-generation sensors

Third-generation glucose biosensors make use of elimination of artificial mediators and the difficulties arise for the efficient electron transfer between the electrodes and glucose oxidase. The best example is organic salt electrodes and mesoporous electrodes with increased surface area and the ability to effectively transfer the electrons [35, 36]. In such a system direct electron transfer occurs between enzymes and electrodes, avoiding the complex mediators.

14.2.7 Different types of glucose biosensors

(i) YSI 23A glucose biosensors:
This glucose biosensor involves an outer carbonate layer and inner enzyme layer. H_2O_2 is produced when glucose enters the enzyme layer and becomes oxidized. The Pt electrode measures the potential difference electrochemically. It is quite expensive as it involves a Pt electrode, high detection voltage etc. A schematic YSI 23A glucose biosensor is displayed in figure 14.3(b).

(ii) Mediator glucose biosensors:
A mediator biosensor device involves small, inexpensive and highly reproducible structures which are capable of detection of glucose oxidase with the help of mediators acting as electron transport entities between glucose oxidase and FAD.

(iii) Current generation of home-use blood glucose biosensors:

The current generation glucose biosensors are displayed in figure 14.4. They make use of a small, compatible, disposable screen printing system, minimizing the required sample size, and the instrument operates automatically.

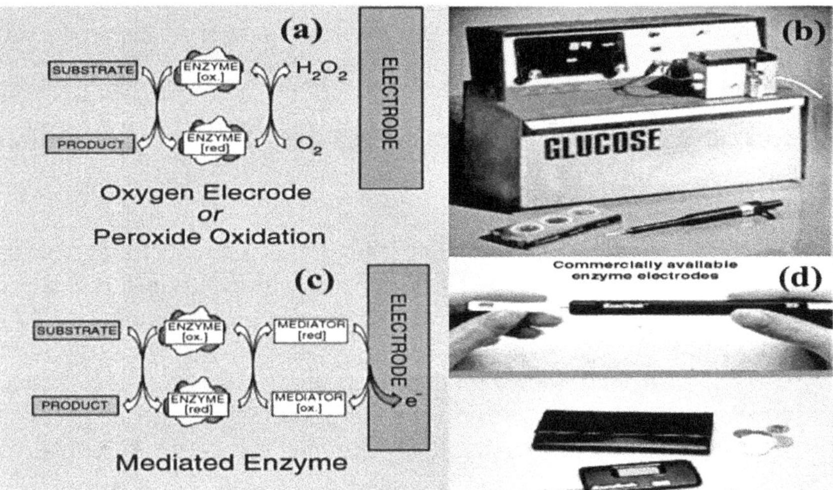

Figure 14.3. (a) Schematic representation of first-generation glucose biosensors; (b) glucose biosensor model number YSI 23A; (c) mediated biosensors; (d) original medisense products. Reprinted from [19], Copyright (2005), with permission from Elsevier.

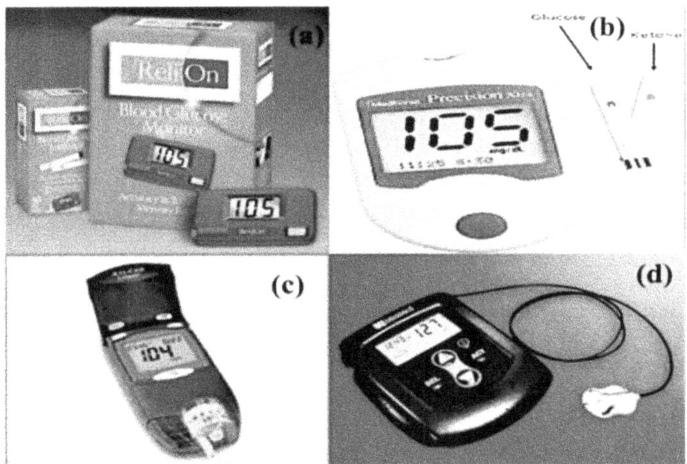

Figure 14.4. (a) Walmart relion blood glucose monitor. (b) Precession extra glucose/ketone monitor. (c) Accu Check Compact Biosensor. (d) MiniMed Implantable glucose biosensor. Adapted from [19]. CC BY 4.0.

(iv) Improved class of glucose biosensors:
An improved and elegant class of glucose biosensors involves elimination of interference in their detection, making use of molecular wires. This can be achieved by modifying the enzyme, transducer or both. Some of the devices make use of a single molecule with delocalized electrons.

(v) Implantable glucose biosensors:
An implantable glucose biosensor involves a component which is a closed loop of glycaemic control systems. Such *in vivo* sensing device measurement

is straightforward but has a few drawbacks such as electrode stability, biocompatibility and calibration of the instrument.

14.2.8 Preparation of the modified electrodes and their electrochemical measurement system

Modification of the surface of the electrode was carried out as follows.

This process is carried out in particular for carbon-based electrode systems:

(i) Using 10 mM H_2O_2 in phosphate buffer 0.1 M having pH = 7, the screen-printed carbon electrodes are used for the cyclic voltammogram of 1 V and −1 V, respectively.

(ii) Carbon-based materials were deposited on the surface of the electrodes as per the standard procedures in the literature [37].

(iii) Nanoparticles were generated as per the standard procedures in the literature [38].

(iv) Immobilization of the GO_x is carried out in specific concentration of the enzyme with 0.05–0.1 M phosphate buffer solution. After the immobilization the remaining solution was discarded and washed with deionized water, then finally washed with 0.1 M phosphate buffer solution.

Electrochemical measurements were carried out using a bio-potentiostat (from Dropsens). The electroactive areas of the electrodes were calculated using the Randles–Sevick equation by performing voltammetric cycles at different scan rates in the presence of 1 mM $Na_4Fe(CN)_6$ in 0.1 of a KCl aqueous solution, previously deoxygenated with nitrogen gas. The diffusion coefficient of the $Na_4Fe(CN)_6$ was used during electrochemical estimation [39]. Electrochemical impedance spectroscopy (EIS) measurements were performed in an AUTOLAB PGSTAT128N potentiostat with an EIS analyzer (Eco Chemie B.V., Utrecht, The Netherlands) using NOVA 2.0 software. The EIS was carried out at 0.15 V in 1 mM $Na_4Fe(CN)_6$ and 0.1 M KCl. Then, a sinusoidal amplitude potential perturbation (5 mV rms) was imposed between 65 kHz and 10 mHz with five points per decade.

14.3 Results and discussion

14.3.1 Glucose sensing mechanism and proposed model (new approaches)

Depending upon the enzyme, transducers used for the determination of glucose oxidase different models have been proposed. Among the different models, two in particular are widely used: (i) charge-based model, and (ii) non-charge-based analytical model. Carbon nanotube filled field-effect transistors (CNTFFETs) were employed for glucose sensing and estimation. For carbon-based, inorganic/organic nano-hybrid-based glucose biosensors make use of the charge-based model. The drift velocity of the charge carrier in the presence of applied voltage is given by the equation

$$V_D = \frac{\mu E}{1 + \frac{E}{E_C}} \tag{14.14}$$

where μ is the mobility of the carriers, E is the electric field, and E_c is the critical electric field under high applied bias. From equation (14.1), the drain current as a function of gate voltage (V_G) and drain voltage (V_D) is obtained as

$$I_D = \beta \frac{2V_{GT}\ V_D \quad - V^2 D}{1 + \frac{V_D}{V_C}}. \tag{14.15}$$

Based on the geometry of the CNTFETs the gate capacitance can be defined as

$$C_G = \frac{C_E\ C_Q}{C_E + C_Q} \tag{14.16}$$

where C_E and C_Q are the electrostatic gate coupling capacitance of the gate oxide and the quantum capacitance of the gated SWCNT, respectively [40–43]. Bare SWCNT FET for different gate voltages without any PBS and glucose concentration is based on equation (14.2). The electrostatic gate capacitance for the C–V can be given as

$$C_E = \frac{2\pi\epsilon}{\ln\left(\frac{4H_{PET}}{d}\right)} L \tag{14.17}$$

14.3.2 Analytical performance and validation procedure for glucose biosensors

In 1994, in order to significantly monitor the inaccuracies of the glucose biosensors and their determination of the efficiencies, the Americans with Disabilities Act of 1990 made the first recommendation for glucose biosensors performance. According to the Act, <10% of the allowable bias from reference methods of glucose concentration is between 1.6 and 22.2 mM l^{-1} [44]. According to US FDA, glucose biosensors should have <20% of the bias and concentration between 1.65 and 22 mM l^{-1}. Glucose monitoring and analytical performance has been validated by healthcare professionals as per the Clinical and Laboratory Standards Institute guidelines. However, The International Organization for Standardization set down the guidelines for evaluating glucose biosensors in real time. In addition, the guidelines prescribe the *in vivo* monitoring of glucose in blood. The guideline points can be summarized as below.

1. Precision or repeatability:
   Samples are collected from venous blood and at least 10 measurements are made, with the mean value of the standard deviation variation coefficients calculated. Five different glucose concentrations spread over 30–400 mg dl^{-1} are collected and kept at 23 °C.
2. Intermediate precision:
   Intermediate precision is the test results conducted with identical test samples but other factors are considered, like calibration, equipment, environmental conditions, time intervals etc. Three glucose samples are used (30–250 mg dl^{-1}),

Table 14.1. Percentage of glucose concentration in milligram per deciliter.

Percentage of samples (%)	Glucose concentration (mg dl^{-1})
05	<50
15	50–80
20	80–120
30	120–200
15	201–300
10	301–400
05	>400

different users with multiple meters are used and the test is conducted for 10 days. Mean of the standard deviation and coefficient of variation for all the samples over 10 measurements are calculated.

3. Accuracy:
   Accuracy is determined for at least 100 different subjects with two different meters over 10 days with capillary blood samples. Samples are measured by two different meters and at least in triplicate. The results obtained from glucose biosensors are compared with the setting results for the measurement of accuracy. Table 14.1 shows the glucose concentration of the samples for the evaluation of accuracy and measurements [45].
4. Linearity:
   Linearity of the results of the glucose biosensors are evaluated by the replicates from the 5–11 samples that are varying in their concentrations. The correlation coefficients and linear regressions are calculated based on the test results of 2–3 for repeatability of the results.
5. User performance:
   User performance is directly dependent on trained users rather than lay users. The results obtained by lay users are compared with the results obtained by validated glucose measurement procedures. At least 50 subjects with varying demography (age, gender, and educational level) are used for obtaining the linear plot. For each investigation second blood samples are collected within 5 min of their interval and test results are fitted to the regression equation.
6. Interference:
   Interference is the material that is electroactive at the measuring potential of the glucose biosensors. A number of interferences developed by FDA include salicylic acid, tetracycline, dopamine, L-DOPA, Methyl-DOPA [46–50] etc. Recent studies on electrochemical interference showed the high efficient detection of various analytes [51–55]. The method also serves applications in the development of adavanced glucose biosensors [56–60].

14.4 Conclusion

This chapter describes the basic principles involved in construction of glucose biosensors and their mechanism of action. The history of transforming glucose biosensors and their first, second and third generations and the limitations of all the classes of glucose biosensors were discussed in detail. The chapter also covers the varieties of commercially available biosensors and their applications as well as limitations. Glucose biosensors and the detection of glucose oxidase and the enzymes utilized for effective measurement require modification of the electrode surface as well as of the electrochemical systems. The different, innovative and new models proposed for glucose biosensors and their utility, and their drawbacks and limitations were discussed in detail. The development of different analytical procedures with and without the involvement of mediators as well as their validation procedure finds glucose biosensors applied in the field of detection of various oxidized and reduced species during effective detection and their quantitative determination. The chapter describes the various possibilities for development of a new class of glucose biosensors, their difficulties in construction and future aspects.

Acknowledgments

All the authors are thankful to Centre for Advanced Materials Technology (CAMT), MSRIT, Bangalore, Karnataka, India, Centre for Nano and Material Sciences (CNMS), Jain University, Bangalore, Karnataka, India and Dr MSSCET, KLE Tech University, Belagavi, Karnataka, India for constant encouragement and support.

Author contributions

Dr Vinayak Adimule conceived the ideas, wrote the manuscript and handled the revision. Dr Santosh Nandi was involved in characterization and interpretation of the drawn out results. Dr Rangappa Keri and Dr Gangadhar Bagihgalli contributed for the revision of the manuscript.

Conflict of interest

All authors declare that they do not have any conflict of interest.

Data availability

Data that supports the findings of the study are openly available.

Funding

All authors declare that they have not received any funding from any source or organization.

References

[1] Wolfbeis O S 2008 Fiber-optic chemical sensors and biosensors *Anal. Chem.* **80** 4269–83

[2] Diamond D 1998 *Principles of Chemical and Biological Sensors* (New York: Wiley)

[3] Sandhu A 2007 Glucose sensing silicon's sweet spot *Nat. Nanotech.* https://doi.org/10.1038/nnano.2007.2

[4] Zhu Z G, Garcia-Gancedo L, Chen C, Zhu X R, Xie H Q, Flewitt A J and Milne W I 2013 Enzyme-free glucose biosensor based on low density CNT forest grown directly on a Si/SiO_2 substrate *Sens. Actuators B Chem.* **178** 586–92

[5] Wen Z, Ci S and Li J 2009 Pt nanoparticles inserting in carbon nanotube arrays: nanocomposites for glucose biosensors *J. Phys. Chem.* C **113** 13482–7

[6] Alwarappan S, Boyapalle S, Kumar A, Li C-Z and Mohapatra S 2012 Comparative study of single, few-, and multilayered graphene toward enzyme conjugation and electrochemical response *J. Phys. Chem.* C **116** 6556–9

[7] Du D, Zou Z, Shin Y, Wang J, Wu H, Engelhard M H, Liu J, Aksay I A and Lin Y 2010 Sensitive immunosensor for cancer biomarker based on dual signal amplification strategy of graphene sheets and multienzyme functionalized carbon nanospheres *Anal. Chem.* **82** 2989–95

[8] Scognamiglio V and Arduini F 2019 The technology tree in the design of glucose biosensors *TrAC, Trends Anal. Chem.* **120** 115642

[9] Idumah C I 2021 Novel trends in conductive polymeric nanocomposites, and bionanocomposites *Synth. Met.* **273** 116674

[10] Aleksandrovskaya A Y, Melnikov P V, Safonov A V, Abaturova N A, Spitsyn B V, Naumova A O and Zaitsev N K 2019 The effect of modified nanodiamonds on the wettability of the surface of an optical oxygen sensor and biological fouling during long-term *in situ* measurements *Nanotechnol. Russ* **14** 389–96

[11] Pan H M, Gonuguntla S, Li S and Trau D 2017 *3.33 Conjugated Polymers for Biosensor Devices* (Amsterdam: Elsevier)

[12] Rehman A and Zeng X 2020 Interfacial composition, structure, and properties of ionic liquids and conductive polymers for the construction of chemical sensors and biosensors: a perspective *Curr. Opin. Electrochem.* **23** 47–56

[13] Ramanavicius S and Ramanavicius A 2021 Conducting polymers in the design of biosensors and biofuel cells *Polymers* **13** 49

[14] Luong J H T, Glennon J D, Gedanken A and Vashist S K 2017 Achievement and assessment of direct electron transfer of glucose oxidase in electrochemical biosensing using carbon nanotubes, graphene, and their nanocomposites *Microchim. Acta* **184** 369–88

[15] Li Y and Shi G 2005 Electrochemical growth of two-dimensional gold nanostructures on a thin polypyrrole film modified ITO electrode *J. Phys. Chem.* B **109** 23787

[16] Chen D, Wang G and Li J 2007 Interfacial bioelectrochemistry: fabrication, properties and applications of functional nanostructured biointerfaces *J. Phys. Chem.* C **111** 2351

[17] Cosnier S 2000 Biosensors based on immobilization of biomolecules by electrogenerated polymer films *Appl. Biochem. Biotechnol.* **89** 127

[18] Elwing H 1998 Protein absorption and ellipsometry in biomaterial research *Biomaterials* **19** 397–406

[19] Newman J D and Turner A P 2005 Home blood glucose biosensors: a commercial perspective *Biosens. Bioelectron.* **20** 2435–53

[20] Rad A S, Mirabi A, Binaian E and Tayebi H 2011 A review on glucose and hydrogen peroxide biosensor based on modified electrode included silver nanoparticles *Int. J. Electrochem. Sci.* **6** 3671–83

[21] Shokuhi Rad A, Mirabi A, Binaian E and Tayebi H 2009 A Review on glucose and hydrogen peroxide biosensor based on modified electrode included silver nanoparticles *Res. J. Biol. Sci.* **4** 1284
[22] Turner A P 2000 Biosensors—sense and sensitivity *Science* **290** 1315–7
[23] Clark L C Jr and Lyons C 1962 Electrode systems for continuous monitoring in cardiovascular surgery *Ann. N. Y. Acad. Sci.* **102** 29–45
[24] Iqbal S S, Mayo M W, Bruno J G, Bronk B V, Batt C A and Chambers J P 2000 A review of molecular recognition technologies for detection of biological threat agents *Biosens. Bioelectron.* **15** 549–78
[25] Habermuller K, Mosbach M and Schuhmann W 2000 Electron-transfer mechanisms in amperometric biosensors *Fresenius J. Anal. Chem.* **366** 560–8
[26] Pearson J E, Gill A and Vadgama P 2000 Analytical aspects of biosensors *Ann. Clin. Biochem.* **37** 119–45
[27] Bartlett P N and Whitaker R G 1987 Strategies for the development of amperometric enzyme electrodes *Biosensors* **3** 359–79
[28] Gorton L and Dominguez E 2002 Electrocatalytic oxidation of NAD(P) H at mediator-modified electrodes *J. Biotechnol.* **82** 371–92
[29] Updike S J and Hicks G P 1967 The enzyme electrode *Nature* **214** 986–8
[30] Liu J and Wang J 2001 Improved design for the glucose biosensor *Food Technol. Biotechnol.* **39** 55–8
[31] Pishko M V, Katakis I, Lindquist S E, Ye L, Gregg B A and Heller A 1990 Direct electrical coomunication between graphite-electrodes and surface andorbed glucsoe-oxicase redox polymer complexes *Angew. Chem. Int. Ed.* **29** 82–9
[32] Juska V B and Pemble M E 2020 A critical review of electrochemical glucose sensing: Evolution of biosensor platforms based on advanced nanosystems *Sensors* **20** 6013
[33] Patolsky F, Weizmann Y and Willner I 2004 Long-range electrical contacting of redox enzymes by SWCNT connectors *Angew. Chem. Int. Ed.* **43** 2113–7
[34] Shichiri M, Kawamori R, Yamasaki Y, Hakui N and Abe H 1982 Wearable artificial endocrine pancreas with needle-tpye glucose *sensor Lancet* **2** 1129–31
[35] Manjunatha J G, Raril C, Hareesha N, Charithra M M, Pushpanjali P A, Tigari G, Ravishankar D K, Mallappaji S C and Gowda J 2020 Electrochemical fabrication of poly (niacin) modified graphite paste electrode and its application for the detection of riboflavin *Open Chem. Eng. J.* **14** 90–8
[36] Manjunatha J G 2020 A promising enhanced polymer modified voltammetric sensor for the quantification of catechol and phloroglucinol *Anal. Bioanal. Electrochem.* **12** 893–903
[37] Manjunatha J G 2018 Highly sensitive polymer based sensor for determination of the drug mitoxantrone *J. Surf. Sci. Technol.* **34** 74–80
[38] Manjunatha J G, Swamy B E and Deraman M 2013 Electrochemical studies of dopamine, ascorbic acid and their simultaneous determination at a poly (rosaniline) modified carbon paste electrode: a cyclic voltammetric study *Anal. Bioanal. Electrochem.* **5** 426–38
[39] Manjunatha J G 2020 Poly (Adenine) modified graphene-based voltammetric sensor for the electrochemical determination of catechol, hydroquinone and resorcinol *Open Chem. Eng. J.* **14** 52–62
[40] Manjunatha J G 2020 A surfactant enhanced graphene paste electrode as an effective electrochemical sensor for the sensitive and simultaneous determination of catechol and resorcinol *Chem. Data Collect.* **25** 100331

[41] Manjunatha J G 2018 A novel voltammetric method for the enhanced detection of the food additive tartrazine using an electrochemical sensor *Heliyon* **4** e00986

[42] Manjunatha J G, Kumara Swamy B E, Deraman M and Mamatha G P 2012 Simultaneous voltammetric measurement of ascorbic acid and dopamine at poly (vanillin) modified carbon paste electrode: a cyclic voltammetric study *Pharm. Chem.* **4** 2489–97

[43] Khan G F, Ohwa M and Wernet W 1996 Design of a stable charge transfer complex electrode for a third-generation amperometric glucose sensor *Anal. Chem.* **68** 2939–45

[44] Palmisano F, Zambonin P G, Centonze D and Quinto M A 2002 disposable, reagentless, third-generation glucose biosensor based on overoxidized poly(pyrrole)/tetrathiafulvalenete-tracyano- quinodimethane composite *Anal. Chem.* **74** 5913–8

[45] González-Sánchez M I, Gómez-Monedero B, Agrisuelas J, Iniesta J and Valero E 2018 Highly activated screen-printed carbon electrode by electrochemical treatment with hydrogen peroxide *Electrochem. Commun.* **91** 36–40

[46] Agrisuelas J, González-Sánchez M I and Valero E 2017 Electrochemical properties of poly (azure A) films synthesized in sodium dodecyl sulfate solution *J. Electrochem. Soc.* **164** G1

[47] Agrisuelas J, González-Sánchez M I and Valero E 2017 Hydrogen peroxide sensor based on *in situ* grown Pt nanoparticles from waste screen-printed electrodes *Sens. Actuators* B **249** 499–505

[48] Anantram M and Leonard F 2006 Physics of carbon nanotube electronic devices *Rep. Prog. Phys.* **69** 507

[49] Tan M L P 2011 Device and circuit-level models for carbon nanotube and grapheme nanoribbon transistors *Thesis* (Cambridge: University of Cambridge, Department of Engineering)

[50] Tan M L P, Lentaris G and Amaratunga G A 2012 Device and circuit-level performance of carbon nanotube field-effect transistor with benchmarking against a nano-MOSFET *Nanoscale Res. Lett.* **7** 467

[51] Manjunatha J G, Deraman M and Basri N H 2015 Electro catalytic detection of dopamine and uric acid at poly (basic blue b) modified carbon nanotube paste electrode *Asian J. Pharm. Clin. Res.* **8** 48–53

[52] Sachith B M, Sandeep S, Nandini S, Nalini S, Karthik C S, Swamy N K, Mallu P and Manjunath J G 2020 Simple sonochemical synthesis of $SrCuO_2$ assisted GCN nanocomposite for the sensitive electrochemical detection of 4-AAP *Surf. Interfaces* **20** 100603

[53] Charithra M M and Manjunatha J G 2021 Fabrication of poly (Evans blue) modified graphite paste electrode as an electrochemical sensor for sensitive and instant riboflavin detection *Mor. J. Chem.* **9** 9–1

[54] Charithra M M, Manjunatha J G, Sreeharsha N, Asdaq S M B and Anwer M 2021 Polymerized carbon nanotube paste electrode as a sensing material for the detection of adrenaline with folic acid *Monatsh. Chem.* **152** 411–20

[55] Tigari G, Manjunatha J G, D'Souza E S and Sreeharsha N 2021 Surfactant and polymer composite modified electrode for the sensitive determination of vanillin in food sample *ChemistrySelect* **6** 2700–8

[56] Prinith N S, Manjunatha J G and Hareesha N 2021 Electrochemical validation of L-tyrosine with dopamine using composite surfactant modified carbon nanotube electrode *J. Iran. Chem. Soc.* **18** 3493–503

[57] Tan M L P 2013 Long channel carbon nanotube as an alternative to nanoscale silicon channels in scaled MOSFETs *J. Nanomater.* **2013** 831252

[58] Lin Y-M, Appenzeller A, Chen Z, Chen Z-G, Cheng H-M and Avouris P 2005 Demonstration of a high performance 40-nm-gate carbon nanotube field-effect transistor *63rd Device Res. Conf. Digest. DRC' (05 2005 1)* pp 113–4
[59] International Organization for Standardization 2003 *In Vitro Diagnostic Test Systems-Requirements for Blood-Glucose Monitoring Systems for Self-Testing in Managing Diabetes Mellitus* (Geneva: International Organization for Standardization)
[60] Heller A and Feldman B 2008 Electrochemical glucose sensors and their applications in diabetes management *Chem. Rev.* **108** 2482–505

IOP Publishing

Chapter 15

Electrochemical sensing devices for estimation of food additives and flavours

Maya Devi, Swetapadma Praharaj and Dibyaranjan Rout

Food additives are substances which are added to food for various technical purposes. But improper use of them poses a threat to human health. Proper estimation of these can reduce the bad impact on health. Of the various reported methods of food additive analysis, electrochemical sensing is a preferred one because of its various advantages. In this chapter a detailed discussion of various electrochemical methods is provided. The recent developments in electrochemical sensors by the use of nanoparticles, modification in electrodes and nature of substrate used in electrodes for effective food analysis is described.

15.1 Introduction

Food additives are organic substances used directly or indirectly in the production, processing, packaging, transportation or storage of food. They are used as: (i) preservatives which slow the spoilage of food and maintain quality of food, (ii) nutrients to enhance the nutritional quality of food, and (iii) agents for improving the taste and appearance of food. The additives which are added to food with a specific purpose are direct additives and the additives to which food is exposed during processing, packaging and storing are indirect additives [1]. The food additives can be mainly divided into six major categories (figure 15.1) according to their function, such as preservatives, nutrient supplement, flavoring agent, colouring agent, texturing agent, and miscellaneous [2]. The different food additives perform various functions. Food preservatives are used to prevent food spoilage from molds, fungi and bacteria and maintain freshness. They are usually found in fruit juices, baked food items, cereals, fruit and vegetables.

Nutritional supplements help to preserve the nutritional value and also help to fortify food with suitable vitamins and minerals. These types of supplements are

doi:10.1088/978-0-7503-5377-9ch15

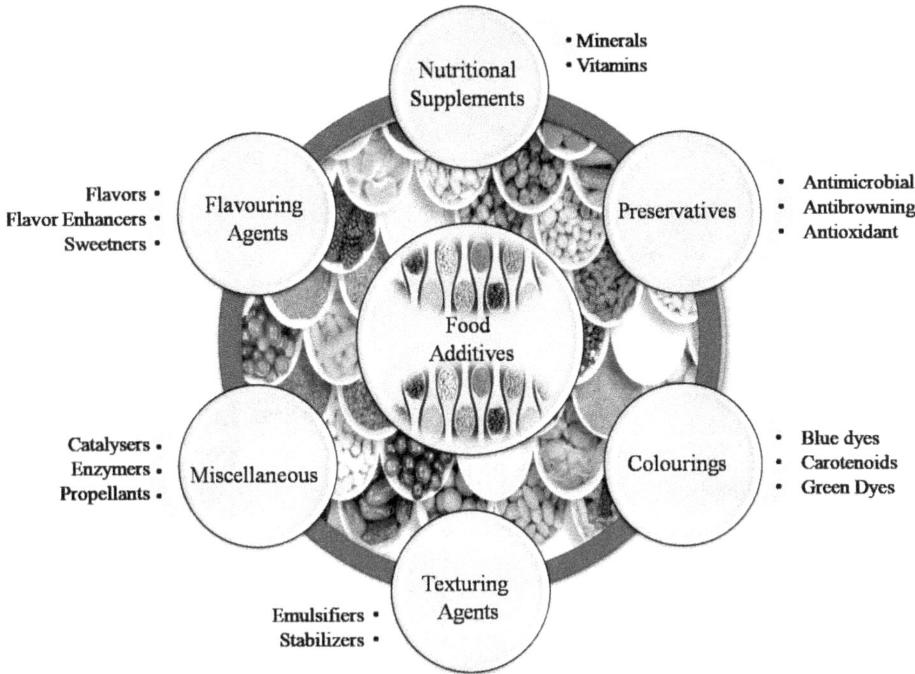

Figure 15.1. Six major types of food additives.

usually found in salt, milk, flour etc. The flavoring agents add specific flavors both natural and synthetic to the food. They are usually used in ice creams, pudding, soft drinks, sauce etc. To compensate the colour loss caused by prolonged exposure to temperature, moisture, and air, coloring additives are added to processed foods. They are also used to provide color to colorless and fun food. Texturing agents are used to provide uniform texture and also help to upgrade the taste of the food. The use of such types of agents is found in jam, jelly, cake, pudding and bread. Other food additives are also added to food to increase the viscosity of food, produce foaming effect and preventing the food from drying out [1]. Food additives improve the nutritional value, help it to stay fresh for a longer period of time and also make the seasonal foods available throughout the year. Along with these advantages, a few of them can have immediate or long-term toxic effects on human health. The immediate effect may include headache and immune response. However, long-term impacts may raise the risk of cardiovascular disease, cancer, and asthma. So, the Scientific Committee on Food does the safety assessment and recommends the food additives for use. For the purpose of safety assessment, it is important to thoroughly study the parameters such as toxicokinetics, subchronic toxicity, chronic toxicity, carcinogenicity, and reproductive and developmental toxicity [3]. As per the recommendation of Codex Alimentarius, an international numbering system is followed for food leveling. It consists of a three digit number. The food additives which were assessed for use in the European Union are assigned a number prefixed

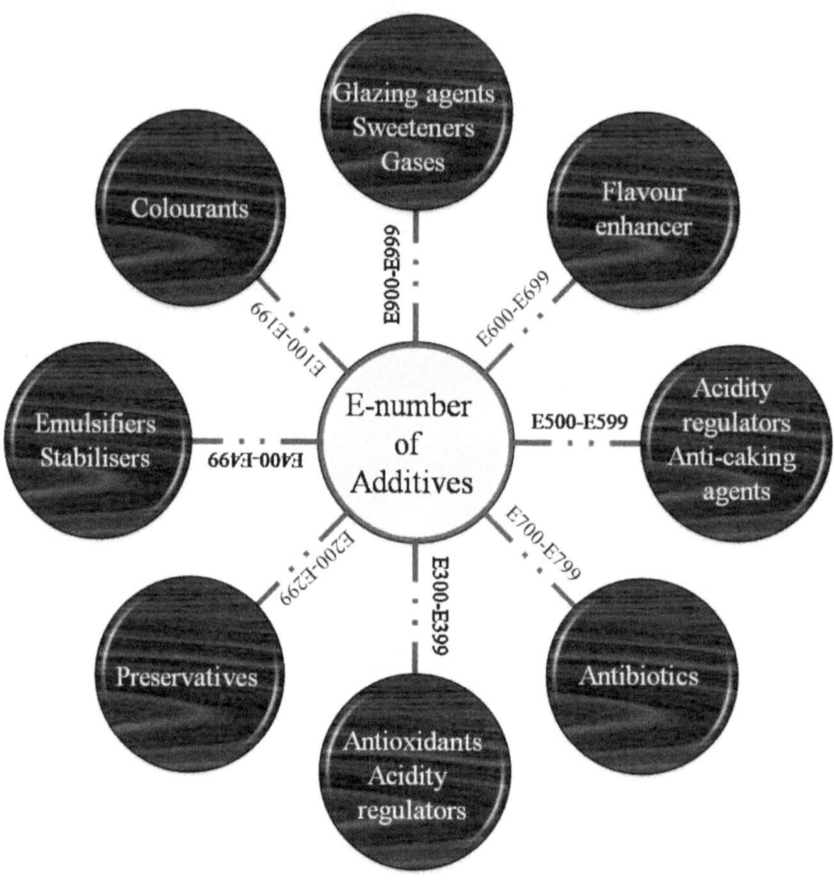

Figure 15.2. Range of E-numbers (unique identification numbers) assigned to different additives.

with 'E' called 'E-number'. Now this uniform numbering scheme is adopted by the Codex Alimentarius Commission for the identification of all additives internationally [5]. The additives with different functions are assigned with a different range of unique identification numbers represented in the figure 15.2 [4].

Each country has their own regulating bodies which determine the additives to be added to food and the safety limit for them. The substances are labeled as 'Generally Recognized as Safe' (GRAS), which includes the views of experts regarding the safety of substances that can be added directly or indirectly to food. The major concern to food safety arises due to the illegal use of food additives. So to improve the situations in food safety and standardize the market level some recommendations on food additives and proper measurement of food additives are suggested. The recommendations are: (i) all non essential food additives should be banned; (ii) food additives recognized as GRAS should be only used in food; (iii) additives with carcinogenic and mutagenic properties used should be labeled properly; (iv) the non-GRAS food items intake should be within acceptable daily intake limit; and (v) food producing and processing units should follow good manufacturing processes prescribed by the regulating bodies [5]. Proper estimation of food additives plays an important role in controlling its impact on human health. This chapter discusses various techniques used for estimation of food additives and flavors. The food additives are assessed in two ways i.e. component identification and content determination. Component identification helps with knowing any illegal use of additives, whereas content determination helps with knowing excessive use of additives. A suitable analytical method is to be used for inspection of food additives.

15.2 Food additives: analysis methods

Different detection methods, including chromatography, spectrophotometry, electrophoresis, optical, and electrochemical approaches, are used to identify food additives [6].

Chromatography is the most traditional method of detecting food additives. Among the types of chromatography, liquid, gas and mass/liquid chromatography techniques are the most commonly used. In chromatography the mixture is applied on a stationary phase and then a pure solvent is allowed to flow on the stationary phase, which carries the components of the mixture separately according to their solubility. In gas chromatography, the stationary state is an immobilized liquid or solid packed in a closed tube, whereas the mobile phase is an inert gas. This method is used to locate volatile compound mixtures or organic compounds that are thermally stable [7]. Here, under a regulated temperature gradient, the sample is converted into vapour and injected into the column, where the molecules are then separated based on the differences in their individual properties. From the column, the sample exits at a steady flow rate until it reaches the detector. The data system shows the detector signals or peaks as a function of time. It is possible to determine the concentration of various food components by analyzing the region beneath the peak [7]. In liquid chromatography, the stationary state is a solid, whereas the mobile phase is a liquid usually composed of organic solvents or buffer solution. The solution is pumped into the column at a constant flux and the different components are differentiated according to affinities between the stationary and mobile phases [8]. Again, based on the columns used, the chromatography techniques can be further classified into different types. In normal phase chromatography, polar substances are utilized as the stationary phase and non-polar or intermediate polar substances as the mobile phase. Similarly, ion exchange resins are utilized in ion-exchange chromatography, whereas polar solvent is employed as the mobile phase in reverse phase chromatography. Sometimes the coupling of high performance liquid chromatography (HPLC) with gas chromatography (GC) is used for determination of different food components. This method has the advantages of high sensitivity and reproducibility. But the limitations of this method are requirement of sample pretreatment, long test period and use of costly instruments.

Spectrophotometry is another method used for analysis of food additives. The spectroscopic techniques are usually classified into four types such as ultraviolet–visible (UV–vis) spectroscopy, infrared and Raman spectroscopy and fluorescence. In these methods the analysis is mainly done by interaction between electromagnetic radiation and the sample. The food additives having different functional groups, which are responsible for absorption and emission of electromagnetic radiation at different wavelengths, can be analyzed by using Lambert–Beer's law in UV–vis spectroscopy. In infrared spectroscopy infrared radiation is passed through the sample, and by observing the peaks at different frequencies information about molecular structure can be obtained. These frequencies corresponding to different normal modes of vibrations, give information about the molecular structure. These methods of food analysis are cost effective, nondestructive and environmental friendly but the major challenge in these methods is the calibration required for each of the products present in the food sample, which is very tedious [9].

In **Raman spectroscopy** the scattering of incident light takes place. Monochromatic light is incident on the molecules of a food sample. Raman spectra are obtained by inelastic collision between the molecules and the incident radiation. The intensity of the scattered light is measured at different frequencies than the incident light. This method can be used for analysis of both solid and liquid food products. This is a very quick method for analysis of food additives but the major limitation in the method is the low analytical sensitivity corresponding to weak Raman scattering [10].

In **fluorescence spectroscopy**, the molecules of the chemical compound present in the food stuff absorb light of specific wavelength and go to the excited energy state. When the molecule is in the lowest vibrational energy level of excited state, it returns to lower energy state by emitting a photon. These wavelengths of excitation and emission depend on the chemical structure and give rise to fluorescence spectra. The chemical compounds with rigid structures having rings can emit radiation of suitable frequency. So although fluorescence technique is more sensitive than UV–vis, very few compounds present in the food stuff can have rigid structure, which limits the application of this technique [11].

Electrophoresis is another method utilized for food additive analysis. In this method the electro-migration of the species occurs in a capillary tube under the action of an electric field. Here the separation of charged particles/ions occurs under the influence of an electric field. An electrophoretic system consists of two electrodes, connected by an electrolyte. Separation of ionic particles occurs due to the difference in their velocity which is related to the mobility of the particle. It is a versatile and efficient method used for the fast and efficient separation of food products. When this method is coupled with HPCL it gives suitable sensitivity and selectivity in food additive analysis [12].

Electrochemical techniques are powerful analytical tools used for measurement of the analytes present in food. Here the detection of analyte is mainly done by the measurement of electric current produced by the chemical reactions in the electrochemical system. An electrochemical sensor has mainly two parts. One is the chemical recognition system and the other is the transducer. Based on the electroanalytical technique the electrochemical sensors can be conductometric, voltammetric or potentiometric [13]. The techniques used in food analysis are polarographic and voltammetric. Currently, polarographic technique is not used in research labs because it requires the stirring of solution after each set of applied

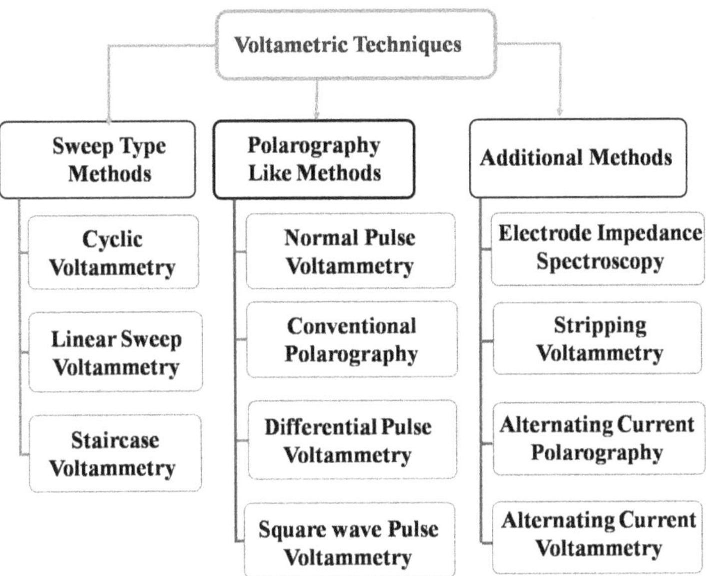

Figure 15.3. (A) list of voltammetric techniques and their classifications.

potential. So it is replaced by modern voltammetric method [14]. In this technique a potential difference is applied between the electrodes which gives rise to a flow of current, flowing through the electrochemical cell. It is a very sensitive tool in detection of a low concentration of organic/inorganic compounds found in food. The different types of voltammetric techniques available are presented in figure 15.3 [15].

Advantages of this method are: (i) it is faster, cheaper and easier to detect food additives in comparison to other methods [16]; (ii) in this method, the effect of interference is low in comparison to chromatographic techniques [17]; (iii) both electrochemically active and inactive compounds can be detected in various modes of voltammetric techniques [18]; and (iv) this technique can be coupled with other techniques easily [19].

15.3 Principles of voltammetric techniques

The voltammetric experiment is conducted in an electrochemical cell, usually consisting of a working electrode, a reference electrode and a counter electrode. The working electrode is one in which the reaction takes place. The potential of the reference electrode is kept constant in comparison to the working electrode. An inert conducting substance is used as counter electrode. An electrolyte is used in which the electrodes are submerged. The electrolyte is mainly responsible for maintenance of ionic strength by controlling the electro-migration effect. In voltammetric technique a potential difference is applied to an electrode which gives rise to a current in the electrochemical cell. So in this technique application of potential difference results in reduction or oxidation of the substance at the surface of the working electrode, giving rise to a current. The effect of applied potential on the rate of reaction and concentration of redox substance at electrode surface, is described by Nernst's equation [20]

$$E = E_0 - \frac{RT}{nF} \ln \frac{C_R^0}{C_O^0}$$

where E_0 is the standard reduction potential for the redox couple, R is molar gas constant, T is absolute temperature, n is number of electrons transferred and F is Farad constant. C_R^0 and C_O^0 are the concentration of the oxidised and reduced agent at the surface. Another relation which relates variation of current, potential and concentration is the Buter–Volmer equation given as

$$\frac{i}{nFA} = k^0 \left\{ C_O^0 \exp\left[-\alpha\theta\right] - C_R^0 \exp\left[(1 - \alpha)\theta\right] \right\}$$

where k^0 is heterogeneous rate constant, α is transfer coefficient, A is area of the electrode and $\theta = nF\frac{E - E_0}{kT}$. From this relation the value of rate constant can be calculated.

Dependence of current on the flux of the material to the surface of the electrode is given by Fick's law. According to this law, $\varphi = -AD_0\left(\frac{\partial C_0}{\partial x}\right)$, where φ is the flux of material and $\frac{\partial C_0}{\partial x}$ is concentration gradient. The faradic current flowing in the cell depends on the flux of the oxidized /reduced agent on the surface of the electrode. The current is a quantitative measure of how fast a material is oxidized or reduced at the electrode surface. This current is mainly due to diffusion. Many factors like concentration of redox species, shape, size and material of the electrode affect the diffusion current. For controlled flow of analyte to the electrode surface the migration of charged ions and convention must be eliminated [20]. In addition to the faradic current, another current called capacitive current also flows due to the charge accumulation on both sides of the metallic electrode and solution interface. The capacitive current flows when surface area changes with time, potential changes with time or both surface area and potential change with time. But in voltammetry, current is measured as a function of potential so rate of potential change is not zero. As faradic and capacitive currents flow together so the measurement gives the sum of both the currents. In the analytical signal, the faradic current is represented as peak and the capacitive current is represented as a sloping line, as shown in figure 15.4 [21].

Some noise also appears due to the fluctuation of electrode and cell. The ratio of faradic current to capacitive current decreases with decrease in concentration of analyte as faradic current is dependent on analyte concentration but capacitive current is independent of this. But if the faradic current is very small in comparison to capacitive current then it limits the determination of analyte. So improvement in ratio of faradic to capacitive current is done by considering a combination of multimode electrode with staircase voltammetry. Keeping the electrode surface area constant, the potential step technique is adopted to increase the faradic-to-capacitive

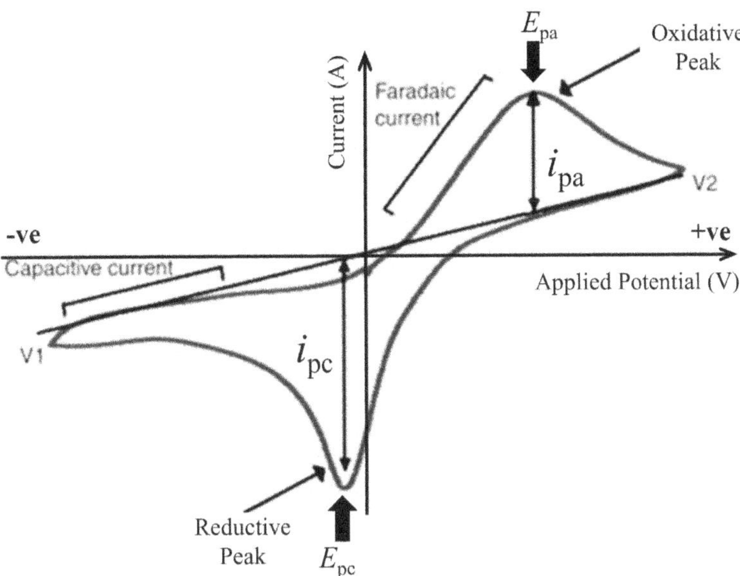

Figure 15.4. (A) schematic diagram of cyclic voltammogram. Reprinted from [21], copyright (2016) with permission from Elsevier.

current ratio. In this technique a suitable choice of current measurement period minimizes the capacitive current [20].

15.4 Types of voltammetric techniques

The various types of voltammetric measurement techniques are classified based on the way of variation of the applied potential. The most commonly used method is cyclic voltammetry (CV) in which the applied voltage on the working electrode is scanned at a constant rate and the faradic current is noted. The voltammogram gives the measurement of peak potential, peak current and half-wave potential. The peak potential gives the idea of electron exchange capability of a molecule [22] and peak current gives the idea of number of electrons exchanged by the antioxidant [23]. In a sample with a greater number of redox species, the integrated area under the oxidation peaks gives the estimation of total amount of exchanged electrons. The signal of the solvent is removed by background subtraction method. This method is adopted to determine the antioxidants in food [24–26]. To increase the sensitivity and speed of sensing, many forms of potential modulation are done, of which three are widely used:

(i) **Normal pulse voltammetry (NPV)**: In this method a series of potential pulses of increasing amplitude is applied. The current measurement is done at the end of each pulse. The voltammogram shows the current in vertical axis and the potential to which it is stepped along the horizontal axis.
(ii) **Differential pulse voltammetry (DPV)** is also used for characterizing the redox behavior of antioxidant. In this method the potential is scanned like

a staircase. The step height and width remain fixed. The measurement is done with an additional pulse [27]. The difference in the measured value of current before and after application of the pulse is taken to minimize the capacitive current. Peak current is used to estimate antioxidant capacity, whereas peak potential is used to identify type of antioxidant. The measurement of different antioxidant is reported by this method [28–30].

(iii) **Square wave voltammetry (SWV)** is another method used in food additive detection. Here the potential scanning is done like a staircase but it is modified by the superposition of a square-shaped pulse. Two potential pulses of equal height and opposite sign are obtained at each step. The current is measured at the end of the two pulses. It has very high sensitivity so it is used in food analysis [31–33].

(iv) For trace analysis a sensitive technique, **stripping voltammetry** is used, which includes two steps i.e. the analyte species is concentrated onto a working electrode, and the preconcentrated analyte is measured by potential scan. Basically, three types of stripping are done. Anodic stripping helps to determine traces of metals. Cathodic stripping helps to trace insoluble salt with mercurous material, and in adsorptive stripping preconcentration a step of analyte is done by adsorption [20].

15.5 Electrochemical sensing devices for estimation of food additives and flavours

Based on the principle of electrochemical sensing the transportable and economical analytic devices are developed. These sensors can be classified into two broad classes: (i) potentiometric, and (ii) voltammetric and amperometric, based on their working. In potentiometric sensors the potential changes according to concentration of analyte, but in voltammetric and amperometric sensors a current close to concentration of electroactive solution is produced due to oxidation–reduction response. The oxidation peaks and high overpotential of the food components is the key in design of sensitive electrochemical sensors (ECS). The sensing depends on the current, potential or charge interaction between analyte and sensing electrode, as shown in figure 15.5 [34].

The electrode of the ECS is made up of materials ranging from noble metals, carbon to conductive polymers. The electrode material plays a very significant role in determination of the sensitivity and selectivity. Noble metals like gold, silver and palladium find applications in ECS. The excellent biocompatibility, stability and conductivity of gold support its use as an electrode. Platinum electrodes are used because they show chemical stability in an oxidizing and reducing environment and also have high catalytic activity. But today the use of nanomaterials plays a crucial role in enhancement of sensitivity, selectivity and stability of the ECS. The modification or functionalization of the electrode surface with nanomaterials amplifies the output signal which is responsible for the increase in sensitivity. Various types of nanomaterials are used for this modification and range from metallic nanoparticles to metal oxide nanoparticles and carbon nanoparticles.

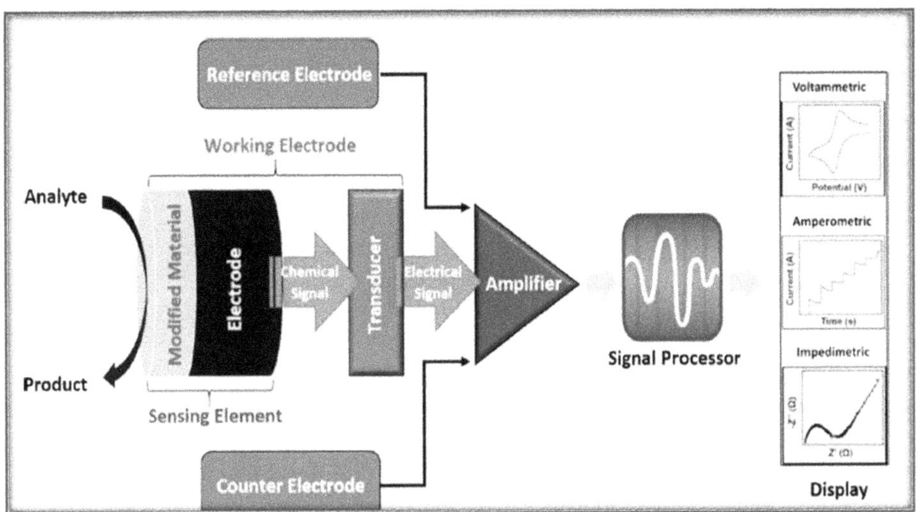

Figure 15.5. Different units of a conventional ECS. Reprinted from [34], copyright (2021), with permission from Elsevier.

Nanoparticles have high surface-to-volume ratio, which is responsible for their enhanced conductivity, tensile strength and chemical reactivity. So they are finding applications in electrochemical sensing. In recent years, gold nanoparticles (AuNPs) find lot of applications in biosensing because of their nontoxicity and inert core. The use of different types of nanoparticles for food analysis in different samples is presented below in table 15.1.

The advent of new technologies like thick and thin film technology, silicon-based technology and photolithographic technologies has played a significant role in development of new miniaturized ECS. For detection and analysis of trace metal elements in natural water stripping technology is used. In this technology the diffusion of heavy metals ions and small complexes occurs at the surface of the electrode, which acts as a barrier to unwanted macromolecules and colloids. So this process is very useful in tracing the heavy metals [47]. In this method the need of mercury electrode, removal of oxygen and forced convection is eliminated. Similarly, the biological sample identification can be done by the use of enzyme electrode [48]. The enzyme electrode can be made by immobilizing the enzyme in the carbon paste matrix. The changes occurring as a result of biocatalytic activity can be traced by the amperometric or potentimetric monitoring. This method is usually used for monitoring the phenolic and organophospate pesticides. Ion selective electrodes are developed which are capable of detecting nitrate pollutants. These electrodes are potentiometric sensors using a selective membrane to minimize matrix interference. The detection limits of these electrodes are very much dependent on the type of internal electrolyte. By choosing an electrolyte with low activity of primary ions and lowering its leakage, the detection limit of the sensor can be reduced significantly to pico range [49, 50]. Some of the modified electrodes, recently reported for food analysis, are presented in table 15.2.

Table 15.1. Different types of nanoparticles for food analysis.

Nanoparticle used	Food sample	Analyte	Limit of detection	References
$ZnO@MnO_2$-rGO nanocomposite		Hydroquinone (HQ), Mono-Tert-butyl hydro-quinone and catechol (CC)	0.0011 μM 0.0012 μM 0.001 μM	[35]
$BiVO_4$ modified g-C_3N_4	Food sample	Tetracycline	1.6 nM	[36]
Gold Platinum core–shell nanoparticle	Friut juice	Carbendazim	1.64 nmol l^{-1}	[37]
Gold nanoparticles and graphdiyne	Kiwi fruit and tomato	Methyl parathion	6.2 pg ml^{-1}	[38]
Gold-ceria nanocomposite	Potato	Chlorpyrifos	0.12 pM	[39]
Gold nanoparticle composite	Water	Paraquat	0.49 pM	[40]
Carbon paste electrode modified with synthesized chitosan nanoparticles and aluminum silicate	Guava leaf	Imidacloprid	1.5×10^{-9} mol l^{-1}	[41]
Cerium oxide catalyzed 1D carbon nanofibers	Real water	Pb (II) and Cu (II)	0.6 ppb 0.3 pb	[42]
Carbon nanotubes functionalized $CoMn_2O_4$	—	Pb^{2+}	0.004 μM.	[43]
Fe_3O_4 magnetic nanoparticles	Apple juice	Esteria Coli	3.5 CFU ml^{-1}	[44]
Silver nanoparticle	Dried *Phyllanthus acidus*	Hg+	8.3×10^{-7} mol l^{-1}	[45]
Silver nanoparticles	Bean, Grape, Apple	Diazinon	7 ng ml^{-1}	[46]

In the standard electrochemical cell, an electrolytic solution is required, which is not suitable for on-site application. So by using such types of devices the analyte can be detected in the laboratory. Now emphasis is given on design of portable ECS consisting of three screen-printed electrodes on an insulated substrate for on-the-spot evaluation of the analyte, as shown in figure 15.6 [80]. The portable ECS works based on (i) CV, (ii) SWV, (iii) DPV, (iv) linear sweep voltammetry, and (v) electrochemical impedance spectroscopy (EIS). The thickness of screen-printed electrodes generally varies from a few microns to 100 microns. This thickness can

Table 15.2. Modified electrode based electrochemical sensors for food additive analysis.

Electrode	Modifier	Sample	Analyte	Method of detection	Linear range	Limit of detection	References
Carboxylated multi-walled carbon nanotubes (MWCNTs) /AuNPs	Single stranded DNA	Raw milk	*Salmonella enterica*	DPV	728.42 μA cm^{-2} ng^{-1}	0.3 pg ml^{-1}	[51]
Graphene quantum dots (GQDs) and AuNPs composite on ITO	2-Aminothiophenol (ATP)	Spiked maize	Aflatoxin B1	CV	382 μA cm^{-2} ng^{-1}	0.008 ng ml^{-1}	[52]
Gum arabic corn flour(GACF)/ invertase-GO_x/	MWCNTs nano-biocomposite/Pt nanoparticles	Fruit juice and vegetable juice	Sucrose	CV	3.109 μA M^{-1}	1×10^{-9} mol l^{-1}	[53]
Mesoporous silica (SBA-15)	Acetylcholinesterase	Soft drinks	Monochrotophus Dimethoate	CV		2.5 ppb 1.5 ppb	[54]
Doped185 poly-(8-aniline-1-naphthalene sulphonic acid) film on Glassy Carbon Electrode	Gold nanoparticle	Yogurt, Roquefort cheese, red wine and beer	Tyrosinase	CV DPV	19 nA cm^{-2} μM^{-1}	0.7 μM	[55]
Polydopamine-gold nanoparticles (PDA-AuNPs)	Exonuclease I (Exo I)	Cauli-flower, cabbage	Malathion	CV EIS DPV	—	5×10^{-1} ng l^{-1}	[56]
Lactate oxidase /carbon (Nafion)	Pt and Pd nanoparticles	Wine	Lactate	CV	3.03 nA mM^{-1} cm^{-2}	0.1 μM	[57]
Molybdenum disulfide (MoS2) /graphene	Gold (Au) nanoparticle	Flour	Gliadin	DPV CV		7 pM	[58]

Glassy carbon	Polyaniline-Fe_3O_4-silver diethyldithiocarbamate (PANi-F-S)	Water	Mercury	CV	1618.86 μA μM^{-1} cm^{-2}	0.051 nM	[59]
Glassy carbon	Cu	Drinking water	Glyphosate	CV DPV	0.62 ± 0.02 μM	0.186 ± 0.004 μM	[60]
Glassy carbon electrodes modified with nano-composites consisting of poly(3,4-ethylenedioxythiophene) polystyrene sulfonate (PEDOT:PSS)	AuNP	Commercial fish and meat	Xanthine (XA).	CV		3×10^{-8}	[61]
Gold screen-printed electrode	Peptide sequences coupled to magnetic nanoparticles	Milk and water	*Staphylococcus aureus* (*S. aureus*),	CV		10^3 CFU ml^{-1}	[62]
Zeolitic imidazole framework-8 (ZIF-8)	$MoTe_2$ Nanoparticle	Spinach and potato	Omethoate (OMT)	CV	10^{-9} g l^{-1}	3.3×10^{-11} g l^{-1}	[63]
Screen-printed MWCNTs gold nanoparticle-based electrodes (Nano-Au/c-MWCNTs)	1-Ethyl-3-(3-dimethylaminopropyl carbodiimide) (EDC) and N-hydroxysuccinimide (NHS)	Fish	Xanthine oxidase (XO)	CV	2.388 μA cm^{-2} μM^{-1}	1.14 nM	[64]
Black phosphorus	Gold Nanoparticles (AuNP)	Apple Juice	Patulin (PAT) and ochratoxin A (OTA)	CV	—	—	[65]
Glassy carbon electrode	Zeolite imidazole skeleton-8 (ZIF-8) and para-sulfonylcalix [3] arene (pSC_4) coated gold nanoparticles		Paraquat (PQ)			0.49 pM	[40]

(*Continued*)

Table 15.2. (*Continued*)

Electrode	Modifier	Sample	Analyte	Method of detection	Linear range	Limit of detection	References
	composite (pSC_4-AuNPs@ZIF-8)						
polyethyleneimine (PEI) improved reduced graphene oxide (rGO)	Silver Nanoparticles (AgNP)	Meat and animal derived food	Arsanilic acid (ASA)	CV	100 ng ml^{-1}	0.38 ng ml^{-1}	[66]
Zn/Ni-ZIF-8–800 derived from bimetal-organic framework	Highly conductive graphene and AuNPs	Beansprouts	6-Benzyl aminopurine (6-BA)	CV EIS	—	0.24 ng ml^{-1}	[67]
Carbon paste electrode	polyarginine	Tap water and candy/chocolate	indigo carmine	CV DPV	0.2–1	25.3 nM	[68]
Carbon nanotube paste electrode	Octyl phenol ethoxylate	Water and blood serum	Phloroglucinol	CV	10–90 μM	0.71 μM	[69]
Graphene paste electrode	Poly adenine	Water	Catechol	CV	2×10^{-6} –1.5×10^{-4} M	0.24 μM	[70]
Carbon nanotube paste electrode	electropolymerisation of alizarin carmine (AC) and immobilization of sodium lauryl sulfate (SLS)		Paracetamol	CV	4.0–100 μM	0.06 μM	[71]
Carbon nanotube paste electrode	Polyglycine	Sewage water	Alizarin Carmine	ESI		0.981 μM	[72]
Carbon nanotube paste electrode	Poly glutamine film-coating		Curcumin	CV DPV	0.4–10 μM	2.79×10^{-8} M	[73]
Graphite paste electrode	Poly niacin	B complex capsule	Riboflavin	LSV	5.0–65.0 μM	0.782 μM	[74]
Graphene paste electrode	Poly tyrosine	Tap water	Catechol	DPV	2–10 μM	0.30 μM	[75]

			Phloroglucinol				
Carbon nanotube paste electrode	Sodium lauryl sulfate and Triton *X*-100 composite surfactant	Cow milk	L-tyrosine	CV	15–160 μM	0.375 μM	[76]
Carbon nanotube electrode	Poly L-methionine	Pharma ceutical formulation	Voltaren (VTN)	CV DPV	2–50 μM	0.10 μM	[77]
Graphene paste electrode	Polyglycine	Ground water Blood sample	Hg (II) Pb (II)	CV		6.6 μM 0.8 μM	[78]
Graphene paste electrode	Sodium dodecyl sulfate	Water	Catechol (CC) and resorcinol (RS)	CV DPV	2–10 μM	106 nM	[79]

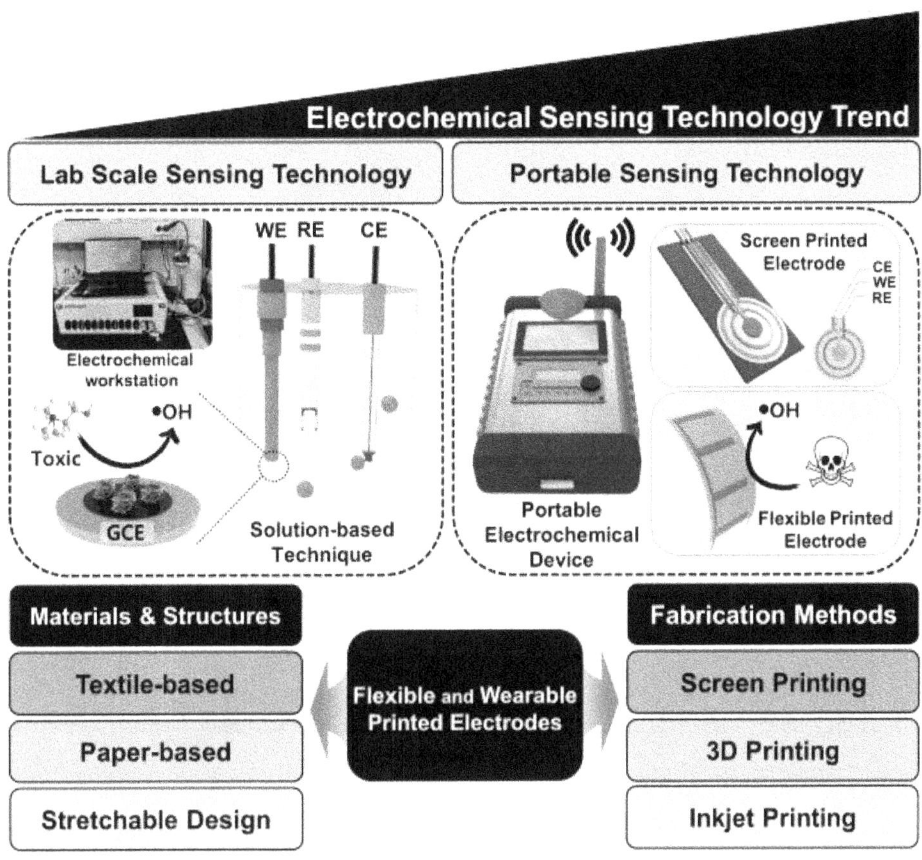

Figure 15.6. Advantages of electrochemical sensing technology and its increasing trend. Reprinted from [80], copyright (2021), with permission from Elsevier.

be varied by controlling the template and amount of printed ink. Screen-printed electrodes are usually made up of commercially available conductive inks which may contain many nonconductive substances. This gives rise to poor conductivity and hence poor electrochemical sensing. To improve the sensing property, the working of ECS can be modulated by use of various nanomaterials.

The used nanomaterials can be metal, carbon or metal oxide-based nanomaterials. Incorporation of these nanomaterials improves the performance of screen-printed electrodes. In ECS, nanoparticles with high conductivity, large surface area and high stability have gained much attention. The incorporation of nanomaterials can be done in two ways: (i) direct addition of them with the printed ink, or (ii) by surface modification of them on the electrode. The different substrates used for screen printing of electrodes are given below in table 15.3.

Screen-printed electrodes can be deposited on the suitable substrate by following various deposition techniques. Many common techniques that are followed for deposition of such electrodes are screen printing, sputtering, thermal evaporation

Table 15.3. Substrates used for screen-printed electrodes.

Material used for fabrication of screen-printed electrode	Characteristic	Fabrication technique	Cost	References
Silicon	• Rigid • Surface modification can be done easily • High temperature resistant • Biocompatible	• Chemical vapor deposition • Photolithography • Dry and wet etching	\$2 kg^{-1}	[81, 82]
Ceramics	• Good thermal, mechanical and electrical properties • Rigid	• Mechanical milling • Laser ablation	\$1–13 kg^{-1}	[83, 84]
Glass	• Transparent • Surface modification can be done easily	• Photolithography • Etching (both dry and Wet) • Laser ablation	\$2–4 kg^{-1}	[81, 82, 85]
Polymethyl methacrylate (PMMA)	• Low water absorption • Transparent • Rigid	• Injection molding • Hot embossing	\$20–25 kg^{-1}	[85, 86]
Cyclic olefin copolymer (COC)	• Rigid • Transparent • Low water absorption	• Injection molding • Hot embossing	<\$ 3 kg^{-1}	[87, 88]
Polycarbonate (PC)	• Transparent • Rigid • Heat resistant	• Injection molding • Hot embossing • Laser ablation	<\$ 3 kg^{-1}	[84, 86]
Whatman chromatography paper	• Homogeneous • Biocompatible • Reproducible	• Screen printing • Inkjet printing • Wax printing • Lithography	\$6 m^{-2}	[90]

and electrodeposition [89]. Of these, the thin film deposition techniques are more sophisticated in comparison to other techniques. Not only the deposition of electrodes, but its integration with the microfluidic state plays a significant role in deciding the selectivity and stability of the electrochemical cell. Along with this the nanomaterial plays a significant role in enhancement of the stability and selectivity of the cell. Usually the nanoparticles can be integrated into the electrode surface by various methods like drop-casting and dip-coating, inkjet printing, field-assisted assay, pressed transfer and screen printing [89].

15.6 Conclusion and future scope

In this chapter it is discussed that the proper synthesis of nanomaterials and functionalization of the electrodes with them by different methods can give rise to more sensitive and stable electrochemical analysis. But a few factors must be addressed in the future:

- Accurate validation studies must be conducted multiple times to increase the number of commercially available sensors for analysis of food additives.
- The impact of environmental constraints and complexity of food samples must be studied properly.
- Stability under actual operating conditions must be studied.
- Study of toxicity and degradation of nanomaterials used must be studied properly.
- Proper integration of research area and information and communications technology must be done for development of portable and sensitive devices for quality control in food processing units.
- The proper technology should be used so that sensitivity of the sensor does not reduce too soon.

References

[1] Awuchi C G, Twinomuhwezi H, Igwe V S and Amagwula I O 2020 Food additives and food preservatives for domestic and industrial food applications *J. Anim. Health* **2** 1–16

[2] Güngörmüş C and Kılıç A 2012 The safety assessment of food additives by reproductive and developmental toxicity studies *Food Additive* ed Y El-Samragy (Intech Open) pp 31–48

[3] Amchova P, Kotolova H and Ruda-Kucerova J 2015 Health safety issues of synthetic food colorants *Regul. Toxicol. Pharm.* **73** 914–22

[4] Wu L, Zhang C, Long Y, Chen Q, Zhang W and Liu G 2022 Food additives: from functions to analytical methods *Crit. Rev. Food Sci. Nutr.* **62** 8497–517

[5] Inetianbor J E, Yakubu J M and Ezeonu S C 2015 Effects of food additives and preservatives on man-a review *Asian J. Sci. Technol.* **6** 1118–35

[6] Yamjala K, Nainar M S and Ramisetti N R 2016 Methods for the analysis of azo dyes employed in food industry–a review *Food Chem.* **192** 813–24

[7] Horwitz W 2010 *Official Methods of Analysis of AOAC International. Volume I, Agricultural Chemicals, Contaminants, Drugs* ed W Horwitz (Gaithersburg, MD: AOAC International) p 1997

[8] Wrolstad R E, Acree T E, Decker E A, Penner M H, Reid D S, Schwarz S J and Sporns P 2005 *Handbook of Food Analytical Chemistry* (Hoboken, NJ: Wiley-Interscience)

[9] Skoog D A, West D M, Holler F J and Crouch S R 2013 *Fundamentals of Analytical Chemistry* (Cengage Learning)

[10] Bumbrah G S and Sharma R M 2016 Raman spectroscopy–basic principle, instrumentation and selected applications for the characterization of drugs of abuse *Egypt. J. Forensic Sci.* **6** 209–15

[11] Albani J R 2006 Fluorescence spectroscopy in food analysis *Encyclopedia of Analytical Chemistry: Applications, Theory and Instrumentation* (Wiley)

[12] Lee J, Song Y S, Sim H J and Kim B 2016 Isotope dilution-liquid chromatography/mass spectrometric method for the determination of riboflavin content in multivitamin tablets and infant formula *J. Food Compos. Anal.* **50** 49–54

[13] Murthy H A, Wagassa A N, Ravikumar C R and Nagaswarupa H P 2022 Functionalized metal and metal oxide nanomaterial-based electrochemical sensors *Functionalized Nanomaterial-Based Electrochemical Sensors* (Woodhead Publishing) pp 369–92

[14] Bard A J and Faulkner L R 2001 *Electrochemical Methods: Fundamentals and Applications* 2nd edn (New York: Wiley)

[15] Parhi S, Dash N, Praharaj S and Rout D 2022 An overview of voltammetric techniques to the present era *Electrochemical Sensors Based on Carbon Composite Materials: Fabrication, Properties and Applications* ed J G Manjunatha (IOP Publishing)

[16] Codognoto L, Zuin V G, De Souza D, Yariwake J H, Machado S A S and Avaca L A 2004 Electroanalytical and chromatographic determination of pentachlorophenol and related molecules in a contaminated soil: a real case example *Microchem. J.* **77** 177–84

[17] Tsai Y C, Coles B A, Holt K, Foord J S, Marken F and Compton R G 2001 Microwave-enhanced anodic stripping detection of lead in a river sediment sample. A mercury-free procedure employing a boron-doped diamond electrode *Electroanalysis* **13** 831–5

[18] Xu Q, Xu C, Wang Q, Tanaka K, Toada H, Zhang W and Jin L 2003 Application of a single electrode, modified with polydiphenylamine and dodecyl sulfate, for the simultaneous amperometric determination of electro-inactive anions and cations in ion chromatography *J. Chromatogr.* A **997** 65–71

[19] Delgado-Zamarreño M M, Bustamante-Rangel M, Sánchez-Pérez A and Carabias-Martínez R 2004 Pressurized liquid extraction prior to liquid chromatography with electrochemical detection for the analysis of vitamin E isomers in seeds and nuts *J. Chromatogr.* A **1056** 249–52

[20] Kounaves S P 1997 Voltammetric techniques *Handbook of Instrumental Techniques for Analytical Chemistry* ed F Settle (Prentice Hall) pp 709–26

[21] Guy O J and Walker K A D 2016 Graphene functionalization for biosensor applications *Silicon Carbide Biotechnology* (Elsevier) pp 85–141

[22] Scholz F 2015 Voltammetric techniques of analysis: the essentials *ChemTexts* **1** 1–24

[23] Haque M A, Morozova K, Lawrence N, Ferrentino G and Scampicchio M 2021 Radical scavenging activity of antioxidants by cyclic voltammetry *Electroanalysis* **33** 23–8

[24] Masek A, Latos-Brozio M, Kałużna-Czaplińska J, Rosiak A and Chrzescijanska E 2020 Antioxidant properties of green coffee extract *Forests* **11** 557

[25] Piljac-Žegarac J, Valek L, Martinez S and Belščak A 2009 Fluctuations in the phenolic content and antioxidant capacity of dark fruit juices in refrigerated storage *Food Chem.* **113** 394–400

[26] Samoticha J, Jara-Palacios M J, Hernández-Hierro J M, Heredia F J and Wojdyło A 2018 Phenolic compounds and antioxidant activity of twelve grape cultivars measured by chemical and electrochemical methods *Eur. Food Res. Technol.* **244** 1933–43

[27] Hoyos-Arbeláez J, Vázquez M and Contreras-Calderón J 2017 Electrochemical methods as a tool for determining the antioxidant capacity of food and beverages: a review *Food Chem.* **221** 1371–81
[28] Hoyos-Arbeláez J, Blandón-Naranjo L, Vázquez M and Contreras-Calderón J 2018 Antioxidant capacity of mango fruit (*Mangifera indica*). An electrochemical study as an approach to the spectrophotometric methods *Food Chem.* **266** 435–40
[29] Leite K C D S, Garcia L F, Lobón G S, Thomaz D, Moreno E K G, Carvalho M F D and Gil E D S 2018 Antioxidant activity evaluation of dried herbal extracts: an electroanalytical approach *Rev. Bras. Farmacogn.* **28** 325–32
[30] Korotkova E, Karbainov Y A and Shevchuk A 2002 Study of antioxidant properties by voltammetry *J. Electroanal. Chem.* **518** 56–60
[31] Novak I, Šeruga M and Komorsky-Lovrić Š 2009 Electrochemical characterization of epigallocatechin gallate using square-wave voltammetry *Electroanalysis* **21** 1019–25
[32] Medeiros R A, Rocha-Filho R C and Fatibello-Filho O 2010 Simultaneous voltammetric determination of phenolic antioxidants in food using a boron-doped diamond electrode *Food Chem.* **123** 886–91
[33] de Macêdo I Y L, Garcia L F, Neto J R O, de Siqueira Leite K C, Ferreira V S, Ghedini P C and de Souza Gil E 2017 Electroanalytical tools for antioxidant evaluation of red fruits dry extracts *Food Chem.* **217** 326–31
[34] Amali R K A, Lim H N, Ibrahim I, Huang N M, Zainal Z and Ahmad S A A 2021 Significance of nanomaterials in electrochemical sensors for nitrate detection: a review *Trends Environ. Anal. Chem.* **31** e00135
[35] Movahed V, Arshadi L, Ghanavati M, Nejad E M, Mohagheghzadeh Z and Rezaei M 2023 Simultaneous electrochemical detection of antioxidants Hydroquinone, Mono-Tert-butyl hydroquinone and catechol in food and polymer samples using ZnO@ MnO_2-rGO nanocomposite as sensing layer *Food Chem.* **403** 134286
[36] Zhao Z, Wu Z, Lin X, Han F, Liang Z, Huang L and Niu L 2023 A label-free PEC aptasensor platform based on g-C_3N_4/$BiVO_4$ heterojunction for tetracycline detection in food analysis *Food Chem.* **402** 134258
[37] Li W, Wang P, Chu B, Chen X, Peng Z, Chu J and Wu D 2023 A highly-sensitive sensor based on carbon nanohorns@ reduced graphene oxide coated by gold platinum core–shell nanoparticles for electrochemical detection of carbendazim in fruit and vegetable juice *Food Chem.* **402** 134197
[38] Xia Z, Zhou Y, Gong Y, Mao P, Zhang N, Yuan C and Xue W 2022 AuNPs and graphdiyne nanocomposite as robust electrocatalyst for methyl parathion detection in real samples *Anal. Sci.* **38** 1513–22
[39] Lakshmi G B V S, Poddar M, Dhiman T K, Singh A K and Solanki P R 2022 Gold-Ceria nanocomposite based highly sensitive and selective aptasensing platform for the detection of the Chlorpyrifos in Solanum tuberosum *Colloids Surf.,* A **653** 129819
[40] Niu Z, Liu Y, Li X, Yan K and Chen H 2022 Electrochemical sensor for ultrasensitive detection of paraquat based on metal–organic frameworks and para-sulfonatocalix [4] arene-AuNPs composite *Chemosphere* **307** 135570
[41] Elbaz G A, Zaazaa H E, Monir H H, Abd El Halim L M and Atty S A 2022 Nano eco-friendly voltammetric determination of pesticide, imidacloprid and its residues in thyme and guava leaves *Sustain. Chem. Pharm.* **29** 100799

[42] Singh S, Pankaj A, Mishra S, Tewari K and Singh S P 2019 Cerium oxide-catalyzed chemical vapor deposition grown carbon nanofibers for electrochemical detection of Pb (II) and Cu (II) *J. Environ. Chem. Eng.* **7** 103250

[43] Bashir N, Akhtar M, Nawaz H Z R, Warsi M F, Shakir I, Agboola P O and Zulfiqar S 2020 A high performance electrochemical sensor for Pb^{2+} ions based on carbon nanotubes functionalized $CoMn_2O_4$ nanocomposite *ChemistrySelect* **5** 7909–18

[44] Wilson D, Materon E M, Ibanez-Redin G, Faria R C, Correa D S and Oliveira O N 2019 Electrical detection of pathogenic bacteria in food samples using information visualization methods with a sensor based on magnetic nanoparticles functionalized with antimicrobial peptides *Talanta* **194** 611–8

[45] Sk I, Khan M A, Ghosh S, Roy D, Pal S, Homechuadhuri S and Alam M A 2019 A reversible biocompatible silver nanoconstracts for selective sensing of mercury ions combined with antimicrobial activity studies *Nano-Struct. Nano-Objects* **17** 185–93

[46] Shrivas K, Sahu S, Sahu B, Kurrey R, Patle T K, Kant T and Ghosh K K 2019 Silver nanoparticles for selective detection of phosphorus pesticide containing π-conjugated pyrimidine nitrogen and sulfur moieties through non-covalent interactions *J. Mol. Liq.* **275** 297–303

[47] Belmont-Hébert C, Tercier M L, Buffle J, Fiaccabrino G C, De Rooij N F and Koudelka-Hep M 1998 Gel-integrated microelectrode arrays for direct voltammetric measurements of heavy metals in natural waters and other complex media *Anal. Chem.* **70** 2949–56

[48] Kulys J J 1989 Amperometric enzyme electrodes in analytical chemistry *Fresenius' Zeitschrift für analytische Chemie* **335** 86–91

[49] Püntener M, Vigassy T, Baier E, Ceresa A and Pretsch E 2004 Improving the lower detection limit of potentiometric sensors by covalently binding the ionophore to a polymer backbone *Anal. Chim. Acta* **503** 187–94

[50] Ceresa A, Bakker E, Hattendorf B, Günther D and Pretsch E 2001 Potentiometric polymeric membrane electrodes for measurement of environmental samples at trace levels: new requirements for selectivities and measuring protocols, and comparison with ICPMS *Anal. Chem.* **73** 343–51

[51] Saini K, Kaushal A, Gupta S and Kumar D 2019 Rapid detection of Salmonella enterica in raw milk samples using Stn gene-based biosensor *3 Biotech.* **9** 1–9

[52] Bhardwaj H, Pandey M K and Sumana G 2019 Electrochemical Aflatoxin B1 immunosensor based on the use of graphene quantum dots and gold nanoparticles *Microchim. Acta* **186** 1–12

[53] Bagal-Kestwal D R and Chiang B H 2019 Platinum nanoparticle-carbon nanotubes dispersed in gum Arabic-corn flour composite-enzymes for an electrochemical sucrose sensing in commercial juice *Ionics* **25** 5551–64

[54] Palanivelu J and Chidambaram R 2019 Acetylcholinesterase with mesoporous silica: Covalent immobilization, physiochemical characterization, and its application in food for pesticide detection *J. Cell. Biochem.* **120** 10777–86

[55] da Silva W, Ghica M E, Ajayi R F, Iwuoha E and Brett C M 2019 Tyrosinase based amperometric biosensor for determination of tyramine in fermented food and beverages with gold nanoparticle doped poly (8-anilino-1-naphthalene sulphonic acid) modified electrode *Food Chem.* **282** 18–26

[56] Xu G, Hou J, Zhao Y, Bao J, Yang M, Fa H and Hou C 2019 Dual-signal aptamer sensor based on polydopamine-gold nanoparticles and exonuclease I for ultrasensitive malathion detection *Sens. Actuators* B **287** 428–36

[57] Shkotova L, Bohush A, Voloshina I, Smutok O and Dzyadevych S 2019 Amperometric biosensor modified with platinum and palladium nanoparticles for detection of lactate concentrations in wine *SN Appl. Sci.* **1** 1–8

[58] Ramalingam S, Elsayed A and Singh A 2020 An electrochemical microfluidic biochip for the detection of gliadin using MoS_2/graphene/gold nanocomposite *Microchim. Acta* **187** 1–11

[59] Hashemi S A, Mousavi S M, Bahrani S, Ramakrishna S and Hashemi S H 2020 Picomolar-level detection of mercury within non-biological/biological aqueous media using ultra-sensitive polyaniline-Fe_3O_4-silver diethyldithiocarbamate nanostructure *Anal. Bioanal. Chem.* **412** 5353–65

[60] del Carmen Aguirre M, Urreta S E and Gomez C G 2019 A Cu^{2+}–Cu/glassy carbon system for glyphosate determination *Sens. Actuators* B **284** 675–83

[61] Khan M Z H, Ahommed M S and Daizy M 2020 Detection of xanthine in food samples with an electrochemical biosensor based on PEDOT: PSS and functionalized gold nanoparticles *RSC Adv.* **10** 36147–54

[62] Eissa S and Zourob M 2020 A dual electrochemical/colorimetric magnetic nanoparticle/peptide-based platform for the detection of *Staphylococcus aureus Analyst* **145** 4606–14

[63] Ding L, Hong H, Xiao L, Hu Q, Zuo Y, Hao N and Wang K 2021 Nanoparticles-doped induced defective ZIF-8 as the novel cathodic luminophore for fabricating high-performance electrochemiluminescence aptasensor for detection of omethoate *Biosens. Bioelectron.* **192** 113492

[64] Sharma N K, Kaushal A, Thakur S, Thakur N, Kumar D and Bhalla T C 2021 Nanohybrid electrochemical enzyme sensor for xanthine determination in fish samples *3 Biotech.* **11** 1–7

[65] Zhao H, Qiao X, Zhang X, Niu C, Yue T and Sheng Q 2021 Simultaneous electrochemical aptasensing of patulin and ochratoxin A in apple juice based on gold nanoparticles decorated black phosphorus nanomaterial *Anal. Bioanal. Chem.* **413** 3131–40

[66] Wang Y, Ma D, Zhang G, Wang X, Zhou J, Chen Y and Wang A 2021 An electrochemical immunosensor based on SPA and rGO-PEI-Ag-Nf for the detection of arsanilic acid *Molecules* **27** 172

[67] Liu Q, Xing Y, Pang X, Zhan K, Sun Y, Wang N and Hu X 2023 Electrochemical immunosensor based on MOF for rapid detection of 6-benzyladenine in bean sprouts *J. Food Compos. Anal.* **115** 105003

[68] Edwin D S S, Manjunatha J G, Raril C, Girish T, Ravishankar D K and Arpitha H J 2021 Electrochemical analysis of indigo carmine using polyarginine modified carbon paste electrode *J. Electrochem. Sci. Eng.* **11** 87–96

[69] Tigari G and Manjunatha J G 2020 Optimized voltammetric experiment for the determination of phloroglucinol at surfactant modified carbon nanotube paste electrode *Instrum. Exp. Tech.* **63** 750–7

[70] Manjunatha J G 2020 Poly (adenine) modified graphene-based voltammetric sensor for the electrochemical determination of catechol, hydroquinone and resorcinol *Open Chem. Eng. J.* **14** 52–62

[71] Manjunatha Charithra M and Manjunatha J G 2020 Electrochemical sensing of paracetamol using electropolymerised and sodium lauryl sulfate modified carbon nanotube paste electrode *ChemistrySelect* **5** 9323–9

[72] Pushpanjali P A and Manjunatha J G 2020 Development of polymer modified electrochemical sensor for the determination of alizarin carmine in the presence of tartrazine *Electroanalysis* **32** 2474–80

[73] Tigari G and Manjunatha J G 2020 Poly (glutamine) film-coated carbon nanotube paste electrode for the determination of curcumin with vanillin: an electroanalytical approach *Monatsh. Chem.* **151** 1681–8

[74] Manjunatha J G, Raril C, Hareesha N, Charithra M M, Pushpanjali P A, Tigari G and Gowda J 2020 Electrochemical fabrication of poly (niacin) modified graphite paste electrode and its application for the detection of riboflavin *Open Chem. Eng. J.* **14** 90–8
[75] Manjunatha J G 2020 A promising enhanced polymer modified voltammetric sensor for the quantification of catechol and phloroglucinol *Anal. Bioanal. Electrochem.* **12** 893–903
[76] Prinith N S, Manjunatha J G and Hareesha N 2021 Electrochemical validation of L-tyrosine with dopamine using composite surfactant modified carbon nanotube electrode *J. Iran. Chem. Soc.* **18** 3493–503
[77] Pushpanjali P A, Manjunatha J G and Srinivas M T 2020 Highly sensitive platform utilizing poly (l-methionine) layered carbon nanotube paste sensor for the determination of voltaren *Flat Chem.* **24** 100207
[78] Raril C and Manjunatha J G 2020 Fabrication of novel polymer-modified graphene-based electrochemical sensor for the determination of mercury and lead ions in water and biological samples *J. Anal. Sci. Technol.* **11** 1–10
[79] Manjunatha J G 2020 A surfactant enhanced graphene paste electrode as an effective electrochemical sensor for the sensitive and simultaneous determination of catechol and resorcinol *Chem. Data Collect.* **25** 100331
[80] Umapathi R, Ghoreishian S M, Sonwal S, Rani G M and Huh Y S 2022 Portable electrochemical sensing methodologies for on-site detection of pesticide residues in fruits and vegetables *Coord. Chem. Rev.* **453** 214305
[81] Ren K, Zhou J and Wu H 2013 Materials for microfluidic chip fabrication *Acc. Chem. Res.* **46** 2396–406
[82] Iliescu C, Taylor H, Avram M, Miao J and Franssila S 2012 A practical guide for the fabrication of microfluidic devices using glass and silicon *Biomicrofluidics* **6** 016505
[83] Fakunle E S and Fritsch I 2010 Low-temperature co-fired ceramic microchannels with individually addressable screen-printed gold electrodes on four walls for self-contained electrochemical immunoassays *Anal. Bioanal. Chem.* **398** 2605–15
[84] Henry C, Lunte S and Santiago J 1999 Ceramic microchips for capillary electrophoresis–electrochemistry *Anal. Commun.* **36** 305–7
[85] Rodrigues R O, Lima R, Gomes H T and Silva A M 2015 Polymer microfluidic devices: an overview of fabrication methods *U. Porto J. Eng.* **1** 67–79
[86] Tsao C W 2016 Polymer microfluidics: Simple, low-cost fabrication process bridging academic lab research to commercialized production *Micromachines* **7** 225
[87] Khanarian G 2001 Optical properties of cyclic olefin copolymers *Opt. Eng.* **40** 1024–9
[88] Li S, Xu Z, Mazzeo A, Burns D J, Fu G, Dirckx M and Chun J H 2008 Review of production of microfluidic devices: material, manufacturing and metrology *MEMS, MOEMS, and Micromachining III* **vol 6993** (SPIE) pp 123–34
[89] Xu Y, Liu M, Kong N and Liu J 2016 Lab-on-paper micro-and nano-analytical devices: fabrication, modification, detection and emerging applications *Microchim. Acta* **183** 1521–42
[90] Wongkaew N, Simsek M, Griesche C and Baeumner A J 2018 Functional nanomaterials and nanostructures enhancing electrochemical biosensors and lab-on-a-chip performances: recent progress, applications, and future perspective *Chem. Rev.* **119** 120–94

IOP Publishing

Chapter 16

Recent advances in electrochemical sensing devices for pharmaceutical and biomedical diagnosis

İpek Kucuk, Selenay Sadak, Özge Selcuk, Cem Erkmen and Bengi Uslu

The most recent advancements in electrochemical sensors created to track drugs and biomarkers used in medication and disease detection are summarized in this chapter. New sensors continue to be developed using various chemical or biological detecting materials. Additionally, mass production technology, which is excellent for the microelectronics industry, enables the development of exceedingly compact, replicable, and inexpensive (disposable) sensor devices. These devices are paired with compact, user-friendly instrumentation powered by microprocessors. Pollution control will surely benefit greatly from 'smart' sensors and other advancements in selective and stable identification elements, such as distant electrodes, molecular devices, multi-parameter sensor arrays, micromachining, and nanotechnology. A development in biological research is the use of electrochemical sensors. Future industries will also have benefits in terms of sensitivity, selectivity, and processing speed. Electrochemical techniques are effective tools for assessing a wide range of targets because they are quick, precise, and non-destructive. The sensitivity has been improved using functional aptamers and nanomaterials, such as carbon nanotubes, graphene, graphene derivatives, and metal nanoparticles. During electrochemical measurement, the target's contact with a certain probe or molecule results in a discernible readout signal.

16.1 Introduction

Therapeutic drug and biomarker monitoring for the diagnosis of diseases has changed how it is used in clinics over time. Early on, measuring and reporting medication concentrations was the only goal. The target concentrations linked to a

doi:10.1088/978-0-7503-5377-9ch16

clinical response were then used to analyze these concentrations. This encouraged physicians and pharmacists to actively use the information to optimize medicine dosage to initiate and sustain a clinical response in the future, in addition to allowing passive estimation of a patient's future reaction [1–3].

Therapeutic drug and biomarker monitoring performed on-site has the potential to enhance patient outcomes and dramatically save medical expenses. Although sensor-based approaches have been under consideration by the scientific community for about 20 years, clinical applications of therapeutic drug and biomarker monitoring have not yet been surpassed and supported, possibly because of the divide between the scientific and clinical communities. Chromatography and spectroscopy as conventional procedures are constrained by a lack of uniformity, expensive equipment, lengthy sample preparation, and lengthy turnaround times. These issues are resolved by the sensors, which also provide a low-cost, user-friendly, *in situ* analytical method to fully realize the potential of therapeutic drug monitoring [3].

Digital health technology is quickly developing, and wearable electrochemical sensors that can allow non-invasive monitoring of chemical indicators are one example. There is a lot of interest in extending this type of sensor technology into other important fields because of recent developments in wearable continuous drug and biomarker monitoring devices [4]. It examines wearable electrochemical sensors to track medicinal medications. This rapidly developing category of drug-sensing wearables responds to the rising desire for individualized medicine for better therapeutic outcomes while minimizing drug side effects and related medical costs. By using patient-specific dosing and monitoring dynamic changes in pharmacokinetic behavior as patients adhere to medicine, doctors and patients can use continuous, non-invasive monitoring of therapeutic medications in bodily fluids to correlate pharmacokinetic features with optimal results. Additionally, law enforcement can use wearable electrochemical drug monitoring devices as effective screening instruments to stop drug trafficking and help on-site forensic investigations. This includes a range of wearable form factors designed for on-site drug misuse screening as well as non-invasive monitoring of therapeutic medicines in various body fluids. With the ultimate goals of supplying exact real-time drug monitoring protocols and autonomous closed-loop platforms for precise dose management and best treatment outcomes, the prospects of such wearable drug monitoring devices are discussed. Finally, gaps and unmet difficulties are highlighted to spur further technology advancements in individualized medicine. Future research and commercialization efforts are projected to be vigorous due to the speed at which advances are now occurring and the large market prospects for such wearable medication monitoring platforms. Current gaps and unmet needs are highlighted to spur technological advancements in individualized therapy in the future. Future research and commercialization efforts are projected to be vigorous due to the speed at which advances are now occurring and the large market prospects for such wearable medication monitoring platforms [5–9].

For the sensitive and targeted detection of novel biomarkers, there has been a significant deal of interest in the development of high-performance electrochemical sensors and biosensor platforms. The electrochemical sensor is appealing to potential multidisciplinary research due to its affordability and ease of downsizing, quick and online monitoring, simultaneous detection capability, etc. It offers several beneficial characteristics. Based on recent research findings, this chapter describes the developments and difficulties for the electrochemical detection of novel biomarkers in biofluids, pharmaceuticals, and biomedical devices. Transition metal nanocomposites' enzymes, antibodies, etc are described as electrode materials for the detection of specific biomarkers in real-world biofluids, early-stage monitoring, and disease-related biomarker identification. These potential electrochemical sensor platforms based on nanomaterials represent the strategy for next-generation sensing devices [10–12].

An electrochemical sensor can be defined as a tool that can detect and quantify important analytes such as organic, inorganic, and biological substances in the sample. Electrochemical sensors have received great attention due to their powerful analytical properties, which encompass the advantages of ease of use, medium cost, energy saving, and portability. They are also fast and performance-optimized analytical tools compared to traditional analytical instruments. An ideal electrochemical sensor should have analytical parameters such as high sensitivity, selectivity, and accuracy, stability [11, 13–15].

The most crucial part of a sensing device is the material used for the electrode because it greatly affects the sensitivity and limit of detections [16]. A wide range of nanomaterials such as metal nanoparticles, carbon nanomaterials, nanocomposites consisting of polymer materials, and metal oxides have been widely used in the design of electrochemical sensors in recent years. Nanomaterials possess large surface area, high surface reactivity, high catalytic efficiency, and strong adsorption capacity. The high surface area/volume ratio of nanomaterials provides an electrochemically active surface area. Nanomaterials have different and enhanced magnetic, optical, electrical, physical, biological, chemical, and mechanical properties compared to macro-sized materials [17].

Thanks to their specific properties, nanomaterials have become very valuable materials used in the structure of electrochemical sensors. The use of nanomaterials in the structure of sensors significantly increases analytical properties such as sensitivity and selectivity. The nanomaterials are used to contribute to improving the performance and efficiency of the sensor [18]. Sensitivity can be defined as the effect of changes in analyte concentration on the measurement results. In other words, the precision indicates the proximity of the measurement results to each other. Nanomaterials also increase the degree of selectivity of measurements. Selectivity is the feature of the developed sensor that shows the distinguishability of the analyte from other materials in the sample. Selectivity demonstrates the sensor's ability to identify and measure analytics. High precision and selectivity provide shorter analytical time and a lower detection limit for analytes [19].

One of the innovations in sensor design is biosensors, developed using biomolecules that have significant effects on living microorganisms. In recent years, advances in this area have attracted great interest from researchers. A biosensor consists of three main elements: a bioreceptor, a transducer, and a signal processing system. Bioelements can be biological molecules such as an enzyme, antibodies, proteins, or nucleic acid, as well as living biological systems such as a cell, tissue, and organism. The bioelement interacts with the analyte, forming a biological response. The resulting changes are converted into an electrical signal that can be measured and amplified by the transducer. The bioreceptor, which is the distinguishing feature of the biosensor, includes a recognition system for analytics. It is responsible for attaching the analyte to the sensor surface [20, 21]. Biosensors have gained great importance due to their advantages in terms of response speed, small sample and reagent consumption, and low energy requirements. On the other hand, biosensors have some limitations, such as low stability and poor repeatability. Biosensors are widely acknowledged as a sensitive detection tool for disease diagnosis as well as the detection of numerous species of pharmaceuticals [22, 23].

In this chapter, the applications of electrochemical sensors in pharmaceutical and biomedical diagnostics will be discussed. Since the use of electrochemical sensors in pharmaceutical and biological molecule analysis is known to be widespread, the advantages of their use, current applications, and future approaches will be discussed.

16.2 Electrochemical sensing devices for pharmaceuticals

Electrochemical methods allow the analysis of the oxidation-reduction properties of substances. All the changes in the pharmacokinetic structure of drug active substances are due to changes in the oxidation-reduction behavior of molecules. In this way, electrochemical techniques have a very important place in monitoring these changes. Electrochemical methods are powerful analytical techniques used in the analysis of many drugs' active ingredients and biological molecules. At the same time, electrochemical techniques are used for the determination of pharmaceutical compounds in dosage forms (tablets, capsules, injections) and biological samples (blood, serum, and urine samples with the addition of real and active ingredients). In recent years, the importance of adsorptive stripping, differential pulse voltammetry (DPV), and square-wave voltammetry (SWV) techniques, which are among the electrochemical techniques, is increasing. Adsorptive stripping methods make it possible to analyze the active substances of the drug at low concentrations [24, 25].

Today, there is a rapidly expanding societal need for point-of-care testing (POCT), or medical tests that may be carried out wherever the patient is, with the capacity to get results instantly and convey them to the clinician for clinical decision-making [4]. Major technological advancements, the internet, the widespread availability of mobile phones, the remarkable advances made in the world of sensors, and the miniaturization of devices make testing possible without the need for in-depth medical knowledge, and they quickly provide results and a doctor's response. Due to their low cost and ability to be adapted to compact hand-held or wearable devices, electrochemical sensors are crucial in this field. The detection of diseases at an early stage, which increases patients' chances of recovery and lowers treatment costs, is one of the toughest difficulties in modern medicine. To do this, it is necessary to discover a way to substitute time-consuming, tedious laboratory analyses that require expensive equipment and skilled specialists in favor of POC devices that execute analysis at or close to the patient's location without the need for sample transportation. The main difficulty these devices face is producing accurate results in only a few minutes without sample preparation. They must be easily transportable, compact, user-friendly, and capable of producing precise, accurate results that even inexperienced patients can use. They can enable remote or rapid clinical choices to enhance patient care, especially in underdeveloped nations without labs and other infrastructure [4, 12, 26–30].

Graphene oxide (GO) is a layer of graphite that contains in its structure various oxygen-containing functional groups, such as carbonyl, carboxyl, and hydroxyl. Reduced graphene oxide (rGO), a two-dimensional carbon material, can be obtained by electrochemical reduction of the oxygen content of GO under suitable conditions [31]. Cadmium sulfide (CdS), a nano-sized II–VI group semiconductor material, provides high sensitivity and selectivity to electrochemical sensors. Lotfi and co-workers developed an electrochemical sensor for the detection of acyclovir, an antiviral agent, using the glassy carbon electrode (GCE) modified with silver nanoparticles (AgNPs)/cadmium sulfide nanowires(CdS NWs)/rGO nanocomposite to prepare AgNPs/CdS NWs/rGO/GCE. In this study, the combination of RG, CdS NWs, and AgNPs produced a strong synergistic effect, which was reported to increase the catalytic ability of the nanomaterials. The presence of amine and hydroxyl groups as electron donors in the acyclovir structure enabled its interaction with cadmium ions and increased the accumulation of the drug on the modified electrode surface. The developed sensor was used for monitoring acyclovir by differential pulse adsorptive anodic stripping voltammetry method. The detection limit was calculated as 3.3 nM at a linear range of 10 nM–4 μM [32].

In recent years, metal nanoparticles have been widely used in the field of electrochemical sensors due to their low cost, high surface area, and chemical stability. However, the aggregation of nanoparticles causes various problems such as reduced conductivity and catalytic activity. Conducting polymers can be used to prevent the aggregation of nanoparticles. At the same time, their synergistic association with metal nanoparticles improves the catalytic properties of the sensor system. Gold nanoparticles (AuNPs) show effective catalytic activity for both oxidation and reduction reactions [33]. A modified carbon paste electrode (CPE) was fabricated by Kummari *et al* for the *in vitro* analysis of the common antiviral drug valacyclovir. Colloidal AuNPs suspension, conductive polymer poly-(3-amino-5-hydroxypyrazole (poly-AHP)), and CPE (AuNPs/poly-AHP/CPE) are used to create a flexible, reusable electrochemical biosensor. On the surface of the cCPE of Hepatit C virus (HCV), a thin poly-AHP film was created along with a uniform distribution of AuNPs, and electrochemical potentiodynamic polymerization of AHP was carried out. Scanning electron microscopy (SEM), energy dispersive x-ray spectroscopy (EDX), x-ray powder diffraction (XRD), and UV-visible spectroscopy were used to characterize the sensor surface. Utilizing cyclic voltammetry (CV), and electrochemical impedance spectroscopy (EIS) the electrochemical effectiveness of the drug for the surface topological/morphological characteristics of the current sensor system was examined. AuNPs/poly-AHP/CPE and DPV are acknowledged as effective sensor systems for the electrochemical determination of valacyclovir.

The anodic peak current showed a linear relationship to valacyclovir in the range of 5–80 nM by applying DPV under ideal conditions, with a detection limit as low as 1.9 nM and an analysis period of 15 s. In the presence of dopamine and serotonin, valacyclovir was specifically detected without interference. It has been demonstrated that the AuNPs/poly-AHP film sensor has good recovery limits and is useful for determining valaciclovir directly from synthetic urine, pharmaceutical formulations, and diluted human serum [34].

Multi-walled carbon nanotubes (MWCNTs) are carbon nanomaterials with various functional groups such as phenol, ester, ether, carboxylic acid, and carbonyl, which have high electrical conductivity, fast electron transfer ability, and significant chemical stability [35]. MWCNTs increase the electroactive surface area of the electrode by interacting with metal nanoparticles. For the first time, Ghanavati *et al* created an electrochemical sensor for the simultaneous, quick electrochemical detection of naproxen and sumatriptan that was decorated with MWCNTs, ZnO, NiO, and Fe_3O_4 nanoparticles on a GCE. The electrooxidation peak currents are significantly increased when the GCE surface is modified with ZnO/NiO/Fe_3O_4/MWCNTs nanocomposite, but the peak potential of naproxen and sumatriptan is shifted to a lower potential, meaning that the target drugs are electro-oxidized on GCE. This indicates that it is simpler than bare GCE on the modified electrode surface. Field emission scanning electron microscopy (FE-SEM), Fourier transform infrared spectroscopy (FT-IR), and XRD were used to study the nanocomposite. Numerous techniques, such as CV, SWV, and EIS, have been used to investigate the enhanced electrode. Outstanding parameters have been optimized, including supporting electrolytes, pH, scanning rate, deposition potential, and deposition duration. The linear detection ranges for naproxen and sumatriptan are 4–350 nM and 6–380 nM, respectively. For naproxen and sumatriptan, the lowest detection limits were determined to be 3 and 2 nM, respectively. The outcome of high-performance liquid chromatography and recovery shows that the method is appropriate for determining naproxen and sumatriptan simultaneously in biological fluids and medicinal substances. The analysis of naproxen and sumatriptan in human urine, serum, and drug samples has been done using the green technique. Naproxen and sumatriptan recovery rates ranged from 97.8% to 102.3% and 95.3%–103.85%, respectively [36].

Metal–organic frameworks (MOFs) are porous polymers that bind organic bridge lignans with metal ions [37]. Using Fe-based MOFs, stable electrochemical sensors with high redox activity can be developed. Due to their high porosity, large surface area, and advanced catalytic activity characteristics, MOFs increase the sensitivity of electrochemical sensing. At the same time, MOFs facilitate the recognition of guest materials thanks to their adjustable pore sizes and can interact privately. When MOFs are used in combination with other metal nanoparticles, their electronic conductivity increases. Platinum nanoparticles (PtNPs) are highly stable nanoparticles used with MOFs. Saeb and Asadpour–Zeynali designed a sensor modified with metal–organic framework and metal nanoparticles (Fe-MOF/PtNPs) for the ultra-trace detection of tinidazole. Tinidazole was initially detected with this sensor, which was made using a simple and direct synthesis technique. Fe-

MOF was first synthesized chemically, and then PtNPs were added to the resulting Fe-MOF /PtNPs/GCE sensor. Synthesized nanocomposite detected using transmission electron microscopy (TEM), Brunauer–Emmett–Teller (BET), Raman spectroscopy, FE-SEM, FT-IR, XRD, UV–vis spectroscopy, EDX, and element map and electrochemical techniques. DPV technique was used under optimum conditions to plot a calibration graph for the determination of a tinidazole reduction peak. The linear range of 0.0196–524.956 M and the detection limit for tinidazole reduction were computed. The findings demonstrate that the manufactured sensor has great stability, good repeatability, and reproducibility when determining tinidazole. The great stability of tinidazole determination in tablets and biological samples is one of this sensor's accomplishments. This specially designed sensor's key benefits include a simple synthesis process, a low detection limit, outstanding selectivity, exceptional stability, repeatability, and good repeatability [38].

Ratiometric sensors were developed as a solution to the problems of reproducibility and low detection robustness of electrochemical sensors. Ratiometric sensors measure the amount of analyte by the ratio of two signals, one from the analyte and one from an internal reference. Examples of internal reference signals are AgNPs. AgNPs enhance the electrocatalytic activity of the electrode. At the same time, the integration of AgNPs into molecularly imprinted polymer (MIP) networks improves electron transfer. MIPs are recognition elements developed to improve electrode sensitivity. Recently, MIPs have attracted attention in the development of electrochemical sensors due to their properties such as high sensitivity, high selectivity, and stability [39]. Perphenazine, a piperazine derivative, is often used as an antiemetic, anti-depressive, and antipsychotic drug in the treatment of schizophrenia psychoses. Liu *et al* designed an electrochemical sensor with a ratiometric approach to quantify perphenazine. Cu-coordinated MIP (Cu-MIP) is electrodeposited onto a flexible porous carbon fabric that has been treated with AgNPs to produce the sensor. Nitrogen and sulfur-doped porous carbon-on-carbon fabric (NSC/CC) was developed and employed as a flexible substrate. Pyrolysis and *in situ* chemical polymerization were used to synthesize NSC/CC. Ag/NSS/CC was made by soaking NSC/CC in 5 ml of $AgNO_3$ solution (2 M) for 1 h at room temperature. The combination was then added to 5 ml of ascorbic acid solution (2 M). As an internal reference, AgNPs generated on NSC/CC were used. Due to the improved adsorptive capacity and electrical characteristics of Cu^{2+} chelation, Cu-MIP exhibited a high level of electrochemical responsiveness; AgNPs can offer a reliable and efficient reference signal for ratiometric measurement. The resultant sensor had strong selectivity and sensitivity as well as acceptable repeatability and anti-interference capabilities. The peak current ratio of perphenazine and AgNPs was shown to be linear to the perphenazine concentration in the range of 1–700 nM ($R^2 = 0.9968$), with a detection limit of 0.43 nM, utilizing DPV to quantitatively determine perphenazine under optimal circumstances. Human serum and pharmaceutical samples were used to test the sensor's applicability, and recovery rates ranging from 92.46% to 104.90% were obtained [40].

Since reductants such as hydrazine and sodium borohydrate used in the preparation of rGO are expensive and highly toxic, researchers have turned to phytochemicals that are lower cost and environmentally friendly reductants. Citrus limetta is an important source of antioxidant phytochemicals such as flavonoids, carotenoids, hesperidin, terpineol, and alkaloids. Gijare *et al* used *citrus limetta* as a reducing agent for rGO synthesis in their study. The Soxhlet extraction method was used, which is a simple and effective method of collecting citrus peel extract. The rGO/TiO_2 composite is prepared using microwave processing. Glucose detection performance was evaluated with rGO/TiO_2/FTO modified electrode. As a result of the electrooxidation of glucose, D-gluconate is produced with H^+ ions (figure 16.1). Thanks to the developed stable and reproducible sensor, a minimum glucose detection limit of 0.32 M and a linearity range of 0.1–12 mM were found by CV technique. The sensor was successfully applied for glucose detection in blood serum [41].

The green synthesis method provides an environmentally friendly and economical approach compared to traditional methods. Plant extracts are the most used materials in the green synthesis of metal nanomaterials. Phytochemicals (alkaloids, flavanoids, terpenoids) found in various parts of plants such as fruits, seeds, and bark are used as reducing agents. Mathad *et al* obtained SNO_2 nanoparticles using cotton seed extract, an agricultural waste, as a reducing agent. rGO is more conductive and electrically stable than GO due to the lack of oxygen-containing functional groups. In this study, glycine, an environmentally friendly reagent, was used for the reduction of GO. The combination of SnO_2 and rGO provided

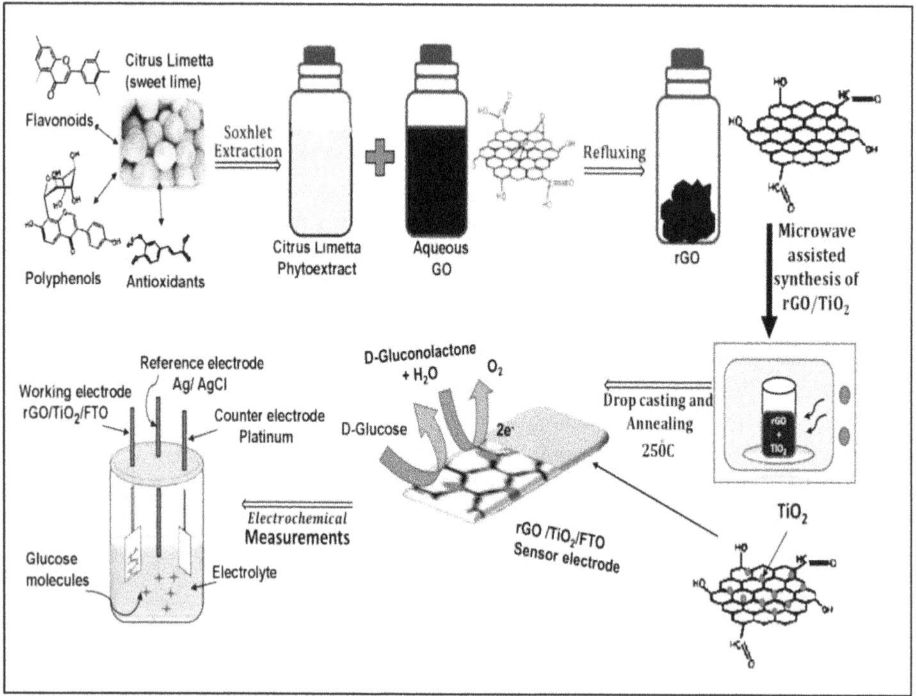

Figure 16.1. Mechanism of reduction of GO particles and rGO/TiO_2/FTO for glucose sensing application. Reprinted with permission from [41] CC BY 4.0.

properties such as high conductivity, biocompatibility, and a chemically stable oxidation state. SEM was used to understand the surface morphology and microstructures of SnO_2@-p-rGO nanocomposite. In figure 16.2(D), (a) rough, porous surface with dense and uniformly dispersed SnO_2 nanoparticles on the rGO structure is visible. Figure 16.2 shows that the stability and interfacial surface area of p-rGO layers is increased. The developed sensor was used for the electrochemical detection of repaglinide, an antidiabetic agent, and the concentration of repaglinide showed linearity in the range of 19.9 nM–14.5 μM. At the same time, the lower limit of detection (LOD) was determined as 0.85 nM by the SWV method. In addition, the sensor was applied to urine and pharmaceutical formulations for repaglinide quantification [42].

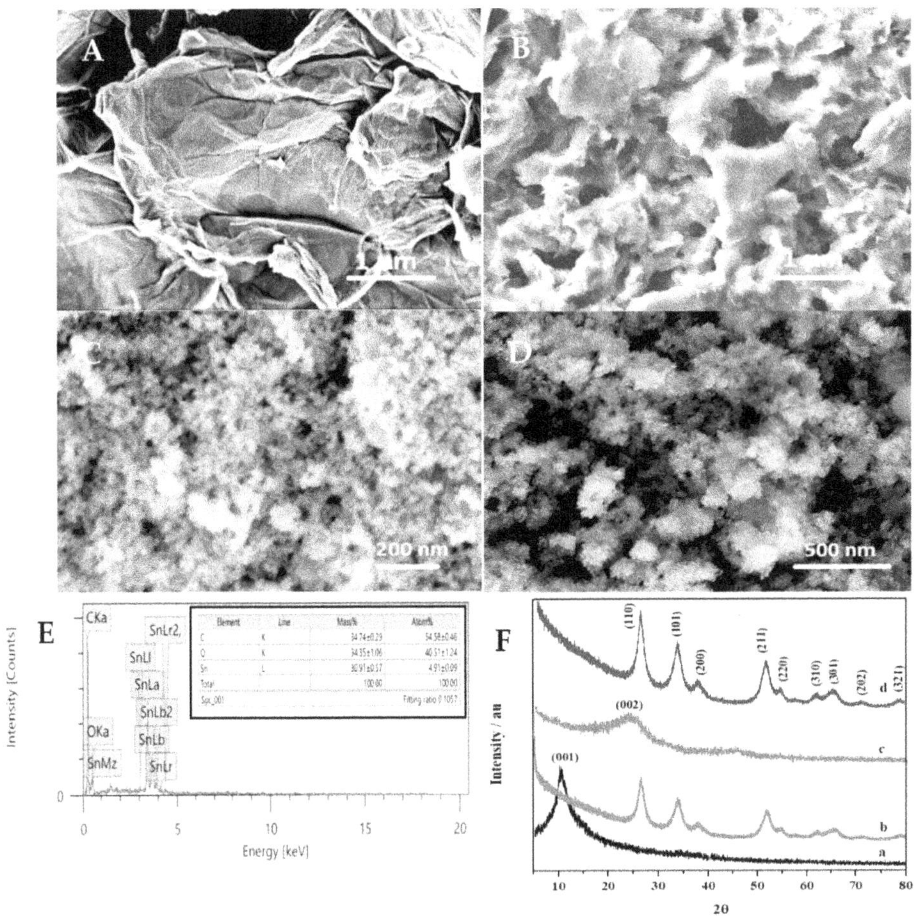

Figure 16.2. SEM images of (a) GO, (b) p-rGO, (c) SnO_2 (inset: particles size distribution histogram), and (d) SnO_2@p-rGO. (e) EDX micrograph of SnO_2@p-rGO; inset graph represents the elemental composition of SnO_2@p-rGO. (f) Powder XRD pattern of GO (a), SnO_2 (b), p-rGO (c), and SnO_2@p-rGO (d). Reprinted with permission from [42] CC BY 4.0.

The latest developments in electrochemical sensors in the field of pharmaceutical drugs are investigated. The selected studies were tabulated by evaluating the electrochemical method, electrode, application, calibration range, and detection limit parameters in table 16.1.

16.3 Electrochemical sensing devices for biomarkers

One of the most promising medical techniques for diagnosing disease is the use of biomarkers. For patients to receive appropriate care, the detection and quantitative analysis of illness biomarkers with high accuracy and sensitivity are crucial. A quantifiable indicator of any physiological state of an organism is the use of biomarkers [103, 104]. Disposable electrodes have been widely used in the scientific world due to their portable, affordable prices, and not requiring cleaning between measurements [105]. Despite their advantages, disposable electrodes generate large amounts of solid waste that is difficult to recycle. Therefore, non-toxic, environmentally friendly solutions and reagents should be used in their production. Considering all these, today, disposable electrode production and modification studies are carried out using environmentally friendly materials. In particular, the use of finger-printed electrodes is preferred in this field [106, 107].

In one study, Nascimento *et al* examined a disposable screen-printed carbon electrode modified with MBs-ACE2/SARS-CoV-2 Spike protein/ACE2-AuNPs bioconjugate. The electrode was developed for the determination of SARS-CoV-2 spike protein. The characterization of the developed electrode was made by using TEM, UV–vis, and EDX. According to the results, it was seen that AuNPs could be successfully used as a redox probe in the electrochemical analysis [108]. Thanks to the oxidation of Au present in AuNPs, a more intense peak current was achieved in the application of DPV compared with the bare electrode. The electrochemical analysis of the SARS-CoV-2 spike protein was made by using DPV. The wide linear range was found to be 0.0009–360.00 fg ml^{-1}, whereas the detection limit was calculated as 0.35 ag ml^{-1}. The results have shown the developed method exhibits a sensitivity of 100.0% and specificity of 93.7% towards SARS-CoV-2 spike protein. Its sensitivity and specificity enable the successful determination of the protein in human saliva samples [109].

Aptamers are single-stranded DNA or RNA molecules that are similar in structure to natural antibodies and are called artificial antibodies. They bind to antibiotics, proteins, and pathogenic bacteria like natural antibodies. Aptamers have some advantages over natural antibodies. Due to their unique structure, they recognize and capture target molecules, they are also easy to prepare, low cost, non-toxic, have a strong binding affinity, and have good thermal stability [110, 111]. In addition, aptasensors are highly susceptible to modification. Another new biosensor was developed for the determination of PfHRP2. Lo *et al* attached methylene blue and PfHRP2 aptamer on the surface of a gold electrode. The characterization of DNA aptamer sequences has shown that aptamer 2106s has a high affinity to bind PfHRP2 thanks to its abundancy, shortness, and interaction capability. The use of SWV method in the detection of PfHRP2 has revealed that

Table 16.1. Recent selected studies for drug detection on electrochemical analysis.

Analyte	Measurement principle	Surface layer content	Electrode	Dynamic range	LOD	Real	References
Perphenazine	DPV	Cu-MIP/Ag/NSC/CC	GCE	0.001–0.7 μM	0.000 43 μM	Human serum, tablet	[40]
Citalopram	DPV	MIP/hNiNS/ AMWCNT@GONRs	GCE	0.5–10 μM, 10–190 μM	0.042 μM	Human urine, human serum, tablet	[43]
Azithromycin	DPV	$Zn_3V_2O_8$/P-rGO	SPCE	0.099–450 μM	0.0067 μM	Human serum, human urine, wastewater	[44]
Tetracycline	DPV	Fe_3O_4@COFs@MIP	SPE	1×10^{-10}–1×10^{-4} M	5.1×10^{-11} M	Water, milk, chicken	[45]
Acetaminophen	DPV	MIP/MoS_2/CNTs	GCE	0.01–300 μM	0.003 μM	Human urine	[46]
Carbendazim	CV	MIP/C-ZIF67@Ni	GCE	4×10^{-13}–1×10^{-9} M	1.35×10^{-13} M	Soil, water	[47]
Milrinone	DPV	nD@Graphite/ZnO	CPE	0.01–7.88 mM	0.007 μM	Human plasma, human serum, urine	[48]
Gemcitabine	DPV	MIP/$CuCo_2O_4$/NCNTs/ FC	GCE	0.1–150 μM	11.3 nM	Ampoule, human serum, human urine	[49]
Methyldopa	DPV	Tb-FeO_3/CuO	CPE	0.04–300 μM	0.009 μM	—	[50]
Paracetamol	LSV	CZTS/MoS_2/CNT	CPE	0.1–1 μM	0.1 μM	Tablet	[51]
Phenobarbital	CV	MIP-PANI-GO	CPE	4.3×10^{-8}–4.3×10^{-7} M	3.3×10^{-8} M	Tablet, human serum, human urine	[52]
Norfloxacin	LSV	MIP/BPNS-AuNP	GCE	0.0001–10 μM	0.012 nM	Milk,tap water, soil	[53]
p-nitrophenol	CV	MIP	GCE	2–400 μM	0.2 μM	Wastewater	[54]
Atorvastatin	DPV	MIP	SPCE	0.05–2 μM	0.049 μM	Water	[55]

(Continued)

Table 16.1. (*Continued*)

Analyte	Measurement principle	Surface layer content	Electrode	Dynamic range	LOD	Real	References
Diclofenac	DPV	MIP	GCE	0.1–0.5 μM	77.5 nM	—	[56]
Tinidazole	DPV	Fe-MOF/PtNPs	GCE	0.02–525 μM	43 nM	Tablet, human plasma, human urine	[38]
Citalopram	DPV	ZIF-8/g-C_3N_4/RGO	CPE	0.009–900.0 μM	0.008μM	Human plasma, human urine	[57]
Selegiline				0.09–900.0 μM	0.014 μM	Tablet	
Nitrofurazone	CV/DPV	MIP-c-MWCNTs-ZIF	GCE	10^{-13}–10^{-6} M	6.7 × 10^{-14} M	Human urine, water	[58]
Norfloxacin	CV	BP-PEDOT:PSS/MIP	GCE	0.001–10 μM	0.25 μM	Milk	[59]
o-nitrophenol	SWV	CoC@ZIF-67@Mn	GCE	0.5–100 μM	0.16 μM	Water	[60]
Omeprazole	DPV	MIP/MWCNTs/Fe_3O_4-CuO-ZnO	CPE	5.0 nM–100 mM	1.5 nM	Water, tablet, human plasma, urine	[61]
Naproxen				5.0 nM–100 μM	1.0 nM		
Methocarbamol				1.0 nM–130 mM	0.7 nM		
Rivaroxaban	DPV	MIP/PVC	GCE	5.4 × 10^{-11}–3.1 × 10^{-3} M	1.002 × 10^{-6} M	Rivarospire® tablets	[62]
Doxorubicin	CV	CDs/CeO_2	SPCE	0.2–20 μM	0.09 μM	—	[63]
Trazodone	DPV	MIP	SPCE	5–80 μM	1.16 μM	Human serum, tap water	[64]
Balofloxacin	SWV	Bi_2O_3/ZnO	GCE	150–1000 nM	40.5 nM	Human serum, tablet	[65]
Silodosin	DPV	MIP/MWCNT	CPE	1 × 10^{-12}–1 × 10^{-3} M	1×10^{-13} M	Human serum, urine Flopadex®, Sympaprost®, eligodosin®	[66]

Dopamine	DPV	MIP	PGE	0.395–3.96 nM	0.193 nM	Human plasma	[67]
Acetaminophen	SWV	NC-fMWCNT-CuN	GCE	0.01–80 μM	0.000 14 μM	—	[68]
Diclofenac				0.05–80 μM	0.000 48 μM		
Chlorpromazine	DPV	DMMIP/Pt/Co_3O_4	GCE	0.005–9 μM	2.6 nM	Tablet, Human serum	[69]
Naproxen	SWV	ZnO/NiO/Fe_3O_4/ MWCNTs	GCE	0.006–18 μM, 18–380 μM	0.003 μM	Human serum, urine, tablet	[36]
Sumatriptan				0.004–17 μM,17–350 μM	0.002 μM		
Vardenafil	DPV	MIP/PtNPs	PtE	1×10^{-12}–5×10^{-8} M	0.2×10^{-12} M	Vardection® Human urine, serum	[70]
	EIS			1×10^{-12}–1×10^{-8} M	0.9×10^{-13} M		
Furazolidone	DPV	CuS-rGO/g-C_3N_4	GCE	0.1–336.4 μM	0.0108 μM	Lake water	[71]
Valacyclovir	DPV	AuNPs/poly-AHP	CPE	5–80 nM	1.19 nM	Tablet, artificial serum, urine	[34]
Levofloxacin	LSV	Co/Ni-MOF	GCE	0.1–500.0 μM	0.022 μM	Eye drop, milk	[72]
Acetaminophen	DPV	MIP/g-C_3N_4 QDs	GCE	1.0×10^{-11}–2.0×10^{-8} M	2.0×10^{-12} M	Paracetamol tablet	[73]
Diclofenac	DPV	MIP	SPCE	0.1–10 μM	70 nM	Tap, river water	[74]
Nevirapine	DPASV	MIP/ErGO	GCE	0.005–400 μM	2 nM	Human serum, tablet	[75]
Thioridazine	DPV	ZIF-67/Bio-MCM-41/ CQD	GCE	0.060–2.47 mM 2.47–69.76 mM	0.031 μM	Human serum, tablet	[76]
Pomalidomide	SWV	MIP/Fe_3O_4/MoS_2	GCE	1.0–20.0 nM	0.33 nM	Human plasma, capsule	[77]

(*Continued*)

Table 16.1. *(Continued)*

Analyte	Measurement principle	Surface layer content	Electrode	Dynamic range	LOD	Real	References
Domperidone	DPV	MIP@ERGO	GCE	0.5–17.2 μM	3.8 nM	—	[78]
Prochlorperazine	DPV	Mn_2O_3/SnO_2	SPCE	0.03–421.45 μM	8 nM	Human serum, urine	[79]
Oxaliplatin	DPV	MIP-Ag@Cu-BDC/N-CNTs	GCE	0.056–200 nM	0.016 nM	Human serum, urine	[80]
Nitrofurazon	DPV	NdO@TC	GCE	0.01–2231 μM	2.7 nM	—	[81]
Famotidine	DPV	Cu_2O/rGO	CPE	0.1–3.0 mM, 3.0–50.0 μM	0.08 μM	Human serum, urine, tablet	[82]
Norfloxacin	DPV	MIP/CoFe-MOFs/AuNPs	GCE	5×10^{-12}–6×10^{-9} M	1.31×10^{-13} M	Milk	[83]
Oxytetracycline	DPV	MOFs@MIPs	MGCE	1×10^{-9}–2×10^{-4} M	4.1×10^{-10} M	Milk	[84]
Atropine	DPV	MSN@ZIF-8	GCE	0.005–9.5 μM	0.98 nM	Human serum, urine, beef, eye drop, cereals	[85]
Linagliptin	DPV AMP	rGO/Poly(β-CD)/Magnetic ZIF-6	GCE	0.03–200 μM 0.02–300 μM	0.01 μM 0.006 μM	Human plasma, urine	[86]
Amoxicillin	CA	Cu/Zn-mMOF	GCE	1.0–205 mM	0.36 μM	Human serum, capsule	[87]
Molnupiravir	CV	rGO	CCE	0.009–4.57 μM	0.003 μM	—	[88]
Levofloxacin	CA	Cu-MOF	SPE	0.1–500	22 nM	Human urine	[89]
Prednisolone	DPV	4-VP/MIP/GO	GCE	1–120 μM	0.004 μM	Human plasma, tablets	[90]
Norfloxacin	DPV	Ni–Co-MOF flexible NF	GCE	0.1–1.5 μM, 2.5–69 μM	0.022 29 μM 0.0944 μM	Human serum, urine	[91]

Dopamine	SWV	CQDs/CuO	GCE	10–180 μM	25.40 μM	Ampoule	[92]
Sertraline	DPV	Cu-MOF/SNDGr	PGE	0.05–2.67 μM	0.038 μM	Human serum	[93]
Favipiravir	DPV	NiS_2NS/BC	GCE	0.42–1100 nM	0.13 nM	Human serum, tablet	[94]
Iodochlorohydroxyquin	DPV	MnO_2/MC	SPCE	0.01–1010 μM	0.001 μM	Human urine, tablet	[95]
Formononetin	DPV	MIP/NG	ITO-PET	0.81–32.19 μM	0.31 μM	Human serum	[96]
Indometacin	DPV	MIP-N-GOQDs-Mo_2C	GCE	10^{-15}–10^{-5}M	9.508×10^{-16} M	Water, tablet	[97]
Amoxicillin	SWV	GO/MIP(ANI)	GCE	1.0×10^{-3}–5.0×10^{-4} M	2.6×10^{-6} M	Human urine, human plasma	[98]
		GO/MIP(MOA)		5.0×10^{-6}–5.0×10^{-4} M	6.1×10^{-7} M		
Sulfamerazine	DPV	MIP/Ni_2P	GCE	0.05–20 μM	0.025 μM	Human urine	[99]
4-acetamidophenol					0.016 μM		
Naproxen	DPV	Poly(r-o-NBA)/GQDs	CPE	1.0–100 nM, 100 nM–2.5μM	0.672 μM	Tablet	[100]
Levodopa	DPV	CB-RuO_2-Nafion	GCE	1–8 μM	17 nM	Madopar® capsule	[101]
Acyclovir	DPV	AgNPs/CdS NWs/RG	GCE	0.01–4 μM, 4–40 μM	0.0033 μM	Tablet, topical cream, human serum	[32]
Enrofloxacin	DPV	AuNPs/CoNi-MOF	GCE	1×10^{-3}–1×10^{-5} M	3.3×10^{-4} M	Meat	[102]

there was a change in MB reduction peak related with the concentration of PfHRP2. The biosensor performed a linear range as 6.67 pM–4.03 nM with a detection limit of 3.73 nM. The sensitivity, selectivity and reliability of the developed sensor led it to be used in the determination of PfHRP2 in real sample analysis like human serum samples [112].

Carbon nanotubes are ultra-light conductive materials that are used as a high-conductivity material in the development of high-sensitivity sensitive sensors that allow biological materials to be modified to surfaces without losing their structure. Multi-walled carbon nanotubes (MWCNTs) and single-walled carbon nanotubes (SWCNTs) are the most used. Particularly, SWCNTs create a large active site for the analyte when modified to the surface of the electrode. In addition, carboxylation with other carbon-based nanomaterials is also carried out to improve the electrochemical properties of the sensor and increase conductivity. SWCNTs with high stability are widely used in electrochemical studies. Si *et al* used a gold electrode modified with MPA/BSA/anti-EGFR/DNA/SWCNT-Fc in their study for the determination of A549 exosomes which is a tumor marker. The developed sensor was characterized with TEM, EDS, and IR. The results have revealed that DNA/SWCNT-Fc composite increased the electron transfer between the electrode and analyte. Moreover, the carboxyl groups attached on the outer sides of the nanotubes led to the conjunction of Fc electroactive labels, thus it ended up with an increase in the signal amplitude. The determination of A549 exosomes was made using SWV. The linear range was calculated as 4.66×10^6–9.32×10^9 exosomes ml^{-1} with a detection limit of 9.38×10^4 exosomes ml^{-1}. Thus, the sensor has high sensitivity, selectivity, and stability. The developed sensor was successfully applied in the determination of A549 exosomes in human serum with an relative standard deviation value of 2.19% which shows the repeatability of the sensor [113].

Graphene quantum dots (GQDs) are nanomaterials that have attracted a lot of attention in recent years [114, 115]. GQDs with oxygen-rich functional groups have a large surface area and are preferred because of their properties such as high conductivity, stability, chemical stability, and good biocompatibility. The chemical properties of these compounds due to the functional groups and π–π bonds in their structures contribute to the widespread use of the compounds [116, 117]. Besides, MIPs are products of molecular imprinting technology that create specific binding sites for the recognition of the target molecule. MIPs are used in drug analysis, biomarker detection of various diseases, bionic sensing, and many other fields. The basis of MIP-based sensors is based on the monitoring of the charge transfer transmitted across the redox probe by the MIP thin film. MIPs have various advantages such as high selectivity, low preparation costs, and enabling rapid and portable sensor development, as well as various disadvantages such as low density of imprinted voids and low conductivity [39, 118–120]. These disadvantages can be eliminated with various modifications. In a study, Santos *et al* modified a screen-printed electrode using functionalized graphene (FG), GQDs and mag@MIP. The developed method was used for the determination of ethinylestradiol. The characterization of (mag@MIP)-GQDs-FG-NF SPE was made by using SEM, EDS and BET techniques and the results have shown that mag@MIP provided a greater surface area for the detection compared with mag@NIP. Thanks to the NF polymer, a better adhesion the (mag@MIP)-GQDs-FG film on the surface of the electrode was accomplished and this improved the physical and chemical stability of the electrode. CV method was used for examining the electrochemical behavior of the electrode. According to the results, an effective electron transfer between ethinylestradiol and the electrode was ensured by GQDs, and FG was used for the modification of the SPE. The analytical analyses of ethinylestradiol using SWV is in the linear range of 10 nM–2.5 μM with a detection limit of 2.6 μM. The ethinylestradiol in river water, serum, and urine samples was successfully determined with recovery between 96%–105% and 97%–104% were achieved in biological and environmental samples [121].

Among the newly modified materials, MOFs play an important role. Among the main reasons for this situation are their high porosity, large surface areas, adaptive active sites, and high conductivity. With many different interactions such as hydrogen bonds, electrostatic interactions, high amounts of analytes are deposited on the sensor surface thanks to MOFs. This situation causes the high preference of MOFs in the development of electrochemical sensors. There are various types of MOFs, but most of them require costly conditions such as high temperature, high pressure, organic solutions to manufacture. Despite all this, this situation can be solved with MOFs produced in mixture such as Ni-MOF, Ce-MOF, and various modifications [122, 123]. In a study by Zhang *et al*, attachment of DpAu, Ni-MOF and AChE on the surface of GCE was carried out for the determination of galantamine. CV and EIS techniques were used for the electrochemical characterization of the developed sensor. The deposition of AuNPs on the electrode surface accelerates the electron transfer. EIS behaviors of the modification on the electrode with D-Ni-BPY and AChE-GA have revealed the enhancement in the resistance due to relatively low electrochemical activity of MOF and AChE-GA. The results demonstrate the successful design of the AChE integrated Ni-MOF biosensor. The detection of galantamine was made by using DPV. A wide linear range of 1×10^{-12}–1×10^{-6} M was calculated with a low detection limit of 0.31 pM. Human serum applications were successfully achieved through utilizing the developed method with recovery from 97.00% to 105.22%. The biosensor has shown high selectivity and sensitivity in human serum sample applications [124].

Magnetic nanoparticles are thought to be the next generation of nanoparticles. Magnetic separation of biological entities has improved illness diagnostics [125]. In a study by Hanoglu *et al*, an SPCE electrode was constructed with using magnetic nanoparticles (MNPs), N-(3-dimethylaminopropyl)-N′-ethylcarbodiimide (EDC), N-hydroxysuccinimide ester (NHS), and 5-methylcytosine antibody (5-mC Ab) for the determination of methylated septin9 (mSEPT9). The surface characterization of the surface of the biosensor was made by performing CV and EIS measurements. As seen in figure 16.3(A), the significant decrease in the CV curves after the modification the electrode's surface with MNP and MNP/EDC/NHS has shown that the massive groups inhibited the electron transfer. Figure 16.3(B) has revealed that there is a growth in the resistance peaks after each hybridization step. The increase in the groups existing on the electrode surface prevented the interference of electrons,

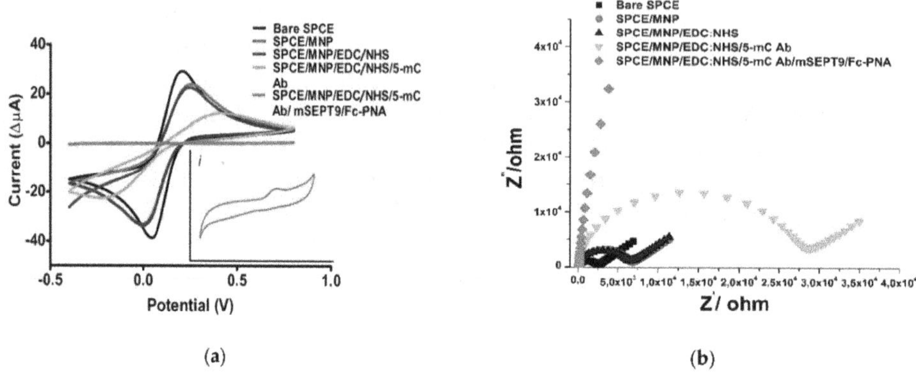

Figure 16.3. Electrochemical surface characterization results of the biosensor. (a) CV peaks and (b) EIS measurements. Reprinted with permission from [126] CC BY 4.0.

which caused an increase in resistance. This is predictable, and the consistency between the EIS results and the CV results indicates that the modification has taken place successfully. According to the DPV results, the detection limit was found to be as low as 0.37% and the concentration range as 25%–100%. The usefulness of the developed electrode makes it possible to detect mSEPT19 in human serums and plasma [126].

Mansuriya *et al* used GQDs and AuNPs for the modification of an screen-printed gold electrode (SPGE). The modified electrode was utilized for the determination of cTnI. The modified electrode was characterized by using SEM, TEM, XRD, and EDX. The biomarker was determined using SWV, CV, EIS and amperometry. For the first time, the detection of an analyte using four different electrochemical techniques was carried out with high sensitivity and rapidity with this immunosensor. Figure 16.4(A) shows that SWV peaks decrease with increasing cTnI concentrations on the anti-cTnI/AuNPs@GQDs/SPGE surface. cTnI showed linearity between 5 and 50 pg ml^{-1} (figure 16.4(B)). To ensure the reproducibility of cTnI, the CV technique has been used for the same concentration range and the results obtained are in harmony (figures 16.4(D) and (E)). The performance of the developed immunosensor, cTnI in human serum using SWV, CV and amperometry methods was also examined [127].

Genosensors enable quick measurement of the analyte using tiny sample amounts, are inexpensive, easy to handle, may be miniaturized, are reusable, and are stable in unfavorable environmental circumstances [128, 129]. Genosensors have a wide range of applications for diagnosing human papillomavirus (HPV) in the lab. The creation of reliable diagnostic tests for HPV infections is a top priority due to the low sensitivity of cytological studies and the high cost of molecular approaches. Approximately 93% of cervical malignancies can be avoided by early HPV infection detection and viral serotype identification in clinical specimens [130]. Using conductive polymer polypyrrole film and AuNPs, Avelino *et al* fabricated a flexible sensor for the electrochemical detection of HPV gene families in cervical samples. HPV gene families are recognized by oligonucleotide sequences. In this study, a

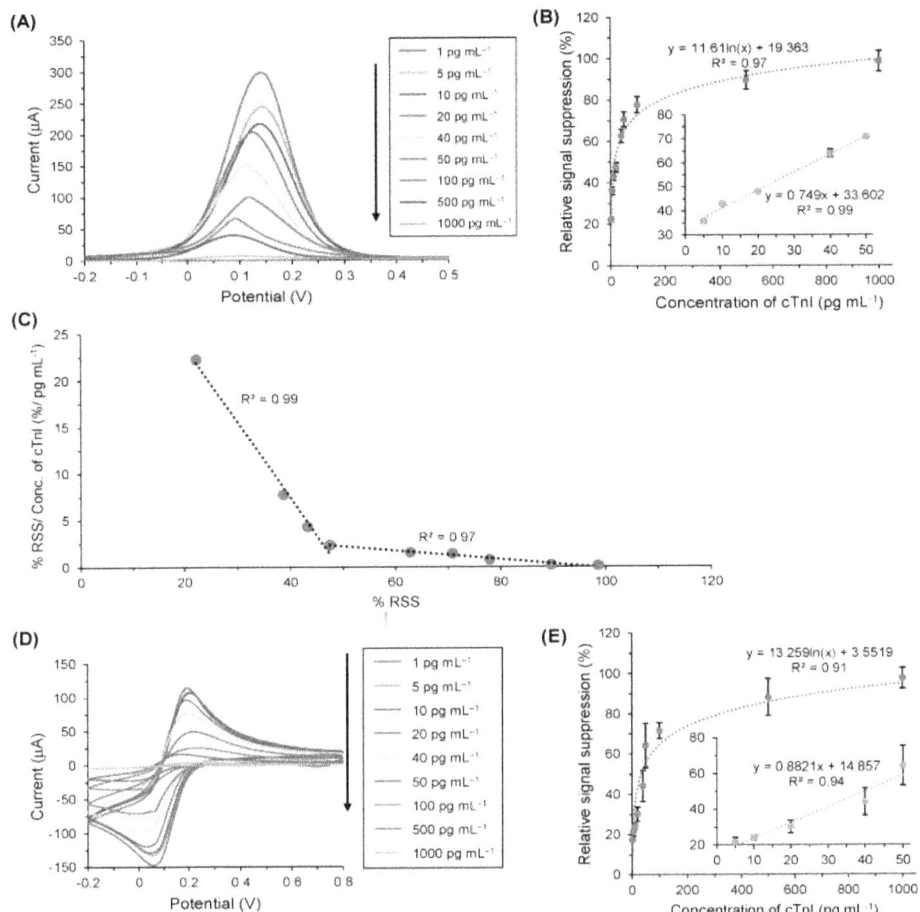

Figure 16.4. (A) Square-wave voltammograms of target binding with cTnI concentrations (1–1000 pg ml^{-1}). (B) Results of concentration-dependent cTnI bioassay using SWV. Inset: linear regression for cTnI (5–50 pg ml^{-1}). (C) Scatchard plot showing a bilinear regime with Kd values of 0.05 pM for concentrations below 40 pg ml^{-1} of cTnI and 0.89 pM for all other concentrations exceeding 40 pg ml^{-1} of cTnI. (D) Cyclic voltammograms of target binding with nine different cTnI concentrations (1–1000 pg ml^{-1}). (E) Results of concentration-dependent cTnI bioassay using CV. Reprinted with permission from [127] CC BY 4.0.

biosensor for the precise detection of HPV genotypes was created using polypyrrole (PPy) sheets and AuNPs. Flexible electrodes based on polyethylene terephthalate (PET) strips coated with indium tin oxide(ITO) were used to create the biosensor. AuNPs and polymeric films were produced using electrosynthesis. To precisely identify HPV gene families, oligonucleotides containing functional amino group modifications are used. On the nanostructured platform, the altered oligonucleotides were chemically immobilized. EIS and CV were employed to analyze electrode change and track molecular hybridization, respectively. After exposing biosensors to plasmid samples and cervical samples, electrochemical changes were seen. Biosensors were based on BSH16 probe, targeting HPV16 at 100 pg l^{-1}–1 fg l^{-1}.

After exposing biosensors to plasmid samples and cervical samples, electrochemical changes were seen. The target HPV16 gene detection biosensor was based on the BSH16 probe, 100 pg l^{-1}–1 fg l^{-1}. With a regression coefficient of 0.98, the LOD and limit of measurement (LOQ) were found to be 0.89 and 2.70 pg l^{-1}, respectively. Cervical tissues underwent screening procedures to assess the sensitivity and specificity of HPV and its viral family. Additionally, the p53 gene's expression, a biomarker for cancer, was observed. In this study, a modular method for label-free HPV family detection and p53 gene monitoring with excellent specificity, selectivity, and sensitivity was successfully created [131]. Biomarker applications carried out in recent years were investigated and selected studies shown in table 16.2 with the parameters of analyte, method, calibration interval and determination limit.

Selected research for recent years are summarized in table 16.2, with the name of biomarker, nanostructure, surface layer types, electrode and some analytical parameters such as LOD or LOQ.

16.4 Conclusion and future prospects

As technology and scientific measures have become increasingly significant in pharmaceutical and therapeutic applications, nanosensors have developed quickly in recent years. The use of nanomaterials in the creation of electrochemical nanosensors has come to light recently. Due to the improved chemical and physical characteristics brought about by discrete nanoelectrode devices or by altering the electrodes' surface with nanomaterials, these devices present a desirable option for improving current electroanalytical techniques in the pharmaceutical industry [11, 25, 180–182].

Research on electrochemical sensors is a promising field. It must be acknowledged that many of the issues in this area continue to revolve around choice. When direct detection is possible in samples, electrochemical sensors' quick analytical speed and ability to detect incredibly small amounts without significantly damaging the sample remain highly desirable properties [108].

Future biosensor developments might result in the development of a cell-friendly analyte for point-of-care and precise medical diagnostics. The introduction of electrochemical sensors has substantially benefited biological research. In addition, processing speed, selectivity, and sensitivity are all advantages that will support future businesses. As a result, sophisticated big systems for disease diagnosis and quality standards of stem cell-based goods can use quick, non-destructive, and customizable electrochemical sensors. The use of arrays to monitor a wide range of organic and inorganic contaminants, as well as the creation of diverse biorecognition materials, industrial advancements in microelectronics, and micro and disposable sensors, are recent innovations in this sector. For monitoring different contaminants, flow injection systems and online systems have also been developed. The characteristics of sensors have also been improved by recent developments in nanomaterials. Building multi-analyte detection systems with arrays of sensors can be helpful for managing pollution as well as for therapeutic and diagnostic monitoring [12].

Table 16.2. Recent selected studies for biomarker detection on electrochemical analysis.

Analyte	Measurement principle	Surface layer content	Electrode	Dynamic range	LOD	Real	References
Spermine	EIS	Gold coated	PCE	10^{-3}–10^{-11} M	10 pM	Synthetic urine, blood plasma samples	[132]
CA19-9, CA72-4	DPV	SNA I/MAL II	Gold electrode	0.005–200.0 U ml^{-1}	0.0032 U ml^{-1}	Tumor markers in human serum or plasma	[133]
LRG1 protein	SWV	BP3-1, BP3-8	Gold electrode	0.025–0.5 μg ml^{-1}	0.025 μg ml^{-1}	Human plasma	[134]
MCF-7	LSV	NGQDs/PHA-L	SPE	5–10^6 cells ml^{-1}	1 cells ml^{-1}	Human serum	[135]
SARS-CoV-2 Spike protein	DPV	MBs-ACE2/SARS-CoV-2 Spike protein/ACE2-AuNPs	SPCE	0.0009–360.00 fg ml^{-1}	0.35 ag ml^{-1}	Human saliva	[109]
Cholesterol	CV	NPG	SPE	50 μM–6 mM	8.36 μM	Human serum	[136]
Indoxyl sulphate	DPV	MIP	ITO	1.0 pM–6.0 mM	0.286 pg ml^{-1}	Human serum, human urine	[137]
5-hydroxyindole-3-acetic acid	DPV	MIP(Py)	GCE	5×10^{-11}–5×10^{-5} M	15×10^{-12} M	Human urine, plasma and serum	[138]
HCV-MPs	DPV	E2-MIP	SPE	0.1–50 ng ml^{-1}	4.6×10^{-4} ng ml^{-1}	Human plasma	[139]
SARS-CoV-2-RBD	EIS	MIP/MP-Au	SPE	2.0–40.0 pg ml^{-1}	0.7 pg ml^{-1}	Human saliva	[140]
MUC1	DPV	MnO@C@AuNPs	GCE	0.001–100 nM0.001–10 nM	0.31 pM0.25 pM	—	[141]
$VEGF_{165}$	DPV	Co-BTC-GO-MOFs	GCE	10^{-13}–10^{-6} M	5.23 pM	Human serum	[142]
Tryptophan	DPV	Cs/Ce-MOF	GCE	0.25–331 μM	0.14 μM	Human serum	[143]
Ethinylestradiol	SWV	(mag@MIP)-GQDs-FG-NF	SPE	10 nM–2.5 μM	2.6 μM	Human serum, human urine, river water	[121]
Riboflavin	DPV	MCOF-AuNPs	GCE	0.005–200 μM	2 nM	Human serum, urine, saliva	[144]

(*Continued*)

Table 16.2. (*Continued*)

Analyte	Measurement principle	Surface layer content	Electrode	Dynamic range	LOD	Real	References
Ascorbic acid	DPV	ZIF-8/PtNPs	GCE	10–2500 μM	5.2 μM	Human serum	[145]
Hydrazine	Amp	ZIF-67/Co Al-LDH	CCE	30–1200 μM	7.71 μM	Water	[146]
Tryptophan	DPV	Grain-CMOF/MWCNT	GCE	0.4–19 μM	0.11 μM	—	[147]
Rutin	AMP	Co-BPDC-MOF	GCE	0.5–1000 μM	0.03 μM	Bromezer® Tablet	[148]
Ephedrine	DPV	MIP/Nafion/MWCNTs	GCE	1.8×10^{-7}–7.5×10^{-5} M	7.2×10^{-8} M	Human saliva	[149]
Histamine	DPV	MIP-Au@Fe-BDC/N,S-GQDs	GCE	0.078–250 μM	0.026 μM	Human serum, canned tuna fish	[150]
Folic acid	SWV	Fe_3O_4@MIP-GO	CPE	2.5–48 μM	0.65 μM	Tablet, water	[151]
Hydroquinone	DPV	MIP/CNTs	GCE	10–100 μM	3.1 μM	—	[152]
Catechol					3.5 μM		
17β-Estradiol	SWV	MIP-GO-AgNP	GCE	10 fM–250 nM	3.01 fM	River water	[153]
Progesterone	DPV	MIP/Pd	CFP	0.1–110 nM	0.05 nM	Injection	[154]
Galanthamine	DPV	AChE@Ni-MOF/DpAu	GCE	1×10^{-12}–1×10^{-6} M	0.31 pM	Human serum	[124]
Cannabidiol	CV	mag-MIP/Gr-UiO-66	SPE	5–100.0 μM	0.05 μM	—	[155]
Protocatechuic acid	DPV	EDAPbCl_4@ZIF-67	GCE	22–337 μM	15 μM	Green tea, Cough syrup	[156]
Quercetin	DPV	Glycosyl/MOF-5-CNFs	GCE	2.0×10^{-7}–1.0×10^{-5} M, 1.0×10^{-5}–2.0×10^{-4} M	8.33×10^{-8} M	—	[157]
Cortisol	DPV	MIP/NiNCs-*N*-CNTs	GCE	10^{-14}–10^{-9} M	2.37×10^{-15} M	Human saliva	[158]
Riboflavin	CV	Ru-CoP/GCN	GCE	0.062–3468.75 μM	1.09 nM	Human serum, urine, mushroom	[159]
Tenofovir	DPV	MIP-Pt@g-C_3N_4/F-MWCNT	SPE	0.005–0.69 μM	0.0030 μM	Biological and water samples	[160]

H_2O_2	CV	Cs/CuO/BP	GCE	0.2–19.8 μm	30 nM	Saliva, gingival crevicular fluid	[161]
Anti-CCP	SWV	(PANI)/MoS_2	Nanogold electrode	0.25–1500 IU ml^{-1}	0.16 IU ml^{-1}	Human serum samples	[162]
A549 exosomes	SWV	DNA/SWCNT-Fc	Gold electrode	4.66×10^6 -9.32×10^9 exosomes ml^{-1}	9.38×10^4 exosomes ml^{-1}	Serum	[113]
DENV NS1	EIS	MIP	C-SPE	50–200 μg l^{-1}	29.3 μg ml^{-1}	Human serum	[163]
PfHRP2	SWV	PfHRP2 aptamer/MB	Gold Electrode	6.67 pM–4.03 nM	3.73 nM	Serum	[112]
BoNT/A	DPV	AuNPs/MB/SNAP-25	Paper based SPE	0.01–10 nM	10 pM	Orange juice	[164]
PSA	SWV	MIP	Gold Electrode	3×10^{-8}–300 ng ml^{-1}	—	Human serum	[165]
NGAL	CV	GF/AuNPs/Anti-LCN2	3D graphene foam	0.05–210 ng ml^{-1}	42 pg l^{-1}	Urine	[166]
Dopamine	Amperometry	Co-NGA	GCE	0.05–200 μM, 200–1500 μM	12 nM	Living cells	[167]
Tyrosine	SWV	CB	SPE	30–500 μM	4.4 μM	Human serum	[168]
circRNA	CV	AuNFs/PNA/MCH/circRNA	CFME	10 fM–1μM	3.29 fM	Human serum	[169]
Lysozyme	DPV	$CuFe_2O_4$/MIP	GCE	0.05–0.8 μg ml^{-1}	1.58×10^{-3} μg ml^{-1}	Urine	[170]
Glucose	DPV	MIP	Au-SPE	0.5–50 μg ml^{-1}	0.59 μg ml^{-1}	Human saliva	[171]
MMP-7	DPV	CNT/GNP	Gold electrode	1×10^{-2}–1×10^{-3} ng ml^{-1}	6 pg ml^{-1}	Human serum, synthetic urine	[172]
N protein	DPV	ds-DNA hybrid	Gold electrode	10 fM–100 nM	1 fM	Throat swabs, blood samples	[111]
HbA1c	SWV	CNT-Nafion@Hb-Nf	SPCE	31.2–500 nM	4.2 nM	Human blood samples	[173]
L-cysteine	CV	$CuFe_2O_4$/rGO-Au	GCE	0.05–0.2 mM	0.383 μM	Human urine	[174]

(*Continued*)

Table 16.2. (*Continued*)

Analyte	Measurement principle	Surface layer content	Electrode	Dynamic range	LOD	Real	References
Nitrite	DPV	$NaNO_2$/2-ATP/PVA-GA/AuNPs	Au-SPE	0.5–50 μg ml^{-1}	4 μM	Exhaled breathe	[175]
MDA-MB-231 cells	DPV	Au-NS-GNPs/83-mer DNA aptamer	Au-SPE	5–2 × 10^6 cell ml^{-1}	2 cell ml^{-1}	Human serum	[110]
NGAL	DPV	Galinstan NPs Re_2O_7	Si electrode	25–650 ng ml^{-1}	2.14 ng ml^{-1}	Human serum	[176]
AβO	LSV	HCR/AgNPs	Gold electrode	1 pM–10 nM	430 fM	Human serum	[41]
P-gp	AMP	AuNPs/ERGO@CNT	GCE	0.01–500 ng ml^{-1}	0.13 ng ml^{-1}	Human serum	[177]
PHB2	SWV	Au/SH-SpA/Ab/SAM/Ab/HRP-Ab	Gold electrode	1.56–50 ng ml^{-1}	0.04 ng ml^{-1}	Human blood samples	[178]
PSMA	DPV	Cys-AuNPs	SPE	0–5 ng ml^{-1}	0.47 ng ml^{-1}	Serum	[179]

Abbreviations

2-ATP	2-Aminothiophenol
4-VP	4-Vinylpyridine
AβO	Amyloid β-peptide oligomer
ACE2	Human angiotensin-converting enzyme 2
AgNPs	Silver nanoparticles
AMP	Amperometry
ANI	Aniline
Anti-CCP	Anti-cyclic citrullinated peptide antibody
Anti-LCN2	Neutrophil gelatinase-associated lipocalin antibody
AuNFs	Gold nanoflowers
AuNPs	Gold nanoparticles
BC	Biomass-derived carbon
β-CD	β-Cyclodextrin
Bi_2O_3	Dibismuth trioxide
Bio-MCM-41	Bio-mobile crystalline material-41
Bont/A	Botulinum neurotoxin A
BP	Black phosphorus
BPDC	Biphenyl-4,4-dicarboxylic acid
BPNS	Black phosphorus nanosheet nanocomposite
BP3-1	Binding protein 3-1
BP3-8	Binding protein 3-8
BTC	Benzene-1,3,5-tricarboxylate
BTN-DAB	Biotinylated detector antibody
C	Carbon
CA	Chronoamperometry
CAB	Specific capture antibody
CA19-9	CA19-9 tumor marker
CA72-4	CA72-4 tumor marker
CB	Carbon black
CC	Carbon cloth
CCE	Carbon ceramic electrode
CdO	Cadmium oxide
CDs	Carbon dots
CDs NWs	Cadmium sulfide nanowires
CeO_2	Cerium oxide
Circ-RNA	Circular-RNA
CMOF	Chiral metal–organic frameworks
COOH-MWCNTs	Carboxylic multi-walled carbon nanotubes
CNTs	Carbon nanotubes
Co Al-LDH	Cobalt-aluminum layered double hydroxide
Co_3O_4	Cobalt tetraoxide
COFs	Covalent organic frameworks
Co-MOFs	Cobalt-based metal–organic framework
COOH-MBs	Micro-sized carboxylic acid magnetic particles
CPE	Carbon paste electrode
CQDs	Carbon quantum dots
Cu_2O	Copper(I) oxide
Cu-BDC	Cu-Benzene dicarboxylic acid

$CuCO_2O_4$	Copper cobaltate
$CuFe_2O_4$	Copper iron oxide
CuNPs	Copper nanoparticle
CuO	Copper oxide
CuS	Copper sulfide
$CuWO_4$	Copper tungsten oxide
CV	Cyclic voltammetry
Cys-AuNPs	Cysteamine-modified gold nanoparticles
CZTS	Copper zinc tin sulfide
DA	Dopamine
DENV NS1	Non-structural protein 1 of dengue virus
DMMIP	Dual-monomer molecularly imprinted polymer
DPASV	Differential pulse anodic stripping voltammetry
DPV	Differential pulse voltammetry
HCV	HCV envelope protein
EDA	Ethylenediammonium
EDC	N-(3-dimethylaminopropyl)-N′-Ethylcarbodiimidehydrochloride
EIS	Electrochemical impedance spectroscopy
erGO	Electrochemically reduced graphene oxide
FC	Ferrocene
Fe_3O_4	Iron oxide
FG	Functionalized graphene
GA	Glutaraldehyde
$G\text{-}C_3N_4$	Graphitic carbon nitride
GCE	Glassy carbon electrode
GCN	Graphitic carbon nitride
GNF	Graphene/nickel foam
GO	Graphene oxide
GoNRs	Graphene oxide nanoribbons
GQDs	Graphene quantum dots
H_2O_2	Hydrogen peroxide
HbA1c	HemoglobinA1c
HCR	Hybridization chain reactor
HCV-MPs	Hepatitis C virüs-mimetic particles
HEMA	2-Hydroxyethyl methacrylate
ITO	Indium tin oxide
LRG1 Protein	Leucine rich alpha-2-glycoprotein 1
LSV	Linear sweep voltammetry
MAASP	N-Methacryloyl-L-aspartic acid
MagNPs	Magnetic nanoparticles
MAL II	Maackia amurensis lectin I
MB	Methylene blue
MBs	Magnetic beads
MC	Mesoporous carbon
MCH	6-Mercaptohexanol
MCF-7	Michigan cancer foundation
MCOF	Melamine based covalent organic framework
MDA-MB-231 Cells	M.D. Anderson metastatic breast 231 cells
MIP	Molecularly imprinted polymer
MOA	Poly-2-methoxyaniline

MMP-7	Matrix metalloproteinase-7
Mn_2O_3	Manganese oxide
MnO_2	Manganese oxide
MnO	Manganese oxide
Mo_2C	Molybdenum carbide
MOF	Metal organic framework
MoS_2	Molybdenum disulfide
MP-Au	Macroporous gold
MSN	Mesoporous silica nanospheres
MUA	11-Mercaptoundecanoic acid
MUC1	Mucin 1
MWCNTs	Multi-walled carbon nanotubes
N	Nitrogen
$NaNO_2$	Sodium nitrite
NC	Nanoclay
N-CNTs	Nitrogen doped carbon nanotubes
NDO	Neodymium oxide
NF	Nickel foam
Nf	Nafion
NG	Nitrogen-doped graphene
NGA	Nitrogen-doped graphene aerogel
NGAL	Neutrophil gelatinase-associated lipocalin
N-GQDs	Nitrogen-doped graphene oxide quantum dots
NHS	N-Hydroxysuccinimide
NiP	Dinickel phosphide
NIP	Non-imprinted sensor
$Ni(OH)_2$	Nickel (II) hydroxide
Ni_2P	Nickel phosphide
$NiCO_2O_4$	Nickel cobaltite
NINCS-N-CNTS	Nickel nanoclusters-loaded nitrogen-doped carbon nanotubes
NiO	Nickel oxide
NiS_2NS	Nickel disulfide nanospheres
NPG	Nanoporous gold
NSC	Nitrogen and sulfur co-doped porous carbon
NsGNPs	Non-spherical gold nanoparticles
o-PD	Ortho-phenylenediamine
PANI	Polyaniline
PBA	Phenylboronic acid
PCE	Plastic chip electrode
Pd	Palladium
PEDOT	Poly(3,4-ethylenedioxythiophene)
PET	Polyethylene terephthalate
PFHRP2	P. Falciparum histidine-rich protein II
PGE	Pencil graphite electrode
p-GP	P-Glycoprotein
PHA-L	Phytohemagglutinin-L
PHB2	Prohibitin 2
PNA	Peptide nucleic acid
Poly(R-O-NBA)	Poly (reduced-O-nitrobenzoic acid)
Poly-AHP	Poly-(3-amino-5-hydroxypyrazole)

P-Rgo	Phosphorous doped reduced graphene oxide
PSA	Prostate specific antigen
Ps-BIL	Thia-bilane structure
PSMA	Prostate-specific membane antigen
PSS	Polystyrene sulfonate
Pt	Platinum nanoparticles
PTE	Platinum electrode
PVA	Polyvinyl alcohol
PVC	Polyvinyl chloride
Py-Co-Pyhis	Hollow nickel nanospheres
Ru-Cop	Ruthenium doped cobalt phosphide
rGO	Reduced graphene oxide
RuO_2	Ruthenium (IV) oxide
S	Sulfur
SAM	Self-assembled monolayer
SARS-Cov-2-RBD	Severe acute respiratory syndrome Coronavirus 2
SCFV	Single chain variable fragment
SH-SPA	Thiol-modified surface protein A
SNA I	Sambucus nigra agglutinin I
SNAP-25	Synaptosomal associated protein
SNDGR	Sulfur–nitrogen co-doped graphene
SNO_2	Tin oxide
SPCE	Screen-printed carbon electrode
Strep-HRP	Streptavidin-peroxidase
SWCNT	Single-walled carbon nanotube
SWV	Square-wave voltammetry
TC	Titanium carbide
UiO-66	Zirconium based metal–organic framework
$VEGF_{165}$	Vascular endothelial growth factor 165
ZIF	Zeolitic imidazolate framework
$Zn_3V_2O_8$	Zinc vanadate
ZnO	Zinc oxide
Γ-Al_2O_3 QDs	Γ-Aluminium oxide quantum dots

References

[1] Irving P M and Gecse K B 2022 Optimizing therapies using therapeutic drug monitoring: current strategies and future perspectives *Gastroenterology* **162** 1512–24

[2] Vermeire S, Dreesen E, Papamichael K and Dubinsky M C 2020 How, when, and for whom should we perform therapeutic drug monitoring? *Clin. Gastroenterol. Hepatol.* **18** 1291–9

[3] Ates H C, Roberts J A, Lipman J, Cass A E G, Urban G A and Dincer C 2020 On-site therapeutic drug monitoring *Trends Biotechnol.* **38** 1262–77

[4] Stranieri A, Venkatraman S, Minicz J, Zarnegar A, Firmin S, Balasubramanian V and Jelinek H F 2022 Emerging point of care devices and artificial intelligence: prospects and challenges for public health *Smart Heal* **24** 100279

[5] Teymourian H, Parrilla M, Sempionatto J R, Montiel N F, Barfidokht A, Van Echelpoel R, De Wael K and Wang J 2020 Wearable electrochemical sensors for the monitoring and screening of drugs *ACS Sens.* **5** 2679–700

[6] Mejia-Salazar J R, Rodrigues Cruz K, Materon Vasques E M and Novais de Oliveira Jr O 2020 Microfluidic point-of-care devices: new trends and future prospects for ehealth diagnostics *Sensors* **20** 1951
[7] Ongaro A E, Ndlovu Z, Sollier E, Otieno C, Ondoa P, Street A and Kersaudy-Kerhoas M 2022 Engineering a sustainable future for point-of-care diagnostics and single-use microfluidic devices *Lab Chip* **22** 3122–37
[8] Roy L, Buragohain P and Borse V 2021 Strategies for sensitivity enhancement of point-of-care devices *Biosens. Bioelectron.* X **10** 100098
[9] He Z, Liu C, Li Z, Chu Z, Chen X, Chen X and Guo Y 2022 Advances in the use of nanomaterials for nucleic acid detection in point-of-care testing devices: a review *Front. Bioeng. Biotechnol.* **10** 1020444
[10] Maduraiveeran G 2022 Metal nanocomposites based electrochemical sensor platform for few emerging biomarkers *Curr. Anal. Chem.* **18** 509–17
[11] Baranwal J, Barse B, Gatto G, Broncova G and Kumar A 2022 Electrochemical sensors and their applications: a review *Chemosensors* **10** 363
[12] Campuzano S, Pedrero M, Yáñez-Sedeño P and Pingarrón J M 2021 New challenges in point of care electrochemical detection of clinical biomarkers *Sens. Actuators* B **345** 130349
[13] Dumitrescu E and Andreescu S 2017 Bioapplications of electrochemical sensors and biosensors *Enzymes as Sensors* **vol 589** ed R B Thompson and C A B T-M in E Fierke (New York: Academic) ch 11 pp 301–50
[14] Privett B J, Shin J H and Schoenfisch M H 2010 Electrochemical sensors *Anal. Chem.* **82** 4723–41
[15] Wang Y, Xu H, Zhang J and Li G 2008 Electrochemical sensors for clinic analysis *Sensors* **8** 2043–81
[16] Ramya M, Senthil Kumar P, Rangasamy G, Uma shankar V, Rajesh G, Nirmala K, Saravanan A and Krishnapandi A 2022 A recent advancement on the applications of nanomaterials in electrochemical sensors and biosensors *Chemosphere* **308** 136416
[17] Maduraiveeran G and Jin W 2017 Nanomaterials based electrochemical sensor and biosensor platforms for environmental applications *Trends Environ. Anal. Chem.* **13** 10–23
[18] Brainina K, Stozhko N, Bukharinova M and Vikulova E 2018 Nanomaterials: electrochemical properties and application in sensors *Phys. Sci. Rev.* **3** 20188050
[19] Huang X, Zhu Y and Kianfar E 2021 Nano biosensors: properties, applications and electrochemical techniques *J. Mater. Res. Technol.* **12** 1649–72
[20] Karunakaran C, Rajkumar R and Bhargava K 2015 Introduction to biosensors *Biosensors and Bioelectronics* ed C Karunakaran, K Bhargava and R B T-B and B Benjamin (Amsterdam: Elsevier) ch 1 pp 1–68
[21] Kurbanoglu S, Uslu B and Ozkan S A 2018 Nanobiodevices for electrochemical biosensing of pharmaceuticals ed A M B T Grumezescu *Nanosructures for the Engineering of Cells, Tissues and Organs* (William Andrew Publishing) ch 8 pp 291–330
[22] Kalambate P K, Rao Z, Dhanjai , Wu J, Shen Y, Boddula R and Huang Y 2020 Electrochemical (bio) sensors go green *Biosens. Bioelectron.* **163** 112270
[23] Hammond J L, Formisano N, Estrela P, Carrara S and Tkac J 2016 Electrochemical biosensors and nanobiosensors *Essays Biochem.* **60** 69–80
[24] Kurbanoglu S and Ozkan S A 2018 Electrochemical carbon based nanosensors: a promising tool in pharmaceutical and biomedical analysis *J. Pharm. Biomed. Anal.* **147** 439–57

[25] Ozkan S A and Uslu B 2016 From mercury to nanosensors: past, present and the future perspective of electrochemistry in pharmaceutical and biomedical analysis *J. Pharm. Biomed. Anal.* **130** 126–40

[26] Chen H, Liu K, Li Z and Wang P 2019 Point of care testing for infectious diseases *Clin. Chim. Acta* **493** 138–47

[27] Sanavio B and Krol S 2015 On the slow diffusion of point-of-care systems in therapeutic drug monitoring *Front. Bioeng. Biotechnol.* **3** 20

[28] Hou Y, Lv C-C, Guo Y-L, Ma X-H, Liu W, Jin Y, Li B-X, Yang M and Yao S-Y 2022 Recent advances and applications in paper-based devices for point-of-care testing *J. Anal. Test.* **6** 247–73

[29] Liu J, Geng Z, Fan Z, Liu J and Chen H 2019 Point-of-care testing based on smartphone: the current state-of-the-art (2017–2018) *Biosens. Bioelectron.* **132** 17–37

[30] Deng J and Jiang X 2019 Advances in reagents storage and release in self-contained point-of-care devices *Adv. Mater. Technol.* **4** 1800625

[31] Dideikin A T and Vul' A Y 2019 Graphene oxide and derivatives: the place in graphene family *Front. Phys.* **6** 149

[32] Lotfi Z, Gholivand M B, Shamsipur M and mirzaei M 2022 An electrochemical sensor based on Ag nanoparticles decorated on cadmium sulfide nanowires/reduced graphene oxide for the determination of acyclovir *J. Alloys Compd.* **903** 163912

[33] Ramalingam V 2019 Multifunctionality of gold nanoparticles: plausible and convincing properties *Adv. Colloid Interface Sci.* **271** 101989

[34] Kummari S, Sunil Kumar V, Yugender Goud K and Vengatajalabathy Gobi K 2022 Nano-Au particle decorated poly-(3-amino-5-hydroxypyrazole) coated carbon paste electrode for in-vitro detection of valacyclovir *J. Electroanal. Chem.* **904** 115859

[35] Arunkumar T, Karthikeyan R, Ram Subramani R, Viswanathan K and Anish M 2020 Synthesis and characterisation of multi-walled carbon nanotubes (MWCNTs) *Int. J. Ambient Energy* **41** 452–6

[36] Ghanavati M, Tadayon F, Basiryanmahabadi A, Torabi Fard N and Smiley E 2023 Design of new sensing layer based on ZnO/NiO/Fe_3O_4/MWCNTs nanocomposite for simultaneous electrochemical determination of naproxen and sumatriptan *J. Pharm. Biomed. Anal.* **223** 115091

[37] Li S, Liu X, Chai H and Huang Y 2018 Recent advances in the construction and analytical applications of metal-organic frameworks-based nanozymes *TrAC - Trends Anal. Chem.* **105** 391–403

[38] Saeb E and Asadpour-Zeynali K 2022 Enhanced electrocatalytic reduction activity of Fe-MOF/Pt nanoparticles as a sensitive sensor for ultra-trace determination of tinidazole *Microchem. J.* **172** 106976

[39] Bozal-Palabiyik B, Erkmen C and Uslu B 2020 Molecularly imprinted electrochemical sensors: analytical and pharmaceutical applications based on ortho-phenylenediamine polymerization *Curr. Pharm. Anal.* **16** 350–66

[40] Liu Y, Xia Y, Tang Y, Chen Y, Cao J, Zhao F and Zeng B 2022 A ratiometric electrochemical sensor based on Cu-coordinated molecularly imprinted polymer and porous carbon supported Ag nanoparticles for highly sensitive and selective detection of perphenazine *Anal. Chim. Acta* **1227** 340301

[41] Gijare M, Chaudhari S, Ekar S, Shaikh S F, Mane R S, Pandit B, Siddiqui M U H and Garje A 2023 Facile green preparation of reduced graphene oxide using citrus limetta-decorated RGO/TiO_2 nanostructures for glucose sensing *Electronics* **12** 294

[42] Mathad A, Korgaonkar K, Jaldappagari S and Kalanur S 2023 Ultrasensitive electrochemical sensor based on SnO2 anchored 3D porous reduced graphene oxide nanostructure produced via sustainable green protocol for subnanomolar determination of anti-diabetic drug, repaglinide *Chemosensors* **11** 50

[43] Aminikhah M, Babaei A and Taheri A 2022 A novel electrochemical sensor based on molecularly imprinted polymer nanocomposite platform for sensitive and ultra-selective determination of citalopram *J. Electroanal. Chem.* **918** 116493

[44] Sharma T S K and Hwa K Y 2022 Architecting hierarchal $Zn_3V_2O_8$/P-RGO nanostructure: electrochemical determination of anti-viral drug azithromycin in biological samples using SPCE *Chem. Eng. J.* **439** 135591

[45] Yang Y, Shi Z, Wang X, Bai B, Qin S, Li J, Jing X, Tian Y and Fang G 2022 Portable and on-site electrochemical sensor based on surface molecularly imprinted magnetic covalent organic framework for the rapid detection of tetracycline in food *Food Chem.* **395** 133532

[46] Ren S, Cui W, Liu Y, Cheng S, Wang Q, Feng R and Zheng Z 2022 Molecularly imprinted sensor based on 1T/2H MoS_2 and MWCNTs for voltammetric detection of acetaminophen *Sens. Actuators* A **345** 113772

[47] Li Y, Chen X, Ren H, Li X, Chen S and Ye B C 2022 A novel electrochemical sensor based on molecularly imprinted polymer-modified C-ZIF67@Ni for highly sensitive and selective determination of carbendazim *Talanta* **237** 122909

[48] Selcuk O, Unal D N, Kanbes Dindar Ç, Süslü İ and Uslu B 2022 Electrochemical determination of phosphodiesterase-3 enzyme inhibitor drug milrinone with nanodiamond modified paste electrode *Microchem. J.* **181** 107720

[49] Hatamluyi B, Sadeghzadeh S, Sadeghian R, Mirimoghaddam M M and Boroushaki M T 2022 A signal on-off ratiometric electrochemical sensing platform coupled with a molecularly imprinted polymer and CuCo2O4/NCNTs signal amplification for selective determination of gemcitabine *Sens. Actuators* B **371** 132552

[50] Baladi M, Amiri M, Akbari Javar H, Mahmoudi-Moghaddam H and Salavati-Niasari M 2022 Green synthesis of perovskite-type TbFeO3/CuO as a highly efficient modifier for electrochemical detection of methyldopa *J. Electroanal. Chem.* **915** 3–10

[51] Chetana S, Kuma N, Choudhary P, Amulya G, Kumar A, Kumar K B and Rangappa D 2022 Cu2ZnSnS4/MoS2/CNT-ternary heterostructures for paracetamol determination *Mater. Chem. Phys.* **294** 126869

[52] Velayati S, Saadati F, Shayani-Jam H, Shekari A, Valipour R and Reza Yaftian M 2022 Fabrication and evaluation of a molecularly imprinted polymer electrochemical nanosensor for the sensitive monitoring of phenobarbital in biological samples *Microchem. J.* **174** 107063

[53] Li G, Qi X, Wu J, Xu L, Wan X, Liu Y, Chen Y and Li Q 2022 Ultrasensitive, label-free voltammetric determination of norfloxacin based on molecularly imprinted polymers and Au nanoparticle-functionalized black phosphorus nanosheet nanocomposite *J. Hazard. Mater.* **436** 129107

[54] Ata S, Feroz M, Bibi I, Mohsin I ul, Alwadai N and Iqbal M 2022 Investigation of electrochemical reduction and monitoring of P-nitrophenol on imprinted polymer modified electrode *Synth. Met.* **287** 117083

[55] Rebelo P, Pacheco J G, Voroshylova I V, Melo A, Cordeiro M N D S and Delerue-Matos C 2022 A simple electrochemical detection of atorvastatin based on disposable screen-printed carbon electrodes modified by molecularly imprinted polymer: experiment and simulation *Anal. Chim. Acta* **1194** 339410

[56] Nguyen D H N, Le Q H, Nguyen T L, Dinh V T, Nguyen H N, Pham H N, Nguyen T A, Nguyen L L, Dinh T M T and Nguyen V Q 2022 Electrosynthesized nanostructured molecularly imprinted polymer for detecting diclofenac molecule *J. Electroanal. Chem.* **921** 116709

[57] Karimi-Harandi M H, Shabani-Nooshabadi M and Darabi R 2022 Simultaneous determination of citalopram and selegiline using an efficient electrochemical sensor based on ZIF-8 decorated with RGO and g-C_3N_4 in real samples *Anal. Chim. Acta* **1203** 339662

[58] Lu H, Liu M, Cui H, Huang Y, Li L and Ding Y 2022 An advanced molecularly imprinted electrochemical sensor based bifunctional monomers for highly sensitive detection of nitrofurazone *Electrochim. Acta* **427** 140858

[59] Li G, Wu J, Qi X, Wan X, Liu Y, Chen Y and Xu L 2022 Molecularly imprinted polypyrrole film-coated poly(3,4-ethylenedioxythiophene):polystyrene sulfonate-functionalized black phosphorene for the selective and robust detection of norfloxacin *Mater. Today Chem.* **26** 101043

[60] Ma F, Jin S, Li Y, Feng Y, Tong Y and Ye B C 2022 Pyrolysis-derived materials of Mn-doped ZIF-67 for the electrochemical detection of o-nitrophenol *J. Electroanal. Chem.* **904** 115932

[61] Vahidifar M, Es'haghi Z, Oghaz N M, Mohammadi A A and Kazemi M S 2022 Multi-template molecularly imprinted polymer hybrid nanoparticles for selective analysis of nonsteroidal anti-inflammatory drugs and analgesics in biological and pharmaceutical samples *Environ. Sci. Pollut. Res.* **29** 47416–35

[62] Abdallah A B, Saher A, Molouk A F S, Mortada W I and Khalifa M E 2022 Applications of electrochemical techniques for determination of anticoagulant drug (Rivaroxaban) in real samples *Biosens. Bioelectron.* **208** 114213

[63] Thakur N, Sharma V, Singh T A, Pabbathi A and Das J 2022 Fabrication of novel carbon dots/cerium oxide nanocomposites for highly sensitive electrochemical detection of doxorubicin *Diam. Relat. Mater.* **125** 109037

[64] Seguro I, Rebelo P, Pacheco J G and Delerue-Matos C 2022 Electropolymerized, molecularly imprinted polymer on a screen-printed electrode—a simple, fast, and disposable voltammetric sensor for trazodone *Sensors* **22** 2819

[65] Ansari S, Ansari M S, Satsangee S P and Jain R 2021 Bi_2O_3/ZnO nanocomposite: synthesis, characterizations and its application in electrochemical detection of balofloxacin as an antibiotic drug *J. Pharm. Anal.* **11** 57–67

[66] Abo Elalaa A S, Abdel-Hamied Abdel-Tawab M, Abdel Ghani N T and El Nashar R M 2022 Computational design and application of molecularly imprinted/MWCNT based electrochemical sensor for the determination of silodosin *Electroanalysis* **34** 1802–20

[67] Shama N A, Aşır S, Ozsoz M, Göktürk I, Türkmen D, Yılmaz F and Denizli A 2022 Gold-modified molecularly imprinted N-methacryloyl-(L)-phenylalanine-containing electrodes for electrochemical detection of dopamine *Bioengineering* **9** 87

[68] Shalauddin M, Akhter S, Basirun W J, Akhtaruzzaman M, Mohammed M A, Rahman N M M A and Salleh N M 2022 Bio-synthesized copper nanoparticle decorated multiwall carbon nanotube-nanocellulose nanocomposite: an electrochemical sensor for the simultaneous detection of acetaminophen and diclofenac sodium *Surf. Interfaces* **34** 102385

[69] Liu Y, Hu X, Xia Y, Zhao F and Zeng B 2022 A novel ratiometric electrochemical sensor based on dual-monomer molecularly imprinted polymer and Pt/Co3O4 for sensitive detection of chlorpromazine hydrochloride *Anal. Chim. Acta* **1190** 339245

[70] Kamal Ahmed R, Saad E M, Fahmy H M and El Nashar R M 2021 Design and application of molecularly imprinted Polypyrrole/Platinum nanoparticles modified platinum sensor for the electrochemical detection of Vardenafil *Microchem. J.* **171** 106771

[71] Bhuvaneswari C and Ganesh Babu S 2022 Nanoarchitecture and surface engineering strategy for the construction of 3D hierarchical CuS-RGO/g-C_3N_4 nanostructure: an ultrasensitive and highly selective electrochemical sensor for the detection of furazolidone drug *J. Electroanal. Chem.* **907** 116080

[72] Jin Y, Xu G, Li X, Ma J, Yang L, Li Y, Zhang H, Zhang Z, Yao D and Li D 2021 Fast cathodic reduction electrodeposition of a binder-free cobalt-doped Ni-MOF film for directly sensing of levofloxacin *J. Alloys Compd.* **851** 156823

[73] Medetalibeyoğlu H 2021 An investigation on development of a molecular imprinted sensor with graphitic carbon nitride (g-C3N4) quantum dots for detection of acetaminophen *Carbon Lett.* **31** 1237–48

[74] Seguro I, Pacheco J G and Delerue-Matos C 2021 Low cost, easy to prepare and disposable electrochemical molecularly imprinted sensor for diclofenac detection *Sensors* **21** 1–11

[75] Hassan Pour B, Haghnazari N, Keshavarzi F, Ahmadi E and Zarif B R 2021 A sensitive sensor based on molecularly imprinted polypyrrole on reduced graphene oxide modified glassy carbon electrode for nevirapine analysis *Anal. Methods* **13** 4767–77

[76] Habibi B, Pashazadeh S, Saghatforoush L A and Pashazadeh A 2021 A thioridazine hydrochloride electrochemical sensor based on zeolitic imidazolate framework-67-functionalized bio-mobile crystalline material-41 carbon quantum dots *New J. Chem.* **45** 14739–50

[77] Afzali M, Mostafavi A, Afzali Z and Shamspur T 2021 Designing a rapid and selective electrochemical nanosensor based on molecularly imprinted polymer on the Fe_3O_4/MoS_2/ glassy carbon electrode for detection of immunomodulatory drug pomalidomide *Microchem. J.* **164** 106039

[78] Kumar D R, Dhakal G, Nguyen V Q and Shim J J 2021 Molecularly imprinted hornlike polymer@electrochemically reduced graphene oxide electrode for the highly selective determination of an antiemetic drug *Anal. Chim. Acta* **1141** 71–82

[79] Koventhan C, Vinothkumar V and Chen S M 2022 Rational design of manganese oxide/tin oxide hybrid nanocomposite based electrochemical sensor for detection of prochlorperazine (Antipsychotic drug) *Microchem. J.* **175** 107082

[80] Mahnashi M H, Mahmoud A M, Alhazzani K, Alanazi A Z, Alaseem A M, Algahtani M M and El-Wekil M M 2021 Ultrasensitive and selective molecularly imprinted electrochemical oxaliplatin sensor based on a novel nitrogen-doped carbon nanotubes/Ag@cu MOF as a signal enhancer and reporter nanohybrid *Microchim. Acta* **188** 124

[81] Srinithi S, Anupriya J, Chen S M and Balakumar V 2022 Ultrasonic fabrication of neodymium oxide@titanium carbide modified glassy carbon electrode: an efficient electrochemical detection of antibiotic drug nitrofurazone *J. Taiwan Inst. Chem. Eng.* **139** 104522

[82] Afruz A, Amiri M and Imanzadeh H 2022 Electrochemical determination of famotidine in real samples using RGO/Cu_2O nanocomposite modified carbon paste electrode *J. Electrochem. Soc.* **169** 016505

[83] Ye C, Chen X, Zhang D, Xu J, Xi H, Wu T, Deng D, Xiong C, Zhang J and Huang G 2021 Study on the properties and reaction mechanism of polypyrrole@norfloxacin molecularly

imprinted electrochemical sensor based on three-dimensional CoFe-MOFs/AuNPs *Electrochim. Acta* **379** 138174

[84] Yang Y, Shi Z, Chang Y, Wang X, Yu L, Guo C, Zhang J, Bai B, Sun D and Fan S 2021 Surface molecularly imprinted magnetic MOFs: a novel platform coupled with magneto electrode for high throughput electrochemical sensing analysis of oxytetracycline in foods *Food Chem.* **363** 130337

[85] Sun J, Zhang L, Liu X, Zhao A, Hu C, Gan T and Liu Y 2021 Rational design of a mesoporous silica@ZIF-8 based molecularly imprinted electrochemical sensor with high sensitivity and selectivity for atropine monitoring *J. Electroanal. Chem.* **903** 115843

[86] Baezzat M R and Shojaei F 2021 Electrochemical sensor based on GCE modified with E-RGO/Poly (B-CD)/magnetic ZIF-67 nanocomposite for the measurement of linagliptin *Diam. Relat. Mater.* **114** 108345

[87] Habibi B, Pashazadeh A and Ali Saghatforoush L 2021 Zn-mesoporous metal-organic framework incorporated with copper ions modified glassy carbon electrode: electrocatalytic oxidation and determination of amoxicillin *Microchem. J.* **164** 106011

[88] Kablan S E, Reçber T, Tezel G, Timur S S, Karabulut C, Karabulut T C, Eroğlu H, Kır S and Nemutlu E 2022 Voltammetric sensor for COVID-19 drug molnupiravir on modified glassy carbon electrode with electrochemically reduced graphene oxide *J. Electroanal. Chem.* **920**

[89] Zhou J, Liu J, Pan P, Li T, Yang Z, Wei J, Li P, Liu G, Shen H and Zhang X 2022 Electrochemical determination of levofloxacin with a Cu–metal–organic framework derivative electrode *J. Mater. Sci., Mater. Electron.* **33** 9941–50

[90] Li Y and Xiong Y 2021 Molecularly imprinted electrochemical sensor for detection of prednisolone in human plasma as a doping agent in sports *Int. J. Electrochem. Sci.* **16** 1–11

[91] Umesh N M, Antolin Jesila J, Wang S F, Vishnu N and Yang Y J 2021 Novel voltammetric detection of norfloxacin in urine and blood serum using a flexible Ni foam based Ni-Co-MOF ultrathin nanosheets derived from Ni-Co-LDH *Microchem. J.* **160** 105747

[92] Elugoke S E, Fayemi O E, Adekunle A S, Mamba B B, Nkambule T T I and Ebenso E E 2022 Electrochemical sensor for the detection of dopamine using carbon quantum dots/copper oxide nanocomposite modified electrode *FlatChem.* **33** 100372

[93] Habibi B, Pashazadeh S, Saghatforoush L A and Pashazadeh A 2021 Direct electrochemical synthesis of the copper based metal-organic framework on/in the heteroatoms doped graphene/pencil graphite electrode: highly sensitive and selective electrochemical sensor for sertraline hydrochloride *J. Electroanal. Chem.* **888** 115210

[94] El-Wekil M M, Hayallah A M, Abdelgawad M A, Abourehab M A S and Shahin R Y 2022 Nanocomposite of gold nanoparticles@nickel disulfide-plant derived carbon for molecularly imprinted electrochemical determination of favipiravir *J. Electroanal. Chem.* **922** 116745

[95] Arumugam B, Nagarajan V, Annaraj J, Balasubramanian K, Palanisamy S, Ramaraj S K and Chiesa M 2022 Synthesis of MnO2 decorated mesoporous carbon nanocomposite for electrocatalytic detection of antifungal drug *Microchem. J.* **182** 107891

[96] Zhang-Peng X, Wei H, Ma J, Li Y, Chen Y, Cui F, Hu F and Du Y 2022 Molecularly imprinted flexible sensor based on nitrogen-doped graphene for selective determination of formononetin *J. Pharm. Biomed. Anal.* **217** 114805

[97] Lu H, Cui H, Duan D, Li L and Ding Y 2021 A novel molecularly imprinted electrochemical sensor based on a nitrogen-doped graphene oxide quantum dot and molybdenum carbide nanocomposite for indometacin determination *Analyst* **146** 7178–86

[98] Yarkaeva Y, Maistrenko V, Dymova D, Zagitova L and Nazyrov M 2022 Polyaniline and poly(2-methoxyaniline) based molecular imprinted polymer sensors for amoxicillin voltammetric determination *Electrochim. Acta* **433** 141222

[99] Zhao M, Sun M, Kang Q, Ma X and Shen D 2022 A differential strategy to enhance the anti-interference ability of electrochemical molecularly imprinted polymers sensors for the determination of sulfamerazine and 4-acetamidophenol *Sens. Actuators* B **366** 131977

[100] Abd-Elsabour M, Abou-Krisha M M, Alhamzani A G and Yousef T A 2022 An effective, novel, and cheap carbon paste electrode for naproxen estimation *Rev. Anal. Chem.* **41** 168–79

[101] Górska A, Paczosa-Bator B and Piech R 2021 Highly sensitive levodopa determination by means of adsorptive stripping voltammetry on ruthenium dioxide-carbon black-nafion modified glassy carbon electrode *Sensors (Switzerland)* **21** 1–12

[102] Wei P, Wang S, Wang W, Niu Z, Rodas-Gonzalez A, Li K, Li L and Yang Q 2022 CoNi bimetallic metal–organic framework and gold nanoparticles-based aptamer electrochemical sensor for enrofloxacin detection *Appl. Surf. Sci.* **604** 154369

[103] Bozal-Palabiyik B, Uslu B and Marrazza G 2019 Nanosensors in biomarker detection *New Developments in Nanosensors for Pharmaceutical Analysis* ed S A Ozkan and A Shah (Amsterdam: Elsevier) pp 327–80

[104] Wang C, Kim J, Zhu Y, Yang J, Lee G-H, Lee S, Yu J, Pei R, Liu G, Nuckolls C *et al* 2015 An aptameric graphene nanosensor for label-free detection of small-molecule biomarkers *Biosens. Bioelectron.* **71** 222–9

[105] Chaloupková Z, Balzerová A, Bařinková J, Medříková Z, Šácha P, Beneš P, Ranc V, Konvalinka J and Zbořil R 2018 Label-free determination of prostate specific membrane antigen in human whole blood at nanomolar levels by magnetically assisted surface enhanced raman spectroscopy *Anal. Chim. Acta* **997** 44–51

[106] Sher M, Faheem A, Asghar W and Cinti S 2021 Nano-engineered screen-printed electrodes: a dynamic tool for detection of viruses *TrAC, Trends Anal. Chem.* **143** 116374

[107] Yáñez-Sedeño P, Campuzano S and Pingarrón J M 2020 Screen-printed electrodes: promising paper and wearable transducers for (bio) sensing *Biosensors* **10** 76

[108] Lee K X, Shameli K, Yew Y P, Teow S-Y, Jahangirian H, Rafiee-Moghaddam R and Webster T J 2020 Recent developments in the facile bio-synthesis of gold nanoparticles (AuNPs) and their biomedical applications *Int. J. Nanomedicine* **15** 275

[109] Nascimento E D, Fonseca W T, de Oliveira T R, de Correia C R S T B, Faça V M, de Morais B P, Silvestrini V C, Pott-Junior H, Teixeira F R and Faria R C 2022 COVID-19 diagnosis by SARS-CoV-2 Spike protein detection in saliva using an ultrasensitive magneto-assay based on disposable electrochemical sensor *Sens. Actuators* B **353**

[110] Akhtartavan S, Karimi M, Sattarahmady N and Heli H 2020 An electrochemical signal-on apta-cyto-sensor for quantitation of circulating human MDA-MB-231 breast cancer cells by transduction of electro-deposited non-spherical nanoparticles of gold *J. Pharm. Biomed. Anal.* **178** 112948

[111] Yu M, Zhang X, Zhang X, Zahra Q ul ain, Huang Z, Chen Y, Song C, Song M, Jiang H, Luo Z *et al* 2022 An electrochemical aptasensor with N protein binding aptamer-complementary oligonucleotide as probe for ultra-sensitive detection of COVID-19 *Biosens. Bioelectron.* **213** 114436

[112] Lo Y, Cheung Y W, Wang L, Lee M, Figueroa-Miranda G, Liang S, Mayer D and Tanner J A 2021 An electrochemical aptamer-based biosensor targeting Plasmodium falciparum histidine-rich protein II for malaria diagnosis *Biosens. Bioelectron.* **192** 113472

[113] Si F, Liu Z, Li J, Yang H, Liu Y and Kong J 2023 Sensitive electrochemical detection of A549 exosomes based on DNA/ferrocene-modified single-walled carbon nanotube complex *Anal. Biochem.* **660** 114971

[114] Krzyczmonik P, Bozal-Palabiyik B, Skrzypek S and Uslu B 2021 Quantum dots-based sensors using solid electrodes *Electroanalytical Applications of Quantum Dot-Based Biosensors* Micro and Nano Technologies B Uslu (Elsevier) ch 3 pp 81–120

[115] Soylemez S, Erkmen C, Kurbanoglu S, Toppare L and Uslu B 2021 Fabrication of quantum dot-polymer composites and their electroanalytical applications *Electroanalytical Applications of Quantum Dot-Based Biosensors* Micro and Nano Technologies B Uslu (Elsevier) ch 8 pp 271–306

[116] Pumera M, Sánchez S, Ichinose I and Tang J 2007 Electrochemical nanobiosensors *Sens. Actuators* B 1195–205

[117] Wackerlig J and Lieberzeit P A A 2015 Molecularly imprinted polymer nanoparticles in chemical sensing-synthesis, characterisation and application *Sens. Actuators* B **207** 144–57

[118] Tarannum N, Hendrickson O D, Khatoon S, Zherdev A V and Dzantiev B B 2020 Molecularly imprinted polymers as receptors for assays of antibiotics *Crit. Rev. Anal. Chem.* **50** 291–310

[119] Pupin R R R, Monteiro G C C, Foguel M V V, Bolzani V S S, Pividori M and Sotomayor M P T P T 2017 Molecularly imprinted polymers (MIP): from the bulk synthesis to hybrid material to classic and new applications *Molecularly Imprinted Polymers (MIPs): Challenges, Uses and Prospects* (Nova Publishers) pp 43–118

[120] Malitesta C, Mazzotta E, Picca R A, Poma A, Chianella I and Piletsky S A 2012 MIP sensors-the electrochemical approach *Anal. Bioanal. Chem.* **402** 1827–46

[121] Santos A M, Wong A, Prado T M, Fava E L, Fatibello-Filho O, Sotomayor M D P T and Moraes F C 2021 Voltammetric determination of ethinylestradiol using screen-printed electrode modified with functionalized graphene, graphene quantum dots and magnetic nanoparticles coated with molecularly imprinted polymers *Talanta* **224** 121804

[122] Cao J, Li X and Tian H 2020 Metal-organic framework (MOF)-based drug delivery *Curr. Med. Chem.* **27** 5949–69

[123] Stock N and Biswas S 2012 Synthesis of metal-organic frameworks (MOFs): routes to various MOF topologies, morphologies, and composites *Chem. Rev.* **112** 933–69

[124] Zhang L, Qiao C, Cai X, Xia Z, Han J, Yang Q, Zhou C, Chen S and Gao S 2021 Microcalorimetry-guided pore-microenvironment optimization to improve sensitivity of Ni-MOF electrochemical biosensor for chiral galantamine *Chem. Eng. J.* **426** 130730

[125] Kudr J, Haddad Y, Richtera L, Heger Z, Cernak M, Adam V and Zitka O 2017 Magnetic nanoparticles: from design and synthesis to real world applications *Nanomaterials* **7** 243

[126] Hanoglu S B, Man E, Harmanci D, Tozan Ruzgar S, Sanli S, Keles N A, Ayden A, Tuna B G, Duzgun O, Ozkan O F *et al* 2022 Magnetic nanoparticle-based electrochemical sensing platform using ferrocene-labelled peptide nucleic acid for the early diagnosis of colorectal cancer *Biosensors* **12** 736

[127] Mansuriya B D and Altintas Z 2021 Enzyme-free electrochemical nano-immunosensor based on graphene quantum dots and gold nanoparticles for cardiac biomarker determination *Nanomaterials* **11** 578

[128] Chao J, Zhu D, Zhang Y, Wang L and Fan C 2016 DNA nanotechnology-enabled biosensors *Biosens. Bioelectron.* **76** 68–79

[129] Babaei A, Pouremamali A, Rafiee N, Sohrabi H, Mokhtarzadeh A and de la Guardia M 2022 Genosensors as an alternative diagnostic sensing approaches for specific detection of virus species: a review of common techniques and outcomes *TrAC, Trends Anal. Chem.* **155** 116686

[130] Shariati M, Ghorbani M, Sasanpour P and Karimizefreh A 2019 An ultrasensitive label free human papilloma virus DNA biosensor using gold nanotubes based on nanoporous polycarbonate in electrical alignment *Anal. Chim. Acta* **1048** 31–41

[131] Avelino K Y P S, Oliveira L S, Lucena-Silva N, Andrade C A S and Oliveira M D L 2021 Flexible sensor based on conducting polymer and gold nanoparticles for electrochemical screening of HPV families in cervical specimens *Talanta* **226** 122118

[132] Luhar S, Ghosh R, Chatterjee P B and Srivastava D N 2022 An impedometric sensor based on boronic acid @ plastic chip electrode for sensitive detection of prostate cancer biomarker spermine *Biosens. Bioelectron.* X **12** 100219

[133] Luo K, Zhao C, Luo Y, Pan C and Li J 2022 Electrochemical sensor for the simultaneous detection of CA72-4 and CA19-9 tumor markers using dual recognition via glycosyl imprinting and lectin-specific binding for accurate diagnosis of gastric cancer *Biosens. Bioelectron.* **216** 114672

[134] Cho C H, Kim J H, Kim J, Yun J W, Park T J and Park J P 2021 Re-engineering of peptides with high binding affinity to develop an advanced electrochemical sensor for colon cancer diagnosis *Anal. Chim. Acta* **1146** 131–9

[135] Tran H L, Dang V D, Dega N K, Lu S M, Huang Y F and Doong R an 2022 Ultrasensitive detection of breast cancer cells with a lectin-based electrochemical sensor using N-doped graphene quantum dots as the sensing probe *Sensors Actuators* B **368** 132233

[136] Wang S, Chen S, Shang K, Gao X and Wang X 2021 Sensitive electrochemical detection of cholesterol using a portable paper sensor based on the synergistic effect of cholesterol oxidase and nanoporous gold *Int. J. Biol. Macromol.* **189** 356–62

[137] Dalal N, Dhiman T K, Lakshmi G B V S, Singh A K, Singh R, Solanki P R and Kumar A 2022 MIP-based sensor for the detection of gut microbiota-derived indoxyl sulphate using PANI-graphene-NiS2 *Mater. Today Chem.* **26** 101157

[138] Moncer F, Adhoum N, Catak D and Monser L 2021 Electrochemical sensor based on MIP for highly sensitive detection of 5-hydroxyindole-3-acetic acid carcinoid cancer biomarker in human biological fluids *Anal. Chim. Acta* **1181** 338925

[139] Antipchik M, Reut J, Ayankojo A G, Öpik A and Syritski V 2022 MIP-based electrochemical sensor for direct detection of hepatitis C virus via E2 envelope protein *Talanta* **250** 123737

[140] Amouzadeh Tabrizi M, Fernández-Blázquez J P, Medina D M and Acedo P 2022 An ultrasensitive molecularly imprinted polymer-based electrochemical sensor for the determination of SARS-CoV-2-RBD by using macroporous gold screen-printed electrode *Biosens. Bioelectron.* **196** 113729

[141] Liu F, Geng L, Ye F and Zhao S 2022 MOF-derivated MnO@C nanocomposite with bidirectional electrocatalytic ability as signal amplification for dual-signal electrochemical sensing of cancer biomarker *Talanta* **239** 123150

[142] Singh S, Numan A, Zhan Y, Singh V, Alam A, Van Hung T and Nam N D 2020 Low-potential immunosensor-based detection of the vascular growth factor 165 (VEGF165) using the nanocomposite platform of cobalt metal-organic framework *RSC Adv.* **10** 27288–96

[143] Zhang L, Sun M, Jing T, Li S and Ma H 2022 A facile electrochemical sensor based on green synthesis of Cs/Ce-MOF for detection of tryptophan in human serum *Colloids Surf. A: Physicochem. Eng. Asp.* **648** 129225

[144] Arul P, Gowthaman N S K, Narayanamoorthi E, Abraham John S and Huang S T 2021 Synthesis of homogeneously distributed gold nanoparticles built-in metal free organic framework: electrochemical detection of riboflavin in pharmaceutical and human fluids samples *J. Electroanal. Chem.* **887** 115143

[145] Ma Y, Zhang Y and Wang L 2021 An electrochemical sensor based on the modification of platinum nanoparticles and ZIF-8 membrane for the detection of ascorbic acid *Talanta* **226** 122105

[146] Habibi B, Pashazadeh A, Pashazadeh S and Ali Saghatforoush L 2022 Electrocatalytic oxidation and determination of hydrazine in alkaline medium through *in situ* conversion thin film nanostructured modified carbon ceramic electrode *J. Electroanal. Chem.* **907** 116038

[147] Wei X, Li L, Lian H, Cao X and Liu B 2021 Grain-like chiral metal-organic framework/multi-walled carbon nanotube composited electrosensing interface for enantiorecognition of tryptophan *J. Electroanal. Chem.* **886** 115108

[148] Sivam T, Gowthaman N S K, Lim H N, Andou Y, Arul P, Narayanamoorthi E and John S A 2021 Tunable electrochemical behavior of dicarboxylic acids anchored Co-MOF: sensitive determination of rutin in pharmaceutical samples *Colloids Surf. A Physicochem. Eng. Asp.* **622** 126667

[149] Jia L, Mao Y, Zhang S, Li H, Qian M, Liu D and Qi B 2021 Electrochemical switch sensor toward ephedrine hydrochloride determination based on molecularly imprinted polymer/nafion-MWCNTs modified electrode *Microchem. J.* **164** 105981

[150] Mahnashi M H, Mahmoud A M, Alhazzani K, AZ A, Algahtani M M, Alaseem A M, Alqahtani Y S A and El-Wekil M M 2021 Enhanced molecular imprinted electrochemical sensing of histamine based on signal reporting nanohybrid *Microchem. J.* **168** 106439

[151] Garcia S M, Wong A, Khan S and Sotomayor M D P T 2021 A simple, sensitive and efficient electrochemical platform based on carbon paste electrode modified with Fe_3O_4@MIP and graphene oxide for folic acid determination in different matrices *Talanta* **229** 122258

[152] Hu C, Huang H, Sun H, Yan Y, Xu F and Liao J 2022 Simultaneous analysis of catechol and hydroquinone by polymelamine/CNT with dual-template molecular imprinting technology *Polymer (Guildf)* **242** 124593

[153] Biyana Regasa M and Nyokong T 2022 Synergistic recognition and electrochemical sensing of 17β-estradiol using ordered molecularly imprinted polymer-graphene oxide-silver nanoparticles composite films *J. Electroanal. Chem.* **922** 116713

[154] Cherian A R, Benny L, George A, Sirimahachai U, Varghese A and Hegde G 2022 Electro fabrication of molecularly imprinted sensor based on Pd nanoparticles decorated poly-(3 thiophene acetic acid) for progesterone detection *Electrochim. Acta* **408** 139963

[155] Tang X, Gu Y, Tang P and Liu L 2022 Electrochemical sensor based on magnetic molecularly imprinted polymer and graphene-UiO-66 composite modified screen-printed electrode for cannabidiol detection *Int. J. Electrochem. Sci.* **17** 1–15

[156] Zhu S, Yang Y, Chen K, Su Z, Wang J, Li S, Song N, Luo S and Xie A 2022 Novel cubic gravel-like EDAPbCl4@ZIF-67 as electrochemical sensor for the detection of protocatechuic acid *J. Alloys Compd.* **903** 163946

[157] Liu Y, Fan J, He F, Li X, Tang T, Cheng H, Li L and Hu G 2021 Glycosyl/MOF-5-based carbon nanofibers for highly sensitive detection of anti-bacterial drug quercetin *Surf. Interfaces* **27** 101488

[158] Duan D, Lu H, Li L, Ding Y and Ma G 2022 A molecularly imprinted electrochemical sensors based on bamboo-like carbon nanotubes loaded with nickel nanoclusters for highly selective detection of cortisol *Microchem. J.* **175** 107231

[159] Shanmugam R, Koventhan C, Chen S M and Hung W 2022 A portable Ru-decorated cobalt phosphide on graphitic carbon nitride sensor: an effective electrochemical evaluation method for vitamin B2 in the environment and biological samples *Chem. Eng. J.* **446** 136909

[160] Mehmandoust M, Soylak M and Erk N 2023 Innovative molecularly imprinted electrochemical sensor for the nanomolar detection of tenofovir as an anti-HIV drug *Talanta* **253** 123991

[161] Wang K *et al* 2022 Non-enzymatic electrochemical detection of H2O2 by assembly of CuO nanoparticles and black phosphorus nanosheets for early diagnosis of periodontitis *Sens. Actuators* B **355** 131298

[162] Selvam S P, Chinnadayyala S R and Cho S 2021 Electrochemical nanobiosensor for early detection of rheumatoid arthritis biomarker: anti- cyclic citrullinated peptide antibodies based on polyaniline (PANI)/MoS2-modified screen-printed electrode with PANI-Au nanomatrix-based signal amplification *Sens. Actuators* B **333** 129570

[163] Siqueira Silva M, Moreira Tavares A P, Leomil Coelho L F, Morganti Ferreira Dias L E, Chura-Chambi R M, Guimarães da Fonseca F, Ferreira Sales M G and Costa Figueiredo E 2021 Rational selection of hidden epitopes for a molecularly imprinted electrochemical sensor in the recognition of heat-denatured dengue NS1 protein *Biosens. Bioelectron.* **191** 113419

[164] Caratelli V, Fillo S, D'Amore N, Rossetto O, Pirazzini M, Moccia M, Avitabile C, Moscone D, Lista F and Arduini F 2021 Paper-based electrochemical peptide sensor for on-site detection of botulinum neurotoxin serotype A and C *Biosens. Bioelectron.* **183** 113210

[165] Mazouz Z, Mokni M, Fourati N, Zerrouki C, Barbault F, Seydou M, Kalfat R, Yaakoubi N, Omezzine A, Bouslema A *et al* 2020 Computational approach and electrochemical measurements for protein detection with MIP-based sensor *Biosens. Bioelectron.* **151** 111978

[166] Danvirutai P, Ekpanyapong M, Tuantranont A, Bohez E, Anutrakulchai S, Wisitsoraat A and Srichan C 2020 Ultra-sensitive and label-free neutrophil gelatinase-associated lipocalin electrochemical sensor using gold nanoparticles decorated 3D graphene foam towards acute kidney injury detection *Sens. Bio-Sensing Res.* **30** 100380

[167] Zou X, Chen Y, Zheng Z, Sun M, Song X, Lin P, Tao J and Zhao P 2022 The sensitive monitoring of living cell-secreted dopamine based on the electrochemical biosensor modified with nitrogen-doped graphene Aerogel/Co3O4 nanoparticles *Microchem. J.* **183** 107957

[168] Fiore L, De Lellis B, Mazzaracchio V, Suprun E, Massoud R, Goffredo B M, Moscone D and Arduini F 2022 Smartphone-assisted electrochemical sensor for reliable detection of tyrosine in serum *Talanta* **237** 122869

[169] Zhang B, Chen M, Cao J, Liang Y, Tu T, Hu J, Li T, Cai Y, Li S, Liu B *et al* 2021 An integrated electrochemical POCT platform for ultrasensitive CircRNA detection towards hepatocellular carcinoma diagnosis *Biosens. Bioelectron.* **192** 113500

[170] Liang A, Tang B, Hou H P, Sun L and Luo A Q 2019 A novel $CuFe_2O_4$ nanospheres molecularly imprinted polymers modified electrochemical sensor for lysozyme determination *J. Electroanal. Chem.* **853** 113465

[171] Diouf A, Bouchikhi B and El Bari N 2019 A nonenzymatic electrochemical glucose sensor based on molecularly imprinted polymer and its application in measuring saliva glucose *Mater. Sci. Eng.* C **98** 1196–209

[172] Palomar Q, Xu X X, Selegård R, Aili D and Zhang Z 2020 Peptide decorated gold nanoparticle/carbon nanotube electrochemical sensor for ultrasensitive detection of matrix metalloproteinase-7 *Sens. Actuators* B **325** 128789

[173] Thiruppathi M, Lee J F, Chen C C and Ho J an A 2021 A disposable electrochemical sensor designed to estimate glycated hemoglobin (HbA1c) level in whole blood *Sens. Actuators* B **329** 129119

[174] Atacan K 2019 $CuFe_2O_4$/reduced graphene oxide nanocomposite decorated with gold nanoparticles as a new electrochemical sensor material for L-cysteine detection *J. Alloys Compd.* **791** 391–401

[175] Diouf A, El Bari N and Bouchikhi B 2020 A novel electrochemical sensor based on ion imprinted polymer and gold nanomaterials for nitrite ion analysis in exhaled breath condensate *Talanta* **209** 120577

[176] Das M, Chakraborty T, Yu Lin C, Fong Lei K and Haur Kao C 2022 Electrochemical detection of acute renal disease biomarker by galinstan nanoparticles interfaced to bilayer polymeric structured dirhenium heptoxide film *Bioelectrochemistry* **147** 108194

[177] Lu T C, Sun Y M, Zhong Y, Lin X H, Lei Y and Liu A L 2023 Electrochemical immunosensor based on AuNPs/ERGO@CNT nanocomposites by one-step electrochemical Co-reduction for sensitive detection of P-glycoprotein in serum *Biosens. Bioelectron.* **222** 115001

[178] Yun Y R, Lee S Y, Seo B, Kim H, Shin M G and Yang S 2022 Sensitive electrochemical immunosensor to detect prohibitin 2, a potential blood cancer biomarker *Talanta* **238** 123053

[179] Kabay G, Yin Y, Singh C K, Ahmad N, Gunasekaran S and Mutlu M 2022 Disposable electrochemical immunosensor for prostate cancer detection *Sens. Actuators* B **360** 131667

[180] Patolsky F and Lieber C M 2005 Nanowire nanosensors *Mater. Today* **8** 20–8

[181] Anker J N and Hall W P 2010 Biosensing with plasmonic nanosensors *Nanoscience and Technology* (World Scientific) pp 308–19

[182] Clark H A, Hoyer M, Philbert M A and Kopelman R 1999 Optical nanosensors for chemical analysis inside single living cells. 1. Fabrication, characterization, and methods for intracellular delivery of PEBBLE sensors *Anal. Chem.* **71** 4831–6

IOP Publishing

Real-Time Applications of Advanced Electrochemical Sensing Devices

Jamballi G Manjunatha

Chapter 17

Electro-analysis of electrochemical sensors for the determination of neurotransmitters

Kübra Turan, Nazlı Şimşek and Gözde Aydoğdu Tığ

Neurotransmitters (NTs) are chemical messengers in the central nervous system that play a particularly significant role in physical and physiological health. Abnormal levels of NTs essential for human health in body fluids have been associated with psychotic, neurodegenerative, and physical diseases such as Parkinson's disease, depression, Alzheimer's disease, dementia, and schizophrenia. The detection of NTs is essential for the diagnosis and treatment of many diseases. It is essential to create reliable sensitive, clinical, and selective assessment techniques to monitor and modulate NTs and screen for these disorders. Electrochemistry is an ideal tool for monitoring NTs due to its ease of use, sensitivity, and portability. This transmission method has sufficient sensitivity to detect NTs in the brain at the nanomolar level. While electroactive NTs such as serotonin (5-HT) and dopamine (DA) can be directly detected, many NTs, including acetylcholine (ACh), glutamate (Glu), and norepinephrine (NE), do not have electroactive properties, in which case electrochemical enzymatic and aptameric and molecular imprinted polymer (MIP)-based biosensors developed for biorecognition elements. This chapter provides an overview of recent advances in the development of electrochemical sensors for the detection of NTs. It aims to summarize common NTs such as acetylcholine, epinephrine, norepinephrine, dopamine, serotonin, tryptamine, glutamate, hydrogen sulfide and nitric oxide, then discusses electrochemical sensing of these NTs and describes the types of NTs. Finally, recent studies targeting NT detection based on electrochemical sensors are reviewed.

17.1 Introduction

One of the biggest problems in biochemical research today is comprehending the numerous and complex functions of the brain. For this reason, many studies are

doi:10.1088/978-0-7503-5377-9ch17

carried out to determine the signals and activities of NTs in the brain. NTs, a chemical that neurons utilize to communicate with one another, operate on muscle cells, or cause glandular cells to respond, were originally identified by Otto Loewi (Tavakolian-Ardakani *et al* 2019). More than 200 NTs entered in neuronal transmissions have been discovered since the discovery of the first one (Aoki *et al* 1986). NTs control and affect many functions, such as mood, memory, learning, heart rate, consciousness, and appetite regulation. The varying concentrations of NTs in the central nervous system cause numerous neurological and physical diseases (Baranwal and Chandra 2018). Determining NTs quantitatively in various human bodily fluids is crucial for diagnosis, therapeutic treatments, and tracking the progression of diseases.

Different analytical techniques have been reported for analyzing NTs, such as chemiluminescence, capillary electrophoresis, fluorimetry, mass spectroscopy, and chromatography. However, the techniques have disadvantages, such as being performed by highly trained personnel, being expensive and time-consuming, and not suitable for on-site monitoring (Si and Song 2018a).

Electrochemical sensing strategies are frequently used to understand brain functions and disorders by analyzing *in vivo* and *in vitro* conditions to assess clinical NTs (Ou *et al* 2019). NT concentrations in biological samples are low (nM), thus requiring methods that are selective, sensitive, reliable, and low-cost (Tavakolian-Ardakani *et al* 2019). The use of electrochemical sensors for the determination of NTs offers an excellent analytical approach with low cost, reliability, sensitivity, and selectiveness. In addition, it can be used as a routine backup test, as electrochemical sensors allow on-site detection and are even suitable for use by untrained personnel (Bucher and Wightman 2015, Naveen *et al* 2017). In addition, the short analysis time allows for the simultaneous determination of two or more NTs. Another advantage of electrochemical sensors is that they provide good reproducibility and accuracy as well as real-time measurements and have a wide linear range with low detection limits (Tavakolian-Ardakani *et al* 2019). Therefore, the use of electrochemical sensors in determining NTs provides a great advantage.

Electrochemical methods have the advantages of simplicity, speed, portability, low cost, and high sensitivity. Moreover, these methods provide excellent analytical performance for the detection of bioactive molecules. Therefore, a voltammetric approach is generally preferred to investigate the redox behavior of various molecules on the electrode. Cyclic voltammetry (CV) has also attracted attention as a suitable technique for the detection of an analyte (Manjunatha *et al* 2021, Tigari and Manjunatha 2020a). Materials such as carbon-based materials, polymers, surfactants, and ionic liquids are frequently used in the development and characterization of electrochemical sensors. Fast, accurate, sensitive, and selective electrochemical techniques for the analysis of biological molecules have been developed in the last decade (Manjunatha 2020). Sensors prepared for NTs have been diversified with many materials to improve their efficiency and detection capabilities.

The main objective of this chapter is to provide information about common NTs such as serotonin, norepinephrine, glutamate, dopamine, acetylcholine, epinephrine, tryptamine, nitric oxide, and hydrogen sulfide and to summarize recent work on sensors that have proven to be effective. The reader will be informed about recently developed sensitive and selective electrochemical approaches to analyze these NTs. This field is vast and rapidly advancing. Therefore, the work done in the last decade is reviewed to inform target selections for future studies. The electrochemical sensors presented are divided into four classes, and each type is focused on separately. NTs and electrochemical sensors are mentioned below.

17.2 Neurotransmitters

Extensible transmitters, both in design and in a broad scope of the invention, are endogenous messengers that are forward-looking and transmitted in a way suitable for widespread use (Hasanzadeh *et al* 2017). NTs control the neurophysiological processes that control the human body, such as emotions, memory, sleep, and other cognitive functions, and are linked to the central nervous system (Banerjee *et al* 2020). Abnormalities in the dysfunction and concentration of NTs are associated with neurodegenerative diseases (attention deficit hyperactivity, epilepsy, Parkinson, Alzheimer's, Huntington's disease, autism spectral disorder), psychotic diseases (depression, schizophrenia, dementia, etc) and other diseases (arrhythmias, glaucoma, congestive heart damage, thyroid hormone deficiency, sudden infant death syndrome, etc) have been associated (Wu *et al* 2012).

Acetylcholine, a widely detected NT, was discovered in 1921 by the Nobel Prize-winning German pharmacologist Otto Loewi (Pradhan *et al* 2014). Although the exact number of NTs in the human body is unknown, more than 200 have been identified since 1921. NTs can be categorized according to their molecular structures (biogenic amines, soluble gases, and amino acid group), functions (inhibitory and stimulant), and effects (slow and fast-acting) (Aoki *et al* 1986). Dopamine, serotonin, epinephrine, norepinephrine, acetylcholine, tryptamine, glutamate, and certain soluble gases (nitric oxide, hydrogen sulfide) are among the NTs.

Because of the importance of these NTs in clinical evaluation, their analysis *in vivo* and *in vitro* is of immense interest for understanding the functioning and disorders of the brain (Li *et al* 2019, Ou *et al* 2019). Table 17.1 shows the types, biological properties, and structure of NTs.

17.3 Types of neurotransmitters

17.3.1 Dopamine

Dopamine (DA) is a monoamine NT from the catecholamine family that is naturally produced in the body and functions in the central nervous system. It is well known to play an essential role in the regulation of body movements and mental activity. At the same time, imbalance in its metabolism affects many systems and organs from the nervous system to the immune system (Manjunatha *et al* 2010, Prinith *et al* 2021). DA is produced by dopaminergic neurons in the substantia nigra and ventral tegmental areas of the brain from tyrosine, an amino acid, which is taken into the brain by an active transport mechanism. Tyrosine is produced from phenylalanine in the liver by phenylalanine hydroxylase. It is then transported to neurons, which are converted into dopamine through a series of reactions. Within

Table 17.1. Some NTs and their structures.

Type of NTs	Biological function	Structure
Dopamine	Responsible for feelings of enjoyment	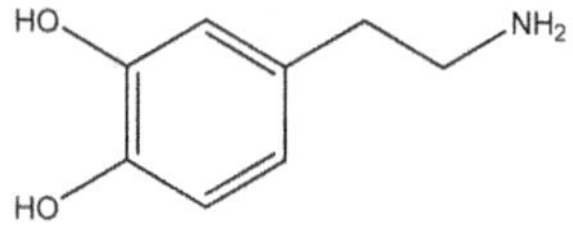 HO, HO, NH_2
Serotonin	Effects mood, sleep, sexuality, appetite, pain regulation	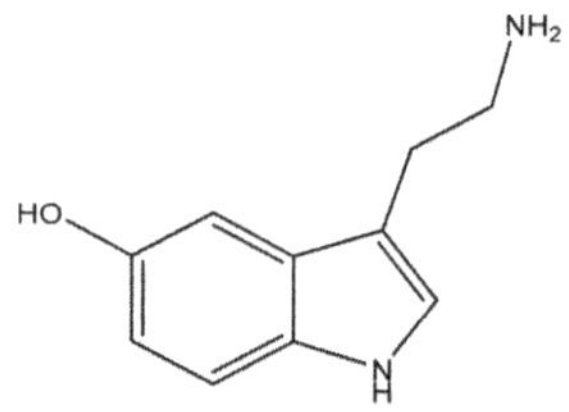 NH_2, HO, N H
Acetylcholine	In charge of learning, thought and memory	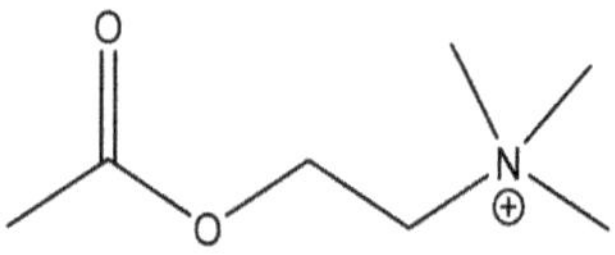 O, O, N ⊕
Epinephrine	High awareness and physical support	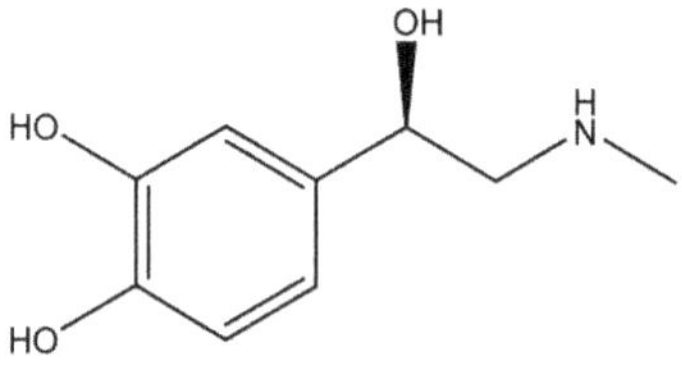 OH, HO, HO, H N
Norepinephrine	Improvement of attention, increased sensitivity	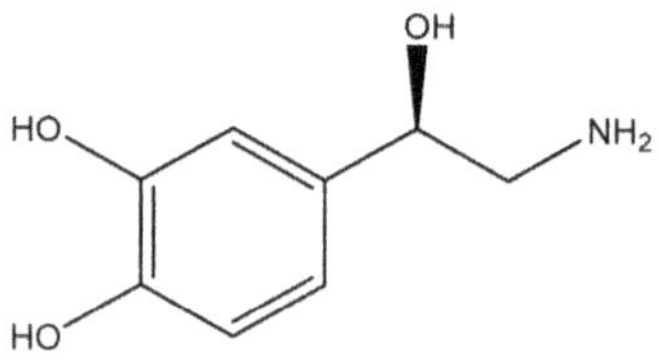 OH, HO, HO, NH_2
Tryptamine	Effective in the gastrointestinal tract and central nervous system	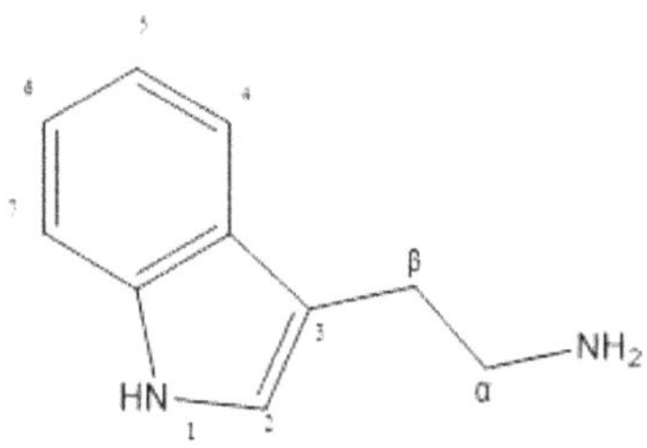 β, α, NH_2, HN

Glutamate	Effective in learning, memory and cognition processes	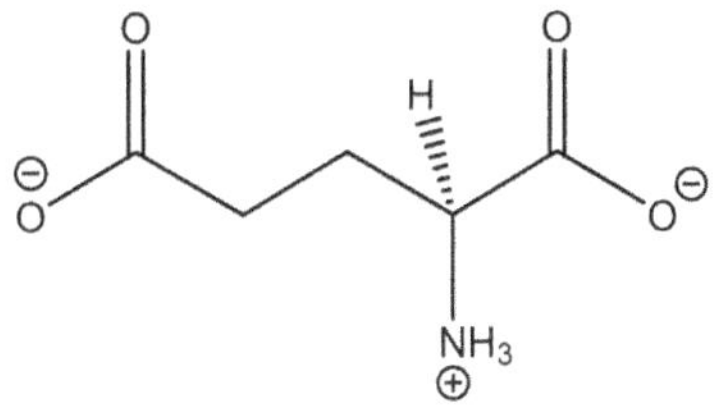
Nitric oxide	Homeostatic functions, cognitive functions, synaptic plasticity and neurosecretion	
Hydrogen sulfide	Neuromodulator in the brain	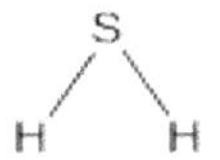

these neurons, tyrosine hydroxylase catalyzes the addition of the –OH group to the meta position of tyrosine to form L-Dopa. L-Dopa is rapidly converted to dopamine in the cytoplasm by Dopa-decarboxylase (Ayano 2016).

DA plays a leading role in inhibiting prolactin production, attention, sleep, mood, behavior, learning, control of nausea and vomiting, and pain. It also contributes to the management of emotion, cognition, and movement. There are five receptors in the body for DA. Its plays a role in the pathophysiology of many diseases, such as Tourette's syndrome, Parkinson, schizophrenia, attention deficit hyperactivity, mood disorders, autism, and obsessive-compulsive disorder, due to their localization in brain regions and various functions (Iversen and Iversen 2007). For this reason, rapid and accurate determination of the concentration of DA in body fluid is crucial for diagnosing diseases.

17.3.2 Serotonin

Serotonin, known as 5-hydroxytryptamine (5-HT), was isolated and characterized by Maurice Rapport and Irvine Page in 1948. 5-HT is found in smooth muscles, gastrointestinal, platelets, central and peripheral nervous systems. In the enterochromaffin cells of the gastrointestinal tract, 90%, 8% in platelets, and 2% in the brain are retained. It gives people a feeling of happiness, vitality, and vitality. When serotonin is released from the brain, blood vessels contract and narrow, and as its level drops, blood vessels expand (Cernat *et al* 2020).

The synthesis of 5-HT takes place in the central nervous and enteric nervous systems. For the synthesis to start in the cells in these systems, tryptophan, which is handled through food and digested, must be taken into the cell by mixing with the blood and circulating. The only tryptophans that can enter the cell are free. Serotonin is produced in the body in two stages. The first stage, tryptophan is converted to 5-hydroxytryptophan by tryptophan hydroxylase, and in the second step, it is decarboxylated to form 5-hydroxytryptamine (5-HT, serotonin) (Mohammad-Zadeh *et al* 2008). 5-HT has essential impression in complex behaviors such as cognition, sleep, appetite, mood, pain control, perception, hormone secretion, motor activities, and temperature regulation. It profoundly impacts various psychopathological and biological processes, such as eating disorders, alcoholism, obsessive-compulsive disorders, anxiety, depression, autism, and epilepsy (Chávez *et al* 2017).

17.3.3 Acetylcholine

Acetylcholine (ACh), an amine chemical compound naturally produced by animals, plants, and fungi, was the first NT identified by British scientist Henry Hallet Dale in 1914 (Francis *et al* 1999). ACh functions as a neuromodulator in both central nervous systems, and the peripheral. ACh is an ester of great biological importance. It is necessary for three main physiological functions; memory, learning, and attention (Chauhan *et al* 2017). In addition to being responsible for carrying nerve impulses from the nerve endings to the organ it affects, or from the nerve ending to the second nerve cell, it is also responsible for forming bioelectrical currents along the nerve and muscle fibers. Various neural disorders, including schizophrenia, progressive dementia Alzheimer's, and Parkinson are associated with decreased ACh levels (Rizzo *et al* 2008). On the other hand, the increase in ACh level causes increased saliva production and reduced heart rate (Mohammadi *et al* 2018). Selective, sensitive, rapid, and accurate determination of ACh in clinical applications is crucial to understanding and treating the physiological and functional aspects of neural disorders caused by changes in body fluids.

17.3.4 Epinephrine

Epinephrine (EPI), known as adrenaline, is one of three types of catechol amines found in the mammalian central nervous system (Zhang *et al* 2002). It plays a crucial role in stimulating a few actions of the sympathetic nervous system, such as psychomotor activity or emotional processes. It is a NT released by the adrenal cortex in the adrenal gland. Changes in EPI levels in biological fluids have been identified as diagnostic symptoms of a variety of diseases. It is also used commercially as a pharmaceutical product in life-threatening situations such as severe allergic reactions and anaphylactic shocks (Sainz *et al* 2020). For these reasons, developing quantitative determination methods to figure out EPI in pharmaceutical and biological samples has become the focus of attention.

17.3.5 Norepinephrine

Norepinephrine (NE), an essential member of the catecholamine NTs secreted by the adrenal medulla in the mammalian central nervous system, plays a vital physiological role in the function of the central nervous, hormonal, reproductive systems, cardiovascular, and renal. NE is synthesized from dopamine by the enzyme dopamine β-hydroxylase. It affects the parts of the brain that focus attention and respond to the environment. Together with EPI, NE forms the basis of the 'fight or flight' response by increasing, blood flow to skeletal muscles, the release of glucose from stores and heart rate (Rosy *et al* 2014).

17.3.6 Tryptamine

Tryptamine (TA), a type of biogenic amine, is a low molecular weight nitrogenous organic compound formed from the enzymatic decarboxylation of free tryptophan and proteins (Meng *et al* 2014, Tatsumi and Ueda 2011). It is believed to be crucial in the coordination of synaptic physiology based on biogenic amines. TA is also found in extremely low concentrations in mammalian brain tissue (Burchett and Hicks 2006). High concentrations of TA in human body fluids can cause symptoms such as respiratory distress, palpitations, nausea, hot flashes, headache, cold sweats, red rash, and high or low blood pressure (Zhang *et al* 2019).

17.3.7 Glutamate

Glutamate (Glu), an anion of glutamic acid that functions as an NT in human body fluids, is one of the chemicals involved in the signal transmission of nerve cells. It was first isolated in 1866 by a German scientist, Karl Ritthausen, in the form of L-glutamic acid, which is isolated from the acid hydrolysate of wheat gluten (Jinap and Hajeb 2010). It is the most abundant in the vertebrate nervous system and the primary excitatory NT in the central nervous system. Glutamate shows its effect by binding and activating cell surface receptors. Although glutamate is a potent neurotoxin, glutamate excitotoxicity is known to be involved in the pathogenesis of many devastating human neurological diseases such as epilepsy, amyotrophic lateral sclerosis, and stroke (Platt 2007, Smith 2000).

17.3.8 Soluble gases

17.3.8.1 Nitric oxide

Nitric oxide (NO) is an unstable, small, and lipid-permeable free radical, as well as an important NT (Snyder and Bredt 1992). NO, a gas with various biological activities produced from arginine by NO synthases, functions by diffusing from the synthesis site to targets in the surrounding cells. To exert its biological effects, NO can form covalent and non-covalent bonds with protein and non-protein targets. Covalent bonds to protein targets are called S-nitrosylation (Donald and Cameron 2021). In addition, NO mediates smooth muscle relaxation, neurotransmission, and modulation of inflammation in several organ systems and pathophysiological conditions.

17.3.8.2 Hydrogen sulfide

Hydrogen sulfide (H_2S), produced from cysteine by cystathionine-β-synthase, two pyridoxal-5′phosphate-dependent enzymes, and cystathionine-γ-lyase, as well as 3-mercaptosulfurtransferase, is the gas that plays a significant role in numerous cellular functions. It can spread across the cell membrane without needing a carrier (Lawrence *et al* 2004, Mathai *et al* 2009, Wang 2002). It has been determined that

H_2S plays critical roles in the central nervous systems, cardiovascular, endocrine, and gastrointestinal, as defined in tissues associated with every function (Brown *et al* 2019). Even low concentrations of H_2S can cause personal distress. In contrast, at higher concentrations, permanent brain damage, loss of consciousness, and even death can result due to the neurotoxic effect of the gas (Shen *et al* 2012).

Due to its importance in clinical NT evaluation, electrochemical sensing strategies are frequently used to understand brain function and disorders by analyzing *in vivo* and *in vitro* conditions (Ou *et al* 2019). The concentrations of NTs in biological samples are low (nM), thus requiring selective, sensitive, reliable, and low-cost electrochemical sensors. In addition, the short analysis time allows the simultaneous determination of two or more NTs. Another advantage of electrochemical sensors is that they provide good repeatability and accuracy, as well as real-time measurements, and have a low detection limit with a wide linear range (Tavakolian-Ardakani *et al* 2019). In this study, the reader will be informed about recently developed electrochemical approaches that can be applied to analyze NTs. Although this field is extensive, it is progressing rapidly. For this reason, studies in the last ten years were examined to inform target choices for future studies.

17.4 Electrochemical analysis of neurotransmitters

Electrochemical techniques are widely used for the detection of species that can readily enter or couple to electrochemical reactions. These techniques have many advantages: excellent sensitivity, easy handling, fast response time, low-cost instruments, good anti-interference, and excellent reproducibility. Many NTs are electrochemically redox active in an aqueous solution (Gao and Gao 2021). Rapid, sensitive, focused, and economical detection of bimolecular analytes important for treatment monitoring and clinical diagnosis may be possible using electrochemical sensing techniques (Hasanzadeh *et al* 2017). Electrochemical techniques are easier to use and faster tools for real-time quantitative analysis of NTs compared to classical techniques (Shadlaghani *et al* 2019). Electrochemical sensors have shown to be efficient and practical for the detection of NTs that are electroactive or can be related to electroactive processes.

NTs are classified into three categories according to their effects (fast-acting and slow-acting), functions (inhibitory, and stimulant), and molecular structure (biogenic amines, soluble gases, and amino acid group) (Banerjee *et al* 2020). Because some NTs such as GABA, glutamate (Glu), acetylcholine (ACh) and norepinephrine (NE) don't have electroactive features, it is required to modify the electrodes with several biological recognition elements rather than their direct electrochemical detection (figure 17.1). The most common classes of indirect NTs detection are electrochemical and aptameric and enzymatic biosensors. These sensors are electrochemical biosensors (Zamani *et al* 2022, Ou *et al* 2019).

17.4.1 Direct detection used sensors

Nonenzymatic electrochemical sensors are also known as direct or enzyme-free electrochemical sensors. They are commonly used for determining NTs and

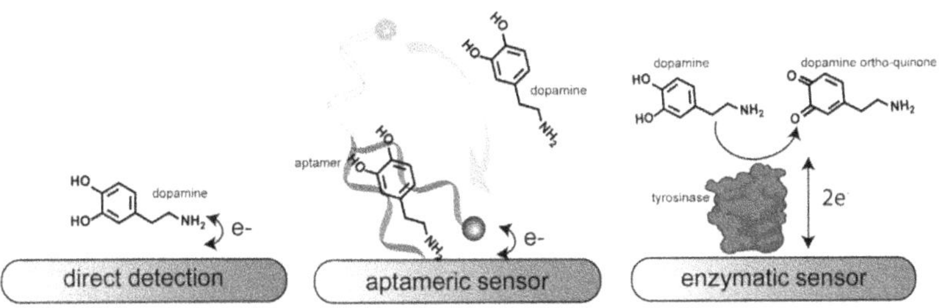

Figure 17.1. Schematic of electrochemical sensors for NT detection. Reproduced from Zamani *et al* (2022). Copyright IOP Publishing Ltd CC BY 3.0.

evaluation of their concentrations in biological and real samples is significant for clinical diagnosis. Direct detection of NTs based on nanocomposite/composite sensors has attracted attention as a good approach for overly sensitive and accurate detection. However, the existence of interfering molecules poses many challenges to their analysis in biological fluids (Tavakolian-Ardakani *et al* 2019). In the following, several diverse types of sensors for the direct detection of NTs are described, as well as electrode structure, electrochemical technique, limits of detection (LODs), linear range, storage stability, sensitivity, real samples, and other important data related to these sensors. In addition, a comparison of these sensors is shown in table 17.2.

Palanisamy *et al* developed a sensitive and selective electrochemical sensor for dopamine based on a glassy carbon electrode (GCE) modified with electrochemically reduced graphene oxide (RGO) and palladium nanoparticles (PdNPs). This modified RGO–PdNPs composite was characterized using SEM, energy dispersive spectroscopy (XPS), and electrochemical impedance spectroscopy (EIS) techniques. The prepared electrode could oxidize DA only at a lower potential than RGO or PdNPs modified electrodes. The RGO–PdNPs nanocomposite due to good conductivity and high surface area allowed efficient DA oxidation. Moreover, RGO–PdNPs composite electrodes displayed more electrocatalytic oxidation for dopamine than those modified with PdNPs and RGO. The sensor's response to DA was linear range of 1–150 μM and with a detection limit of 0.233 μM. The proposed sensor was applied to commercial dopamine injection samples (Palanisamy *et al* 2013).

Electrochemical sensors take advantage of their fast detection rates and low cost, making them useful for routine analysis of biomolecules such as DA. In this study, electrochemical sensors modified with self-assembled gold nanoparticles (SA-AuNPs) with an electrochemical sensor fabricated with monodisperse AuNPs with a size of 8.5 ± 1.7 nm were used to achieve better signal stability between the same group of electrodes for detecting DA. Dopamine could be sensitively detected in detection limit of 0.04 μM and the linear range of 1–50 μM. This research could lead to the development of industrialized nanosensors with stable and nanoscopic properties (Qin *et al* 2021).

Zhu *et al* present a novel method to monitor the release of dopamine and serotonin *in vivo*. A graphene-iron-tetrasulfophthalocyanine (GR-FeTSPc)-coated carbon fiber microelectrode (GR-FeTSPc/CFME) was developed to detect 5-HT and DA selectively and simultaneously in mouse brain. The electrocatalytic activity

Table 17.2. Electrochemical sensors used to detection direct of NTs.

Sensor	Sensor platform	NTs	Sample	LOD	Linear range	References
Direct	RGO–PdNPs/GCE	Dopamine	Commercial DA injection	0.233 μM	1–150 μM	Palanisamy *et al* (2013)
	SA-AuNPs/ITO	Dopamine	—	0.04 μM	1–50 μM	Qin *et al* (2021)
	GR-FeTSPc/CFME	Dopamine	Mouse brain	50 nM	0.05–50 μM	Zhu *et al* (2018)
	AuNR@ZIF-8	Dopamine	Human serum albumin	0.03 μM	0.1–50 μM	Zhao *et al* (2021)
	AuNR@ZIF-8	Serotonin	Human serum albumin	0.007 μM	0.1–25 μM	Zhao *et al* (2021)
	GR-FeTSPc/CFME	Serotonin	Mouse brain	20 nM	0.1–100 μM	Zhu *et al* (2018)
	MCSNP/SPE	Acetylcholine	ACh ampule, urine, and serum	20.0 nM	0.1–500 μM	Mohammadi *et al* (2018)
	GNRs/GCE	Epinephrine	Commercial EPI injection	2.1 μM	6.4–100 μM	Sainz *et al* (2020)
	ZnO/MWCNTs/GCE	Epinephrine	Spiked human blood serum and adrenaline titrate injection samples	0.016 μM	0.4–2.4 μM	Shaikshavali *et al* (2020)
	CTAB-SnO_2/GCE	Epinephrine	Human urine samples	10 nM	0.1–250 μM	Lavanya and Sekar (2017)
	CTAB-SnO_2/GCE	Norepinephrine	Human urine samples	6 nM	0.1–300 μM	Lavanya and Sekar (2017)
	AuILCCDCPE	Norepinephrine	Human urine samples	3.12 nM	0.05–10 μM	Atta *et al* (2020)
	GCE	Tryptamine	Food sample	0.8 nM	47–545 nM	Costa *et al* (2016)
	Nafion/AuNPS/GNRs/CPE	Nitric oxide		0.04 μM	0.39–2.34 μM 2.34–104.7 μM	Wenninger *et al* (2021)
	Ti_3AlC_2MXene/GCE	Hydrogen sulfide	Bovine serum samples	16.0 nM	100 nM–300 μM	Liu *et al* (2021)

RGO: reduced graphene oxide; NPs: nanoparticles; GCE: glassy carbon electrode; MCSNP: magnetic core–shell manganese ferrite nanoparticles; CTAB: cetyltrimethylammonium bromide; CFME: coated carbon fiber microelectrode; AuNR@ZIF-8: Gold Nanorods Zeolitic Imidazolate Framework-8; GR-FeTSPc: graphene-iron-tetrasulfophthalocyanine; AuILCCDCPE: gold nanoparticles/ionic liquid crystal/β-cyclodextrin modified carbon paste electrode; Ti_3AlC_2MXene: two-dimensional transition metal carbonitrides.

of GR-FeTSPc/CFME for the oxidation of 5-HT and DA is enhanced. An oxidation peak at 0.36 V observed in GR-FeTSPc/CFME only for 5-HT, while for 5-HT and DA, respectively, reversible redox peaks with redox peak potentials of 0.10 and 0.15 V were observed. Differential pulse voltammetry (DPV) was used to determine 5-HT and DA simultaneously. The LODs were 20 and 50 nM for 5-HT and DA, respectively. In addition, the proposed microsensor exhibited great selectivity in homovanillic acid and AA medium. The experimental results demonstrated that this sensor could simultaneously detect the release of DA and 5-HT *in vivo* in the mouse brain, as well as monitor the time variation of the two biomolecules' concentration (Zhu *et al* 2018).

In the last decade, the production of electrochemical nonenzymatic ACh biosensors has increased rapidly. In this study, Mohammadi *et al* developed a novel sensor based on a modified electrochemical screen-printed electrode (SPE) using magnetic core–shell nanoparticles to determine acetylcholine rapidly and sensitively. The electrochemical behavior was examined by CV. The modified electrode showed a significant increase in peak current and a reduction in the potential of approximately 130 mV. The reproducibility of the measurements was assessed by recording responses to 50.0 μM ACh on the same day and with four different improved sensors prepared identically. Under optimized conditions, the DPV has a detection limit of 0.02 μM and an operating range of 0.1–500.0 μM. Finally, the accuracy of the system was evaluated with ampoules, urine, and serum samples (Mohammadi *et al* 2018).

In another study, a GCE modified with zigzag-like graphene nanospheres (GNRs) synthesized by a solution-based chemical route was used to generate a novel electrochemical sensor for EPI detection. AFM, SEM, EIS, and Raman spectroscopy were conducted for characterization. To determine EPI, the sensor used the reduction peak corresponding to the epinephrinephrinochrome–leucoepinephrinochrome transition ($E = -0.25$ V) to minimize interferences in place of the oxidation peak ($E = +0.6$ V), which is usually used in the literature as an analytical signal. The sensor structure is more straightforward than in the literature, and the detection potential ($E = -0.25$ V) is sufficient to avoid interferences. Using DPV, a linear range from 6.4–100 μM was obtained and an LOD of 2.1 μM. The sensor's applicability has been demonstrated with satisfactory results for EPI detection in pharmaceutical samples sensors (Sainz *et al* 2020).

In their study, Lavanya and Sekar developed an overly selective and sensitive electrochemical sensor for the simultaneous detection of EPI and NE, catecholamine NTs, based on a modified GCE using cetyltrimethylammonium bromide (CTAB)-SnO_2 nanoparticles synthesized by microwave irradiation (shown in figure 17.2). Throughout the study, characterization was performed HRTEM, XRD, FTIR, UV and CV techniques. The prepared CTAB-SnO_2/GCE showed two sharps, well-defined reduction peaks and excellent electrocatalytic activity for EPI and NE with a significant potential difference of 0.354 V in pH 5.0 phosphate buffer (PBS). Using the voltammetric method, the sensor could detect target molecules individually or simultaneously. Under optimum conditions, the lowest LODs of 10 and 6 nM and a wide concentration range of 0.1–250 and 0.1–300 μM were obtained for EPI and

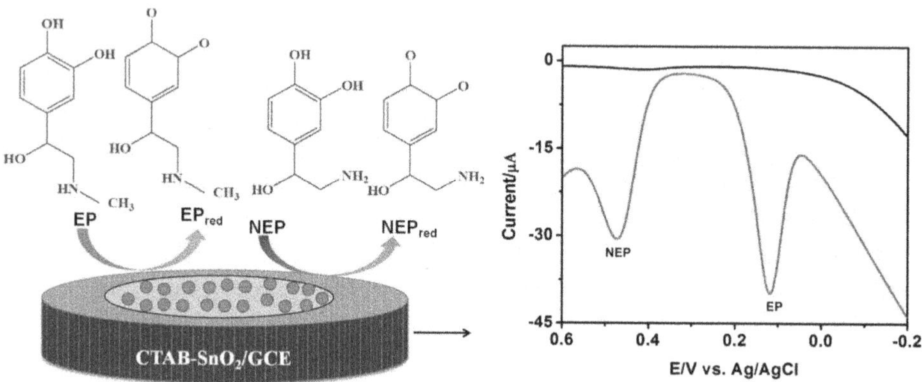

Figure 17.2. Possible reaction mechanism for electro-reduction of EP and NEP at the CTAB-SnO_2/GCE. Reprinted from Lavanya and Sekar (2017), copyright (2017), with permission from Elsevier.

NE, respectively. Furthermore, the developed CTAB-SnO_2/GCE sensor was found to have excellent selectivity in the presence of biological interferences such as uric acid and ascorbic acid in interference studies, stability, and good reproducibility. The developed electrode has been successfully applied to the simultaneous determination of EPI and NE in human urine samples with satisfactory recoveries. Therefore, it can be used as a valuable tool for selective and sensitive simultaneous analysis of EPI and NE in clinical and research laboratories (Lavanya and Sekar 2017).

In the current study by Atta *et al* a novel electrochemical sensor for the sensitive and direct detection of NE based on AuNPs/ionic liquid crystal/β-cyclodextrin modified carbon paste electrode (AuILCCDCPE) was effectively optimized and developed. AuILCCDCPE sensor demonstrated a good electrochemical response towards NE in human urine in the linear ranges of 0.05–10 μM with a low LOD of 3.12 nM in under optimal conditions. Moreover, the sensor has many advantages, such as anti-interference ability in the presence of other compounds, good stability, and high sensitivity (9.44 $\mu A\ \mu M^{-1}$). The developed sensor has a remarkable extensibility for electrochemical applications in testing additional drugs. It is believed to be useful for application in pharmaceutical quality control laboratories and other medical applications (Atta *et al* 2020).

Costa *et al* investigated the electrochemical oxidation of tryptamine, an indole-derived biogenic amine, using GCE and square wave adsorptive stripping voltammetry (SWAdSV) with a focus on trace-level detection in food products. Electrochemical oxidation of tryptamine occurred in two consecutive steps using different voltammetric techniques on GCE in acidic media. The oxidation mechanism of TA was proposed and the proton and electron numbers for the oxidation reactions were estimated by DPV and over a wide pH range. The SWAdSV method with a low detection limit of 0.8 nM and a linear response range of 47–545 nM was developed and successfully applied for the determination of TA in food samples such as tomato, banana, gorgonzola cheese, mozzarella cheese, chicken, and pepper sausage. Furthermore, the method is cost-effective, selective, accurate, simple, rapid,

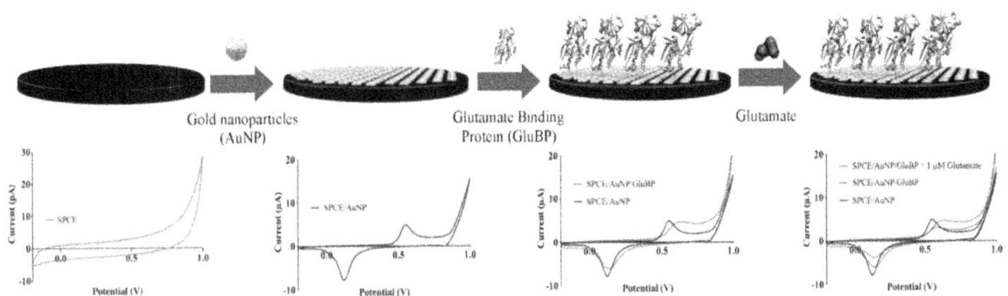

Figure 17.3. Production stages of the GluBP/AuNP/SPCE. Reprinted from Zeynaloo *et al* (2021), Copyright (2021), with permission from Elsevier.

and sensitive and offers a convenient alternative for routine laboratory analysis (Costa *et al* 2016).

Zeynaloo *et al* created a novel platform for directly measuring glutamate levels by covalent immobilizing a genetically engineered periplasmic glutamate binding protein (GluBP) on gold nanoparticle modified screen-printed carbon electrodes (GluBP/AuNP/SPCE) in shown figure 17.3. The produced GluBP-based biosensor, enzyme-free, is highly efficient in measuring Glu concentrations at pH 7.40 with relatively low detection limit, and fast response. The developed biosensor for glutamate detection was evaluated by CV. A LOD of 0.15 μM and linear range of 0.1–0.8 μM was determined from dose–response plots relating the electrochemical signal to the Glu concentration. The high selectivity of the glutamate biosensor towards co-interfering compounds was confirmed. The developed biosensor system can potentially be used as a platform for glutamate detection in biological samples (Zeynaloo *et al* 2021).

Various immune response, physiological and pathophysiological processes or neurotransmission depend on changes in nitric oxide (NO) concentration. Therefore, direct, and rapid detection of this gas is highly necessary. In this study, a novel electrochemical sensor was proposed for fast and sensitive detection of NO at 0.72 V versus Ag/AgCl based on a carbon paste electrode (CPE) modified with AuNPs, GNRs and Nafion; AuNPs were used as a catalyst to keep NO oxidation at a lower potential, and GNRs were used to increase electrical conductivity, and the surface area. The voltammetric investigations with the optimized sensor were carried out in 0.1 M phosphate buffer, pH 7.4, and at room temperature. Based on DPV measurements, a good linear analytical signal was obtained in the range of 0.39–2.34 μM and 2.34–104.7 μM, respectively. High reproducibility was achieved for these ranges. The detection limit was 40 nM, while the quantification limit was 130 nM. Further investigation into physiological interferences revealed no discernible impact, giving the sensor good selectivity. The results show that this new sensor provides a quick and easy approach to determining low concentrations of NO in liquids using direct voltammetry. Additionally, this sensor is distinguished by quick measurement, excellent biocompatibility, and simple construction (Wenninger *et al* 2021).

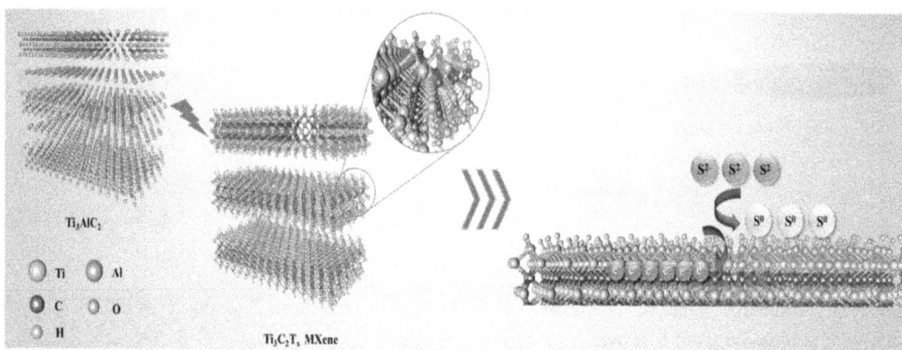

Figure 17.4. Scheme of separation of Al layers from Ti_3AlC_2MXene phase and the electron transfer mechanism of S_2 on the electrode (Liu *et al* 2021) John Wiley & Sons. [© 2021 Wiley-VCH GmbH].

Graphene is widely used as an electrode material due to its high specific surface area, high electrical conductivity, fast electron transfer rate and mechanical properties (Raril and Manjunatha 2020). At the same time, as an alternative to this material in the literature, two-dimensional metal–carbon structures can also be used for electrochemical sensor fabrication. Liu *et al* preferred the layered structure of two-dimensional transition metal–carbon/nitride, an excellent potential biosensing material due to its many advantages. For this purpose, a graphene-like system composed of multilayered $Ti_3C_2T_x$ MXene was successfully synthesized by a simple hydrogen fluoride etching method, and a nonenzymatic H_2S electrochemical sensor was prepared (figure 17.4). The obtained results showed that the electrochemical $Ti_3C_2T_x$ MXene/GCE sensor has a high sensitivity for H_2S detection, a wide detection range of 100 nM–300 μM and a low detection limit of 16.0 nM. In conclusion, this study presents new research on the direct electrochemical sensing of H_2S and the fabrication of a novel biosensing platform for MXene materials (Liu *et al* 2021).

17.4.2 Aptameric biosensors

Aptamers are manufactured RNA or single-stranded DNA oligomers that are relatively short and have a specified nucleotide sequence intended to house a target molecule, such as dopamine. Aptamers shift from a quite simple chain structure to a complex 3D shape that wraps around their target once they are bonded to it. Aptamers are particularly appealing for a wide range of applications due to their combination of high selectivity and stability of biological systems and, perhaps, for large-scale synthesis of artificial systems (Abu-Ali *et al* 2020).

While functionalized electrodes can enhancement surface area, a ligand is sometimes needed to improve selectivity and enrich the detection compound on the electrode surface. To that end, aptamer-based electrochemical biosensors are the most used materials (Liu and Liu 2021). Because of its ease of use, portability, and sensitivity, electrochemistry is an excellent tool for monitoring NTs (Banerjee *et al* 2020). In this chapter, diverse electrochemical sensors have been obtained and defined for various targets based on aptamers and the combination of electrodes.

Chen *et al* presented an electrochemical aptasensor for label-free determination of dopamine by electrochemical deposition of an Au nanocomposite with a ribonucleic acid (RNA) aptamer and RGO. 5′-SH-(CH_2)6-GUCUCUGUGUGCGCCAGA-GACAGUGGGGCAGAUAUGGGCCAGCACAGA AUGAGGCCC-3′ were the sequences of the RNA aptamer. This aptamer was immobilized on the AuNPs surface with thiol-gold bonds, and determination of DA was determined by the specific interaction of this aptamer. Direct electrochemical analysis was possible due to the redox activity of the analyte. CV and SEM were used to analyze the nanocomposite's structure and conductivity. The nanocomposite-modified electrode was found to improve the current response significantly. The linear operating range for the electrochemical aptasensor was determined to be from 0.5 to 20 μM, with an LOD of 0.13 μM. Additionally, it was shown to provide high selectivity, stability, and in the determination of DA in human serum (Chen *et al* 2016).

Geng *et al* successfully prepared an electrochemical aptasensor for the selective detection of serotonin based on DNA aptamer conformation folding, shown in figure 17.5. The effect of MB-labeled regions on the analytical performances of the developed aptasensors is discussed by comparing the sensitivity of aptasensors labeled as MB at the aptamer intermediate with the sensitivity of MB-labeled aptasensors at the 3′ end of the aptamer. DNA aptamer sequences are terminal-labeled MB:(5′–3′) HS-GACTGGTAGGCAGATAGGGGAAGCTGATTTCG ATGCGTGGGTC-MB and intermediate-labeled MB:(5′–3′) HS-GACTGGTAG GCAGATAGGGGAAGCTGA TT (MB)T CGATGCGTGGGT C. Due to specific conformational changes caused by the aptamer, the sensitivity of the intermediate-labeled aptasensor was found to be much higher than the terminal-labeled aptasensor. The proposed aptasensors exhibited high sensitivity and a fast electrochemical response for determining 5-HT. A detection limit of 0.017 fM with a linear range of 1 pM–10 nM was calculated for 5-HT by the intermediate-labeled aptasensor, under optimal experimental conditions. Furthermore, the developed aptasensor has reusable and reproducible properties. This sensor showed good selectivity for the detection of serotonin in rat cerebrospinal fluid diluted approximately 100-fold in real sample analysis (Geng *et al* 2021).

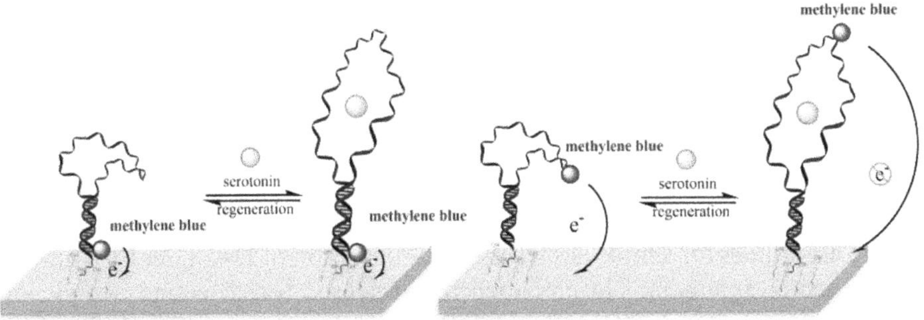

Figure 17.5. The schematic diagram of the DNA aptamer sensors with MB molecule. Reprinted from Geng *et al* (2021), Copyright (2021), with permission from Elsevier.

A PEDOT:PSS-based organic aptamer functionalized electrochemical transistor (OECT) functionalized with aptamers was developed to detect the existence of EPI molecules by Saraf *et al* (2018). A schematic representation of the aptamer-based OECT device is shown in figure 17.6. The present method relies on immobilizing aptamers, which have a high affinity for binding to EPI, on the gate electrode. The addition of EPI to the system allows the aptamers' negative charges to be scanned, and EPI's oxidation generates a faradaic current. The inclusion of aptamers increased the sensitivity and specificity to the EPI molecule. The reason for the very low LOD of 90 pM of the current aptasensor is electrooxidation of the molecule in the vicinity of the gate electrode and the high binding affinity of aptamers to epinephrine. The combined impact of these two processes lowers the total channel current as shown in the current–time curve and transfer characteristics. In addition, to determine selectivity properties of the OECT sensor, experiments were performed against commonly used interfering agents such as DOPAC, ascorbic acid, DA, etc. It was found that there was no reduction in current, indicating the high specificity of the sensor. By incorporating aptamers into the transistor, a sensor has been prepared to show 90 pM, the lowest detection limit for EPI, comparable to the normal physiological level. The current approach provides a simple and rapid method to detect EPI, but further work is required to test the reusability and reproducibility of the proposed sensor. The current work provides the groundwork for improving a wide range of individualized point-of-care testing devices for therapeutic and curative applications (Saraf *et al* 2018).

Glutamate is the primary excitatory NT in the central nervous system. To diagnose pathological conditions in the brain, it is essential to determine the release and levels of glutamic acid and investigate signal transduction. In a study by Wu *et al*, glutamic acid-sensitive DNA aptamers were discovered via the Capture-SELEX process (figure 17.7). This process has isolated the glutamic acid specialized sequences from a DNA library. The short aptamer sequence (12 μM dissociation constant of 1d04) modified with ferrocene redox probes and a thiol binding group to promote amperometric Glu detection had been used in developing an aptameric electrochemical sensor. For this, a shortened aptamer sequence called Glu1 was immobilized on a gold electrode surface via Au-thiol bonds and tagged with a ferrocene redox tag at its 3′ end. Sensor performance was evaluated by alternating current voltammetry using 6-mercapto-1-hexanol as filler. The Glu1 aptasensor showed good selectivity for Glu in ten-fold diluted human serum, and a detection limit of 0.0013 pM, and comprehensive detection ranges from 0.01 pM to 1 nM. This performance is sensitive enough to identify physiological levels of glutamate in a variety of *in vitro* and *in vivo* samples. The aptasensor, which can discriminate between amino acid analogs like GABA and glutamic acid as well as between the interfering electroactive NT DA, may offer an alternative to current sensor systems (Wu *et al* 2022).

There are studies for simple electrochemical detection of DA, Glu, EPI, and 5-HT prepared using aptamers. No electrochemical sensor for detecting acetylcholine, norepinephrine, nitric oxide, or hydrogen sulfide has been reported. But optical sensors are available for these NTs. In addition, a comparison of these sensors is shown in table 17.3.

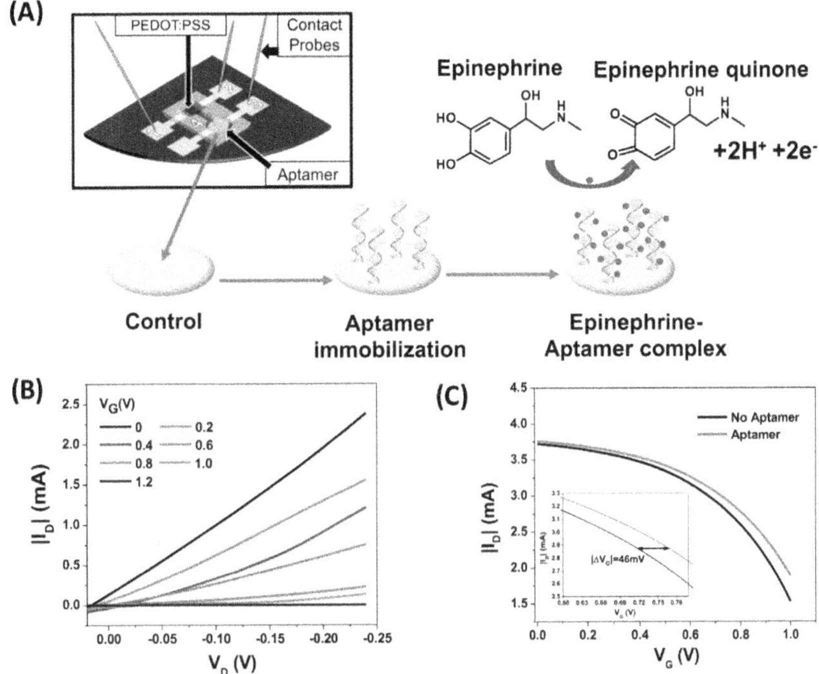

Figure 17.6. (A) Schematic illustration of the fabricated aptamer-based OECT device on a silicon wafer. The source, drain and gate electrodes are labelled as 1, 2 and 3, respectively. (B) Output characteristics (ID versus VD) of the aptamer-based OECT device measured in pure PBS solution in the absence of epinephrine at VG values of 0, 0.2, 0.4, 0.6, 0.8 and 1.0 V (from top to bottom). (C) Transfer characteristics (ID versus VG) of the OECT device measured in PBS solution before and after immobilization of the aptamers on the gate electrode at VD = +0.1 V. Reprinted from Saraf *et al* (2018). Copyright 2018, with permission from Elsevier.

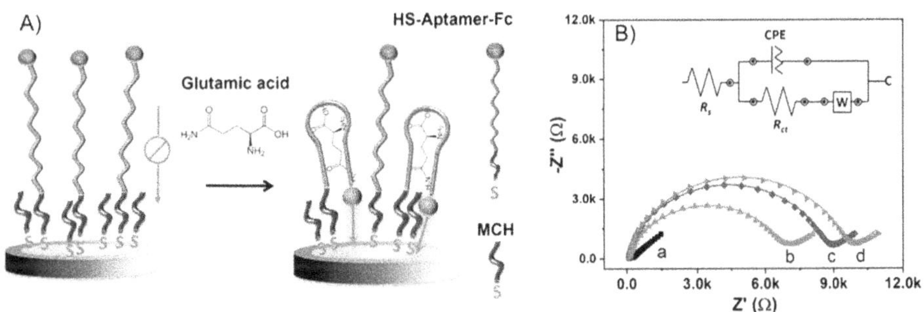

Figure 17.7. A Scheme of the label-free aptasensor working principle for the detection of glutamic acid via ferrocene- and thiol-tagged aptamer receptors. B EIS data recorded in 5.0 mM $[Fe(CN)_6]^{3-/4-}$, 10 mM PBS, 150 mM NaCl, 15 mM KCl at (a) bare gold rod electrode (AuR), (b) AuR+aptamer, (c) AuR+aptamer +MCH, and (d) after adding 1 μM glutamic acid. Solid lines are fitting results according to the equivalent circuitry model in the inset. Reprinted by permission from Springer Nature Customer Service Centre GmbH: [Springer-Verlag GmbH Germany, part of Springer Nature] [Analytical and Bioanalytical Chemistry] Wu *et al* (2022), copyright (2022).

Table 17.3. Electrochemical sensors used to detection aptameric of NTs.

Biosensor	Sensor platform	Aptamer	NT	Sample	LOD	Linear range	References
Aptameric	RGO-AuNPs/GCE	RNA	Dopamine	Human serum	0.13 μM	0.5–20 μM	Chen *et al* (2016)
	Gold nanostructure	5′ end thiolated dopamine aptamer	Dopamine	Human serum	0.01 nM	0.163–20 nM	Taheri *et al* (2018)
	Aptamer/Au	DNA	Serotonin	Rat cerebrospinal fluid samples	0.017 fM	1 pM–10 nM	Geng *et al* (2021)
	OECT	EPI binding thiolated32-mer aptamers	Epinephrine	—	90 pM	—	Saraf *et al* (2018)
	AuR/aptamer/MCH	ssDNA, 1d04 aptamer	Glutamate	Human serum, artificial CSF	0.0013 pM	0.01 pM–1 nM	Wu *et al* (2022)

AuR: gold rod electrode; MCH: 6-mercapto-1-hexanol; OECT: organic electrochemical transistor.

17.4.3 Enzymatic biosensors

Turning non-electroactive species into an electroactive analyte is necessary to detect them electrochemically. Many non-electroactive analytes can benefit from the electrochemically detectable signal produced by enzyme-based sensors, which are produced by processes. They are typically very selective and allow for lower detection limits. NT detection in bodily samples such as cerebrospinal fluid, urine blood, and serum has been made possible with enzyme-based electrochemical biosensors and sensors. Additionally, it has been shown that saliva samples can be used to perform some basic neurochemical assays; these biosensors constitute a potential population screening tool and detection of unrecognized people at extremely early stages of neurological disease development (Tavakolian-Ardakani *et al* 2019)

Enzyme-based electrochemical sensors are often a preferred NT sensing approach due to the non-electroactivity of some NTs. However, they also have some disadvantages. Enzymatic biosensors can be damaged by instability caused by enzyme denaturation or can be expensive due to the high cost of enzymes (Shadlaghani *et al* 2019).

Using the CTAC aggregation as a template, the Fe_3O_4 nanoparticles were coated with SiO_2 using the Stober method, and they were also covered with mesoporous SiO_2 to create Fe_3O_4@SiO_2@vmSiO_2 microsphere. Figure 17.8(a) shows the immobilized laccase (Lac) for DA determination, which was covalently bound to a modified microsphere and stabilized on a GCE surface (Fe_3O_4@SiO_2@vmSiO_2-Lac/GCE). EIS and CV looked into the biosensor's electrochemical characteristics. The immobilized Lac biosensor exhibits a linear range of 1.5–75 μM, a low detection limit of 0.177 μM, strong DA electrocatalytic activity, high selectivity, and sensitivity. The immobilized Lac biosensor detected DA in drug injection with a 98.7%–100.5% recovery rate (Li *et al* 2018).

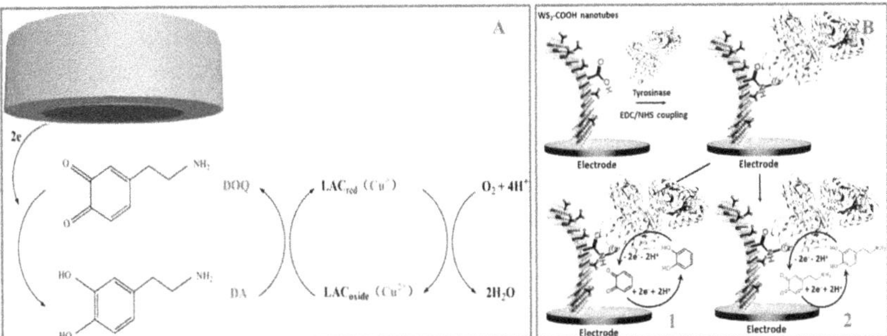

Figure 17.8. (a) Schematic representation for electrochemical detection of DA by laccase-immobilized Fe_3O_4@SiO_2@vmSiO_2-LAC/GCE as described in Li *et al* (2018). Reprinted by permission from Springer Nature Customer Service Centre GmbH: [Springer Science Business Media, LLC, part of Springer Nature] [Journal of Materials Science] [Li *et al* 2018], Copyright 2018. (b) Sketch of the functionalization of WS_2 modified glassy carbon electrodes with the enzyme tyrosinase via a standard EDC/NHS coupling reaction. These modified bioelectrodes served in the detection of catechol (1) and dopamine (2) at 0.2 V versus Ag+/Ag. Reproduced from Palomar *et al* (2020) with permission of The Royal Society of Chemisty.

Tyrosinase (Ty) has been successfully used in the fabrication of various electrochemical biosensors for the determination of dopamine. Immobilization of Ty is carried out using various protocols such as carbon paste immobilization, adsorption, covalent bonding, conducting polymer films, chemical grafting on polymers, entrapment, and cross-linking (Lete *et al* 2017).

More effective immobilization of the tyrosinase enzyme was achieved using WS_2 nanotubes functionalized with carboxylic acid functionalities (WS_2–COOH) to develop an electrochemical biosensor for dopamine and catechol shown in figure 17.8(b). To ensure reproducibility of the deposits, the nanotubes were placed in GCE using a dispersion filtration transfer technique. The K_m, Michaelis Menten constant, of the system is lower than the K_m of the tyrosinase in solution. This is because the ability to access of the active site of enzyme and efficient mass transfer of the target molecule are high. Due to the electrostatic interactions between the amine function of DA and the –COOH groups of the nanotubes, accumulation of the substrate was observed in dopamine determination and signal capture was improved at low DA levels. The overall sensor performance for dopamine was characterized by sensitivities of 6.2 ± 0.7 mA l mol^{-1} and 3.4 ± 0.4 mA l mol^{-1} and linear ranges of 0.5–10 μM and 10–40 μM. Nevertheless, these WS_2–COOH nanotubes are a potential nanomaterial for biosensing and sensing applications due to their easy handling and minimal environmental impact (Palomar *et al* 2020).

Baluta *et al* developed fast, simple, sensitive, and selective biosensor systems for detecting serotonin and dopamine using two different approaches. Miniature NT biosensors were prepared by immobilization of Lac in the electroactive layer of a Pt electrode (Pt-E) coated with poly(2,6-bis(3,4-ethylenedioxythiophene)-4-methyl-4-octyl-dithienosylol) for 5-HT detection (System A). Another biosensor was designed with an Au electrode that was modified with electroconductive polymer poly(2,6-bis (selenophen-2-yl)-4-methyl-4-octyl-dithienosylol) via immobilization of horseradish peroxidase (HRP) for DA detection. System A is based on Pt-E/bisEDOTTDTSi/ Lac for 5-HT, and system B is based on Au-E/bisSeDTSi/Pox for DA. Electrochemical measurements were performed using DPV and CV techniques. Their catalytic oxidation achieved the detection procedure of these NTs to reactive quinone derivatives in the presence of enzymes. Under optimized conditions, the analytical performance was determined with detection limits of 48 and 73 nM and over a wide linear range of 0.1–200 μM for 5-HT and DA biosensors, respectively. It showed high selectivity and reproducibility. Furthermore, the method was effectively performed for NTs determination in the presence of interfering compounds such as uric acid, ascorbic acid, and L-cysteine. The properties of both systems allow convenient, long-term techniques, simple, and stable for NT detection and show promise as great analytical bio-tools (Baluta *et al* 2020a).

In a study by Deng *et al*, for the simultaneous detection of glutamate and acetylcholine, an electrochemical biosensor array using a working electrode made of nanoporous pseudo carbon paste (nano-PPCPE) was created. Glutamate oxidase (GluOx) and acetylcholinesterase (AChE) enzymes were used in the preparation of the glutamate sensor and acetylcholine sensor, respectively. The nano-PPCPE is an amalgam of polystyrene microspheres, polymerized pyrrole, and graphite powder packed inside glass tubes with a copper wire. Two anodic peaks associated with Glu and ACh were seen on the cyclic voltammogram, which was produced at a scan rate of 0.05 V s^{-1}. The sensor's ability to discriminate between the two NTs and the fact that it can detect both glutamate and acetylthiocholine iodide simultaneously suggests that the multimodal detection of NTs is conceivable. For Glu and ACh, an LOD of 2.5×10^{-7} M with a linear range of 5×10^{-7}–1×10^{-5} M and an LOD of 1.5×10^{-7} M with a linear range of 5×10^{-6}–2×10^{-4} M were demonstrated (Deng *et al* 2013).

In recent years, multi-walled carbon nanotubes (MWCNTs) have been extensively functionalized for electrocatalytic and sensing applications of bioactive and electroactive molecules. They are frequently used in the electroanalysis of bioactive molecules due to their unique properties such as accurate and stable voltammograms, outstanding electronic behavior, chemical stability, low background current, high surface-to-volume ratio, high electrical conductivity, long potential fields, compatibility, distinctive electrical properties, and high electron transfer ability, simple preparation, and easy surface modification. Their use in MWCNT-based electrochemical sensor fabrication exhibits enhanced electroanalytical properties, including improved sensitivity, wide potential range, stubby background current, resistance to surface fouling, and reduction of overpotential (Hareesha *et al* 2021, Pushpanjali *et al* 2020, Pushpanjali and Manjunatha 2020, Tigari and Manjunatha 2020b).

In another paper, a novel amperometric bienzymatic GCE/CS-MWCNTs-Fe_3O_4NP biosensor based on chitosan (CS), MWCNTs, and iron-oxide nanoparticles (Fe_3O_4NPs), GCE was fabricated for acetylcholine determination. The biosensor was prepared with EDC-NHS by covalent binding of AChE and choline oxidase (ChOx) on the surface of the GCE/CS-MWCNTs-Fe_3O_4NPs electrode. The intended use of CS is to immobilize AChE and ChOx. Nanocomposites were characterized by EIS, CV and SEM. Electrochemical measurements are based on detecting enzymatically produced hydrogen peroxide (H_2O_2). The linear ranges of the biosensor were 0.02–0.111 μM and 0.111–1.87 μM, while the detection limit was calculated as 0.61 nM. In addition, the amperometric biosensor showed long-term stability, repeatability, reproducibility, high sensitivity, and good selectivity. The fabricated biosensor was used to determine ACh levels in serum samples. The significance of this work is that it provides an effective immobilization platform and demonstrates a novel and simple strategy for developing a multienzyme-based biosensor for ACh determination (Bolat *et al* 2017).

Gopal *et al* developed a sensitive, suitable, and simple electrochemical biosensor to quantify EPI. For this purpose, a Ty/MWCNTs/GCE biosensor platform was prepared by modification of GCE with MWCNTs followed by immobilization of Ty enzyme. The electrochemical behavior of EPI for the sensor platform was studied and a redox mechanism was predicted. The characterization was performed by electrochemical techniques such as Tafel plot studies, CV, and EIS. The effect of buffer pH on the electrochemical redox behavior of EPI was investigated, and the optimum pH value of 7.0 was determined. Electrochemical kinetic parameters were also evaluated. Hill coefficient and K_m^{app} values were estimated to examine the kinetic behavior of Ty enzyme. K_m^{app} was figured out as 0.159 mM, and Ty enzyme has high catalytic activity. The practical efficiency of the proposed Ty/MWCNT/GCE biosensor was verified regarding reproducibility and stability. Finally, this biosensor is evaluated for practical application in quantifying EPI in human serum samples (Gopal *et al* 2020).

Baluta *et al* developed and characterized a novel NE biosensing assay. A miniaturized Au-E biosensor was designed and fabricated by immobilizing tyrosinase on carbon nanoparticles and an electroactive cysteamine layer covering the gold electrode (shown in figure 17.9). It utilized the catalytic oxidation of NE to quinone measured by techniques such as DPV and CV. The sensor exhibited excellent parameters, such as a sensitivity of 4.2 μA mM^{-1} cm^{-2}, a wide linear range from 1–200 μM, a detection limit of 196 nM. It showed good selectivity when examined with a wide range of compounds such as 4tBC, EPI, AA, DA, CYS, and UA. It has also been successfully applied to real samples (Baluta *et al* 2020b).

In a different work, Ganesana *et al* improved a Glu biosensor with CS as a matrix for the immobilization of glutamate oxidase enzyme on a platinum electrode surface. A poly-o-phenylenediamine (PPD) layer was electropolymerized on a 50 μm Pt wire. GluOx in a CS matrix was dripped onto the electrode surface. This platform was coated with ascorbate oxidase to eliminate the effect due to extracellular ascorbic acid levels, which are significantly higher in real brain tissue samples. L-Glu responses were recorded at +0.6 V constant potential against Ag/AgCl. The biosensor has detection limit of 44 nM, a high sensitivity of 0.097 ± 0.001 nA μM^{-1}, a linear range from 5 to 150 μM, and storage stability of 7 days. Therefore, this biosensor shows promise for future neuroscience applications (Ganesana *et al* 2019).

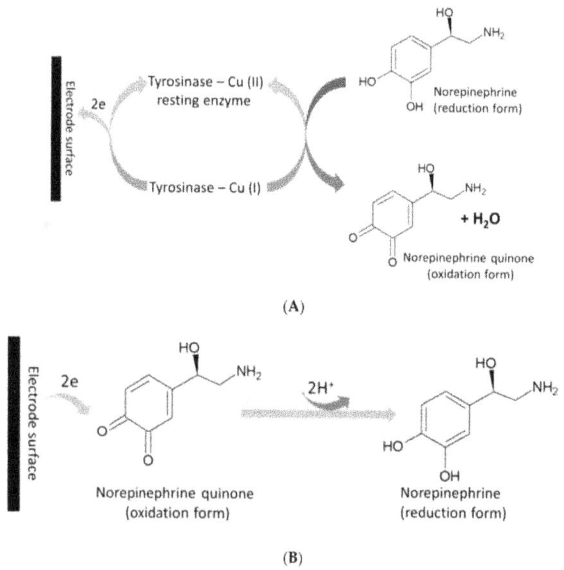

Figure 17.9. Schematic representation of electron transfer between Ty and Au-modified electrode (A) anodic reaction and (B) cathodic reaction. Reproduced from Baluta *et al* (2020b) CC BY 4.0.

In a study performed by Wang *et al*, for the detection of nitric oxide, the AuNPs-decorated MIL-101(Fe) (APPPM(Fe)) nanocomposite created using the iron-based metal organic framework (MIL-101(Fe)), which was created via a straightforward solvothermal-synthesis (shown in figure 17.10). Due to the multilayer construction method, the amount of AuNPs loaded onto APPPM(Fe) increased and produced an excellent biocompatible interface and good conductivity. High-loading amounts of HRP were immobilized on APPPM(Fe) by the strong Au–S bond and the porous structure of MIL-101(Fe) while supporting natural bioactivity of HRP. With a low detection limit of 0.01 μM and a large linear range from 0.033 to 5370 μM, the built-in biosensor demonstrated outstanding electrocatalysis activity for detecting NO and showed exceptional stability, reproducibility and specificity, according to electrochemical kinetics. According to a study of real sample data, the suggested biosensor provides quick and easy detection of nitric oxide in serum samples. As a result, the current technology offers a chance for implementation in NO clinic disease diagnosis and treatment (Wang *et al* 2022).

An enzymatic biosensing system for H_2S detection based on the inhibition of ascorbate oxidase (AOx) activity was developed. Optimum glutaraldehyde (GA) concentration of 17.5% (w/v) and optimum pH 7 were optimized. Between the H_2S concentration and the decrease in concentration of oxygen was observed a linear relationship. Inhibition of ascorbate oxidase by H_2S was found using a Clark electrode as a transducer in a batch system. It proved to be a simple method to monitor the biological reaction before and after H_2S injection. The biosensing platform stability was determined as a response to H_2S concentration under optimal operating conditions. A detection limit of 0.5 mg l^{-1} and the H_2S concentration range of 1–15 mg l^{-1} were calculated (Amoabediny *et al* 2014). In addition, a comparison of these sensors is shown in table 17.4.

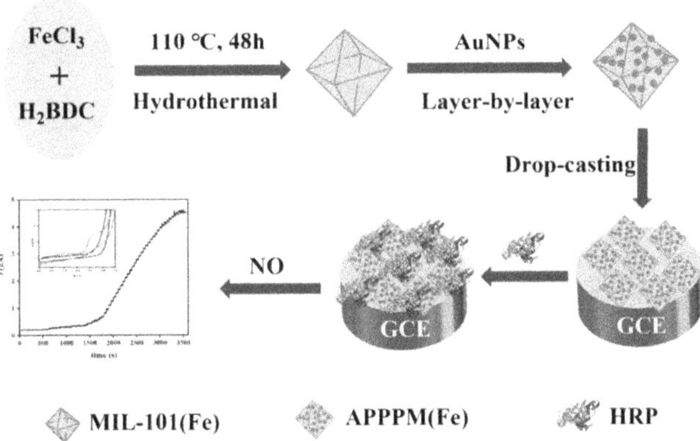

Figure 17.10. Schematic representations of the preparation of APPPM(Fe) and the construction of the nitric oxide biosensor. Reprinted by permission from Springer Nature Customer Service Centre GmbH: [Springer-Verlag GmbH Austria, part of Springer Nature] [Microchimica Acta] from Wang *et al* (2022), copyright (2022).

Table 17.4. Electrochemical sensors used to detection enzymatic of NTs.

Biosensor	Sensor platform	Enzyme	NT	Sample	LOD	Linear range	References
Enzymatic	Fe_3O_4@SiO_2@vmSiO_2	Laccase	Dopamine		0.177 μM	1.5–75 μM	Li *et al* (2018)
	WS_2–COOH nanotube/GCE	Tyrosinase	Dopamine	—	—	0.5–10 μM	Palomar *et al* (2020)
	Au-E/bisSeDTSi/Pox	Horseradish peroxidase	Dopamine	Human serum	73 nM	0.1–200 μM	Baluta *et al* (2020a)
	Pt-E/bisEDOTDTSi/Lac	Laccase	Serotonin	Human serum	48 nM	0.1–200 μM	Baluta *et al* (2020a)
	MWCNTs-Fe_3O_4NPs-CS	Acetylcholine esterase-choline oxidase	Acetylcholine	Serum sample	0.61 nM	0.02–0.111 μM 0.111–1.87 μM	Bolat *et al* (2017)
	Ty/MWCNTs/GCE	Tyrosinase	Epinephrine	Spiked PBS and human serum sample	0.51 μM	0.6–100 μM	Gopal *et al* (2020)
	Au-E/Cys/CDs/Ty	Tyrosinase	Norepinephrine	Synthetic urine	196 nM	1–200 μM	Baluta *et al* (2020b)
	Pt-Ir/PPD/GlutOx/AsOx/BSA	Glutamate oxidase	Glutamate	Rat brain	0.044 nM	5–150 μM	Ganesana *et al* (2019)
	HRP/APPPM(Fe)/GCE	Horseradish peroxidase	Nitric oxide	Serum			Wang *et al* (2022)
	PGE/ [SWCNTs/(DOPE: DOTAP:DSPE-PEG)/NOR]	Nitric oxide reductase	Nitric oxide	Biological system	0.13 μM	0.44–9.09 μM	Gomes *et al* (2019)
	Clark oxygen electrode	Ascorbate oxidase	Hydrogen sulfide	—	0.5 mg l^{-1}	1–15 mg l^{-1}	Amoabediny *et al* (2014)

bisEDOTDTSi: poly[2,6-bis(3,4-ethylenedioxythiophene)-4-methyl-4-octyl-dithienosilole; bisSeDTSi: poly[2,6-bis(selenophen-2-yl)-4-methyl-4-octyl-dithienosilole; Pox: horse-radish peroxidase; Ty: Tyrosinase; Au-E: gold electrode; Cys:Cysteamine; CDs: carbon dots, GlutOx: glutamate oxidase; AsOx: ascorbate oxidase; PPD: poly-o-phenylenediamine; DOPE: 1,2-di-(9Z-octadecenoyl)-sn-glycero-3-phosphoethanolamine; DOTAP: 1,2-di-(9Z-octadecenoyl)-3-trimethylammonium-propane; DSPE-PEG: 1,2-distearoyl-sn-glycero-3-phosphoethanolamine—polyethylene glycol; NOR: nitric oxide reductase.

17.4.4 Molecularly imprinted polymer-based sensor/biosensors

Molecular recognition is an essential step for artificial biological mimicry systems. Molecular imprinting techniques can separate and detect many molecules, such as metal ions, microbial cells, bacteriophage, protein, amino acids, drugs, pesticides, food additives, organic pollutants, and biomolecules, regardless of size and shape. Molecular imprinting is also frequently used in electrochemical sensors with lower detection limits, selectivity, and sensitivity. The preparation steps of conventional electrochemistry methods are easy and economical at the same time. Combining these features with molecular imprinting has led to the emergence of new and exciting molecularly imprinted electrochemical sensors (MIECSs). Molecularly imprinted polymer (MIP) sensors' advantages over conventional electrochemical and traditional instrumental techniques are miniaturization, high chemical and mechanical stability, high sensitivity and selectivity, reusability and automation. With high selectivity and sensitivity, nM detection limits can be achieved using lower sample volumes, such as 5–20 ml. The concentration of the target analyte can be obtained from the measured current. These sensors can achieve greater selectivity and sensitivity by changing the working electrode surface (Ann Maria *et al* 2021, Suryanarayanan *et al* 2010, Yáñez-Sedeño *et al* 2017). By combining aptamers or MIPs with electrochemistry, new products with fast detection and high selectivities can be obtained. MIP-based electrochemical sensors are used in many application areas. A few studies are presented in table 17.5.

Li *et al* have fabricated a novel electrochemical sensor for the simultaneous determination of DA and uric acid (UA) based on nanoporous gold leaf (NPGL) dual-templated (MIP). The NPGL provided an expanded specific surface area and an enhanced electron transfer rate for the sensor. The MIP layer was electrochemically polymerized on the NPGL layer in the presence of monomers and double templates to ensure specific identification towards DA and UA. Under optimum conditions for DA, a detection limit of 0.3 μM and a good linear range of 2.0–180 μM at an operating potential of 0.15 V (versus Ag/AgCl) were determined. The responses remained stable at 96% even after 30 days of storage. Simultaneous real sample determination of UA and DA was performed with bovine serum. The dual-template MIP sensor demonstrated sufficient robustness, selectivity, reproducibility and accuracy. It is concluded that this study offers a cost-effective, facile, reliable, and rapid method for the simultaneous quantification of two or more chemicals/biomolecules (Li *et al* 2020).

In Si and Song's study, they proposed electrochemical sensors for detecting multiple NTs using MIPs. DA, EPI and NE were specifically chosen as model NTs, and three sensors were developed to detect each analyte. o-phenylenediamine (o-PD) and pyrrole (PPy) were used as functional monomers to prepare the MIP sensor. DA, NE, and EPI were detected by applying DPV to the developed selective MIP-based sensor. Furthermore, the selectivity of each analyte was determined. It was shown that the MIP sensors have higher sensitivity than the non-imprinted NIP sensors. The detection limits of all developed MIP sensors were found to be different but less than 1.3×10^{-5} M. The selectivity of the norepinephrine sensor was found to be lower. Cross-reaction testing confirmed that the developed technique could be

Table 17.5. Electrochemical sensors used to MIP-based detection of NTs.

Biosensor	Sensor platform	Functional Monomer	NT	Sample	LOD	Linear range	References
MIP	MIP/NPGL/GCE	—	Dopamine	Bovine serum	0.3 μM	2.0–180 μM	Li *et al* (2020)
	AuNPs@MIPs/GCE	PANI	Serotonin	Human blood serum	11.7 nM	0.2–10.0 μM	Xue *et al* (2014)
	Fe_3O_4@Au@SiO_2_MIP/GPE	PVP, TEOS,	Serotonin	Pharmaceutical capsules and urine samples	0.002 μM	0.01–1000 μM	Amatatongchai *et al* (2019)
	MIP/AuNPs/GCE	3-TBA	Epinephrine	Injection sample	76 nM	90 nM–100 μM	Liu and Kan (2019)
	MIP/SWNTs/GCEs	Silica sol	Norepinephrine	Blood sample	33.3 nM	99.9 nM–15.0 μM	Wang *et al* (2017)
	MIPs/HA-MWCNTs/PPy-SG/GCE	PPy	Tryptamine	Cheese and lactobacillus beverage sample	74 nM	90 nM–70 μM	Xing *et al* (2012)

PPy: polypyrol; NPGL: nanoporous gold leaf; PANI: Polyaniline; GPE: graphite electrode; PVP: poly (vinylpyrrolidone); TEOS: tetraethoxysilane, PTMOS: phenyltrimethoxysilane; 3-TBA: 3-tiyofen boronik asit.

used as a multi-analyte sensing platform capable of detecting multiple NTs simultaneously from a single sample solution. This technique can be further extended to measure other electroactive NT species and has enormous potential for simultaneous monitoring of different chemical species (Si and Song 2018b). Multi-analyte detection platform is shown in figure 17.11.

For sensitive and targeted 5-HT detection, a bilayer membrane sensing interface based on RGO/polyaniline (PANI) nanocomposites was developed. It is embedded with AuNPs and MIPs. Figure 17.12 shows a schematic representation of the preparation of sensor. Electrostatic adsorption was used first to bind the protonated anilines to RGO sheets, and then CV was used to form the RGO/PANI membrane on the bare electrode. This procedure was used to develop the RGO/PANI nanocomposites. In the presence of 5-HT and p-amino thiophenol, functionalized gold nanoparticles (F-AuNPs) were synthesized to fabricate MIPs embedded with AuNPs on modified electrodes (AuNPs@MIPs) membrane. UV–vis and Raman spectroscopy were used to evaluate the materials prepared for this work, while electrochemical techniques SEM, and EDS characterized the membranes. In addition, the obtained sensor provided a remarkable selectivity towards ascorbic acid and other interferences. A detection limit (11.7 nM) and linear concentration range values (0.2–10.0 μM) were determined. Further, the resulting biomimetic sensor was used to detect serotonin in human serum samples (Xue *et al* 2014).

In another study, Liu and Kan prepared a novel MIP based on the electrochemical sensor for EPI detection by modifying MIP/AuNPs composite on the surface of GCE. 3-thiophene boronic acid (3-TBA) monomer was used for the electropolymerization of MIP in the existence of EPI. The modified AuNPs and conductive poly(3-TBA) matrix increased the electrode's surface area and improved the sensor's sensitivity. SEM, CV, and EIS characterization procedures were carried out. The prepared sensor demonstrated selective recognition capabilities and sensitive sensing against EPI. Under optimum conditions, the sensor can sensitively detect EPI with a linear concentration range 90 nM–100 μM and LOD of 76 nM. The sensor showed dual recognition capability to EPI and can recognize EPI from its analogs. The sensor was successfully applied to analyze EPI in injection samples (Liu and Kan 2019).

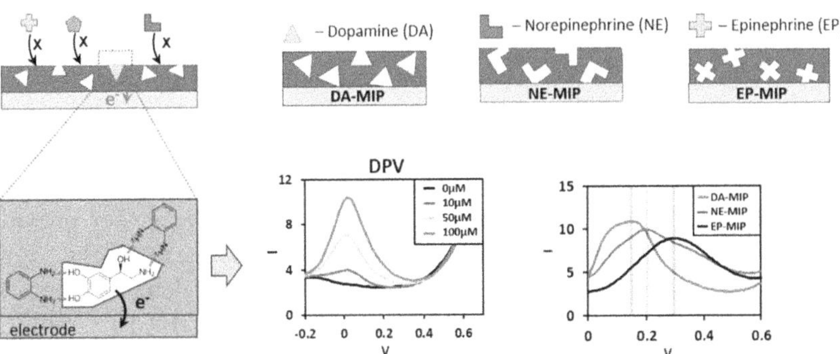

Figure 17.11. Schematic representation of MIP-based sensor for multi-analyte detection. Reprinted from Si and Song (2018b), copyright (2018), with permission from Elsevier.

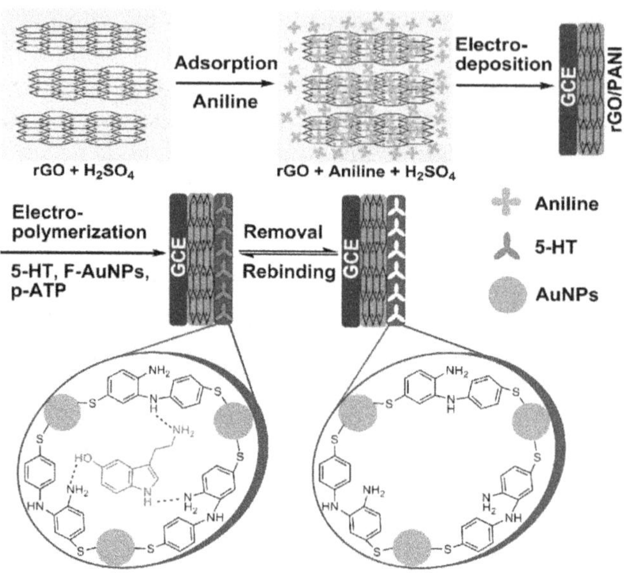

Figure 17.12. Schematic representation for the preparation of the AuNPs@MIES. Reprinted from Xue *et al* (2014). Copyright (2014), with permission from Elsevier.

Wang *et al* constructed a novel MIP/SWNTs/GCEs sensor for norepinephrine detection using the sol–gel method. A selective and sensitive imprinted electrochemical sensor was created by antimony-doped tin oxide (ATO)-silica composite sol on a GCE modified with single-walled carbon nanotubes (SWCNTs). SWCNTs enhanced the transfer of electrons and biological molecules on the electrode, increasing the sensitivity of the MIP sensor with its electrochemical signals. The surface morphology of the sensors was characterized by SEM and the optimal conditions were determined. As determined by CV, the proposed electrochemical sensor had an LOD of 33.3 nM and showed a good linear relationship concentration in the range of 99.9 nM–15 μM. Furthermore, the sensors were effectively applied to detect NE concentration in human blood serum samples (Wang *et al* 2017).

Xing *et al* reported an MIP based on the electrochemical sensor for the sensitive detection of tryptamine based on polypyrrole-sulfonated graphene/hyaluronic acid-MWCNTs (PPy-SG/HA-MWCNTs). MIPs were synthesized by electropolymerization technique using tryptamine as a template- and para-aminobenzoic acid (pABA) (monomer). The key parameters affecting the performance of the MIP sensor were investigated and optimized. The surface properties of the prepared sensors were tested by chronoamperometry and characterized by CV. The sensor provided TA determination with good selectivity from samples containing interfering tyramine, dopamine, and tryptophan. Under optimum conditions, the detection limit for TA was 7.4×10^{-8} mol l^{-1} and the linear range was 9.0×10^{-8}–7.0×10^{-5} mol l^{-1}. This MIP-based electrochemical sensor was successfully used for the determination of TA in real food samples (Xing *et al* 2012).

17.5 Conclusion

This chapter presents information about common NTs, and electrochemical sensors developed for these NTs are discussed. Neurological diseases affect people's quality of life and cause social and economic costs. Early detection of these diseases is important both socially and economically. When these diseases occur, there are changes in the values of some NTs. Altered concentrations of NTs in the central nervous system cause numerous neurological and physical disorders. In short, the analysis of NTs facilitates diagnosing central nervous system diseases. Electrochemical sensors/biosensors are a promising technique for the detection of NTs. The use of electrochemical sensors for the determination of NTs offers low cost, reliability, sensitivity, and selectivity. Electrochemical sensors provide many important analytical features and represent promising candidates for future clinical diagnostics. Therefore, sensors and biosensors developed for the sensitive and selective detection of NTs provide significant advantages in diagnosing and monitoring these diseases. According to research results in the literature, electrochemical sensors based on the recognition of NTs can be produced in different forms. Even sensor platforms that allow the co-detection of several NTs can be constructed. The fact that different methods can prepare sensors facilitates the production of sensors with more suitable features in terms of rapid detection, portability, sensitivity, accuracy, cost-effectiveness, functionality, and miniaturization. Most electrochemical methods provide a quick, highly sensitive and selective analysis of NTs.

Despite some challenges, electrochemical sensors are the best method for detecting NTs in real samples and clinical laboratories. These challenges are expected to be overcome soon. Consequently, developing sensors that enable automated, rapid, or *in situ* detection analysis for NTs is still ongoing.

Acknowledgments

Kübra Turan was supported by The Scientific and Technological Research Council of Türkiye (TÜBİTAK) with a 2218-National Postdoctoral Research Fellowship Program scholarship in the 121C426 number project.

References

Abu-Ali H, Cansu O, Davis F, Walch N and Nabok A 2020 Electrochemical aptasensor for detection of dopamine *Chemosensors* **8** 28

Amatatongchai M, Sitanurak J, Sroysee W, Sodanat S, Chairam S, Jarujamrus P, Nacapricha D and Lieberzeit P A 2019 Highly sensitive and selective electrochemical paper-based device using a graphite screen-printed electrode modified with molecularly imprinted polymers coated Fe_3O_4@Au@SiO_2 for serotonin determination *Anal. Chim. Acta* **1077** 255–65

Amoabediny G, poor N Z M, Baniasadi L, Omidi M, Yazdian F, Attar H, heydarzadeh A, Zarami A S H and Sheikhha M H 2014 An inhibitory enzyme electrode for hydrogen sulfide detection *Enzyme Microb. Technol.* **63** 7–12

Ann Maria C G, Varghese A and Nidhin M 2021 Recent advances in nanomaterials based molecularly imprinted electrochemical sensors *Critical Reviews in Analytical Chemistry* (London: Taylor and Francis Ltd) p 10

Aoki I, Shirane K, Tokimoto T and Nakagawa K 1986 Separation of fine particles using rotating tube with alternate flow *Rev. Sci. Instrum.* **57** 2859–61

Atta N F, Galal A, El-Ads E H and Galal A E 2020 Efficient electrochemical sensor based on gold nanoclusters/carbon ionic liquid crystal for sensitive determination of neurotransmitters and anti-Parkinson drugs *Adv. Pharm. Bull.* **10** 46–55

Ayano G 2016 Dopamine: Receptors, functions, synthesis, pathways, locations and mental disorders: Review of literatures *J. Ment. Disord. Treat.* **2** 2–5

Baluta S, Lesiak A and Cabaj J 2020a Simple and cost-effective electrochemical method for norepinephrine determination based on carbon dots and tyrosinase *Sensors (Switzerland)* **20** 1–13

Baluta S, Zając D, Szyszka A, Malecha K and Cabaj J 2020b Enzymatic platforms for sensitive neurotransmitter detection *Sensors (Switzerland)* **20** 423

Banerjee S, McCracken S, Faruk Hossain M and Slaughter G 2020 Electrochemical detection of neurotransmitters *Biosensors* **10** 101

Baranwal A and Chandra P 2018 Clinical implications and electrochemical biosensing of monoamine neurotransmitters in body fluids, *in vitro*, *in vivo*, and *ex vivo* models *Biosens. Bioelectron.* **121** 137–52

Bolat E Ö, Tığ G A and Pekyardımcı Ş 2017 Fabrication of an amperometric acetylcholine esterase-choline oxidase biosensor based on MWCNTs-Fe_3O_4NPs-CS nanocomposite for determination of acetylcholine *J. Electroanal. Chem.* **785** 241–8

Brown M D, Hall J R and Schoenfisch M H 2019 A direct and selective electrochemical hydrogen sulfide sensor *Anal. Chim. Acta* **1045** 67–76

Bucher E S and Wightman R M 2015 Electrochemical analysis of neurotransmitters *Annu. Rev. Anal. Chem.* **8** 239–61

Burchett S A and Hicks T P 2006 The mysterious trace amines: protean neuromodulators of synaptic transmission in mammalian brain *Prog. Neurobiol.* **79** 223–46

Cernat A, Ştefan G, Tertis M, Cristea C and Simon I 2020 An overview of the detection of serotonin and dopamine with graphene-based sensors *Bioelectrochemistry* **136** 107620

Chauhan N, Chawla S, Pundir C S and Jain U 2017 An electrochemical sensor for detection of neurotransmitter-acetylcholine using metal nanoparticles, 2D material and conducting polymer modified electrode *Biosens. Bioelectron.* **89** 377–83

Chávez J L, Hagen J A and Kelley-Loughnane N 2017 Fast and selective plasmonic serotonin detection with aptamer-gold nanoparticle conjugates *Sensors (Switzerland)* **17** 681

Chen T, Tang L, Yang F, Zhao Q, Jin X, Ning Y and Zhang G J 2016 Electrochemical determination of dopamine by a reduced graphene oxide–gold nanoparticle-modified glassy carbon electrode *Anal. Lett.* **49** 2223–33

Costa D J E, Martínez A M, Ribeiro W F, Bichinho K M, di Nezio M S, Pistonesi M F and Araujo M C U 2016 Determination of tryptamine in foods using square wave adsorptive stripping voltammetry *Talanta* **154** 134–40

Deng Y, Wang W, Ma C and Li Z 2013 Fabrication of an electrochemical biosensor array for simultaneous detection of L-glutamate and acetylcholine *J. Biomed. Nanotechnol.* **9** 1378–82

Donald J A and Cameron M S 2021 Nitric oxide *Handbook of Hormones: Comparative Endocrinology for Basic and Clinical Research* **vol 29** (New York: Academic) pp 1083–6

Francis P T, Palmer A M, Snape M and Wilcock G K 1999 The cholinergic hypothesis of Alzheimer's disease: a review of progress *J. Neurol. Neurosurg. Psychiatry* **66** 137–47

Ganesana M, Trikantzopoulos E, Maniar Y, Lee S T and Venton B J 2019 Development of a novel micro biosensor for *in vivo* monitoring of glutamate release in the brain *Biosens. Bioelectron.* **130** 103–9

Gao L L and Gao E Q 2021 Metal–organic frameworks for electrochemical sensors of neurotransmitters *Coord. Chem. Rev.* **434** 213784

Geng X, Zhang M, Long H, Hu Z, Zhao B, Feng L and Du J 2021 A reusable neurotransmitter aptasensor for the sensitive detection of serotonin *Anal. Chim. Acta* **1145** 124–31

Gomes F O, Maia L B, Loureiro J A, Pereira M C, Delerue-Matos C, Moura I, Moura J J G and Morais S 2019 Biosensor for direct bioelectrocatalysis detection of nitric oxide using nitric oxide reductase incorporated in carboxylated single-walled carbon nanotubes/lipidic 3 bilayer nanocomposite *Bioelectrochemistry* **127** 76–86

Gopal P, Narasimha G and Reddy T M 2020 Development, validation and enzyme kinetic evaluation of multi walled carbon nano tubes mediated tyrosinase based electrochemical biosensing platform for the voltammetric monitoring of epinephrine *Process Biochem.* **92** 476–85

Hareesha N, Manjunatha J G, Amrutha B M, Pushpanjali P A, Charithra M M and Prinith Subbaiah N 2021 Electrochemical analysis of indigo carmine in food and water samples using a poly(glutamic acid) layered multi-walled carbon nanotube paste electrode *J. Electron. Mater.* **50** 1230–8

Hasanzadeh M, Shadjou N and Guardia M de la 2017 Current advancement in electrochemical analysis of neurotransmitters in biological fluids *TrAC Trends Anal. Chem.* **86** 107–21

Iversen S D and Iversen L L 2007 Dopamine: 50 years in perspective *Trends Neurosci.* **30** 188–93

Jinap S and Hajeb P 2010 Glutamate. Its applications in food and contribution to health *Appetite* **55** 1–10

Lavanya N and Sekar C 2017 Electrochemical sensor for simultaneous determination of epinephrine and norepinephrine based on cetyltrimethylammonium bromide assisted SnO_2 nanoparticles *J. Electroanal. Chem.* **801** 503–10

Lawrence N S, Deo R P and Wang J 2004 Electrochemical determination of hydrogen sulfide at carbon nanotube modified electrodes *Anal. Chim. Acta* **517** 131–7

Lete C, Lakard B, Hihn J Y, del Campo F J and Lupu S 2017 Use of sinusoidal voltages with fixed frequency in the preparation of tyrosinase based electrochemical biosensors for dopamine electroanalysis *Sens. Actuators* B **240**

Li N, Nan C, Mei X, Sun Y, Feng H and Li Y 2020 Electrochemical sensor based on dual-template molecularly imprinted polymer and nanoporous gold leaf modified electrode for simultaneous determination of dopamine and uric acid *Microchim. Acta* **187** 496

Li X, Tian A, Wang Q, Huang D, Fan S, Wu H and Zhang H 2019 An electrochemical sensor based on platinum nanoparticles and mesoporous carbon composites for selective analysis of dopamine *Int. J. Electrochem. Sci.* **14** 1082–91

Li Z, Zheng Y, Gao T, Liu Z, Zhang J and Zhou G 2018 Fabrication of biosensor based on core–shell and large void structured magnetic mesoporous microspheres immobilized with laccase for dopamine detection *J. Mater. Sci.* **53** 7996–8008

Liu F and Kan X 2019 Conductive imprinted electrochemical sensor for epinephrine sensitive detection and double recognition *J. Electroanal. Chem.* **836** 182–9

Liu X, He L, Li P, Li X and Zhang P 2021 A direct electrochemical H_2S sensor based on Ti_3 C_2T_x MXene *ChemElectroChem.* **8** 3658–65

Liu X and Liu J 2021 Biosensors and sensors for dopamine detection *View* **2** 20200102

Manjunatha J G 2020 Poly (adenine) modified graphene-based voltammetric sensor for the electrochemical determination of catechol, hydroquinone and resorcinol *Open Chem. Eng. J.* **14** 52–62

Manjunatha J G, Kumara Swamy B E, Gilbert O, Mamatha G P and Sherigara B S 2010 Sensitive voltammetric determination of dopamine at salicylic acid and TX-100, SDS, CTAB modified carbon paste electrode *Int. J. Electrochem. Sci.* **5** 682–95

Manjunatha J G, Raril C, Hareesha N, Charithra M M, Pushpanjali P A, Tigari G, Ravishankar D K, Mallappaji S C and Gowda J 2021 Electrochemical fabrication of poly (niacin) modified graphite paste electrode and its application for the detection of riboflavin *Open Chem. Eng. J.* **14** 90–8

Mathai J C, Missner A, Kügler P, Saparov S M, Zeidel M L, Lee J K and Pohl P 2009 No facilitator required for membrane transport of hydrogen sulfide *Proc. Natl Acad. Sci.* **106** 16633–8

Meng X, Guo W, Qin X, Liu Y, Zhu X, Pei M and Wang L 2014 A molecularly imprinted electrochemical sensor based on gold nanoparticles and multiwalled carbon nanotube-chitosan for the detection of tryptamine *RSC Adv.* **4** 38649–54

Mohammad-Zadeh L F, Moses L and Gwaltney-Brant S M 2008 Serotonin: a review *J. Vet. Pharmacol. Ther.* **31** 187–99

Mohammadi S Z, Beitollahi H and Tajik S 2018 Nonenzymatic coated screen-printed electrode for electrochemical determination of acetylcholine *Micro Nano Syst. Lett.* **6** 9

Naveen M H, Gurudatt N G and Shim Y B 2017 Applications of conducting polymer composites to electrochemical sensors: a review *Appl. Mater. Today* **9** 419–33

Ou Y, Buchanan A M, Witt C E and Hashemi P 2019 Frontiers in electrochemical sensors for neurotransmitter detection: Towards measuring neurotransmitters as chemical diagnostics for brain disorders *Anal. Methods* **11** 2738–55

Palanisamy S, Ku S and Chen S M 2013 Dopamine sensor based on a glassy carbon electrode modified with a reduced graphene oxide and palladium nanoparticles composite *Microchim. Acta* **180** 1037–47

Palomar Q, Gondran C, Lellouche J P, Cosnier S and Holzinger M 2020 Functionalized tungsten disulfide nanotubes for dopamine and catechol detection in a tyrosinase-based amperometric biosensor design *J. Mater. Chem.* B **8** 3566–73

Platt S R 2007 The role of glutamate in central nervous system health and disease—a review *Vet. J.* **173** 278–86

Pradhan T, Jung H S, Jang J H, Kim T W, Kang C and Kim J S 2014 Chemical sensing of neurotransmitters *Chem. Soc. Rev.* **43** 4684–713

Prinith N S, Manjunatha J G and Hareesha N 2021 Electrochemical validation of L-tyrosine with dopamine using composite surfactant modified carbon nanotube electrode *J. Iran. Chem. Soc.* **18** 3493–503

Pushpanjali P A and Manjunatha J G 2020 Development of polymer modified electrochemical sensor for the determination of alizarin carmine in the presence of tartrazine *Electroanalysis* **32** 2474–80

Pushpanjali P A, Manjunatha J G and Srinivas M T 2020 Highly sensitive platform utilizing poly (L-methionine) layered carbon nanotube paste sensor for the determination of voltaren *FlatChem.* **24** 100207

Qin X, Zhang J, Shao W, Liu X, Zhang X, Chen F, Qin X, Wang L, Luo D and Qiao X 2021 Modification of electrodes with self-assembled, close-packed AuNPs for improved signal reproducibility toward electrochemical detection of dopamine *Electrochem. Commun.* **133** 107161

Raril C and Manjunatha J G 2020 Fabrication of novel polymer-modified graphene-based electrochemical sensor for the determination of mercury and lead ions in water and biological samples *J. Anal. Sci. Technol.* **11** 3

Rizzo S *et al* 2008 Benzofuran-based hybrid compounds for the inhibition of cholinesterase activity, β amyloid aggregation, and Aβ neurotoxicity *J. Med. Chem.* **51** 2883–6

Rosy , Chasta H and Goyal R N 2014 Molecularly imprinted sensor based on o-aminophenol for the selective determination of norepinephrine in pharmaceutical and biological samples *Talanta* **125** 167–73

Sainz R *et al* 2020 Chemically synthesized chevron-like graphene nanoribbons for electrochemical sensors development: Determination of epinephrine *Sci. Rep.* **10** 1–11

Saraf N, Woods E R, Peppler M and Seal S 2018 Highly selective aptamer based organic electrochemical biosensor with pico-level detection *Biosens. Bioelectron.* **117** 40–6

Shadlaghani A, Farzaneh M, Kinser D and Reid R C 2019 Direct electrochemical detection of glutamate, acetylcholine, choline, and adenosine using non-enzymatic electrodes *Sensors (Switzerland)* **19** 447

Shaikshavali P, Madhusudana Reddy T, Venu Gopal T, Venkataprasad G, Kotakadi V S, Palakollu V N and Karpoormath R 2020 A simple sonochemical assisted synthesis of nanocomposite (ZnO/MWCNTs) for electrochemical sensing of Epinephrine in human serum and pharmaceutical formulation *Colloids Surf.* A **584** 124038

Shen X, Peter E A, Bir S, Wang R and Kevil C G 2012 Analytical measurement of discrete hydrogen sulfide pools in biological specimens *Free Radical Biol. Med.* **52** 2276–83

Si B and Song E 2018a Recent advances in the detection of neurotransmitters *Chemosensors* **6** 1–24

Si B and Song E 2018b Molecularly imprinted polymers for the selective detection of multi-analyte neurotransmitters *Microelectron. Eng.* **187–8** 58–65

Smith Q R 2000 Transport of glutamate and other amino acids at the blood-brain barrier *J. Nutr.* **130** 1016–22

Snyder S H and Bredt D S 1992 Biological roles of nitric oxide *Sci. Am.* **266** 68–77

Suryanarayanan V, Wu C T and Ho K C 2010 Molecularly imprinted electrochemical sensors *Electroanalysis* **22** 1795–811

Taheri R A, Eskandari K and Negahdary M 2018 An electrochemical dopamine aptasensor using the modified Au electrode with spindle-shaped gold nanostructure *Microchem. J.* **143** 243–51

Tatsumi H and Ueda T 2011 Ion transfer voltammetry of tryptamine, serotonin, and tryptophan at the nitrobenzene/water interface *J. Electroanal. Chem.* **655** 180–3

Tavakolian-ArdakaniHosu , Cristea , Mazloum-Ardakani and Marrazza 2019 Latest trends in electrochemical sensors for neurotransmitters: a review *Sensors* **19** 2037

Tigari G and Manjunatha J G 2020a Optimized voltammetric experiment for the determination of phloroglucinol at surfactant modified carbon nanotube paste electrode *Instrum. Exp. Tech.* **63** 750–7

Tigari G and Manjunatha J G 2020b Poly(glutamine) film-coated carbon nanotube paste electrode for the determination of curcumin with vanillin: an electroanalytical approach *Monatsh. Chem.* **151** 1681–8

Wang R 2002 Two's company, three's a crowd: can H_2S be the third endogenous gaseous transmitter? *FASEB J.* **16** 1792–8

Wang Y, Zhou Y, Chen Y, Yin Z, Hao J, Li H and Liu K 2022 Simple and sensitive nitric oxide biosensor based on the electrocatalysis of horseradish peroxidase on AuNPs@metal–organic framework composite-modified electrode *Microchim. Acta* **189** 162

Wang Z, Wang K, Zhao L, Chai S, Zhang J, Zhang X and Zou Q 2017 Tin oxide-silica composite sol on a glassy carbon electrode modified by single-walled carbon nanotubes for detection of norepinephrine *Mater. Sci. Eng.* C **80** 180–6

Wenninger N, Bračič U, Kollau A, Pungjunun K, Leitinger G, Kalcher K and Ortner A 2021 Development of an electrochemical sensor for nitric oxide based on carbon paste electrode modified with Nafion, gold nanoparticles and graphene nanoribbons *Sens. Actuators* B **346**

Wu C, Barkova D, Komarova N, Offenhäusser A, Andrianova M, Hu Z, Kuznetsov A and Mayer D 2022 Highly selective and sensitive detection of glutamate by an electrochemical aptasensor *Anal. Bioanal. Chem.* **414** 1609–22

Wu L, Feng L, Ren J and Qu X 2012 Electrochemical detection of dopamine using porphyrin-functionalized graphene *Biosens. Bioelectron.* **34** 57–62

Xing X, Liu S, Yu J, Lian W and Huang J 2012 Electrochemical sensor based on molecularly imprinted film at polypyrrole-sulfonated graphene/hyaluronic acid-multiwalled carbon nanotubes modified electrode for determination of tryptamine *Biosens. Bioelectron.* **31** 277–83

Xue C, Wang X, Zhu W, Han Q, Zhu C, Hong J, Zhou X and Jiang H 2014 Electrochemical serotonin sensing interface based on double-layered membrane of reduced graphene oxide/polyaniline nanocomposites and molecularly imprinted polymers embedded with gold nanoparticles *Sens. Actuators* B **196** 57–63

Yáñez-Sedeño P, Campuzano S and Pingarrón J M 2017 Electrochemical sensors based on magnetic molecularly imprinted polymers: a review *Anal. Chim. Acta* **960** 1–17

Zamani M, Wilhelm T and Furst A L 2022 Perspective—electrochemical sensors for neurotransmitters and psychiatrics: steps toward physiological mental health monitoring *J. Electrochem. Soc.* **169** 047513

Zeynaloo E, Yang Y-P, Dikici E, Landgraf R, Bachas L G and Daunert S 2021 Design of a mediator-free, non-enzymatic electrochemical biosensor for glutamate detection *Nanomed. Nanotechnol. Biol. Med.* **31** 102305

Zhang D, Wang Y, Geng W and Liu H 2019 Rapid detection of tryptamine by optosensor with molecularly imprinted polymers based on carbon dots-embedded covalent-organic frameworks *Sens. Actuators* B **285** 546–52

Zhang H M, Zhou X L, Hui R T, Li N Q and Liu D P 2002 Studies of the electrochemical behavior of epinephrine at a homocysteine self-assembled electrode *Talanta* **56** 1081–8

Zhao L, Niu G, Gao F, Lu K, Sun Z, Li H, Stenzel M, Liu C and Jiang Y 2021 Gold nanorods (AuNRs) and zeolitic imidazolate framework-8 (ZIF-8) core–shell nanostructure-based electrochemical sensor for detecting neurotransmitters *ACS Omega* **6** 33149–58

Zhu M, Zeng C, Ye J and Sun Y 2018 Simultaneous *in vivo* voltammetric determination of dopamine and 5-hydroxytryptamine in the mouse brain *Appl. Surf. Sci.* **455** 646–52

IOP Publishing

Real-Time Applications of Advanced Electrochemical Sensing Devices

Jamballi G Manjunatha

Chapter 18

Electrochemical devices for determination of amino acids

Parvin Abedi Ghobadloo, Arezou Taghvimi, Samin Hamidi and Sevinc Kurbanoglu

In recent decades, much research has been done about electrochemistry, which has led to identifying biological molecules. Among these compounds, amino acids (AAs) have attracted the attention of many scientists and researchers. According to successive research, AAs have been identified as the most important compounds that play a significant role in human health concerning their functional influences in many diseases. Also, the determination of AAs is so important in brain diseases too. Due to their importance in many medicinal and therapeutic fields, many different methods were used to identify them, such as liquid chromatography, GC, etc. Nevertheless, new and innovative methods were replaced due to their time-consuming feature, expensiveness, poor selectivity, and the need to prepare materials before identifying. Electrochemical sensors and biosensors are the techniques that overcome these disadvantages. They are simple to use, selective, sensitive, and time-saving. This chapter explains AAs, their structure, and the sensors and biosensors used for their discrimination and determination of them.

18.1 Introduction

18.1.1 Definition of amino acids

In recent years, AAs have attracted more attention than before because of their critical roles in metabolism and well-being. AAs have been found in foods, pharmaceutical products, humans, animals, plants, microorganisms, and the environment, so they significantly affect life [1–3]. The differences in their carbon numbers, structure, and chemical groups give them different physiological, chemical, nutritional, and biochemical properties [4, 5].

doi:10.1088/978-0-7503-5377-9ch18

There is an idea that AAs names have been derived because of their discoveries, properties such as taste, appearance, chemical structure, synthesis processes, and their isolation source [6, 7]. Organic compounds containing amino and acid groups are defined as AAs (figure 18.1) [8].

According to the Greek alphabet, different carbon atoms in AAs are named in the order. Classification of AAs is according to the location of the core structural functional groups. α, β, γ, δ, ε AAs are named because of the connection of the amino group to the α, β, γ, δ, ε carbon, respectively [9, 10]. However, there are hundreds of AAs in Nature; alpha AAs are widespread, and 22 of them can be found in the genetic code. In addition, AAs can be classified by polarity, ionization, and side chain group type (aliphatic, acyclic, aromatic, containing hydroxyl or sulfur, etc) [11].

In 1851 the optical activity and rotate plane-polarized light of natural AAs were witnessed by Louis Pasteur. Almost ten years later, in 1908, the first introduction of L and D isomers was defined by Emil Fischer. His definition of AA isomers was valuable because their structure can demonstrate stereochemical information without three-dimensional representation. According to the contract AAs, AAs exist in two isomeric forms: D and L AAs. If the NH_2 agent attached to the alpha carbon is on the left side, it is named L-type. When the NH_2 moiety is placed on the alpha carbon's right side, this AA is called D-type (figure 18.2). Unlike natural sugars, which are D-type, natural AAs are all L-type [12]. D-type AAs are found in plants, animals, processed foods, and microorganisms.

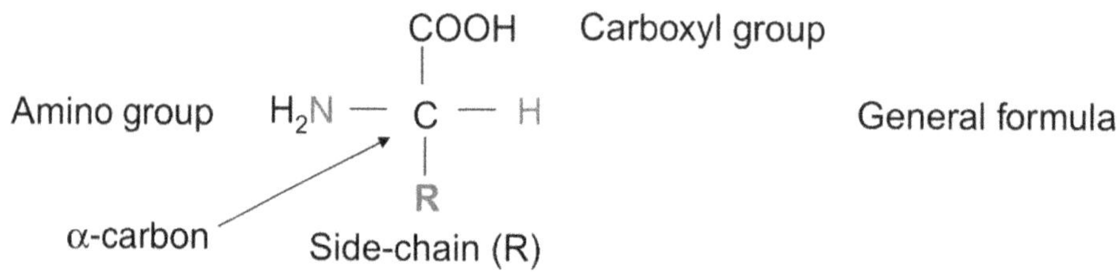

Figure 18.1. Chemical structure of the AA glycine with the side chain R = H. Reprinted from [8], copyright (2012) with permission from Elsevier.

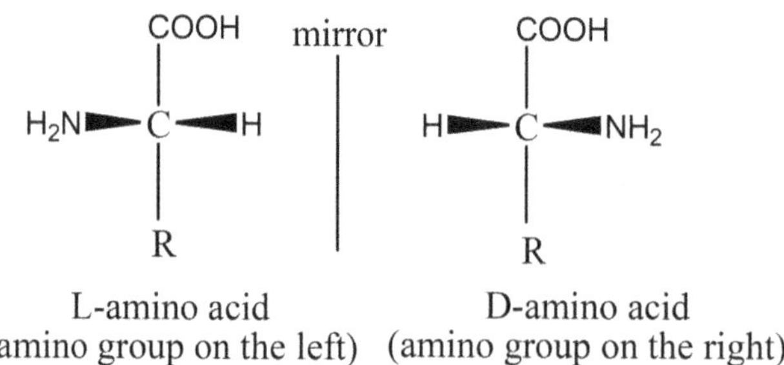

Figure 18.2. L- and D-AAs structure.

In 2012 Bischoff and Schlüter pointed out that proteinogenic AAs can have both L and D isomers, excluding glycine, which was introduced as the simplest AA in Nature with a single H (hydrogen) in its side chain [8].

In the 19th century, D-AAs were considered a type of AAs that could be synthesized. However, in 1927 unexpectedly, the existence of D-AAs was identified in octopins for the first time, although before particular recognition of them, they were reported sporadically in animals and plants from 1884. In addition, after almost eight years, W A Jacobs and L C Craig found another D-AA (D-proline) obtained from a group of fungi called ergot. After these discoveries, a large number of D-AAs were determined in fungi and bacteria such as D-alanine, D-aspartate, D-glutamate, D-leucine, D-valine, and D-phenylalanine and so forth [13]. Hence, the existence of D-AAs in microbes, bacteria, animals, and plants (D-alanine, D-aspartate, D-glutamate, D-proline, D-serine, and D-tryptophan) [14] and foods have illustrated the importance of AAs in biochemistry and physiology [15].

18.1.2 Determination of amino acids

AAs participate in protein and fatty acids synthesis and many other biological processes; they are also significant in several diseases, such as type 2 diabetes, kidney disease, liver disease, and cancer. Therefore, scientists tried to find methods for the recognition of AAs. Thus, their importance has extended the analytical techniques for measuring AAs. High-performance liquid chromatography (HPLC), colorimetric, fluorimetric detection, molecular imprinted polymer methods, and capillary electrophoresis (CE) are examples of these methods [16–18].

18.1.2.1 Electrochemical sensors and biosensors

Electrochemical devices are one instrumental method developed recently for detecting AAs. They are based on constructing sensors and biosensors [19]. Electrochemical sensors are highly regarded and reliable due to their accuracy, high speed, as well as low cost, simplicity of operation, and portability [20–22]. A clear example of them is glucose sensors used for diabetic patients [23–25]. Electrochemical sensors are very reliable and successful in bioanalysis [26, 27]. In the last decade, electrochemical sensors have been greatly improved due to their size, matrix, and electroactive materials. As a classification of electrochemical sensors, electrochemical biosensors —enzyme-based biosensors, DNA-based biosensors, immunosensors, and microbial biosensors—have also been introduced, and the selectivity of biochemical reactions is due to the high selectivity of these sensors [28–32]. When a biological molecule is fixed on sensitive electrochemical transducers such as amperometric ones, its performance can be compared with optical sensors-chemiluminescence based on a biosensor. Physiological measurement at the laboratory level and *in vivo* is considered an important principle. Electrochemical sensors can be used practically, having solidity, biocompatibility, and non-toxicity [33–36].

Biosensors were developed almost 40 years ago with the platinum electrode stabilized with glucose oxidase for measuring glucose in blood samples [37–40]. Enhancements were announced in the quality of this sensor's membrane matrix and the type of enzyme or antibody immobilization [21]. Biosensors used today to identify AAs are an integral and essential part of science and technology. It can be said that this method is a combination of important branches of science and technology, which was later called a branch of bioelectronics. A biosensor is defined as a 'self-contained integrated device' by the International Union of Pure and Applied Chemistry (IUPAC), which uses a biological compound in direct spatial contact with a transduction element for quantifying [41–43]. Biological substances such as enzymes, antibodies, or analytes form biosensors, which are very selective due to their ability to tailor interactions specific to analytes by biological detection elements on the sensor substrate with a particular binding desire to the target molecule. Therefore, since the biological material is in contact with a suitable transducer, the biochemical signals are altered into continuous or non-continuous electrical signals, which can then be estimated. The choice of a biological component and transducer depends on the analyte or substance one wants to analyze. It can be said that biosensors have applications in various fields, such as agriculture, food, food preservation, medicine, etc. Today, biosensors are classified as the most important field of chemistry because they can save one's time and energy in sample preparation because they have omitted sample preparation [37–40, 44].

18.2 Electrochemical assay of amino acids using sensors

García-Carmona *et al* developed a method for the detection of branched-chain AAs: vertically aligned nickel nanowires-based electrochemical sensor (v-NiNWs). Characterization of v-NWs was adjusted by template electrodeposition protocol and polycarbonate membrane. They used gold film as a working electrode and the electrodeposition process for sputtering one side of the membrane. Also, a Pt wire was introduced as a counter, and Ag/AgCl was a reference electrode. Then, the electrodeposition of nickel and copper nanowires was developed at -1.0 V on the membrane pores. After that, they cut it into a suitable shape, and the membrane was plunged into dichloromethane. In the next stage, it is cleaned with water, isopropanol, and ethanol. The adhesive tape, a non-conductive substrate that contains orientated nanowires, was directly stuck into the ceramic substrate actuating as a working electrode. Finally, conductive silver paint (Electrolube, UK) was used to create the electric contact and an epoxy protective overcoat (242-SB, ESL Europe) for the isolation. In addition, metallic oxides were created into the electrodic surface before each experiment by applying a potential of $+0.70$ V in 0.1 mM NaOH for 150 s. The detection limit and linear range were 8 μM and 25–700 μM, respectively. In this work, the capability of copper and nickel-based (v-NWs) electrochemical transducer were explored. The study showed that (v-NiNWs) were the best transducer for detecting AAs (figure 18.3). It was developed to rapidly detect AAs for non-invasive screening of maple syrup urine disease. This method was introduced as simple, fast in response, and time-saving [45].

Nickel, in many studies, was introduced as a non-enzymatic glucose sensor which declared it as an essential material. Fleishman explored its sensing mechanism for the first time, and $Ni(OH)_2$ and NiOOH were mentioned as catalytic sensing species.

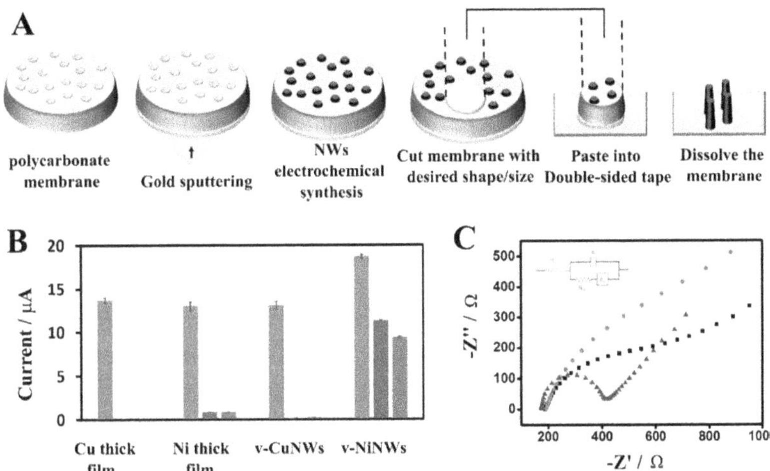

Figure 18.3. (A) Scheme of the fabrication of v-NWs as exclusive transducers for AAs detection. (B) Nickel and copper-based electrochemical sensors response toward target AAs detection. Tyr (green), Phe (blue), Leu (red). (C) Nyquist diagrams using 5 mM $Fe(CN)_6^{4-/3-}$ in 0.1 M KCl. SPCE (blue), v-NiNWs (black), and 110-Ni (red). (Inset: equivalent electrical circuit diagram.) Reprinted with permission from [45] John Wiley & Sons. Copyright 2021 Wiley-VCH GmbH.

Their reversible mechanism ($NiOOH \rightarrow Ni(OH)_2 \rightarrow NiOOH$) is the essential feature of its usage in electrochemical sensors. Thus, in 2018, a nickel-based electrochemical sensor for detecting AAs such as phenylalanine and glycine was tested in urine by Petrialia *et al* in their investigation, the VLSI technology was used for developing this sensor. First, a silicon oxide layer (first passivation layer) grew to separate electrodes from the substrate. Second, ejection of a nickel film on the electrode areas operated. Next, the deposition of the second passivation layer separated the first metallization (Ni) from the second one (Au). The etch carried out the connection of the first and the second metallization. The second metallization (Au) was then sputtered and lithographically defined in complementary electrode regions and contact areas. The final device is composed of four electrochemical cells, each one containing three planar microelectrodes, a working electrode (WE) in nickel, a counter (CE), and a reference (RE) electrode made in gold. The detection methods were cyclic voltammetry (CV) and chronoamperometry. This paper's linear range and sensitivity were 30–150 μM and 0.21 $\mu A\ \mu M^{-1}\ cm^{-2}$, respectively. Therefore, a Ni electrode can be considered an easy way to determine AAs [46].

For the identification of chiral AAs in pharmaceutical products, biochemistry, and medical science, the fabrication of a useful and simple way is necessary. In recent years there has been in electrochemistry because of its time-saving and cheap properties. Therefore, Yongxin Tao *et al* constructed an Alfa-cyclodextrin (α-CD) based chiral sensing device to identify isomers of tyrosine (Try). Firstly, α-CD was modified with Cu^{2+} and CS (chitosan) to establish coordination-driven self-assembly. Construction of The CS/Cu_2-α-CD self-assembly illustrated an accurate percentage of D-tyrosine in racemic solution; also, it showed the specific determination of Try. According to the information mentioned above, it can be said that this method can encourage the building of new chiral sensing methods for the identification of chiral compounds [47].

Lijun Zangh *et al* created a sensor to determine α-AAs. They introduced organic electrochemical transistors (OECTs) with gate electrodes modified with molecularly imprinted polymer (MIP) films. For developing the device, MIP film was deposited on the surface of the Au electrode by CV according to the conditions stated in this paper. Three-electrode cell systems were utilized in this study: Au was the working electrode, saturated calomel electrode (SCE) was mentioned as the reference electrode, and Pt wire was the counting electrode. After that, methyl alcohol/acetic acid solution was used to extract template molecules from the polymer film and cleaned with water. The modified electrode by L-tryptophan MIP film was achieved. L-trp MIP/Au was comprised of other electrodes (D-Trp MIP/Au, D/L-Tyr MIP/Au electrodes, and NIP/Au without template). The detection methods were CV and electrochemical impedance spectroscopy (EIS). This method was a highly selective and sensitive chiral recognition biosensor for D/L-tryptophan (D/L-Trp) and D/L-tyrosine (D/L-Tyr), demonstrated in figure 18.4. As can be seen from the picture, the OECT sensor was modified with MIP film related to the size and structure of the analytes. Thus, other AAs could not cause any disruption. The linear response range, sensitivity, and limit of detection (LOD) were from 300 nM to 10 μM, 3.19 and 3.64 μA μM^{-1}, and 2 nM and 30 nM for L-Trp and L-Tyr, respectively [48].

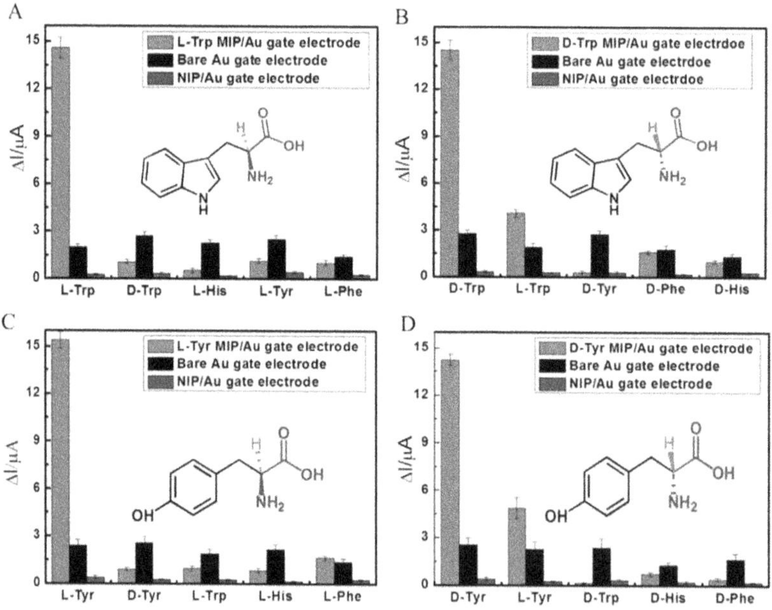

Figure 18.4. The current change of OECTs with (A) L-Trp MIP/Au, (B) D-Trp MIP/Au, (C) L-Tyr MIP/Au, and (D) D-Tyr MIP/Au electrodes as gate electrodes with the addition of 3 μM of different AAs in PBS (pH 7.4). Reprinted from [48], copyright (2018) with permission from Elsevier.

Wenting Liang *et al* characterized a chiral sensor for enantiomeric recognition by coupling three-dimensional graphene with hydroxypropyl-β-cyclodextrin (3D-G/HP-β-CD), and they investigated the utilization and selectivity of this sensor for recognition of tryptophan (Trp) isomers (enantiomers). Firstly, the aluminum powder was used to polish the bare glassy carbon electrode (GCE), then in an ultrasonic bath, it was cleaned three times with ethanol and highly purified water and dried in air. Finally, 5 ml of 3D-G/HP-β-CD and 3D-G were added to the surface of GCE and dried by the infrared lamp to prepare 3D-G/HP-b-CD/GCE and 3D-G/GCE. Differential pulse voltammetry (DPV) was used to recognize Trp enantiomers. Moreover, bare GCE and 3D-G/GCE controlled the experiments [49]. Bare GCE was not useful because both L- and D-Trp overlapped, and they could not identify them. Although the peaks can be identified in the second one, it is not enough for exact and efficient determination of the isomers. They witnessed that their organized sensor is an excellent choice for the detection of AAs (L, D-Trp). Therefore, according to their investigation, 3D-G/HP-β-CD can be considered a chiral selector. This work's linear range and detection limits were 0.5–175 mM and 9.6–38 nM, respectively [49].

An MIP/reduced graphene oxide (rGO) composite was designed by Weihua Zheng *et al* for the detection of tyrosine in serum and urine samples. The reduced graphene oxide was prepared in four steps. First, 4 mg of GO, in double distilled water consisting of poly diallyl dimethylammonium chloride (PDDA) (0.2 ml) diffused by ultrasonic. Second, ascorbic acid, a reducing power, was added and refluxed (2 h at 90 °C). In the third step, it was centrifuged three times for isolation and removal of PDDA. In the end, for producing rGO suspension, the product was redispersed in a mixture of water and ethanol. MIP electropolymerized on the rGO-modified GCE using a novel monomer of 2-amino-5-mercapto-1, 3, 4-thiadiazole and the template of tyrosine. This sensor showed a specific area for the identification of tyrosine. The method used in this study was DPV, and the respective amount of linear range and LOD were 0.1–400 μM and 0.046 μM in optimal conditions [50].

Guzel Ziyatdinova *et al* developed a (poly (p-coumaric acid)/MWNT/GCE) for the recognition of L-cysteine in optimal conditions in human urine. Like other mentioned works, polishing and cleaning GCE with alumina and acetone was the first step, followed by adding multi-walled carbon nanotubes (MWCNTs) on the surface of the GCE and drying by evaporation. Then, p-coumaric acid was electropolymerized by potentiodynamic electrolysis on the surface of MWNT/GCE. Modified MWNT/GCE by p-coumaric acid is the effective sensor that could identify the expected AA (L-cysteine). DPV was used to determine it, and the analytical range and detection limit were 7.5–1000 μM and 1.1 μM, respectively. It is claimed that the organized sensor is very selective to determine the L-cysteine in the presence of glucose, uric and ascorbic acids, dopamine, L-tyrosine, and S-containing substances (homocysteine, glutathione, L-methionine, L-cystine, and α-lipoic acid) [51].

In 2019, Quanguo He *et al* tested an electrochemical sensor for rapid and sensitive detection of AA Trp in human serum samples, which was Cu_2O-ERGO/GCE. They claimed this sensor could also be used to determine Trp in real samples. They prepared cuprous oxide and reduced graphene oxide simply. Cuprous oxide nanoparticles (Cu_2ONPs) and GO were synthesized first, then mixed and ultrasonicated to produce Cu_2O-GO. After that, cleaning of GCE and dropping of Cu_2O-GO on its surface constructed Cu_2O-GO/GCE, which reduced and gave Cu_2O-ERGO/GCE. The method they used to detect Trp was square-wave voltammetry (SWV). The respective amount of linear range and detection limits were 0.02–20 μM and 0.01 μM, respectively. According to this study, it is simple to prepare, and selectivity is an important sensor characteristic [52].

Recently, many scientists have introduced helical structures because they can be used for the chiral parting of racemic AAs. Although in Nature, helical structures such as α-helical peptides, double-helical nucleic acids, and triple-helical collagens can widely be found, the synthesis of artificial ones has increased via the self-assembly of small synthetic molecules. Therefore, a chiral helical self-assembly sensing device was constructed by Jiapei Yang *et al.* First, the chiral (1R,2R)-N, N-didodecylcyclohexane-1,2-dicarboxamide (DDC) was synthesized according to the conditions mentioned in this work, then DDC was self-assembled by alcohol. In addition, the intermolecular hydrogen bond and solvent polarity effect of DDC self-assembly created a chiral helical structure that effectively recognized Trp. Furthermore, DDC self-assembly dropped on the surface of GCE; therefore, GCE (DDC/GCE) was modified. A three-electrode cell was used for electrochemical experiments, which comprised DDC/GCE, saturated calomel electrode (SCE), and platinum plate as the working electrode, reference electrode, and counter electrode, respectively. First, the solution of L- and D-Trp was operated to plunge the DDC/GCE into it. After that, the combination of DPVs of L- and D-Trp with DDC self-assembly evaluated the sensor's power for the determination of AAs in the potential range of 0.4–1.2 V, respectively [53].

According to the studies, gold nanoparticles (AuNPs) have catalytic features. They can increase the electron transfer rate; therefore, in 2019, Isabela A Mattioli *et al* fabricated a new electrochemical device to detect tryptophan (EGPU-tAuNP). For developing this device, the first electrodeposition of AuNP onto the surface of a graphite-polyurethane composite electrode (EGPU) was started according to the data explained in this work. For the next step, CV was used for electrochemical treatment in NaOH. Moreover, the experimental conditions were optimized, then recognition of Trp was started by DPV in synthetic urine and commercial poly-AAs supplements. The linear range and LOD were 0.6–2.0 μM and 5.3×10^{-8} M (0.053 μM), respectively [54] (figure 18.5).

Wenqin Yao *et al* constructed an electrochemical device for detecting AA (L-Trp) in real samples (human urine, AA injection, and human serum). For the organization of the sensor, they modified Bare GCE with Ni-ZIF-8/N S-CNTs/CS and Ni-ZIF-8/N S-CNTs/CS/GCE introduced as a sensor that can identify the considered AA. DPV, CV, chronocoulometry (CC), and EIS techniques were used, between them, CV showed the best performance. The linear range and LOD were 5–850 and 0.69 μM. This work also determined other compounds, but the AA condition is reported here [55].

Recent years have seen various methods for detecting AAs polymers as the most attractive ones. Between them, metal–organic frameworks (MOFs) as porous

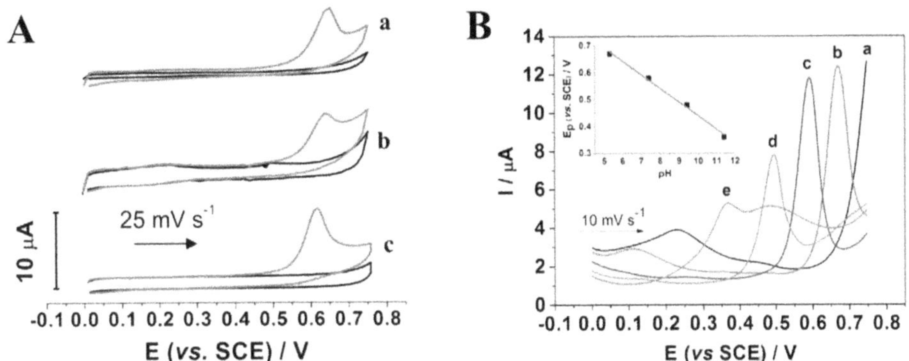

Figure 18.5. Cyclic voltammograms obtained in the presence of 100.0 μM Trp using (a) EGPU, (b) EGPU-AuNP, and (c) EGPU-tAuNP in 0.10 M Britton–Robinson buffer (pH 7.4) from 0.0 to 0.75 V versus SCE ($\nu = 25$ mV s^{-1}) (solid) and blank (dash). (B) DPV voltammograms for EGPU-tAuNP in the presence of 50.0 μM Trp in 0.10 M. Britton–Robinson buffer at different pH values: (a) 3.3, (b) 5.3, (c) 7.4, (d) 9.4, and (e) 11.4. Parameters: $\nu = 10$ mV s^{-1}; $a = 50$ mV. Inset: Ep (versus SCE) versus pH plot. Reprinted from [54], copyright (2019) with permission from Elsevier.

coordination polymers have drawn researchers' interest in using electrochemical sensors. However, according to some investigations, besides their advantages, such as large surface areas and high crystallization power, they have demerits too. For instance, MOFs are assumed to be limited materials when they used as electrodes or electrocatalysts. After a while, some scientists declared that if their frameworks' design and surface modification were reasonable, they could also utilize them as electrochemical biosensors. Therefore, in 2021, Xiaofeng Wei *et al* constructed a sensor that was the L-histidine-regulated zeolitic imidazolate framework (L-His-ZIF). Zeolite imidazole framework (ZIF) is another type of MOF. In their study, they first synthesized ZIFs; then, L-His-ZIF was prepared by dissolving L-His in water and mixed with triethylamine. Next, as stated in the paper, $Zn(NO_3).6H_2O$ and MIM (2-methylimidazole) is dissolved in methanol in different amounts. After that, $ZN(NO_3).6H_2O$ solution was added to the MIM and L-His solutions mixture. After incubation, centrifugation, and washing in methanol three times, the L-His-ZIF achieved was a white crystal. In the second phase, the L-His-ZIF interface was fabricated by rinsing the working electrode (GCE) with Al_2O_3, pure water, ethanol, and nitric acid, and drying with N_2. Further, the synthesized L-His-ZIF was weighed and added to the DMF solution to achieve the L-His-ZIF solution. Next, the mentioned solution was dropped onto the surfaces of the electrode. Finally, the organization of the L-His-ZIF interface was performed. Figure 18.6 shows its preparation scheme. They announced that this electrochemical sensor could identify the enantiomers of glutamate by electrochemical impedance spectroscopy (EIS). The LOD and linear range were 0.1–50 and 0.06 nM, respectively. It can be said that by introducing different interfaces of MOFs, which have high power in the applications of biochemical and pharmaceutical analyses, they can be used to identify enantiomeric chiral [56].

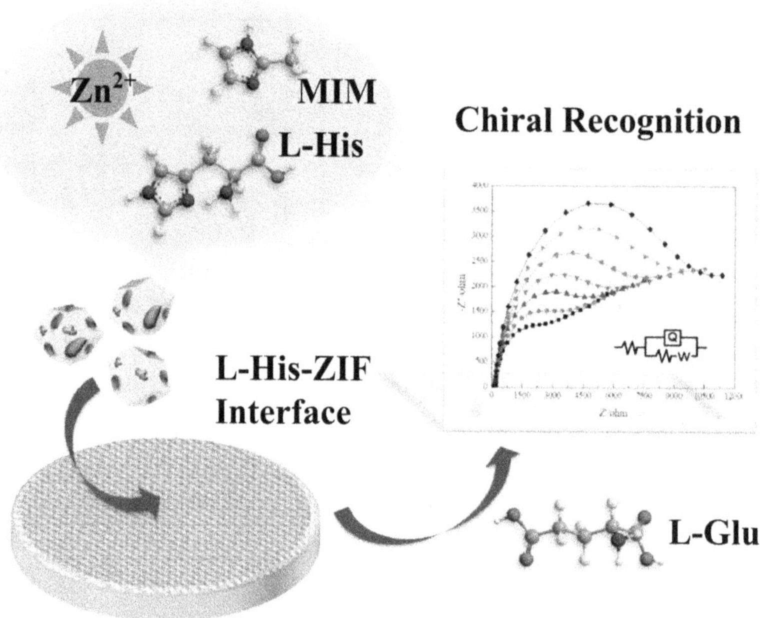

Figure 18.6. The scheme of L-His-ZIF interface for enantioselective identification of L-glutamate. Reprinted from [56], copyright (2021) with permission from Elsevier.

The synthesis of tetra-4-{(E)-[(8-aminonaphthalen-1-yl)imino]methyl}-2-methoxyphenol Co(II) phthalocyanine (CoTANImMMPPc) was used for the production of the new electrochemical sensor by Mounesh *et al.* They introduced the (CoTANImMMPPc/MWCNTs/GC) sensor for discovering L-alanine and L-arginine. To begin with, they developed ANImMMP and CoTANImMMPPc. M-vanillin, diaminonaphthalene, and methanol were utilized to synthesize the novel Schiff base ligand and the mixture was stirred under a nitrogen atmosphere; H_2SO_4 was added. Later, stirring for 6 h, refluxing under a vacuum, cleaning with water, recrystallization by methanol, and purification by column chromatography was performed. Moreover, by dissolving CoTCAPc, K_2CO_3, and DCC catalyst in DMF, the CoTANImMMPPc complex was obtained. Accordingly, stirring of the mixture was performed, and next, the ANImMMP was added to the reaction mixture, and the solution was stirred again for 46 h. After these steps, a dark green precipitate was achieved, followed by washing with cold and warm water and hexane (figure 18.7). Eventually, rubbing and rinsing of the GCE surface were performed by alumina Slurry, water, and acetone, respectively. Additionally, CoTANImMMPPcand nafion binder was ultrasonicated for 30 min for their dispersion in DMF. the GCE was modified with the CoTANImMMPPc material by dropping technique. Finally, the CoTANImMMPPc material settled on the GCE electrode, then was dried at 25 °C. It is interesting to note that the CoTANImMMPPc-MWCNT electrode was prepared in the same conditions using MWCNTs/CNTs. Therefore, these electrodes were used for the discovery of L-Ala and L-Arg. The methods used

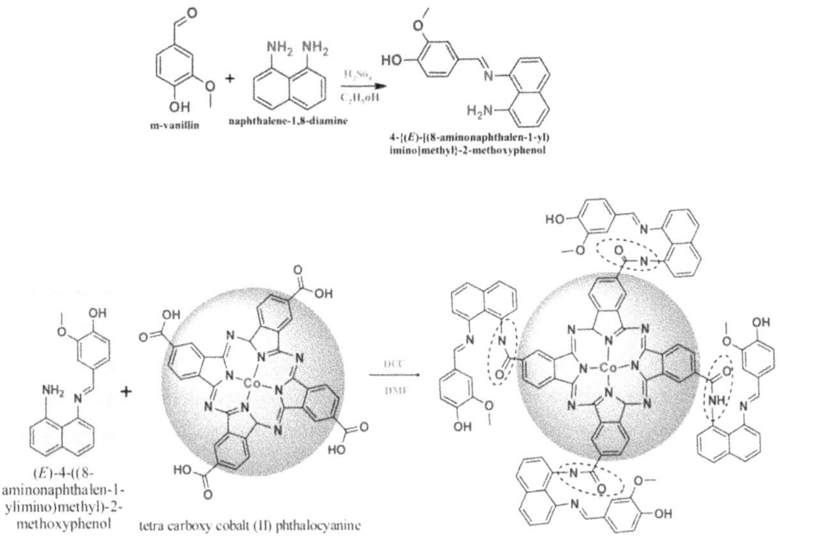

Figure 18.7. The steps of developing the ANImMMP and CoTANImMMPPc. Reproduced from [57], with permission from the Royal Society of Chemistry, CC BY 3.0.

for detecting L-alanine and L-arginine were CV, DPV, and CA, with a linear range between 50 and 500 nM, and the limit of detections in CV were 1.5 and 1.2, respectively. Although, DPV and CA demonstrated a linear range from 50 to 500 and LOD of 1.8 and 2.3 nM. Therefore, CoTANImMMPPc-MWCNTs/GCE illustrated excellent analytical performance and can be applied as a sensor for the detection of L-alanine and L-arginine[57].

Nowadays, using the smartphone as an electrochemical sensor has become common because it is easy to analyze, process, and exchange data. In 1991 carbon nanotubes were uncovered, and carbonaceous-based nanomaterials, because of their useful effect in selectivity and sensitivity considered an essential part of electrochemical studies. Among these nanomaterials, carbon black, according to past studies, was so cheap and selective. Thus, Luca Fiore *et al* decided to characterize a smartphone-assisted electrochemical device in which a carbon black (CB)-based screen-printed electrode (SPE) was used for exact sensing of L-tyrosine in serum (figure 18.8). For the construction of SPEs, a polyester was used that was transparent and flexible. After that, they used graphite-based ink and silver-based ink as the working, counter electrodes, and reference electrodes for creating a three-electrode cell. For the next step, CB is dispersed based on the explained conditions. Then, modification of SPEs with CB dispersion via immersing approaches was achieved. For identification of tyrosine SWV was utilized by SPEs coupled with potentiostat EmStat3 Blue connected to a smartphone. This sensor is declared to be simple, cheap, and sensitive; in addition, it does not need any preparation before use. More importantly, it can discriminate tyrosine from another AA in serum samples. The line range and detection limits were 30–500 and 4.4 μM, respectively (figure 18.8) [58].

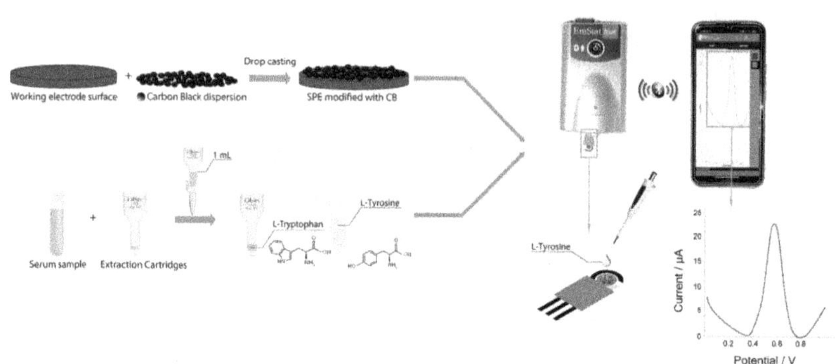

Figure 18.8. Scheme of the smartphone-assisted electrochemical sensor combined with sample pre-treatment for tyrosine detection in serum. Reprinted from [58], copyright (2022) with permission from Elsevier.

18.3 Electrochemical assay of amino acids using biosensors

In 2018 Kambiz Varmira *et al*, organized an electrochemical biosensor (enzymatic biosensing) for the identification of L-tyrosine in food samples like eggs, cheese, and yogurt, which are rich in L-tyrosine. Firstly, chitosan-1-ethyl-3-methylimidazolium bis(trifluoromethylsulfonyl) imide/graphene-MWCNTs-IL/GCE (Ch-IL/Gr-MWCNTs-IL/GCE) was prepared by firstly polishing GCE, and secondly dropping Gr-MWCNTs-IL solution onto the surface of GCE, and thirdly dropping Ch-IL onto the surface of Gr-MWCNTs-IL/GCE. Secondly, palladium—platinum (PdPt) bimetallic alloy nanoparticles were synthesized such Ch-IL/Gr-MWCNTs-IL/GCE was a working electrode in an electrochemical cell. In this cell, there was a solution comprising H_2PtCl_6, $PdCl_2$, and KCl. The reference electrode was Ag/AgCl and the counter electrode was Pt wire. It was then scanned by CV. Thirdly, PdPt NPs/Ch-IL/Gr-MWCNTs-IL/GCE was absorbed in a glutaraldehyde (GA) solution. Finally, TyrH solution was dropped onto the surface of the PdPt NPs/Ch-IL/Gr-MWCNTs-IL/GCE figure 18.9 [59].

TyrH/PdPt NPs/Ch-IL/Gr-MWCNTs-IL/GCE tyrosine hydroxylase/palladium–platinum bimetallic alloy nanoparticles/chitosan-1-ethyl-3-methylimidazolium bis (trifluoromethylsulfonyl) imide/graphene-MWCNTs-IL/GCE was the biosensor used for detection of L-tyrosine. The linear range and LOD were 0.01×10^{-9}–8.0×10^{-9} mol l^{-1} and 8.0×10^{-9}–160.0×10^{-9} mol l^{-1} and 0.009×10^{-9} mol l^{-1}, respectively. This method was fast, sensitive, selective, and reliable for recognizing L-tyrosine [59].

In 2020, a non-mediator-style electrochemical biosensor based on D-AA oxidase (DAAO) was constructed for the determination of chiral D-AA by Tingting Tian *et al.* In their study, for the fabrication of a non-mediator-style D-AA biosensor, DAAO, CNTs, and nafion were used as a chiral selector, a signal amplifier, and a protective agent, respectively. For the fabrication of the enzyme electrode, they first blended CNTs and DAAO solutions and achieved DAAO/CNTs, then GCE was modified by dropping a suitable amount of the above-mentioned solutions and nafion solution. The examined samples were urine and milk, which showed

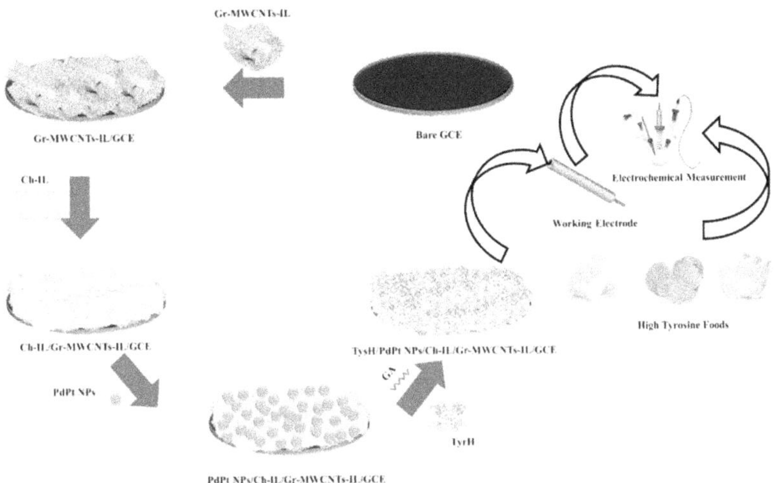

Figure 18.9. The steps of developing a biosensor for the determination of L-tyrosine. Reprinted from [59], copyright (2018) with permission from Elsevier.

satisfactory results. In this study, they achieved that this biosensor is highly selective against enantiomer intrusion; therefore, it can recognize D-alanine in an alanine racemic mixture. Furthermore, they solved the problem of enantiomer recognition by this biosensor because all chiral sensors that existed before can identify a single isomer, for example, L- or D-isomer. It means that those sensors needed more differences for the identification of mixture solutions [60].

In living organisms such as humans, there are basic AAs that have demonstrated indispensable roles in immune systems and cardiovascular. L-arginine is one of them that has a key role in diseases. Thus, in this work, Ashish Kumar Singh *et al* created a quick and selective biosensor (ADI/Fe_3O_4/MWCNT/PANI/Nf/GCE) for discovering stated AAs in leukemic blood samples. They modified GCE by PANI/MWCNTs/Fe_3O_4. Before their study, the bio-enzymatic approaches have reported for sensing L-arginine, but they tried a new single enzymatic way for recognition and claimed that it is simple and sensitive toward arginine. Preparing this selective biosensor was done in five steps that were purification of enzyme arginine deiminase (ADI), iron oxide (Fe_3O_4) nanoparticles synthesis, MWCNTs preparation, the Fe_3O_4/MWCNT/PANI-nafion nanocomposite film construction and the enzyme ADI immobilization, respectively. In the first stage, the *Pseudomonas putida* MTCC1313 bacteria grew in arginine deiminase for 24 h, then the cell biomass was harvested developed by centrifugation and phosphate buffer solution (PBS) used for cleaning. Next, crude enzyme was extracted by sonication and purified using the salt precipitation method. In the second stage, $Fe(NO)_3$, $Fe(NO_3)_2.9H_2O$, and NaOH were dissolved in the water independently, mixing the solutions. After that, the represented product was cleaned with ethyl alcohol and pure water. Finally, the sample was dried and the powder of iron oxide nanoparticles was achieved, in addition, scanning electron microscopy (SEM) was used for the fabrication of its size and morphology. In the third step, MWCNTs with carboxylic group functionalized according to the conditions stated in this study and Fourier transform infrared spectroscopy (FTIR) was utilized to confirm its existence. In the fourth step, the Fe_3O_4/MWCNT/PANI-nafion was constructed by the electropolymerization method on the GCE. Next, the electrochemical cell consisted of aniline, carboxylated MWCNTs, and iron oxide nanoparticles utilized for Plunging nafion-modified GCE. In the last step, the glutaraldehyde cross-linker was deposited on the surface of the cleaned Fe_3O_4/MWCNT/PANI-nafion-modified electrodes in water first. Secondly, it was dried and washed again in pure water to remove the unbound glutaraldehyde. Later, the purified enzyme was immobilized on the glutaraldehyde-treated nanocomposite-modified electrode by drop-casting on the surface and incubating for 2 h. Finally, deionized water was used for cleaning the enzyme-modified Fe_3O_4/MWCNT/PANI-nafion nanocomposite film electrode and kept in pH 7.2 phosphate buffer until use. In their study, CV, DPV, and CA were used to identify AA L-arginine. The linear range and detection limit values were 100–1000 and 20 μM [61].

In 2021 Siba Moussa *et al* organized a selective biosensor based on DAAO. In this paper, they used three various DAAO enzymes: wild-type yeast DAAO (RgDAAO WT) and variants human DAAO W209R and yeast M213G and they constructed three biosensors for determination of D-serine and D-alanine. These biosensors have different selectivity of AAs. First, enzyme preparation was carried out according to the conditions declared in the paper. Then, calibration and regeneration as well as storage of biosensors were developed. First, they calibrated (*in vitro*) using chronoamperometry in 0–50 μM D-serine or D-alanine in PBS (0.01 M, pH 7.4) in 15 min, 0.5 V versus Ag/AgCl. Next, to determine their storage power, they were increased from −20 °C dry to room temperature. Furthermore, they were immersed in the D-serine solution to test their regeneration power and capability. Finally, the cross-linking process was done because it had effects on the DAAO structure and function. Organization of biosensors was carried out with variable amounts of enzymes: RgDAAO WT, RgDAAO M213G, and hDAAO W209R, with 56.8, 48, and 45 mg ml^{-1}, respectively. Standard solutions of D-serine and L-alanine were used for the calibration of these biosensors. They showed that they have various selectivity and sensitivity toward AAs [62]. From the above-mentioned information, all steps had a great effect on the selectivity and sensitivity of the biosensors but also on their structural changes. Their study stated that enzymatic biosensors (DAAO biosensors) could be used to understand D-AA function and their roles in neurological disease [62]. Some selected studies are tabulated in table 18.1.

18.4 Conclusion

As mentioned above, use of sensors and biosensors have replaced traditional methods such as chromatography, spectrophotometry, colorimetric, titrimetric, etc. Studies related to sensors and biosensors in the last few years that have been used to identify AAs are reported in table 18.1. Studies of electrochemical devices

Table 18.1. Electrochemical sensors and biosensors for determination of AAs.

Amino acid	Method	Electrode	Linear range	Limit of detection	Matrix	References
Cysteine	DPV	FCPA	0.25–100 μM	0.14 μM	Blood	[63]
Tyrosine Phenylalanine	FS	*E. coli* DH5a cells	5–100 μM	4.9 μM 3.7 μM	Urine	[64]
L-tryptophan	SWV	E-AB	0.1–10 mM	NS	Urine	[65]
Phenylalanine	SWV	E-AB	90 nM–7 μM	0.4×10^{-6} M	Blood	[66]
Tryptophan	NS	$SrTiO_3$@rGO	0.03–917.3 μM	0.0071 μM	Urine and blood serum samples	[67]
Phenylalanine	CV, DPV, EIS	MIP (4-[(4-methacryloyloxy) phenylazo] benzoic acid)/ MWCNT	$0.5–3 \times 10^{-6}$ M	0.2086×10^{-6} M	NS	[68]
L-phenylalanine D-phenylalanine	CV, DPV	Perylene-functionalized graphene/β-CD	$0.01–5 \times 10^{-3}$ M	0.08×10^{-9} M 0.2×10^{-9} M	NS	[69]
Phenylalanine	CV, DPV	β-CD/CNTs@rGO	$0.2–13.0 \times 10^{-6}$ M	0.08×10^{-6} M	NS	[70]
Phenylalanine	CV, CA	ZIF-67 Encapsulated PtPd Alloy Nanoparticle (PtPd@ZIF-67)	$5–500 \times 10^{-6}$ M	20×10^{-9} M	NS	[71]
Phenylalanine	Potentiometry	MIP polyvinyl chloride	$1 \times 10^{-8}–1 \times 10^{-4}$ M	5×10^{-9} M	Blood serum	[72]
Phenylalanine	CA	L-Phe-MIP PPy/Ag	$0.1–50 \times 10^{-3}$ M	1.39×10^{-3} M 27.7×10^{-3} M	NS	[73]
L-tyrosine	CV, LSV	Al-CuSe-NPs/SPCE	0.15–10 μM	0.04 μM	Pharmaceutical samples	[74]
L-tyrosine	Amperometry	CuO/β-CD/Nf/GCE	0.01–0.100 μM	0.0082 μM	Human blood serum, Foods and drinks, Urine	[75]
L-tyrosine	DPV	MW/Nf/GCE	2–120 μM	0.8 μM	Serum	[76]
L-tyrosine	DPV	MIP/pTH/Au@ZIF-67/GCE	0.01–4 μM	0.000 79 μM	Serum	[77]
D-alanine	DPV	Fe_3O_4@Au@Ag@CuxO NPs	0.1–10 000 nM	0.052 nM	NS	[78]
Tyrosine Tryptophan	DPV	PDPA/PGE	20–1000 μM 20–1400 μM	NS	Serum	[79]
Tyrosine	DPV	β-CD/AgNPs/GCE	NS	NS	NS	[80]

(Continued)

Table 18.1. (*Continued*)

Amino acid	Method	Electrode	Linear range	Limit of detection	Matrix	References
Tryptophan	CV DPV	MnWO4/rGO/GCE	0.001–120 μM	4.4 nM	Milk	[81]
5-Hydroxy tryptophan (5-HTP)	Amperometry	ePAD/3D-BIA	0.165–50 μM	50 nM	Food/pharmaceutical samples	[82]
Dopamine	DPV	npoly(3,4-ethylene dioxythiophene) (PEDOT)	1–1000 μM	NS	Milk	[83]
Lysine						
Histidine						
Cysteine						
Tyrosine						
L-glutamate	Amperometry	AuNPs/AP/MCH	1.00×10^{-10}–1.00×10^{-3} M	1.00×10^{-10} M	Serum	[84]
L-glutamate	Amperometry	RuO_2–ZnO NPs/GCE	0.1 nM–0.01 mM	96.0 pM	NS	[85]
Tryptophan	DPV	Carbon Cloth	20–6000 μM	7.69 μM	Human serum	[86]
Tryptophan	DPV	(PImox) film/GCE	3–464 μM	0.770 μM	Human urine	[87]
Trptophan	SWV	Cabon black nanoballs (CBBN)/ CNTs/GCE	0.025–4.8 μM	0.011 μM	Milk, human urine	[88]
L-tyrosine	DPV	D-CNT@PPy@Pt NPs@β-CD/SPE	3–30 μM	0.107 nM	NS	[89]
Tryptophan			19.6–196 μM	0.133 nM		
L-tyrosine	SWV	SS-CS/GCE	10–100 μM	350 nM	NS	[90]
L-tyrosine	CV DPV	CuS NS/CS/F-MWCNTs/GCE	0.08–1.0 μM	4.9 nM	Pig serum	[91]
L-tyrosine	DPV	GSH-Cu/Pt/GCE	NS	NS	NS	[92]
L-tyrosine	SWV	CS-GalN/GCE	10–1000 μM	650 nM	NS	[93]
D-tyrosine				860 nM		
Tryptophan	CV	PPy/FeCN/SPCE	3.3×10^{-7}–1.06×10^{-5} M	1.05×10^{-7} M	Pharmaceutics	[94]
Tyrosine	CV	GC/CNT/PEDOT/NF/Crown	0.06–20 × 10^{-9} M	0.429×10^{-9} M	Biological Fluids	[95]
Tryptophan	CV SWV	PT-Ag/L-Try/GCE	200×10^{-9}–400×10^{-3} M	20×10^{-9} M	Food samples (Soybeans Extract)	[96]
Phenylalanie	DPV	Sensor with Gum Arabic based polyurethane modification	200–1000 × 10^{-6} M	48.01×10^{-6} M	NS	[97]

Phenylalanie	DPV	Sensor with p-Toluene Sulfonic Acid Modified Pt Electrode	2–2000 $\times 10^{-6}$ M	0.59 $\times 10^{-6}$ M	NS	[98]
Phenylalanie	CV LSV	PPy-β-CD/GCE	0.1–0.75 $\times 10^{-3}$ M	NS	NS	[99]
Tyrosine	SWV	EB-Ppy-BSA/GCE	100 $\times 10^{-9}$–800 $\times 10^{-6}$ M	8.8 $\times 10^{-9}$ M	NS	[100]
Tyrosine	CV SWV	EB-PPy/MGA	0.4–600 $\times 10^{-6}$ M	85 $\times 10^{-9}$ M	Real samples	[101]
Tyrosine	CV SWV	MIP PPy/AuE	5.0 $\times 10^{-9}$–2.5 $\times 10^{-8}$ M	2.5 $\times 10^{-9}$ M	NS	[102]
Tryptophan	CV	MIOPPy/pABSA/GCE	NS	NS	NS	[103]
Tryptophan	DPV, CV	3DCu(*x*)O-ZnO NPs/PPy/RGO	0.053–480 $\times 10^{-6}$ M	0.016 $\times 10^{-6}$ M	Human blood	[104]
Tyrosine	CV, DPV	Polythreonine-modified graphite-carbon nanotube paste electrode	9.6 $\times 10^{-7}$ M	2.9 $\times 10^{-7}$ M	Pharmaceutical sample	[105]
L-tyrosine D-tyrosine L-tryptophan D-tryptophan	DPV	L/D-DHCNT@PPy@AuNPs @L/D-Cys	NS	1.88 $\times 10^{-1}$ M 5.72 $\times 10^{-1}$ M 0.012 M 0.14 M	NS	[106]
Tryptophan	DPV	MIP-QCM biosensor	15.2–750 ng ml^{-1}	0.73 ng ml^{-1}	NS	[107]
Tryptophan	CV	MIP-MWCNTs/GCE	2 nM–0.2 μM, 0.2–10 μM, 10–100 μM	1.0 $\times 10^{-9}$ M	Human serum	[108]

Abbreviations: 3D: three-dimensional, Ag: silver dendrites composite, Au: gold nanoparticles, AuE: gold electrode, BSA: bovine serum albumin, CBBN: carbon black nanoballs, CNTs: carbon nanotubes, CS: chitosan, CS-GaIN: abis-aminosaccharides composite, EB: electron beam, FCPA: ferrocene carbamate phenyl acrylate, GCE: glassy carbon electrode, GSH: glutathione, L/D-DHCNT: left-/right-handed double helix carbon nanotubes, MGA: modified gum acacia, MIOPPy: molecularly imprinted overoxidized polypyrrole, MIP: molecularly imprinted polyaniline, MWCNT: multi-walled carbon nanotubes, NF: nafion, NPs: nanoparticals, pABSA: p-aminobenzene sulfonic acid, PDA: polydopamine, PEDOT: poly 3–4-ethylene dioxythiophene, PPy: polypyrrole, PT: polythiophene, PTH: polythionine, QCM: quartz crystal microbalance, rGO: reduced graphene oxide, SPCE: carbon screen-printed electrode, ZIF-67: zeolitic imidazolate framework-67 composite, β-CD: β-cyclodextrin.

showed that they are very suitable for identifying AAs. Because AAs play an important role in human health. Electrochemical sensors and biosensors combined with nanomaterials and polymers have played an important role in identifying AAs. According to the studies conducted on AAs, bare electrodes to the most complex electrochemical devices have been investigated. As described in the above-mentioned articles, the linear range and detection limit of AAs by these electrodes have been investigated. It has been concluded that modified sensors and biosensors have shown better performance for detecting AAs in different samples. They can be considered selective, sensitive, cheap, simple to use, and time-saving electrochemical bio(sensors), compared with past detection devices. Electrochemical devices showed that they can be used in different samples such as urine, blood, milk, juice, pharmaceutical samples, serum, etc to detect AAs. These methods show that new ways and sensors will be produced to identify AAs in the future. It is expected that in the future, with the wide spread of electrochemical devices, it will be possible to identify even the smallest amount of AAs for use in human health.

References

[1] Elango R, Ball R O and Pencharz P B 2008 Individual amino acid requirements in humans: an update *Curr. Opin. Clin. Nutr. Metab. Care* **11** 34–9

[2] Li P, Yin Y L, Li D *et al* 2007 Amino acids and immune function *Br. J. Nutr.* **98** 237–52

[3] Hall R 2018 Bioresources amino acids and proteins *Fish Nutrition* (New York: Academic) pp 143–79

[4] Agostinelli E 2020 Biochemical and pathophysiological properties of polyamines *Amino Acids* **52** 111–7

[5] Beaumont M and Blachier F 2020 Amino acids in intestinal physiology and health *Amino Acids in Nutrition and Health*Advances in Experimental Medicine and Biology *ed G Wu (Springer)*

[6] Closs E I, Boissel J P, Habermeier A and Rotmann A 2006 Structure and function of cationic amino acid transporters (CATs) *J. Membr. Biol.* **213** 67–77

[7] Yang Z, Htoo J K and Liao S F 2020 Methionine nutrition in swine and related monogastric animals: beyond protein biosynthesis *Anim. Feed Sci. Technol.* **268** 114608

[8] Bischoff R and Schlüter H 2012 Amino acids: chemistry, functionality and selected non-enzymatic post-translational modifications *J. Proteomics* **75** 2275–96

[9] Wu G 2013 *Amino Acids: Biochemistry and Nutrition* ed G Wu (CRC Press)

[10] Hu S, He W and Wu G 2022 Hydroxyproline in animal metabolism, nutrition, and cell signaling *Amino Acids* **54** 513–28

[11] Kumar J, Rajalakshmi Mohanraj H, Christian P K and David J T 2019 Controlling the electronic properties of zigzag graphene nanoribbon using amino acids and oxygen molecule-A first principles DFT study *Appl. Surf. Sci.* **494** 627–34

[12] Pundir C S, Lata S and Narwal V 2018 Biosensors for determination of D and L-amino acids: a review *Biosens. Bioelectron.* **117** 373–84

[13] Berg C P 1953 Physiology of the D-amino acids *Physiol. Rev.* **33** 145–89

[14] Grishin D V, Zhdanov D D, Pokrovskaya M V and Sokolov N N 2020 D-amino acids in nature, agriculture and biomedicine *Front. Life Sci.* **13** 11–22

[15] Marcone G L, Rosini E, Crespi E and Pollegioni L 2020 D-amino acids in foods *Appl. Microbiol. Biotechnol.* **104** 555–74
[16] Song Y, Xu C, Kuroki H *et al* 2018 Recent trends in analytical methods for the determination of amino acids in biological samples *J. Pharm. Biomed. Anal.* **147** 35–49
[17] Sarkar P, Tothill I E, Setford S J and Turner A P F 1999 Screen-printed amperometric biosensors for the rapid measurement of L- and D-amino acids *Analyst* **124** 865–70
[18] Choi M M F and Wong P S 2002 Application of a datalogger in biosensing: a glucose biosensor *J. Chem. Educ.* **79** 982
[19] Dinu A and Apetrei C 2020 A review on electrochemical sensors and biosensors used in phenylalanine electroanalysis *Sensors* **20**
[20] Iftikhar F J, Shah A, Akhter M S *et al* 2019 Introduction to nanosensors In: *New Developments in Nanosensors for Pharmaceutical Analysis* (Amsterdam: Elsevier) pp 1–46
[21] Stefan R-I 2001 *Electrochemical Sensors in Bioanalysis* (Boca Raton, FL: CRC Press)
[22] Ronkainen N J, Halsall H B and Heineman W R 2010 Electrochemical biosensors *Chem. Soc. Rev.* **39** 1747–63
[23] Yarman A, Kurbanoglu S and Scheller F F W 2020 Noninvasive biosensors for diagnostic biomarkers *Commercial Biosensors and Their Applications: Clinical, Food, and Beyond* (Amsterdam: Elsevier) p 167
[24] Hammond J L, Formisano N, Estrela P *et al* 2016 Electrochemical biosensors and nanobiosensors *Essays Biochem.* **60** 69–80
[25] Kim J, Campbell A S and Wang J 2018 Wearable non-invasive epidermal glucose sensors: a review *Talanta* **177** 163–70
[26] Baiulescu G E 2000 *Quality and Reliability in Analytical Chemistry* (CRC Press)
[27] Štulík K 1999 Challenges and promises of electrochemical detection and sensing *Electroanalysis* **11** 1001–4
[28] Mollarasouli F, Kurbanoglu S, Ozkan S A S A *et al* 2019 The role of electrochemical immunosensors in clinical analysis *Biosensors* **9** 1–19
[29] Soylemez S and Kurbanoglu S 2021 Enzyme-based electrochemical nanobiosensors using quantum dots *Electroanalytical Applications of Quantum Dot-Based Biosensors* (Elsevier) pp 307–39
[30] Kurbanoglu S, Uslu B and Ozkan S A S A 2018 Nanobiodevices for electrochemical biosensing of pharmaceuticals *Nanostructures for the Engineering of Cells, Tissues and Organs: From Design to Applications* (William Andrew Publishing) pp 291–330
[31] Kurbanoglu S, Erkmen C and Uslu B 2020 Frontiers in electrochemical enzyme based biosensors for food and drug analysis *TrAC Trends Anal. Chem.* **124** 1–24
[32] Ahangari H, Kurbanoglu S, Ehsani A and Uslu B 2021 Latest trends for biogenic amines detection in foods: enzymatic biosensors and nanozymes applications *Trends Food Sci. Technol.* **112** 75–87
[33] Majeed S, Naqvi S T R, ul Haq M N and Ashiq M N 2022 Electroanalytical techniques in biosciences: conductometry, coulometry, voltammetry, and electrochemical sensors *Analytical Techniques in Biosciences* (Elsevier) pp 157–78
[34] Bakker E and Qin Y 2006 Electrochemical sensors *Anal. Chem.* **78** 3965–84
[35] Wang Y, Xu H, Zhang J and Li G 2008 Electrochemical sensors for clinic analysis *Sensors* **8** 2043–81
[36] Magner E 1998 Trends in electrochemical biosensors *Analyst* **123** 1967–70

[37] Vetterl V and Hason S 2005 Electrochemical properties of nucleic acid components *Electrochemistry of Nucleic Acids and Non-Conjugated Proteins* ed E Palecek, F W Scheller and J Wang (Amsterdam: Elsevier) pp 18–60
[38] Scheller F, Schubert F, Pfeiffer D *et al* 1989 Research and development of biosensors. A review *Analyst* **114** 653–62
[39] Scheller F and Schubert F 1992 *Biosensors* (Elsevier)
[40] Renneberg R, Scheller F W and Yarman A 2018 Biosensors *Bioanalytics: Analytical Methods and Concepts in Biochemistry and Molecular Biology* ed F Lottspeich and J Engels (New York: Wiley)
[41] Kissinger P T 2005 Biosensors—a perspective *Biosens. Bioelectron.* **20** 2512–6
[42] Turner A P F A 2013 Biosensors: sense and sensibility *Chem. Soc. Rev.* **42** 3184–96
[43] Scheller F and Schubert F 1992 *Techniques and Instrumentation in Analytical Chemistry: Biosensors* (Amsterdam: Elsevier Science)
[44] Mishra P and Sahu P P 2022 *Biosensors in Food Safety and Quality: Fundamentals and Applications* (CRC Press)
[45] García-Carmona L, González M C and Escarpa A 2018 Electrochemical on-site amino acids detection of maple syrup urine disease using vertically aligned nickel nanowires *Electroanalysis* **30** 1505–10
[46] Petralia S, Sciuto E L, Messina M A *et al* 2018 Miniaturized and multi-purpose electrochemical sensing device based on thin Ni oxides *Sens. Actuators* B **263** 10–9
[47] Tao Y, Chu F, Gu X *et al* 2018 A novel electrochemical chiral sensor for tyrosine isomers based on a coordination-driven self-assembly *Sens. Actuators* B **255** 255–61
[48] Zhang L, Wang G, Xiong C *et al* 2018 Chirality detection of amino acid enantiomers by organic electrochemical transistor *Biosens. Bioelectron.* **105** 121–8
[49] Liang W, Rong Y, Fan L *et al* 2018 3D graphene/hydroxypropyl-β-cyclodextrin nanocomposite as an electrochemical chiral sensor for the recognition of tryptophan enantiomers *J. Mater. Chem.* C **6** 12822–9
[50] Zheng W, Zhao M, Liu W *et al* 2018 Electrochemical sensor based on molecularly imprinted polymer/reduced graphene oxide composite for simultaneous determination of uric acid and tyrosine *J. Electroanal. Chem.* **813** 75–82
[51] Ziyatdinova G, Kozlova E and Budnikov H 2018 Selective electrochemical sensor based on the electropolymerized p-coumaric acid for the direct determination of l-cysteine *Electrochim. Acta* **270** 369–77
[52] He Q, Tian Y, Wu Y *et al* 2019 Electrochemical sensor for rapid and sensitive detection of tryptophan by a cu2o nanoparticles-coated reduced graphene oxide nanocomposite *Biomolecules* **9** 176
[53] Yang J, Wu D, Fan G C *et al* 2019 A chiral helical self-assembly for electrochemical recognition of tryptophan enantiomers *Electrochem. Commun.* **104** 106478
[54] Mattioli I A, Baccarin M, Cervini P and Cavalheiro É T G 2019 Electrochemical investigation of a graphite-polyurethane composite electrode modified with electrodeposited gold nanoparticles in the voltammetric determination of tryptophan *J. Electroanal. Chem.* **835** 212–9
[55] Yao W, Guo H, Liu H *et al* 2020 Highly electrochemical performance of Ni-ZIF-8/ N S-CNTs/CS composite for simultaneous determination of dopamine, uric acid and L-tryptophan *Microchem. J.* **152** 104357

[56] Wei X, Chen Y, He S *et al* 2021 L-histidine-regulated zeolitic imidazolate framework modified electrochemical interface for enantioselective determination of L-glutamate *Electrochim. Acta* **400** 139464
[57] Mounesh , Sharan Kumar T M, Praveen Kumar N Y *et al* 2021 Novel Schiff base cobalt(ii) phthalocyanine with appliance of MWCNTs on GCE: enhanced electrocatalytic activity behaviour of α-amino acids *RSC Adv.* **11** 16736–46
[58] Fiore L, De Lellis B, Mazzaracchio V *et al* 2022 Smartphone-assisted electrochemical sensor for reliable detection of tyrosine in serum *Talanta* **237** 122869
[59] Varmira K, Mohammadi G, Mahmoudi M *et al* 2018 Fabrication of a novel enzymatic electrochemical biosensor for determination of tyrosine in some food samples *Talanta* **183** 1–10
[60] Tian T, Liu M, Chen L *et al* 2020 D-amino acid electrochemical biosensor based on D-amino acid oxidase: mechanism and high performance against enantiomer interference *Biosens. Bioelectron.* **151** 111971
[61] Singh A K, Sharma R, Singh M and Verma N 2020 Electrochemical determination of L-arginine in leukemic blood samples based on a polyaniline-multiwalled carbon nanotube —magnetite nanocomposite film modified glassy carbon electrode *Instrum. Sci. Technol.* **48** 400–16
[62] Moussa S, Murtas G, Pollegioni L and Mauzeroll J 2021 Enhancing electrochemical biosensor selectivity with engineered d-amino acid oxidase enzymes for d-serine and d-alanine quantification *ACS Appl. Bio Mater.* **4** 5598–604
[63] Balamurugan T S T, Huang C H, Chang P C and Huang S T 2018 Electrochemical molecular switch for the selective profiling of cysteine in live cells and whole blood and for the quantification of aminoacylase-1 *Anal. Chem.* **90** 12631–8
[64] Lin C, Jair Y C, Chou Y C *et al* 2018 Transcription factor-based biosensor for detection of phenylalanine and tyrosine in urine for diagnosis of phenylketonuria *Anal. Chim. Acta* **1041** 108–13
[65] Idili A, Gerson J, Parolo C *et al* 2019 An electrochemical aptamer-based sensor for the rapid and convenient measurement of l-tryptophan *Anal. Bioanal. Chem.* **411** 4629–35
[66] Idili A, Parolo C, Ortega G and Plaxco K W 2019 Calibration-free measurement of phenylalanine levels in the blood using an electrochemical aptamer-based sensor suitable for point-of-care applications *ACS Sens.* **4** 3227–33
[67] Govindasamy M, Wang S F, Pan W C *et al* 2019 Facile sonochemical synthesis of perovskite-type SrTiO3 nanocubes with reduced graphene oxide nanocatalyst for an enhanced electrochemical detection of α-amino acid (tryptophan) *Ultrason. Sonochem.* **56** 193–9
[68] Sajini T, John S and Mathew B 2019 Tailoring of photo-responsive molecularly imprinted polymers on multiwalled carbon nanotube as an enantioselective sensor and sorbent for L-PABE *Compos. Sci. Technol.* **181** 107676
[69] Niu X, Yang X, Mo Z *et al* 2019 Perylene-functionalized graphene sheets modified with β-cyclodextrin for the voltammetric discrimination of phenylalanine enantiomers *Bioelectrochemistry* **129** 189–98
[70] Yi Y, Zhang D, Ma Y *et al* 2019 Dual-signal electrochemical enantiospecific recognition system via competitive supramolecular host-guest interactions: the case of phenylalanine *Anal. Chem.* **91** 2908–15

[71] Xu X, Ji D, Zhang Y *et al* 2019 Detection of phenylketonuria markers using a ZIF-67 encapsulated PtPd alloy nanoparticle (PtPd@ZIF-67)-based disposable electrochemical microsensor *ACS Appl. Mater. Interfaces* **11** 20734–42

[72] Bangaleh Z, Sadeghi H B, Ebrahimi S A and Najafizadeh P 2019 A new potentiometric sensor for determination and screening phenylalanine in blood serum based on molecularly imprinted polymer *Iran. J. Pharm. Res.* **18** 61–71

[73] Ou S H, Pan L S, Jow J J *et al* 2018 Molecularly imprinted electrochemical sensor, formed on Ag screen-printed electrodes, for the enantioselective recognition of D and L phenylalanine *Biosens. Bioelectron.* **105** 143–50

[74] Murtada K, Salghi R, Ríos A and Zougagh M 2020 A sensitive electrochemical sensor based on aluminium doped copper selenide nanoparticles-modified screen printed carbon electrode for determination of L-tyrosine in pharmaceutical samples *J. Electroanal. Chem.* **874** 114466

[75] Karthika A, Rosaline D R, Inbanathan S S R *et al* 2020 Fabrication of Cupric oxide decorated β-cyclodextrin nanocomposite solubilized Nafion as a high performance electrochemical sensor for L-tyrosine detection *J. Phys. Chem. Solids* **136** 109145

[76] Li Z Y, Gao D Y, Wu Z Y and Zhao S 2020 Simultaneous electrochemical detection of levodapa, paracetamol and l-tyrosine based on multi-walled carbon nanotubes *RSC Adv.* **10** 14218–24

[77] Chen B, Zhang Y, Lin L *et al* 2020 Au nanoparticles @metal organic framework/polythionine loaded with molecularly imprinted polymer sensor: Preparation, characterization, and electrochemical detection of tyrosine *J. Electroanal. Chem.* **863** 114052

[78] Liu H, Shao J, Shi L *et al* 2020 Electroactive NPs and D-amino acids oxidase engineered electrochemical chiral sensor for D-alanine detection *Sens. Actuators* B **304** 127333

[79] Krishnan R G and Saraswathyamma B 2021 Simultaneous resolution and electrochemical quantification of tyrosine and tryptophan at a poly (diphenylamine) modified electrode *J. Electrochem. Soc.* **168** 027509

[80] Wu S, Wang H, Wu D *et al* 2021 Silver nanoparticle driven signal amplification for electrochemical chiral discrimination of amino acids *Analyst* **146** 1612–9

[81] Sundaresan R, Mariyappan V, Chen S and Keerthi M 2021 Colloids and surfaces a : physicochemical and engineering aspects electrochemical sensor for detection of tryptophan in the milk sample based on MnWO 4 nanoplates encapsulated RGO nanocomposite *Colloids Surf. A: Physicochem. Eng. Asp.* **625** 126889

[82] Eduardo da Silva Ferreira M, Canhete de Moraes N, Souza Ferreira V *et al* 2022 A novel 3D-printed batch injection analysis (BIA) cell coupled to paper-based electrochemical devices: a cheap and reliable analytical system for fast on-site analysis *Microchem. J.* **179** 107663

[83] Li L, Chen L and Chen Z 2022 High throughput sensing of multiple amino acids with differential pulse voltammetry measurement *Anal. Biochem.* **647** 1–6

[84] Wang W, He Y, Gao Y *et al* 2022 A peptide aptamer based electrochemical amperometric sensor for sensitive L-glutamate detection *Bioelectrochemistry* **146** 108165

[85] Alam M M, Uddin M T, Asiri A M *et al* 2020 Fabrication of selective L-glutamic acid sensor in electrochemical technique from wet-chemically prepared RuO_2 doped ZnO nanoparticles *Mater. Chem. Phys.* **251** 123029

[86] Salve M, Amreen K, Pattnaik P K and Goel S 2022 Carbon cloth-based electrochemical device for specific and sensitive detection of ascorbic acid and tryptophan *IEEE Sens. J.* **22** 6072–9
[87] He Q, Liu J, Feng J *et al* 2020 Sensitive voltammetric sensor for tryptophan detection by using polyvinylpyrrolidone functionalized graphene/GCE *Nanomaterials* **10** 125
[88] Wu B, Xiao L, Zhang M *et al* 2021 Facile synthesis of dendritic-like CeO_2/rGO composite and application for detection of uric acid and tryptophan simultaneously *J. Solid State Chem.* **296** 122023
[89] Ning G, Wang H, Fu M *et al* 2022 Dual signals electrochemical biosensor for point-of-care testing of amino acids enantiomers *Electroanalysis* **34** 316–25
[90] Zou J, Chen X Q, Zhao G Q *et al* 2019 A novel electrochemical chiral interface based on the synergistic effect of polysaccharides for the recognition of tyrosine enantiomers *Talanta* **195** 628–37
[91] Zhu Q, Liu C, Zhou L *et al* 2019 Highly sensitive determination of L-tyrosine in pig serum based on ultrathin CuS nanosheets composite electrode *Biosens. Bioelectron.* **140** 111356
[92] Ye Q, Yin Z Z, Wu H *et al* 2020 Decoration of glutathione with copper-platinum nanoparticles for chirality sensing of tyrosine enantiomers *Electrochem. Commun.* **110** 106638
[93] Zou J and Yu J G 2019 Chiral recognition of tyrosine enantiomers on a novel bis-aminosaccharides composite modified glassy carbon electrode *Anal. Chim. Acta* **1088** 35–44
[94] Dinu A and Apetrei C 2021 Development of a novel sensor based on polypyrrole doped with potassium hexacyanoferrate (Ii) for detection of l-tryptophan in pharmaceutics *Inventions* **6** 56
[95] Atta N F, Ahmed Y M and Galal A 2018 Layered-designed composite sensor based on crown ether/Nafion®/polymer/carbon nanotubes for determination of norepinephrine, paracetamol, tyrosine and ascorbic acid in biological fluids *J. Electroanal. Chem.* **828** 11–23
[96] GunaVathana S D, Thivya P, Wilson J and Peter A C 2020 Sensitive voltammetric sensor based on silver dendrites decorated polythiophene nanocomposite: selective determination of L-Tryptophan *J. Mol. Struct.* **1205** 127649
[97] Alışık F, Burç M, Titretir Duran S *et al* 2021 Development of Gum-Arabic-based polyurethane membrane-modified electrodes as voltammetric sensor for the detection of phenylalanine *Polym. Bull.* **78** 4699–719
[98] Alışık F, Burç M, Köytepe S and Titretir Duran S 2020 Preparation of molecularly imprinted electrochemical L-phenylalanine sensor with p-toluene sulfonic acid modified Pt electrode *J. Electrochem. Soc.* **167** 167508
[99] Shishkanova T V, Habanová N, Řezanka M *et al* 2020 Molecular recognition of phenylalanine enantiomers onto a solid surface modified with electropolymerized pyrrole-β-cyclodextrin conjugate *Electroanalysis* **32** 767–74
[100] Ramya R, Muthukumaran P and Wilson J 2018 Electron beam-irradiated polypyrrole decorated with Bovine serum albumin pores: simultaneous determination of epinephrine and L-tyrosine *Biosens. Bioelectron.* **108** 53–61
[101] Dhananjayan N, Jeyaraj W and Karuppasamy G 2019 Interactive studies on synthetic nanopolymer decorated with edible biopolymer and its selective electrochemical determination of L-tyrosine *Sci. Rep.* **9** 1–12

[102] Ermiş N and Tinkiliç N 2018 Preparation of molecularly imprinted polypyrrole modified gold electrode for determination of tyrosine in biological samples *Int. J. Electrochem. Sci.* **13** 2286–98

[103] Gong L, Li S, Yin Z *et al* 2021 Enantioselective recognition of tryptophan isomers with molecularly imprinted overoxidized polypyrrole/poly(p-aminobenzene sulfonic acid) modified electrode *Chirality* **33** 176–83

[104] Ghanbari K and Bonyadi S 2018 An electrochemical sensor based on reduced graphene oxide decorated with polypyrrole nanofibers and zinc oxide-copper oxide p-n junction heterostructures for the simultaneous voltammetric determination of ascorbic acid, dopamine, paracetamol, and tryptoph *New J. Chem.* **42** 8512–23

[105] Chenthattil Raril , Manjunatha J G, Ravishankar D K *et al* 2020 Validated electrochemical method for simultaneous resolution of tyrosine, uric acid, and ascorbic acid at polymer modified nano-composite paste electrode *Surf. Eng. Appl. Electrochem.* **56** 415–26

[106] Zhang Q, Fu M, Lu H *et al* 2019 Novel potential and current type chiral amino acids biosensor based on L/D-handed double helix carbon nanotubes@polypyrrole@Au nanoparticles@L/D-cysteine *Sens. Actuators* B **296** 126667

[107] Prabakaran K, Jandas P J, Luo J *et al* 2021 Molecularly imprinted poly(methacrylic acid) based QCM biosensor for selective determination of L-tryptophan *Colloids Surf. A: Physicochem. Eng. Asp.* **611** 125859

[108] Wu Y, Deng P, Tian Y *et al* 2020 Rapid recognition and determination of tryptophan by carbon nanotubes and molecularly imprinted polymer-modified glassy carbon electrode *Bioelectrochemistry* **131** 107393

www.ingramcontent.com/pod-product-compliance
Lightning Source LLC
Chambersburg PA
CBHW082122210326
41599CB00031B/5840

9780750353786